Contents

3

Polynomials 167

4

Linear Equations 255

5

Integer Exponents 339

9

Quadratic Equations 565

10

Graphing Linear and Quadratic Equations 599

11

Solving Systems of Equations 637

12

Inequalities 689

Preface

The main purpose of this textbook is to help students make the transition from arithmetic to algebra as smoothly and pleasantly as possible. The similarities between arithmetic and algebra are constantly stressed to help students gain confidence and insight into algebra. When each algebraic topic is introduced, a relationship is shown to arithmetic or to a previous algebraic topic. Knowing this relationship should give students a better understanding of algebra. It is the author's contention that when students see how mathematics fits together, they will gain more understanding and appreciation for mathematics. As a college student, it is not enough to simply learn the rote operations of a subject. This approach may take more time and effort on the part of the instructor and the student; however, the author feels that the rewards are well worth the additional investment.

This textbook is presented in a way that students have probably not seen previously. Most students should be capable of mastering the material presented without becoming anxious or bored. There is a continuous flow from one mathematical concept to the next, and new concepts are not presented without prior preparation. A fictitious student, Charlie, is introduced to ask pertinent questions and to give the reader an occasional break from the rigors of the mathematics.

Formal names, definitions, and rules have been kept to a minimum throughout the textbook. One purpose of this textbook is not for the student to merely memorize a host of new and strange terms but to better understand the structure of basic algebra—and even to feel good about it. A large number of new mathematical names, definitions, and rules can make mathematics seem foreign and frightening to many students, so their use has been minimized throughout the textbook. There will be time later to introduce additional names, definitions, and rules after the basics of algebra have been learned and the student feels somewhat comfortable in this new environment.

This textbook is the second in a series and is designed for a one semester or one quarter course in elementary algebra. Each chapter contains practice problems to be worked as the material is covered so that the student will feel comfortable with each concept before progressing to the next concept. The answers to the practice problems are listed at the end of each chapter. In addition, there are numerous exercises at the end of each section as well as review problems at the end of each chapter. These problems are grouped according to difficulty. The answers to the odd-numbered problems are listed in the back of the textbook. An application section containing practical problems related to the material is given at the end of some sections and chapters. The answers to the odd-numbered application problems are also listed in the back of the textbook.

There is also a section entitled "Useful Information" at the end of each chapter. It contains the key words, phrases, and concepts of the chapter as well as other information the student might need when working through the chapter. This section is more extensive than a summary; in fact, it sometimes contains information not found in the body of the text.

Chapter 1 is designed to help the student learn how the whole number system is organized using expanded notation. It emphasizes more than just performing arithmetic operations on whole numbers. It also introduces some of the similarities between arithmetic and algebra so the student feels more comfortable about starting algebra. When comparisons are made in the chapter between arithmetic and algebra, they are set apart from the regular text. This does not mean that the comparisons are optional. They are very important to future success in working algebraic problems and should receive careful attention. However, it is not recommended that the student be tested on the algebraic concepts at this time. One intent of introducing these comparisons is to reduce student anxiety concerning algebra; testing at this time would only increase this anxiety.

Chapter 3 is the first chapter on algebra and is a very important chapter. The algebra problems are consistently related to what has already been learned and applied in arithmetic. Again, the similarity between algebra and arithmetic is constantly stressed. The introduction of each algebraic topic is related to previous work; thus, no topic is introduced without some background.

The author has used the material in this textbook with great success in his classes since 1979. The author is confident that students will experience some of the same success in their mathematics classes.

ACKNOWLEDGMENTS

Special thanks to Don Kern, Ed.D. mathematics education, for his assistance in the preparation of this textbook. He read the entire manuscript and made many valuable suggestions for improving the material.

REVIEWERS

Julia Brown
Atlantic Community College

Roe V. Hurst
Central Virginia Community College

Sally Copeland
Johnson County Community College

Ned W. Schillow
Lehigh County Community College

William J. Roberts
Plymouth State College

ACCURACY CHECK REVIEWERS

Joan E. Bell
North Eastern State University

Sue L. Korsak
New Mexico State University

J. Terry Wilson
San Jacinto College

TO THE INSTRUCTOR

In your annotated edition, you will find some notes to the instructor at the beginning of each chapter. These notes point out some of the unique features of the chapter and what the author hopes these features will help the students accomplish. I hope that you will find these notes to be helpful. If you have any suggestions on how to better present the material or improve the book in any way, please send them to me personally at Mobile College, P.O. Box 13220, Mobile, Alabama 36613 or to Wm. C. Brown Publishers.

CHAPTER 1

Whole Numbers

The objective of this chapter is:

1. To show you how to add, subtract, multiply, and divide whole numbers in expanded notation so that your understanding of the decimal number system will be improved.

When the shortcuts of arithmetic are eliminated, you should see that arithmetic and algebra are basically the same. In fact, *Webster's New Collegiate Dictionary* defines algebra as "a generalization of arithmetic in which letters representing numbers are combined according to the rules of arithmetic."* When arithmetic procedures are presented in this chapter, algebra will frequently be compared to arithmetic. As you can see, Webster's definition of algebra certainly justifies these comparisons. One of the goals of this textbook is to present algebra in such a way that algebra will indeed be a natural extension of arithmetic. So, as you work through the first chapter of this textbook, keep in mind that you are not just learning arithmetic but laying the foundation for algebra.

Let me introduce you to Charlie, a friend that will also be taking the course. Perhaps you will find that Charlie will have many of the same questions and concerns about the material being covered that you will have. We hope that through Charlie you can identify with many of the questions students typically ask about mathematics. Now, here's what Charlie looks like:

Hi! My name is Charlie.

1.1 HOW THE WHOLE NUMBER SYSTEM IS ORGANIZED

The different symbols, called **digits,** used in forming whole numbers are

0, 1, 2, 3, 4, 5, 6, 7, 8, 9

Let's examine the process of counting using the symbols 0, 1, 2, 3, 4, 5, 6, 7, 8, 9 and the concept of grouping. We will count chairs in the illustrations.

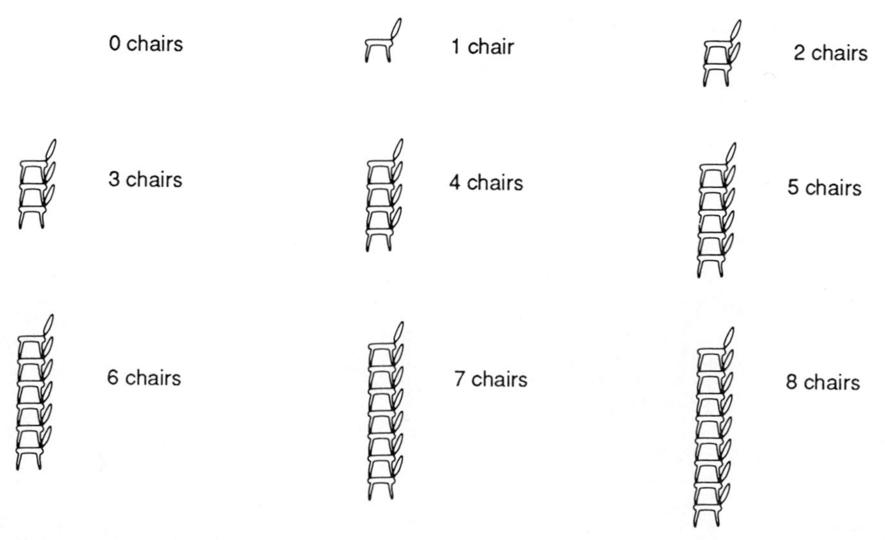

*By permission. From Webster's Ninth New Collegiate Dictionary © 1988 by Merriam-Webster Inc., publisher of the Merriam-Webster® dictionaries.

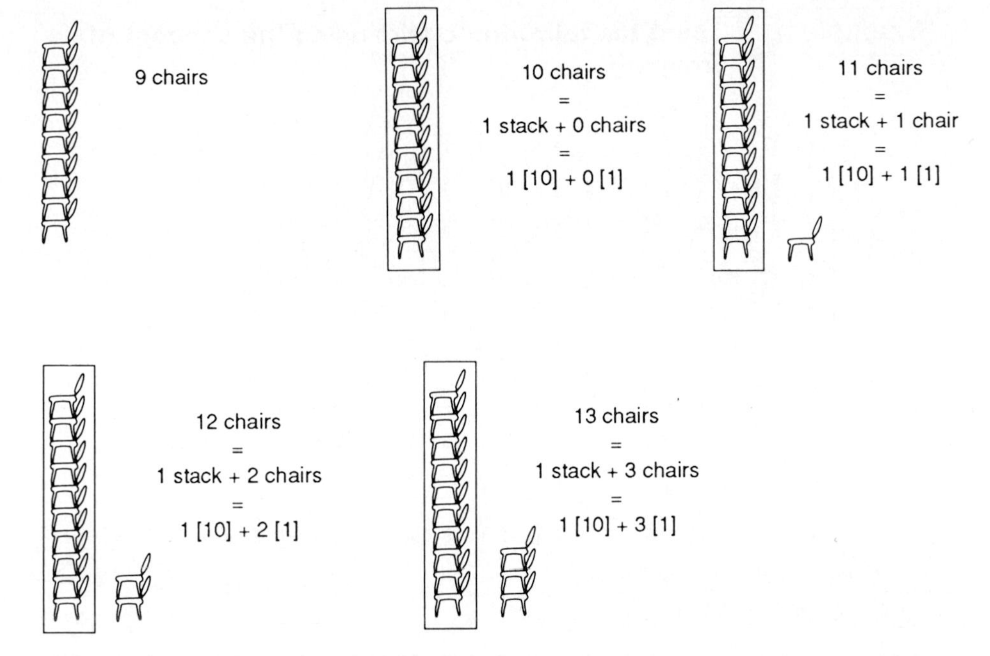

Notice that whenever you have more than nine individual chairs, you can start to group them into stacks and use the single digits to count the stacks—until you get more than nine stacks. Then you can group the stacks together into a larger group called a truckload.

Practice 1: Count the following chairs using the concept of grouping as shown above.

1. ____________________

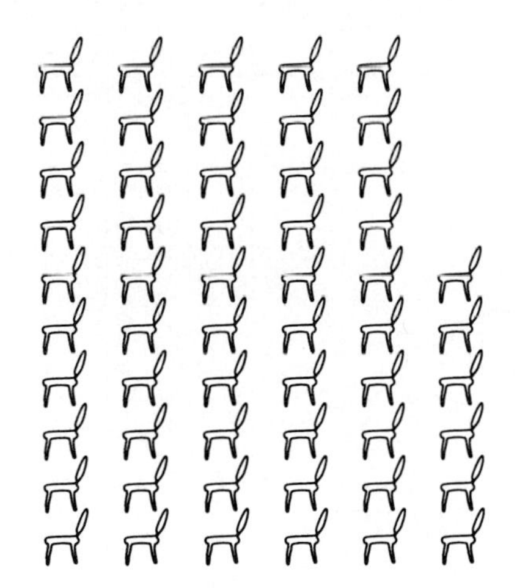

Example 1 **Count the following chairs using the concept of grouping.**

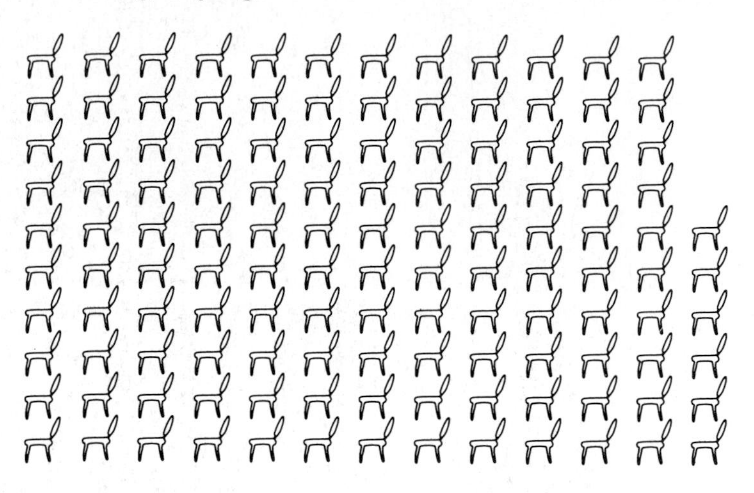

Step 1: Group the chairs.

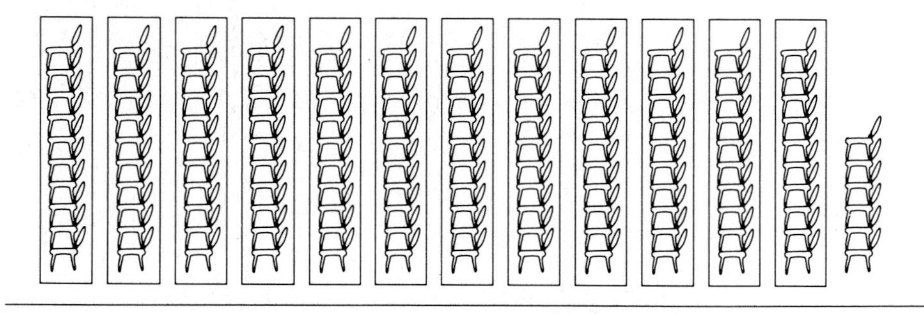

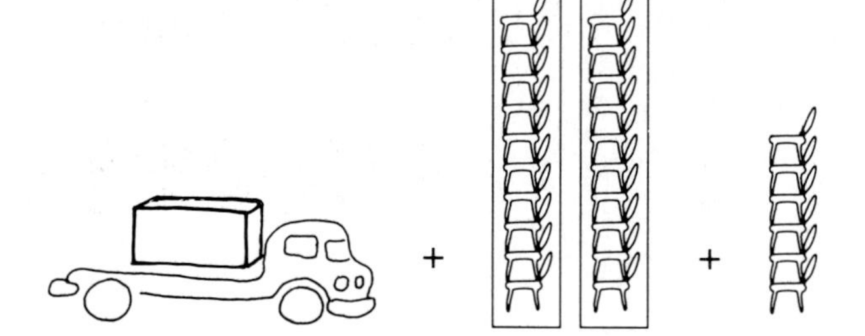

Step 2: Count the various groups.

$$1 \text{ truck} + 2 \text{ stacks} + 6 \text{ chairs} = 126 \quad \text{or}$$

$$1(10^2) + 2(10) + 6(1) = 126$$

Thus, a truckload of chairs is ten groups of ten chairs or 100 chairs. ■

Practice 2: Count the following chairs using the concept of grouping.

A N S W E R

2. ____________

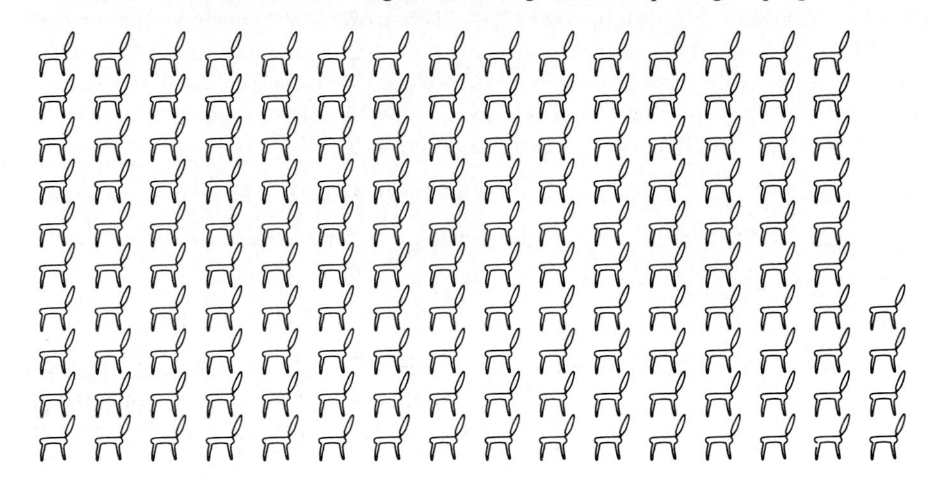

By combining the **digits**

0, 1, 2, 3, 4, 5, 6, 7, 8, 9

and using the concept of grouping you can form the **whole numbers**

0, 1, 2, 3, 4, 5, 6, 7, 8, 9, 10, 11, 12, 13, . . .

The three dots at the end of the sequence mean that the numbers continue indefinitely. These dots are necessary because it is impossible to list all the whole numbers.

Example 2 **Write the whole number 325 showing the groups of tens.**

Solution:

$$325 = 300 + 20 + 5$$

$$325 = 3(100) + 2(10) + 5(1)$$

$$325 = 3[10(10)] + 2(10) + 5(1)$$

$$325 = 3(10^2) + 2(10) + 5(1)$$

Since 5(1) equals 5, 325 can be written as

$$3(10^2) + 2(10) + 5$$

When a number is written showing the groups of tens as in Example 2, it is written in **expanded notation.** For example,

$$3(10^2) + 2(10) + 5$$

is written in **expanded notation** and

325

is written in **short notation.** ■

I don't understand what 10^2 represents.

A N S W E R

Well, Charlie, the whole number system uses groups of tens to form numbers. The symbols used are the digits 0, 1, 2, 3, 4, 5, 6, 7, 8, 9. Therefore, there can be at most 9 of any one thing without starting a new group. For example, there can be at most nine individual chairs, nine stacks of chairs, or nine truckloads of chairs. If there are nine chairs and one chair is added, then the individual chairs become one stack (group) of ten chairs which is represented by 10.

You can have 10, 11, 12, 13, . . . , 18, 19, 20, 21, . . . , 87, 88, 89, 90, . . . , 100 chairs. Where,

$$\begin{aligned} \text{100 chairs} &= \text{1 truckload} + \text{0 stacks} + \text{0 chairs} \\ &= \text{1 truckload of chairs} \\ &= \text{ten stacks of ten chairs} \\ &= 10(10) \text{ chairs} \\ &= 10^2 \text{ chairs} \end{aligned}$$

Thus, 10^2 means that we have gone through the grouping process twice. In fact, that is what the 2 in 10^2 tells us.

Example 3 **Write 524 in expanded notation.**

Solution:

$$\begin{aligned} 524 &= 5(100) + 2(10) + 4 \\ &= 5(10^2) + 2(10) + 4 \end{aligned}$$

■

3. ______________

4. ______________

5. ______________

6. ______________

7. ______________

8. ______________

Practice 3: Write 382 in expanded notation.

Practice 4: Write $2(10^2) + 3(10) + 8$ in short notation.

Practice 5: Write 605 in expanded notation.

Practice 6: Write $8(10^2) + 0(10) + 3$ in short notation.

Practice 7: Write 400 in expanded notation.

Practice 8: Write $6(10^2) + 2(10) + 0$ in short notation.

Example 4 **Write 8462 in expanded notation.**

Solution:

$$\begin{aligned} 8462 &= 8000 + 400 + 60 + 2 \\ &= 8(1000) + 4(100) + 6(10) + 2 \\ &= 8[10(100)] + 4(100) + 6(10) + 2 \\ &= 8\{10[10(10)]\} + 4[10(10)] + 6(10) + 2 \\ &= 8(10^3) + 4(10^2) + 6(10) + 2 \end{aligned}$$

Looking from right to left.

2 represents 2 ones

6 represents 6 groups of tens

4 represents 4 groups of ten groups of tens

8 represents 8 groups of ten groups of ten groups of tens ■

Thus, 10^3 means that we have gone through the grouping process three times. In fact, that is what the 3 in 10^3 tells us. So 10^3 means that we have grouped the chairs into stacks and the stacks into truckloads and the truckloads into a larger group called a warehouse. Now,

$$\begin{aligned} \text{Stack} &= 10 \text{ chairs} \\ \text{Truckload} &= 10 \text{ stacks} \\ &= 10(10) \text{ chairs} \\ &= 10^2 \text{ chairs} \\ &= 100 \text{ chairs} \\ \text{Warehouse} &= 10 \text{ trucks} \\ &= 10[10(10)] \text{ chairs} \\ &= 10^3 \text{ chairs} \\ &= 1000 \text{ chairs} \end{aligned}$$

Practice 9: Write 2573 in expanded notation.

Practice 10: Write $3(10^3) + 0(10^2) + 0(10) + 7$ in short notation.

Practice 11: Write 7050 in expanded notation.

Practice 12: Write $8(10^3) + 0(10^2) + 0(10) + 0$ in short notation.

A N S W E R

9. ____________

10. ____________

11. ____________

12. ____________

Why do we need to use expanded notation? It sure seems like a lot of trouble.

I am glad you asked that question, Charlie. Expanded notation, which shows the grouping concept of whole numbers, is very helpful in understanding how the decimal number system works. Many areas of mathematics that you will study later are also based on the grouping concept and cannot be written in short notation. For example, the algebraic expression

$$3x^2 + 2x + 5$$

does not appear to be related to 325. However, when we write 325 as

$$3(10^2) + 2(10) + 5$$

you should see some similarity. We will further explore this similarity in later sections. Also, when using different units of measurements such as inches and feet we must show the groups. There is no easy way to write

3 yards + 2 feet + 5 inches or

3 hours + 2 minutes + 5 seconds

in short notation.

EXERCISE 1.1

Write the following whole numbers in expanded notation.

A

1. 32
2. 47
3. 39
4. 87
5. 93
6. 48
7. 67
8. 17
9. 40
10. 66
11. 33
12. 11

B

13. 235
14. 405
15. 749
16. 454
17. 360
18. 555
19. 448
20. 522
21. 386
22. 407
23. 870
24. 900

C

25. 3609
26. 4620
27. 8300
28. 9999
29. 8088
30. 5893
31. 5800
32. 34,050
33. 33,010
34. 54,731
35. 88,888
36. 80,000

1.2 ADDING WHOLE NUMBERS

In order to add whole numbers, we must have some standard procedure or rule to follow. The rule for addition will be defined for you by the addition table for whole numbers.

Addition Table

+	0	1	2	3	4	5	6	7	8	9
0	0	1	2	3	4	5	6	7	8	9
1	1	2	3	4	5	6	7	8	9	10
2	2	3	4	5	6	7	8	9	10	11
3	3	4	5	6	7	8	9	10	11	12
4	4	5	6	7	8	9	10	11	12	13
5	5	6	7	8	9	10	11	12	13	14
6	6	7	8	9	10	11	12	13	14	15
7	7	8	9	10	11	12	13	14	15	16
8	8	9	10	11	12	13	14	15	16	17
9	9	10	11	12	13	14	15	16	17	18

When you need to add any two whole numbers from 0 through 9, look at the addition table to find the sum.

Example 1 **Add the following whole numbers using the addition table.**

$$7 + 6$$

+	0	1	2	3	4	5	**6**	7	8	9
0	0	1	2	3	4	5	6	7	8	9
1	1	2	3	4	5	6	7	8	9	10
2	2	3	4	5	6	7	8	9	10	11
3	3	4	5	6	7	8	9	10	11	12
4	4	5	6	7	8	9	10	11	12	13
5	5	6	7	8	9	10	11	12	13	14
6	6	7	8	9	10	11	12	13	14	15
7	7	8	9	10	11	12	**13**	14	15	16
8	8	9	10	11	12	13	14	15	16	17
9	9	10	11	12	13	14	15	16	17	18

Looking at the addition table we see that

$$7 + 6 = 13$$

Without the addition table, it would be difficult to add; however, the table only shows the addition of any two whole numbers 0 through 9. Addition involving a whole number larger than 9 is performed using several steps. ■

Basic Rule for Addition: Only like things can be added.

For example, apples can be added to apples and oranges can be added to oranges, but apples cannot be added to oranges without renaming them both as fruits and thus making them like things.

3 apples + 5 apples = 8 apples

3 oranges + 5 oranges = 8 oranges

3 apples + 5 oranges = 3 apples + 5 oranges since only like things can be added.

In a similar manner, only like groups in expanded notation can be added together. Expanded notation will be used throughout this chapter. Through the use of expanded notation, you should gain a better understanding of the structure of the decimal number system. Also, you should be better prepared to handle many of the algebra topics discussed later in this textbook.

Example 2 **Add the following whole numbers using expanded notation. Check your answer by adding in short notation.**

$$232 + 324$$

Step 1: Write the numbers in vertical form using expanded notation.

$$\begin{aligned} 232 &= \ 2(10^2) + 3(10) + 2(1) \\ +\ \underline{324} &= +\ \underline{3(10^2) + 2(10) + 4(1)} \end{aligned}$$

A N S W E R

If we break the example down into its parts, we see that there are actually three additions in this example. Starting right to left:

$$\begin{array}{r} 2(1) \\ +\ 4(1) \\ \hline \end{array} \quad \textbf{Addition 1} \qquad \begin{array}{r} 3(10) \\ +\ 2(10) \\ \hline \end{array} \quad \textbf{Addition 2} \qquad \begin{array}{r} 2(10^2) \\ +\ 3(10^2) \\ \hline \end{array} \quad \textbf{Addition 3}$$

Step 2: Add the groups of ones.

$$\begin{array}{rcr} 232 & = & 2(10^2) + 3(10) + \mathbf{2(1)} \\ +\ \underline{324} & = & +\ 3(10^2) + 2(10) + \mathbf{4(1)} \\ \hline \mathbf{6} & & \mathbf{6(1)} \end{array}$$

Step 3: Add the groups of tens.

$$\begin{array}{rcr} 232 & = & 2(10^2) + \mathbf{3(10)} + 2(1) \\ +\ \underline{324} & = & +\ 3(10^2) + \mathbf{2(10)} + 4(1) \\ \hline 56 & & \mathbf{5(10)} + 6(1) \end{array}$$

Step 4: Add the groups of 10^2.

$$\begin{array}{rcr} 232 & = & \mathbf{2(10^2)} + 3(10) + 2(1) \\ +\ \underline{324} & = & +\ \mathbf{3(10^2)} + 2(10) + 4(1) \\ \hline 556 & & \mathbf{5(10^2)} + 5(10) + 6(1) \end{array}$$

Note that like things were added together (groups of ones were added to groups of ones, groups of tens to groups of tens, and ten groups of tens to ten groups of tens). **It was possible to use the addition table since we did not add any numbers larger than 9 (2 + 4 = 6, 3 + 2 = 5, and 2 + 3 = 5).** ■

1. ____________________

Practice 1: Add the following whole numbers using expanded notation. Check your answer by adding in short notation.

235 + 354

2. ____________________

Practice 2: Add the following whole numbers using expanded notation. Check your answer by adding in short notation.

354 + 545

3. ____________________

Practice 3: Add the following whole numbers using expanded notation. Check your answer by adding in short notation.

247 + 632

Now, let's look at an example in algebra and compare it with its equivalent in arithmetic.

Example 3

Arithmetic	**Algebra**
$232 + 324$	$(2x^2 + 3x + 2) + (3x^2 + 2x + 4)$
Step 1: Write the numbers in vertical form using expanded notation.	*Step 1:* Write in vertical form.
$\begin{array}{rcr} 232 & = & 2(10^2) + 3(10) + 2 \\ +\ 324 & = & +\ 3(10^2) + 2(10) + 4 \\ \hline \end{array}$	$\begin{array}{r} 2x^2 + 3x + 2 \\ +\ 3x^2 + 2x + 4 \\ \hline \end{array}$
There are three additions. Starting on the right:	There are three additions. Starting on the right:
$\begin{array}{r} 2 \\ +\ 4 \\ \hline \end{array}$ **Addition 1**	$\begin{array}{r} 2 \\ +\ 4 \\ \hline \end{array}$ **Addition 1**
$\begin{array}{r} 3(10) \\ +\ 2(10) \\ \hline \end{array}$ **Addition 2**	$\begin{array}{r} 3x \\ +\ 2x \\ \hline \end{array}$ **Addition 2**
$\begin{array}{r} 2(10^2) \\ +\ 3(10^2) \\ \hline \end{array}$ **Addition 3**	$\begin{array}{r} 2x^2 \\ +\ 3x^2 \\ \hline \end{array}$ **Addition 3**
Step 2: Add the groups of ones.	*Step 2:* Add the groups of ones.
$\begin{array}{rcr} 232 & = & 2(10^2) + 3(10) + \mathbf{2} \\ +\ 324 & = & +\ 3(10^2) + 2(10) + \mathbf{4} \\ \hline \mathbf{6} & & \mathbf{6} \end{array}$	$\begin{array}{r} 2x^2 + 3x + \mathbf{2} \\ +\ 3x^2 + 2x + \mathbf{4} \\ \hline \mathbf{6} \end{array}$
Step 3: Add the groups of tens.	*Step 3:* Add the groups of x's.
$\begin{array}{rcr} 232 & = & 2(10^2) + \mathbf{3(10)} + 2 \\ +\ 324 & = & +\ 3(10^2) + \mathbf{2(10)} + 4 \\ \hline 56 & & \mathbf{5(10)} + 6 \end{array}$	$\begin{array}{r} 2x^2 + \mathbf{3x} + 2 \\ +\ 3x^2 + \mathbf{2x} + 4 \\ \hline \mathbf{5x} + 6 \end{array}$
Step 4: Add the groups of 10^2.	*Step 4:* Add the groups of x^2.
$\begin{array}{rcr} 232 & = & \mathbf{2(10^2)} + 3(10) + 2 \\ +\ 324 & = & +\ \mathbf{3(10^2)} + 2(10) + 4 \\ \hline 556 & & \mathbf{5(10^2)} + 5(10) + 6 \end{array}$	$\begin{array}{r} \mathbf{2x^2} + 3x + 2 \\ +\ \mathbf{3x^2} + 2x + 4 \\ \hline \mathbf{5x^2} + 5x + 6 \end{array}$ ■

Wait a minute! Except for the x's and tens, there is no real difference between the arithmetic and algebra.

The additions are basically the same, Charlie. The only difference is in one case groups of tens are added and in the other case groups of x's are added. However, the item being added does not affect the outcome. The outcome depends on how many things are added.

For example,

$$3x + 2x = 5x$$

$$3 \text{ apples} + 2 \text{ apples} = 5 \text{ apples}$$

$$3(10) + 2(10) = 5(10)$$

Algebra may not be so bad after all.

Let's look at another addition example.

Example 4 **Add the following whole numbers using expanded notation and the concept of regrouping and carrying. Check your answer by adding in short notation.**

$$453 + 379$$

Step 1: Write the numbers in vertical form using expanded notation.

$$\begin{array}{rcr} 453 & = & 4(10^2) + 5(10) + 3 \\ +\ \underline{379} & = & +\ \underline{3(10^2) + 7(10) + 9} \end{array}$$

There are three additions to be performed.

Step 2: Add the groups of ones.

$$\begin{array}{rcr} \mathbf{453} & = & 4(10^2) + 5(10) + \mathbf{3} \\ +\ \underline{\mathbf{379}} & = & +\ \underline{3(10^2) + 7(10) + \mathbf{9}} \\ \mathbf{12} & & \mathbf{12} \end{array}$$

Step 3: Regroup the sum and carry.

$$12 = 1(10) + 2$$

We have regrouped the 12 ones into 1(10) and 2 ones. Now carry the 1(10) to the ten's column.

$$\begin{array}{rcr} \mathbf{1} & & \mathbf{1(10)} \\ 453 & = & 4(10^2) + 5(10) + 3 \\ +\ \underline{379} & = & +\ \underline{3(10^2) + 7(10) + 9} \\ 2 & & 2 \end{array}$$

ANSWER

Step 4: Add the groups of tens, including the 1(10) that was carried.

$$\begin{array}{rcl} \mathbf{1} & & \mathbf{1(10)} \\ \mathbf{453} = & 4(10^2) + & \mathbf{5(10)} + 3 \\ + \underline{\mathbf{379}} = & + \underline{3(10^2) +} & \underline{\mathbf{7(10)} + 9} \\ \mathbf{132} & & \mathbf{13(10)} + 2 \end{array}$$

Step 5: Regroup the sum and carry.

$$\begin{aligned} 13(10) &= 130 \\ &= 1(100) + 30 \\ &= 1[10(10)] + 3(10) \\ &= 1(10^2) + 3(10) \end{aligned}$$

We have regrouped the 13(10) into $1(10^2)$ and 3(10). Now carry the $1(10^2)$ to the 10^2 column.

$$\begin{array}{rcl} 11 & \mathbf{1(10^2)} & 1(10) \\ 453 = & 4(10^2) + & 5(10) + 3 \\ + \underline{379} = & + \underline{3(10^2) +} & \underline{7(10) + 9} \\ 32 & & 3(10) + 2 \end{array}$$

Step 6: Add the groups of 10^2, including the $1(10^2)$ that was carried.

$$\begin{array}{rcl} \mathbf{11} & \mathbf{1(10^2)} & 1(10) \\ \mathbf{453} = & \mathbf{4(10^2)} + & 5(10) + 3 \\ + \underline{\mathbf{379}} = & + \underline{\mathbf{3(10^2)} +} & \underline{7(10) + 9} \\ \mathbf{832} & \mathbf{8(10^2)} + & 3(10) + 2 \end{array}$$

This example points out that *it is sometimes necessary to regroup and carry when adding.* ■

Practice 4: Add the following whole numbers using expanded notation and the concept of regrouping and carrying. Check your answer by adding in short notation.

$$542 + 329$$

4. ____________________

Practice 5: Add the following whole numbers using expanded notation and the concept of regrouping and carrying. Check your answer by adding in short notation.

$$586 + 372$$

5. ____________________

Practice 6: Add the following whole numbers using expanded notation and the concept of regrouping and carrying. Check your answer by adding in short notation.

$$759 + 686$$

6. ____________________

A N S W E R

7. ____________________

8. ____________________

Practice 7: Add the following whole numbers using expanded notation and the concept of regrouping and carrying. Check your answer by adding in short notation.

$$807 + 596$$

Practice 8: Add the following whole numbers using expanded notation and the concept of regrouping and carrying. Check your answer by adding in short notation.

$$596 + 79$$

This is one area of addition where algebra is perhaps easier than arithmetic since *numbers are not regrouped or carried in algebra.* Let's look at an example.

Good! I'm getting tired of carrying these numbers.

Example 5

Arithmetic

$$278 + 356$$

Step 1: Write the numbers in vertical form using expanded notation.

$$\begin{array}{rcr} 278 & = & 2(10^2) + 7(10) + 8 \\ +\ \underline{356} & = & +\ \underline{3(10^2) + 5(10) + 6} \end{array}$$

There are three additions to be performed.

Step 2: Add the groups of ones.

$$\begin{array}{rcr} 278 & = & 2(10^2) + 7(10) + \mathbf{8} \\ +\ \underline{356} & = & +\ \underline{3(10^2) + 5(10) + \mathbf{6}} \\ \mathbf{14} & & \mathbf{14} \end{array}$$

Algebra

$$(2x^2 + 7x + 8) + (3x^2 + 5x + 6)$$

Step 1: Write in vertical form.

$$\begin{array}{r} 2x^2 + 7x + 8 \\ +\ \underline{3x^2 + 5x + 6} \end{array}$$

There are three additions to be performed.

Step 2: Add the groups of ones.

$$\begin{array}{r} 2x^2 + 7x + \mathbf{8} \\ +\ \underline{3x^2 + 5x + \mathbf{6}} \\ \mathbf{14} \end{array}$$

Step 3: Regroup and carry.

$$\begin{array}{rcl} \mathbf{1} & & \mathbf{1(10)} \\ 278 & = & 2(10^2) + 7(10) + 8 \\ +\ 356 & = & +\ 3(10^2) + 5(10) + 6 \\ \hline 4 & & 4 \end{array}$$

Step 4: Add the groups of tens, including the 1(10) that was carried.

$$\begin{array}{rcl} \mathbf{1} & & \mathbf{1(10)} \\ 278 & = & 2(10^2) + \mathbf{7(10)} + 8 \\ +\ 356 & = & +\ 3(10^2) + \mathbf{5(10)} + 6 \\ \hline \mathbf{134} & & \mathbf{13(10)} + 4 \end{array}$$

Step 5: Regroup and carry.

$$\begin{array}{rcl} \mathbf{1}1 & & \mathbf{1(10^2)} \quad 1(10) \\ 278 & = & 2(10^2) + 7(10) + 8 \\ +\ 356 & = & +\ 3(10^2) + 5(10) + 6 \\ \hline 34 & & 3(10) + 4 \end{array}$$

Step 6: Add the groups of 10^2, including the $1(10^2)$ that was carried.

$$\begin{array}{rcl} \mathbf{1}1 & & \mathbf{1(10^2)} \quad 1(10) \\ \mathbf{2}78 & = & \mathbf{2(10^2)} + 7(10) + 8 \\ +\ \mathbf{3}56 & = & +\ \mathbf{3(10^2)} + 5(10) + 6 \\ \hline \mathbf{6}34 & & \mathbf{6(10^2)} + 3(10) + 4 \end{array}$$

Step 3: Add the groups of x's.

$$\begin{array}{r} 2x^2 + \mathbf{7x} + 8 \\ +\ 3x^2 + \mathbf{5x} + 6 \\ \hline \mathbf{12x} + 14 \end{array}$$

Step 4: Add the groups of x^2.

$$\begin{array}{r} \mathbf{2x^2} + 7x + 8 \\ +\ \mathbf{3x^2} + 5x + 6 \\ \hline \mathbf{5x^2} + 12x + 14 \end{array}$$

■

The answers are not the same!

That's right, Charlie. The answers are different since numbers cannot be regrouped or carried in algebra. In algebra, the groups are groups of x's; thus, it is not known how many of an item it takes to make a group. In the case of chairs, it is not known how many chairs it will take to make a stack. Therefore, *it is not possible to regroup and carry in algebra.*

How many chairs do I put in a stack?

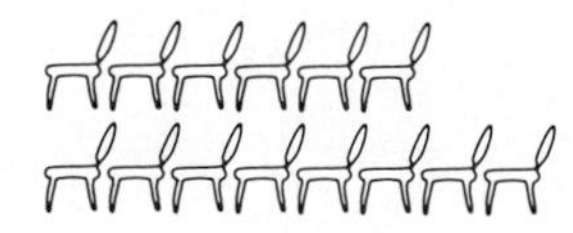

x chairs.

But how many is x?

We don't know how many x is, Charlie.

EXERCISE 1.2

Add the following whole numbers in each problem using expanded notation. Check your answers by adding in short notation.

A

1. 43 + 4

2. 25 + 3

3. 5 + 54

4. 25 + 7

5. 19 + 6

6. 18 + 30

7. 73 + 25

8. 57 + 80

9. 83 + 76

10. 68 + 38

11. 76 + 67

12. 97 + 86

13. 59 + 78

14. 75 + 79

15. 77 + 99

B

16. 136 + 42

17. 367 + 80

18. 507 + 89

19. 378 + 48

20. 354 + 76

21. 573 + 400

22. 587 + 322

23. 648 + 472

24. 938 + 500

25. 572 + 269

26. 693 + 952

27. 780 + 590

28. 808 + 787

29. 809 + 396

30. 955 + 688

C

31. 2486 + 4300

32. 5738 + 2587

33. 4792 + 8010

34. 6462 + 3750

35. 7009 + 6099

36. 7897 + 6786

37. 8695 + 67,497

38. 17,848 + 7895

39. 48,944 + 86,843

APPLICATIONS

Work the following problems using short notation. When an asterisk appears by a problem, you may want to use your calculator.

1. Find the perimeter (the sum of the lengths of the sides) of the following triangle.

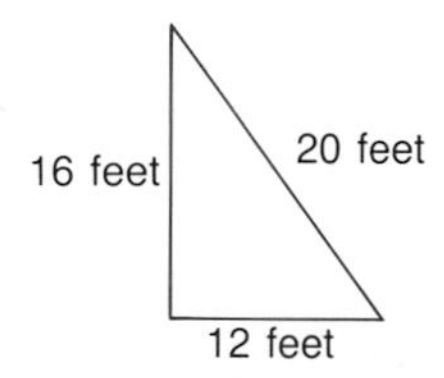

***2.** Find the perimeter (the sum of the lengths of the sides) of the following rectangle.

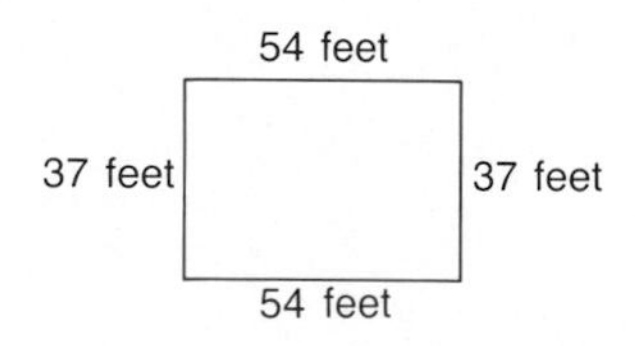

***3.** Joyce wants to fence her yard. The dimensions of the yard are shown in the figure below. How many feet of fence will Joyce need to buy?

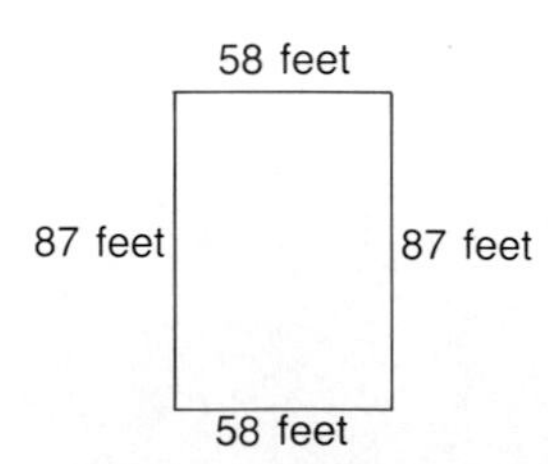

4. When Joyce found the number of feet to fence her entire yard in the preceding problem, she decided to fence only a small part of the backyard. The dimensions Joyce wants to fence are shown in the figure below. How many feet of fence will Joyce need to buy in order to fence this section of her yard?

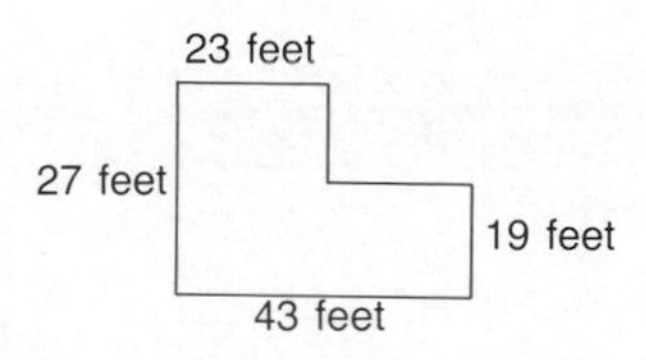

Wait a minute! We don't have all the dimensions.

That's true, Charlie. However, you can use the dimensions given to find the missing ones.

***5.** Sam wants to install a light at point B by running an electrical wire from point A to point B along the path shown in the figure below. How many feet of electrical wire will Sam need to buy?

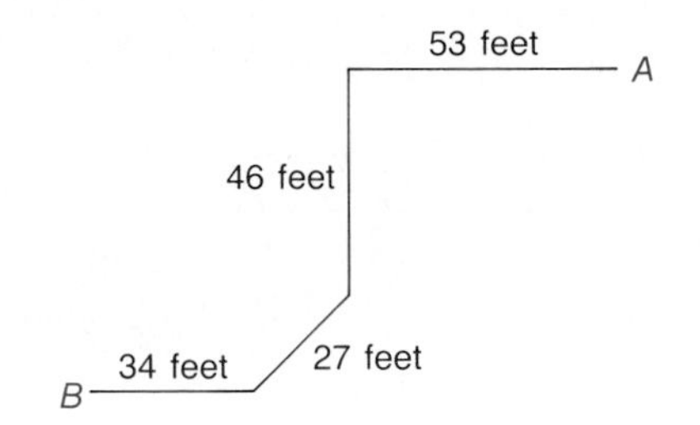

1.3 SUBTRACTING WHOLE NUMBERS

In Section 1.2, the addition table was used to add whole numbers. Since subtraction is the inverse of addition, you can use the addition table to subtract whole numbers by just using the table in reverse.

> **Definition of Subtraction.** If a, b, and c represent whole numbers, then $a - b = c$ if and only if $b + c = a$.

So, the subtraction

$$a - b = ?$$

becomes the addition

$$b + ? = a$$

Written in vertical form, the subtraction

$$\begin{array}{r} a \\ -\ b \\ \hline ? \end{array} \quad \text{becomes the addition} \quad \begin{array}{r} b \\ +\ ? \\ \hline a \end{array}$$

For example,

$$8 - 5 = ?$$

is changed to

$$5 + ? = 8$$

Now, looking back at the addition table

$$5 + 3 = 8$$

thus,

$$8 - 5 = 3$$

> **Basic Rule for Subtraction:** Only like things can be subtracted.

For example, apples can be subtracted from apples and oranges can be subtracted from oranges, but apples cannot be subtracted from oranges.

8 apples − 5 apples = 3 apples

8 oranges − 5 oranges = 3 oranges

8 oranges − 5 apples = 8 oranges − 5 apples since only like things can be subtracted.

Just as in addition, the following subtraction problems will be worked using expanded notation.

Example 1 **Subtract the following whole numbers using expanded notation. Check your answer by subtracting in short notation.**

557 − 235

Step 1: Write the numbers in vertical form using expanded notation.

$$\begin{array}{rrl} 557 & = & 5(10^2) + 5(10) + 7(1) \\ -\underline{235} & = & -\underline{2(10^2) + 3(10) + 5(1)} \end{array}$$

There are really three subtractions within this example.

$$\begin{array}{r} 7(1) \\ -\underline{5(1)} \end{array} \quad \textbf{Subtraction 1} \qquad \begin{array}{r} 5(10) \\ -\underline{3(10)} \end{array} \quad \textbf{Subtraction 2} \qquad \begin{array}{r} 5(10^2) \\ -\underline{2(10^2)} \end{array} \quad \textbf{Subtraction 3}$$

Step 2: Subtract the groups of ones.

$$\begin{array}{rrr} 557 & = & 5(10^2) + 5(10) + \mathbf{7(1)} \\ -\underline{235} & = & -\underline{2(10^2) + 3(10) + \mathbf{5(1)}} \\ \mathbf{2} & & \mathbf{2(1)} \end{array}$$

Step 3: Subtract the groups of tens.

$$\begin{array}{rrr} 557 & = & 5(10^2) + \mathbf{5(10)} + 7(1) \\ -\underline{235} & = & -\underline{2(10^2) + \mathbf{3(10)} + 5(1)} \\ \mathbf{22} & & \mathbf{2(10)} + 2(1) \end{array}$$

Step 4: Subtract the groups of 10^2.

$$\begin{array}{rrr} \mathbf{557} & = & \mathbf{5(10^2)} + 5(10) + 7(1) \\ -\underline{\mathbf{235}} & = & -\underline{\mathbf{2(10^2)} + 3(10) + 5(1)} \\ \mathbf{322} & & \mathbf{3(10^2)} + 2(10) + 2(1) \end{array}$$

■

Only like things are subtracted; groups of ones are subtracted from groups of ones, groups of tens from groups of tens, and ten groups of tens from ten groups of tens. Also, don't be confused by the plus signs in the expanded notation. They are used to show that the groups on each row are connected. You subtract between rows, not within each row.

A N S W E R

Practice 1: Subtract the following whole numbers using expanded notation. Check your answer by subtracting in short notation.

$$578 - 453$$

1. ____________

Practice 2: Subtract the following whole numbers using expanded notation. Check your answer by subtracting in short notation.

$$765 - 543$$

2. ____________

Practice 3: Subtract the following whole numbers using expanded notation. Check your answer by subtracting in short notation.

$$938 - 726$$

3. ____________

Now, let's look at an example in algebra and compare it with its equivalent in arithmetic.

Example 2

Arithmetic	Algebra
$557 - 235$	$(5x^2 + 5x + 7) - (2x^2 + 3x + 5)$
Step 1: Write the numbers in vertical form using expanded notation.	*Step 1:* Write in vertical form.
$\begin{array}{rcr} 557 & = & 5(10^2) + 5(10) + 7 \\ -\ 235 & = & -\ 2(10^2) + 3(10) + 5 \\ \hline \end{array}$	$\begin{array}{r} 5x^2 + 5x + 7 \\ -\ 2x^2 + 3x + 5 \\ \hline \end{array}$
There are three subtractions.	There are three subtractions.
$\begin{array}{r} 7 \\ -\ 5 \\ \hline \end{array}$ **Subtraction 1**	$\begin{array}{r} 7 \\ -\ 5 \\ \hline \end{array}$ **Subtraction 1**
$\begin{array}{r} 5(10) \\ -\ 3(10) \\ \hline \end{array}$ **Subtraction 2**	$\begin{array}{r} 5x \\ -\ 3x \\ \hline \end{array}$ **Subtraction 2**
$\begin{array}{r} 5(10^2) \\ -\ 2(10^2) \\ \hline \end{array}$ **Subtraction 3**	$\begin{array}{r} 5x^2 \\ -\ 2x^2 \\ \hline \end{array}$ **Subtraction 3**

Step 2: Subtract the groups of ones.

$$\begin{array}{rrr} 557 & = & 5(10^2) + 5(10) + \mathbf{7} \\ -\ 235 & = & -\ 2(10^2) + 3(10) + \mathbf{5} \\ \hline \mathbf{2} & & \mathbf{2} \end{array}$$

Step 3: Subtract the groups of tens.

$$\begin{array}{rrr} 557 & = & 5(10^2) + \mathbf{5(10)} + 7 \\ -\ 235 & = & -\ 2(10^2) + \mathbf{3(10)} + 5 \\ \hline 22 & & \mathbf{2(10)} + 2 \end{array}$$

Step 4: Subtract the groups of 10^2.

$$\begin{array}{rrr} 557 & = & \mathbf{5(10^2)} + 5(10) + 7 \\ -\ 235 & = & -\ \mathbf{2(10^2)} + 3(10) + 5 \\ \hline 322 & & \mathbf{3(10^2)} + 2(10) + 2 \end{array}$$

Step 2: Subtract the groups of ones.

$$\begin{array}{r} 5x^2 + 5x + \mathbf{7} \\ -\ 2x^2 + 3x + \mathbf{5} \\ \hline \mathbf{2} \end{array}$$

Step 3: Subtract the groups of x's.

$$\begin{array}{r} 5x^2 + \mathbf{5x} + 7 \\ -\ 2x^2 + \mathbf{3x} + 5 \\ \hline \mathbf{2x} + 2 \end{array}$$

Step 4: Subtract the groups of x^2.

$$\begin{array}{r} \mathbf{5x^2} + 5x + 7 \\ -\ \mathbf{2x^2} + 3x + 5 \\ \hline \mathbf{3x^2} + 2x + 2 \end{array}$$

■

It looks like I will be able to do algebra if I can do arithmetic.

That's correct, Charlie. If you can do arithmetic, you are on your way to being able to do algebra. The only difference is instead of groups of tens being subtracted, groups of x's are subtracted. The items being subtracted do not affect the outcome. The outcome depends on how many things are subtracted.

For example,

$$5x - 3x = 2x$$

$$5 \text{ apples} - 3 \text{ apples} = 2 \text{ apples}$$

$$5(10) - 3(10) = 2(10)$$

You know, I may learn to like algebra.

Here is a slightly more difficult subtraction example.

Example 3 **Subtract the following whole numbers using expanded notation and the concept of borrowing and regrouping. Check your answer by subtracting in short notation.**

$$635 - 378$$

<u>*Step 1:*</u> Write the numbers in vertical form using expanded notation.

$$\begin{array}{rcrl} 635 & = & & 6(10^2) + 3(10) + 5 \\ -\ \underline{378} & = & - & \underline{3(10^2) + 7(10) + 8} \end{array}$$

There are three subtractions to be performed.

<u>*Step 2:*</u> Subtract the groups of ones.

$$\begin{array}{rcrl} 63\mathbf{5} & = & & 6(10^2) + 3(10) + \mathbf{5} \\ -\ \underline{37\mathbf{8}} & = & - & \underline{3(10^2) + 7(10) + \mathbf{8}} \end{array}$$

At this time we cannot subtract 8 from 5; thus, we borrow and regroup.

<u>*Step 3:*</u> Borrow and regroup. A group of tens is borrowed and regrouped to form 10 ones. Since we already have 5 ones, this gives us a total of 15 ones. However we now have only 2 tens left.

$$\begin{array}{rcrl} \mathbf{2\ 15} & & & \qquad\quad \mathbf{2(10)} \quad \mathbf{15} \\ 6\,\cancel{3}\,\cancel{5} & = & & 6(10^2) + \cancel{3(10)} + \cancel{5} \\ -\ \underline{3\ 7\ 8} & = & - & \underline{3(10^2) + 7(10) + 8} \end{array}$$

<u>*Step 4:*</u> Subtract the groups of ones, including the 10 ones that were borrowed.

$$\begin{array}{rcrl} 2\ \mathbf{15} & & & \qquad\quad 2(10) \quad \mathbf{15} \\ 6\,\cancel{3}\,\cancel{5} & = & & 6(10^2) + \cancel{3(10)} + \cancel{5} \\ -\ \underline{3\ 7\ \mathbf{8}} & = & - & \underline{3(10^2) + 7(10) + \mathbf{8}} \\ \mathbf{7} & & & \qquad\qquad\qquad\quad \mathbf{7} \end{array}$$

<u>*Step 5:*</u> Subtract the groups of tens, omitting the 1 ten that was borrowed. The second subtraction has now changed since one group of tens was borrowed and regrouped into 10 ones in order to do the first subtraction.

$$\begin{array}{rcrl} \mathbf{2}\ 15 & & & \qquad\quad \mathbf{2(10)} \quad 15 \\ 6\,\cancel{3}\,\cancel{5} & = & & 6(10^2) + \cancel{3(10)} + \cancel{5} \\ -\ \underline{3\ \mathbf{7}\ 8} & = & - & \underline{3(10^2) + \mathbf{7(10)} + 8} \\ 7 & & & \qquad\qquad\qquad\quad 7 \end{array}$$

At this time we cannot subtract 7 from 2; thus, we borrow and regroup.

<u>*Step 6:*</u> Borrow and regroup. We borrow and regroup one of the groups of 10^2 into 10 groups of tens [$10^2 = 10(10)$]. We now have $10 + 2 = 12$ groups of tens.

$$\begin{array}{rcrl} \mathbf{12} & & & \qquad\quad \mathbf{12(10)} \\ 5\,\cancel{2}\,15 & & & \mathbf{5(10^2)} \quad \cancel{2(10)} \quad 15 \\ \cancel{6}\,\cancel{3}\,\cancel{5} & = & & \cancel{6(10^2)} + \cancel{3(10)} + \cancel{5} \\ -\ \underline{3\ 7\ 8} & = & - & \underline{3(10^2) + 7(10) + 8} \\ 7 & & & \qquad\qquad\qquad\quad 7 \end{array}$$

<u>*Step 7:*</u> Subtract the groups of tens including the 10 tens that were borrowed.

$$\begin{array}{rcrl} \mathbf{12} & & & \qquad\quad \mathbf{12(10)} \\ 5\,\cancel{2}\,15 & & & 5(10^2) \quad \cancel{2(10)} \quad 15 \\ \cancel{6}\,\cancel{3}\,\cancel{5} & = & & \cancel{6(10^2)} + \cancel{3(10)} + \cancel{5} \\ -\ \underline{3\ \mathbf{7}\ 8} & = & - & \underline{3(10^2) + \mathbf{7(10)} + 8} \\ \mathbf{5}\,7 & & & \qquad\qquad \mathbf{5(10)} + 7 \end{array}$$

A N S W E R

Step 8: Subtract the groups of 10^2, omitting the $1(10^2)$ that was borrowed.

$$
\begin{array}{rcrcrcr}
12 & & & & 12(10) & & \\
\mathbf{5}\,\cancel{2}\,15 & & \mathbf{5(10^2)} & & \cancel{2(10)} & & 15 \\
\cancel{6\,3\,5} & = & \cancel{6(10^2)} & + & \cancel{3(10)} & + & \cancel{5} \\
-\ \underline{\mathbf{3}\,7\,8} & = & -\ \mathbf{3(10^2)} & + & 7(10) & + & 8 \\
\hline
\mathbf{2}\,5\,7 & & \mathbf{2(10^2)} & + & 5(10) & + & 7
\end{array}
$$

This example points out that *it is sometimes necessary to borrow and regroup when subtracting.* ■

4. ________________

Practice 4: Subtract the following whole numbers using expanded notation and the concept of borrowing and regrouping. Check your answer by subtracting in short notation.

$$852 - 635$$

5. ________________

Practice 5: Subtract the following whole numbers using expanded notation and the concept of borrowing and regrouping. Check your answer by subtracting in short notation.

$$737 - 485$$

6. ________________

Practice 6: Subtract the following whole numbers using expanded notation and the concept of borrowing and regrouping. Check your answer by subtracting in short notation.

$$465 - 98$$

7. ________________

Practice 7: Subtract the following whole numbers using expanded notation and the concept of borrowing and regrouping. Check your answer by subtracting in short notation.

$$536 - 328$$

Just as numbers cannot be regrouped or carried in algebra, *numbers cannot be regrouped or borrowed in algebra.*

I'm glad. Borrowing is a good way to lose friends.

Example 4

Arithmetic

$$723 - 356$$

<u>*Step 1:*</u> Write the numbers in vertical form using expanded notation.

$$\begin{array}{rcr} 723 & = & 7(10^2) + 2(10) + 3 \\ -\ \underline{356} & = & -\ \underline{3(10^2) + 5(10) + 6} \end{array}$$

There are three subtractions to be performed.

<u>*Step 2:*</u> Subtract the groups of ones.

$$\begin{array}{rcr} 72\mathbf{3} & = & 7(10^2) + 2(10) + \mathbf{3} \\ -\ \underline{35\mathbf{6}} & = & -\ \underline{3(10^2) + 5(10) + \mathbf{6}} \end{array}$$

This subtraction can be worked the same way as Example 3.

Algebra

$$(7x^2 + 2x + 3) - (3x^2 + 5x + 6)$$

<u>*Step 1:*</u> Write in vertical form.

$$\begin{array}{r} 7x^2 + 2x + 3 \\ -\ \underline{3x^2 + 5x + 6} \end{array}$$

There are three subtractions to be performed.

<u>*Step 2:*</u> Subtract the groups of ones.

$$\begin{array}{r} 7x^2 + 2x + \mathbf{3} \\ -\ \underline{3x^2 + 5x + \mathbf{6}} \end{array}$$

It appears one of the x's must be regrouped. But how many is an x? This question cannot be answered because x can be any number. This particular example will have to be worked later in the book after negative numbers have been introduced. ■

Example 5 **Subtract the following whole numbers using expanded notation and the concept of borrowing and regrouping. Check your answer by subtracting in short notation.**

$$502 - 327$$

<u>*Step 1:*</u> Write the numbers in vertical form using expanded notation.

$$\begin{array}{rcr} 502 & = & 5(10^2) + 0(10) + 2 \\ -\ \underline{327} & = & -\ \underline{3(10^2) + 2(10) + 7} \end{array}$$

<u>*Step 2:*</u> Since at this time we cannot subtract 7 from 2, we will borrow and regroup. There are no groups of tens to borrow. Thus, it becomes necessary to borrow and regroup a group of 10^2. We take one of the groups of 10^2 and regroup it into 10 groups of tens. We now have

$$\begin{array}{rcr} \mathbf{4\ 10} & & \mathbf{4(10^2)} \quad \mathbf{10(10)} \quad\quad \\ \cancel{5}\,\cancel{0}\,2 & = & \cancel{5(10^2)} + \cancel{0(10)} + 2 \\ -\ \underline{3\,2\,7} & = & -\ \underline{3(10^2) + 2(10) + 7} \end{array}$$

A N S W E R

Step 3: We now borrow and regroup one of the groups of tens.

$$\begin{array}{r} \mathbf{9} \quad\quad \\ 4\ \not{10}\ \mathbf{12} \\ \not{5}\ \not{0}\ \not{2} \\ -\ 3\ \ 2\ \ 7 \\ \hline \end{array} = \begin{array}{r} \mathbf{9(10)} \quad\quad \\ 4(10^2) \quad \not{10(10)} \quad \mathbf{12} \\ \not{5(10^2)} + \not{0(10)} + \not{2} \\ -\ 3(10^2) + 2(10) + 7 \\ \hline \end{array}$$

Step 4: Subtract 7 from 12.

$$\begin{array}{r} 9 \quad\quad \\ 4\ \not{10}\ \mathbf{12} \\ \not{5}\ \not{0}\ \not{2} \\ -\ 3\ \ 2\ \ \mathbf{7} \\ \hline \mathbf{5} \end{array} = \begin{array}{r} 9(10) \quad\quad \\ 4(10^2) \quad \not{10(10)} \quad \mathbf{12} \\ \not{5(10^2)} + \not{0(10)} + \not{2} \\ -\ 3(10^2) + 2(10) + \mathbf{7} \\ \hline \mathbf{5} \end{array}$$

Step 5: Subtract 2(10) from 9(10).

$$\begin{array}{r} \mathbf{9} \quad\quad \\ 4\ \not{10}\ 12 \\ \not{5}\ \not{0}\ \not{2} \\ -\ 3\ \ \mathbf{2}\ \ 7 \\ \hline \mathbf{7}\ \ 5 \end{array} = \begin{array}{r} \mathbf{9(10)} \quad\quad \\ 4(10^2) \quad \not{10(10)} \quad 12 \\ \not{5(10^2)} + \not{0(10)} + \not{2} \\ -\ 3(10^2) + \mathbf{2(10)} + 7 \\ \hline \mathbf{7(10)} + 5 \end{array}$$

Step 6: Subtract $3(10^2)$ from $4(10^2)$.

$$\begin{array}{r} 9 \quad\quad \\ \mathbf{4}\ \not{10}\ 12 \\ \not{5}\ \not{0}\ \not{2} \\ -\ \mathbf{3}\ \ 2\ \ 7 \\ \hline \mathbf{1}\ \ 7\ \ 5 \end{array} = \begin{array}{r} 9(10) \quad\quad \\ \mathbf{4(10^2)} \quad \not{10(10)} \quad 12 \\ \not{5(10^2)} + \not{0(10)} + \not{2} \\ -\ \mathbf{3(10^2)} + 2(10) + 7 \\ \hline \mathbf{1(10^2)} + 7(10) + 5 \end{array}$$

■

8. ______________

Practice 8: Subtract the following whole numbers using expanded notation and the concept of borrowing and regrouping. Check your answer by subtracting in short notation.

$$703 - 486$$

9. ______________

Practice 9: Subtract the following whole numbers using expanded notation and the concept of borrowing and regrouping. Check your answer by subtracting in short notation.

$$800 - 576$$

EXERCISE 1.3

Subtract the following whole numbers in each problem using expanded notation. Check your answer by subtracting in short notation.

A

1. 29 − 8

2. 35 − 3

3. 46 − 5

4. 31 − 8

5. 42 − 9

6. 36 − 7

7. 58 − 23

8. 46 − 15

9. 67 − 36

10. 74 − 45

11. 55 − 49

12. 57 − 38

13. 60 − 43

14. 67 − 59

15. 66 − 49

B

16. 446 − 46

17. 573 − 40

18. 634 − 23

19. 572 − 38

20. 643 − 97

21. 638 − 365

22. 892 − 384

23. 538 − 269

24. 438 − 385

25. 703 − 538

26. 957 − 689

27. 700 − 399

28. 803 − 594

29. 777 − 699

30. 700 − 400

C

31. 7693 − 5739

32. 7430 − 3891

33. 4702 − 3547

34. 8432 − 6378

35. 7603 − 5607

36. 9053 − 4987

37. 8003 − 5706

38. 7002 − 5994

39. 8000 − 5599

APPLICATIONS

Work the following problems using short notation. When an asterisk appears by a problem, you may want to use your calculator.

1. John wants to buy a stereo system for $536. He only has $329. How much more money does John need in order to buy the stereo system?

2. George and Mary plan to make a canvas tent. It takes 83 yards of canvas to construct the tent. The local store has only 37 yards of canvas; however, the store manager is willing to order the remaining canvas. How much canvas needs to be ordered?

3. In Chemistry 111, it is necessary to have a total of 525 points to receive a grade of A for the semester. Jane has a total of 439 points prior to the last test of the semester. How many points will Jane need on her last test to receive an A in chemistry?

***4.** Last year there was a total of 382,704 acres of land under cultivation in Elmore County. This year 27,896 acres have been removed from cultivation. How many acres of land are currently under cultivation in Elmore County?

***5.** A new car costs $9823 with automatic transmission and air conditioning. Sam wants to buy the same model car without the automatic transmission and air conditioning. If the automatic transmission costs $785 and the air conditioner costs $593, how much will Sam's car cost?

1.4 MULTIPLYING WHOLE NUMBERS

Working with whole numbers, mathematicians have found that repeated additions such as the following occur quite often.

$$5 + 5 + 5 + 5 + 5 + 5$$

$$7 + 7 + 7 + 7 + 7 + 7 + 7 + 7 + 7$$

ANSWER

They decided that it would be convenient to have a shorter notation for this type of addition, so the following notation was introduced.

$$5 + 5 + 5 + 5 + 5 + 5 = 6 \times 5$$

$$7 + 7 + 7 + 7 + 7 + 7 + 7 + 7 + 7 = 9 \times 7$$

Now, 6×5 represents the addition of 6 fives, and 9×7 represents the addition of 9 sevens. This special form of addition is called **multiplication.**

A similar problem soon occurred with multiplication. Repeated multiplications such as the following began to occur quite often.

$$5 \times 5 \times 5 \times 5 \times 5 \times 5$$

$$7 \times 7 \times 7 \times 7 \times 7 \times 7 \times 7 \times 7 \times 7$$

When you solve one problem, you sometimes create another problem.

That's true, Charlie. However, they solved the multiplication problem in the same way they solved the addition problem. They decided to use the following notation

$$5 \times 5 \times 5 \times 5 \times 5 \times 5 = 5^6$$

Now, 5^6 represents the multiplication of 6 fives. The number 5 is called the **base** and the number 6 is called the **exponent.** Thus,

$$7 \times 7 \times 7 \times 7 \times 7 \times 7 \times 7 \times 7 \times 7 = 7^9$$

where 7 is the base and 9 is the exponent.

1. ______ Practice 1: Write the following addition as a multiplication.

$$2 + 2 + 2 + 2 + 2 + 2$$

2. ______ Practice 2: Write the following multiplication using an exponent.

$$2 \times 2 \times 2 \times 2 \times 2 \times 2$$

3. ______ Practice 3: Write the following multiplication as an addition.

$$5 \times 9$$

4. ______ Practice 4: Write the following without using an exponent.

$$9^5$$

5. ______ Practice 5: Write the following addition as a multiplication.

$$4 + 4 + 4$$

6. ______ Practice 6: Write the following multiplication using an exponent.

$$4 \times 4 \times 4$$

7. ______ Practice 7: Write the following multiplication as an addition.

$$7 \times 2$$

Practice 8: Write the following without using an exponent.

$$2^7$$

Practice 9: Write the following addition as a multiplication.

$$7 + 7 + 7 + 7 + 7 + 7 + 7$$

Practice 10: Write the following multiplication using an exponent.

$$7 \times 7 \times 7 \times 7 \times 7 \times 7 \times 7$$

Practice 11: Write the following multiplication as an addition.

$$3 \times 5$$

Practice 12: Write the following without using an exponent.

$$5^3$$

A N S W E R

8. ______

9. ______

10. ______

11. ______

12. ______

The multiplication table for whole numbers can be constructed using the addition table for whole numbers; however, we will not do the construction of the multiplication table. We will just write the table.

Multiplication Table

×	0	1	2	3	4	5	6	7	8	9
0	0	0	0	0	0	0	0	0	0	0
1	0	1	2	3	4	5	6	7	8	9
2	0	2	4	6	8	10	12	14	16	18
3	0	3	6	9	12	15	18	21	24	27
4	0	4	8	12	16	20	24	28	32	36
5	0	5	10	15	20	25	30	35	40	45
6	0	6	12	18	24	30	36	42	48	54
7	0	7	14	21	28	35	42	49	56	63
8	0	8	16	24	32	40	48	56	64	72
9	0	9	18	27	36	45	54	63	72	81

When we multiply two whole numbers that are not larger than 9, we look at the multiplication table to find the product (answer).

Example 1 **Multiply the following whole numbers using the multiplication table.**

$$7 \times 5$$

×	0	1	2	3	4	**5**	6	7	8	9
0	0	0	0	0	0	0	0	0	0	0
1	0	1	2	3	4	5	6	7	8	9
2	0	2	4	6	8	10	12	14	16	18
3	0	3	6	9	12	15	18	21	24	27
4	0	4	8	12	16	20	24	28	32	36
5	0	5	10	15	20	25	30	35	40	45
6	0	6	12	18	24	30	36	42	48	54
7	0	7	14	21	28	**35**	42	49	56	63
8	0	8	16	24	32	40	48	56	64	72
9	0	9	18	27	36	45	54	63	72	81

Solution: Looking at the multiplication table we find that $7 \times 5 = 35$. ■

Without the multiplication table it would be difficult to multiply; however, the table only shows the multiplication of any two whole numbers 0 through 9. Multiplication involving a whole number larger than 9 is performed using several steps. In order to multiply whole numbers larger than 9, we will need the use of three properties.

Commutative Property of Multiplication. If a and b represent whole numbers, then $a \times b = b \times a$.

For example,

$$2 \times 3 = 6$$

and

$$3 \times 2 = 6$$

Thus,

$$2 \times 3 = 3 \times 2$$

Associative Property of Multiplication. If a, b, and c represent whole numbers, then $(a \times b) \times c = a \times (b \times c)$.

For example,

$$(3 \times 2) \times 5 = 6 \times 5 = 30$$

and

$$3 \times (2 \times 5) = 3 \times 10 = 30$$

Thus,

$$(3 \times 2) \times 5 = 3 \times (2 \times 5)$$

Distributive Property of Multiplication over Addition. If a, b, and c represent whole numbers, then $a \times (b + c) = (a \times b) + (a \times c)$.

For example,

$$2 \times (3 + 5) = 2 \times 8 = 16$$

and

$$(2 \times 3) + (2 \times 5) = 6 + 10 = 16$$

Thus,

$$2 \times (3 + 5) = (2 \times 3) + (2 \times 5)$$

Example 2 **Multiply the following whole numbers using expanded notation. Check your answer by multiplying in short notation.**

$$3 \times 32$$

Step 1: Write the numbers in expanded notation.

$$3 \times [3(10) + 2]$$

Step 2: Apply the distributive property of multiplication over addition.

$$\begin{aligned} 3 \times [3(10) + 2] &= [3 \times 3(10)] + (3 \times 2) \\ &= 9(10) + 6 \end{aligned}$$

Thus, $3 \times 32 = 96$.

Now, let's do the same example using a vertical format.

A N S W E R

Step 1: Write the numbers in vertical form.

$$\begin{array}{r} 32 \\ \times\ 3 \\ \hline \end{array}$$ **Short Notation**

$$\begin{array}{rr} & 3(10) + 2 \\ \times & 3 \\ \hline \end{array}$$ **Expanded Notation**

Step 2:

Multiply 3 times 2.

$$\begin{array}{r} 32 \\ \times\ 3 \\ \hline 6 \end{array}$$

Multiply 3 times 2.

$$\begin{array}{rr} & 3(10) + 2 \\ \times & 3 \\ \hline & 6 \end{array}$$

Step 3:

Multiply 3 times 3.

$$\begin{array}{r} 32 \\ \times\ 3 \\ \hline 96 \end{array}$$

Multiply 3 times 3(10).

$$\begin{array}{rr} & 3(10) + 2 \\ \times & 3 \\ \hline & 9(10) + 6 \end{array}$$ ■

By working the following practice problems using expanded notation, you should gain a better understanding of arithmetic as well as better preparation for algebra.

Practice 13: Multiply the following whole numbers using expanded notation. Check your answer by multiplying in short notation.

2 × 32

13. ______________

Practice 14: Multiply the following whole numbers using expanded notation. Check your answer by multiplying in short notation.

3 × 21

14. ______________

Practice 15: Multiply the following whole numbers using expanded notation. Check your answer by multiplying in short notation.

4 × 22

15. ______________

Let's look at a similar example in algebra.

Example 3

Arithmetic	**Algebra**
2×24	$2 \times (2x + 4)$
Step 1: Write the numbers in vertical form using expanded notation.	*Step 1:* Write in vertical form.
$\begin{array}{rcr} 24 & = & 2(10) + 4 \\ \times\ \underline{2} & = \times & \underline{\qquad 2} \end{array}$	$\begin{array}{r} 2x + 4 \\ \times\ \underline{\qquad 2} \end{array}$
Apply the distributive property and multiply.	Apply the distributive property and multiply.
Step 2: Multiply 2 times 4.	*Step 2:* Multiply 2 times 4.
$\begin{array}{rcr} 24 & = & 2(10) + \mathbf{4} \\ \times\ \underline{\mathbf{2}} & = \times & \underline{\qquad \mathbf{2}} \\ \mathbf{8} & & \mathbf{8} \end{array}$	$\begin{array}{r} 2x + \mathbf{4} \\ \times\ \underline{\qquad \mathbf{2}} \\ \mathbf{8} \end{array}$
Step 3: Multiply 2 times 2(10).	*Step 3:* Multiply 2 times $2x$.
$\begin{array}{rcr} 24 & = & \mathbf{2(10)} + 4 \\ \times\ \underline{\mathbf{2}} & = \times & \underline{\qquad \mathbf{2}} \\ 48 & & \mathbf{4(10)} + 8 \end{array}$	$\begin{array}{r} \mathbf{2x} + 4 \\ \times\ \underline{\qquad \mathbf{2}} \\ \mathbf{4x} + 8 \end{array}$ ■

It looks like using expanded notation in arithmetic will help me with algebra.

Example 4 **Multiply the following whole numbers using expanded notation and the concept of regrouping and carrying. Check your answer by multiplying in short notation.**

$$6 \times 34$$

Step 1: Write the numbers in vertical form using expanded notation.

$$\begin{array}{rcr} 34 & = & 3(10) + 4 \\ \times\ \underline{6} & = \times & \underline{\qquad 6} \end{array}$$

Apply the distributive property and multiply.

ANSWER

Step 2: Multiply 6 times 4.

$$\begin{array}{r} 34 \\ \times\ 6 \\ \hline 24 \end{array} = \begin{array}{r} 3(10) + 4 \\ \times\ \ \ 6 \\ \hline 24 \end{array}$$

Step 3: Regroup and carry.

$$24 = 2(10) + 4$$

We have regrouped the 24 ones into 2(10) and 4 ones. Now carry the 2(10) to the 10 column.

$$\begin{array}{r} 2\ \\ 34 \\ \times\ 6 \\ \hline 4 \end{array} = \begin{array}{r} 2(10)\ \ \ \ \ \ \\ 3(10) + 4 \\ \times\ \ \ 6 \\ \hline 4 \end{array}$$

Step 4: Multiply 6 times 3(10) and then add the 2(10) that was carried.

$$6 \times 3(10) = 18(10)$$

$$18(10) + 2(10) = 20(10)$$

$$\begin{array}{r} 2\ \\ 34 \\ \times\ 6 \\ \hline 204 \end{array} = \begin{array}{r} 2(10)\ \ \ \ \ \ \\ 3(10) + 4 \\ \times\ \ \ 6 \\ \hline 20(10) + 4 \end{array}$$

Step 5: Regroup and carry.

$$\begin{aligned} 20(10) &= 200 \\ &= 200 + 0 \\ &= 2(100) + 0(10) \\ &= 2(10^2) + 0(10) \end{aligned}$$

We have regrouped the 20(10) into $2(10^2)$ and 0(10). Now carry the $2(10^2)$ to the 10^2 column.

$$\begin{array}{r} 2\ \\ 34 \\ \times\ 6 \\ \hline 204 \end{array} = \begin{array}{r} 2(10)\ \ \ \ \ \ \\ 3(10) + 4 \\ \times\ \ \ 6 \\ \hline 2(10^2) + 0(10) + 4 \end{array}$$

■

Practice 16: Multiply the following whole numbers using expanded notation. Check your answer by multiplying in short notation.

$$5 \times 31$$

16. ______

Practice 17: Multiply the following whole numbers using expanded notation. Check your answer by multiplying in short notation.

$$7 \times 34$$

17. ______

A N S W E R

18. ____________

Practice 18: Multiply the following whole numbers using expanded notation. Check your answer by multiplying in short notation.

$$6 \times 87$$

19. ____________

Practice 19: Multiply the following whole numbers using expanded notation. Check your answer by multiplying in short notation.

$$7 \times 426$$

Example 5 **Multiply the following whole numbers using expanded notation and the concept of regrouping and carrying. Check your answer by multiplying in short notation.**

$$24 \times 35$$

Step 1: Write the numbers in vertical form using expanded notation.

$$\begin{array}{rcr} 35 & = & 3(10) + 5 \\ \times\ \underline{24} & = & \times\ \underline{2(10) + 4} \end{array}$$

Apply the distributive property and multiply.

Step 2: Multiply 4 times 5.

$$\begin{array}{rcr} 35 & = & 3(10) + \mathbf{5} \\ \times\ \underline{24} & = & \times\ \underline{2(10) + \mathbf{4}} \\ \mathbf{20} & & \mathbf{20} \end{array}$$

Step 3: Regroup and carry.

$$20 = 2(10) + 0$$

We have regrouped the 20 ones into 2(10) and 0 ones. Now carry the 2(10) to the ten's column.

$$\begin{array}{rcr} \mathbf{2} & & \mathbf{2(10)} \\ 35 & = & 3(10) + 5 \\ \times\ \underline{24} & = & \times\ \underline{2(10) + 4} \\ 0 & & 0 \end{array}$$

Step 4: Multiply 4 times 3(10) and then add the 2(10) that was carried.

$$4 \times 3(10) = 12(10)$$

$$12(10) + 2(10) = 14(10)$$

$$\begin{array}{rcr} \mathbf{2} & & \mathbf{2(10)} \\ 35 & = & \mathbf{3(10)} + 5 \\ \times\ \underline{24} & = \times & \underline{2(10) + \mathbf{4}} \\ \mathbf{140} & & \mathbf{14(10)} + 0 \end{array}$$

Step 5: Regroup and carry.

$$\begin{aligned} 14(10) &= 140 \\ &= 100 + 40 \\ &= 1(100) + 4(10) \\ &= 1[10(10)] + 4(10) \\ &= 1(10^2) + 4(10) \end{aligned}$$

We have regrouped the 14(10) into $1(10^2)$ and 4(10). Now carry the $1(10^2)$ to the 10^2 column.

$$\begin{array}{rcr} 2 & & 2(10) \\ 35 & = & 3(10) + 5 \\ \times\ \underline{24} & = & \underline{\times\ 2(10) + 4} \\ \mathbf{140} & & \mathbf{1(10^2)} + 4(10) + 0 \end{array}$$

Step 6: Multiply 2(10) times 5 and then place the product in the appropriate column.

$$\begin{array}{rcr} 35 & = & 3(10) + \mathbf{5} \\ \times\ \underline{\mathbf{2}4} & = & \underline{\times\ \ \mathbf{2(10)} + 4} \\ 140 & & 1(10^2) + 4(10) + 0 \\ \mathbf{10} & & \mathbf{10(10)} \end{array}$$

Step 7: Regroup and carry.

$$\begin{aligned} 10(10) &= 100 \\ &= 1(100) + 0(10) \\ &= 1[10(10)] + 0(10) \\ &= 1(10^2) + 0(10) \end{aligned}$$

We have regrouped the 10(10) into $1(10^2)$ and 0(10). Now carry the $1(10^2)$.

$$\begin{array}{rcr} \mathbf{1} & & \mathbf{1(10^2)} \\ 35 & = & 3(10) + 5 \\ \times\ \underline{24} & = & \underline{\times\ 2(10) + 4} \\ 140 & & 1(10^2) + 4(10) + 0 \\ 0 & & 0(10) \end{array}$$

A N S W E R

Step 8: Multiply 2(10) times 3(10) and then add the $1(10^2)$ that was carried. Then place the result in the appropriate column.

$$2(10) \times 3(10) = 6[10(10)] = 6(10^2)$$

$$6(10^2) + 1(10^2) = 7(10^2)$$

$$\begin{array}{rr} \mathbf{1} & \mathbf{1(10^2)} \\ 35 = & \mathbf{3(10)} + 5 \\ \times \underline{\mathbf{24}} = & \underline{\times\ \mathbf{2(10)} + 4} \\ 140 & 1(10^2) + 4(10) + 0 \\ 70 & \mathbf{7(10^2)} + 0(10) \end{array}$$

Step 9: Add the columns.

$$\begin{array}{rr} 35 = & 3(10) + 5 \\ \times \underline{24} = & \underline{\times\ 2(10) + 4} \\ \mathbf{140} & \mathbf{1(10^2) + 4(10) + 0} \\ \underline{\mathbf{70}} & \underline{\mathbf{7(10^2) + 0(10)}} \\ \mathbf{840} & \mathbf{8(10^2) + 4(10) + 0} \end{array}$$

■

20. ____________________

Practice 20: Multiply the following whole numbers using expanded notation. Check your answer by multiplying in short notation.

$$24 \times 53$$

21. ____________________

Practice 21: Multiply the following whole numbers using expanded notation. Check your answer by multiplying in short notation.

$$32 \times 43$$

22. ____________________

Practice 22: Multiply the following whole numbers using expanded notation. Check your answer by multiplying in short notation.

$$65 \times 70$$

Practice 23: Multiply the following whole numbers using expanded notation. Check your answer by multiplying in short notation.

$$30 \times 43$$

A N S W E R

23. ________________

Let's look at another algebra example.

Example 6

Arithmetic

$$23 \times 21$$

Step 1: Write the numbers in vertical form using expanded notation.

$$\begin{array}{rcr} 23 & = & 2(10) + 3 \\ \times\ \underline{21} & = & \times\ \underline{2(10) + 1} \end{array}$$

Apply the distributive property and then multiply.

Step 2: Multiply 1 times 3.

$$\begin{array}{rcr} \mathbf{23} & = & 2(10) + \mathbf{3} \\ \times\ \underline{\mathbf{21}} & = & \times\ \underline{2(10) + \mathbf{1}} \\ \mathbf{3} & & \mathbf{3} \end{array}$$

Step 3: Multiply 1 times 2(10).

$$\begin{array}{rcr} \mathbf{23} & = & \mathbf{2(10)} + 3 \\ \times\ \underline{\mathbf{21}} & = & \times\ \underline{2(10) + \mathbf{1}} \\ 23 & & \mathbf{2(10)} + 3 \end{array}$$

Step 4: Multiply 2(10) times 3 and place the product in the appropriate column.

$$\begin{array}{rcr} 23 & = & 2(10) + \mathbf{3} \\ \times\ \underline{\mathbf{21}} & = & \times\ \underline{\mathbf{2(10)} + 1} \\ 23 & & 2(10) + 3 \\ \mathbf{6} & & \mathbf{6(10)} \end{array}$$

Algebra

$$(2x + 3) \times (2x + 1)$$

Step 1: Write in vertical form.

$$\begin{array}{r} 2x + 3 \\ \times\ \underline{2x + 1} \end{array}$$

Apply the distributive property and then multiply.

Step 2: Multiply 1 times 3.

$$\begin{array}{r} 2x + \mathbf{3} \\ \times\ \underline{2x + \mathbf{1}} \\ \mathbf{3} \end{array}$$

Step 3: Multiply 1 times $2x$.

$$\begin{array}{r} \mathbf{2x} + 3 \\ \times\ \underline{2x + \mathbf{1}} \\ \mathbf{2x} + 3 \end{array}$$

Step 4: Multiply $2x$ times 3 and place the product in the appropriate column.

$$\begin{array}{r} 2x + \mathbf{3} \\ \times\ \underline{\mathbf{2x} + 1} \\ 2x + 3 \\ \mathbf{6x} \qquad \end{array}$$

Step 5: Multiply 2(10) times 2(10) and then place the product in the appropriate column.

$$\begin{array}{rcr} 23 & = & 2(10) + 3 \\ \times\ \underline{21} & = & \times\ \underline{\mathbf{2(10)} + 1} \\ 23 & & 2(10) + 3 \\ \mathbf{46} & & \mathbf{4(10^2)} + 6(10) \end{array}$$

Step 6: Add the columns.

$$\begin{array}{rcr} 23 & = & 2(10) + 3 \\ \times\ \underline{21} & = & \times\ \underline{2(10) + 1} \\ \mathbf{23} & & \mathbf{2(10) + 3} \\ \underline{\mathbf{46}} & & \underline{\mathbf{4(10^2) + 6(10)}} \\ \mathbf{483} & & \mathbf{4(10^2) + 8(10) + 3} \end{array}$$

Step 5: Multiply $2x$ times $2x$ and then place the product in the appropriate column.

$$\begin{array}{r} 2x + 3 \\ \times\ \underline{\mathbf{2x} + 1} \\ 2x + 3 \\ \mathbf{4x^2} + 6x \end{array}$$

Step 6: Add the columns.

$$\begin{array}{r} 2x + 3 \\ \times\ \underline{2x + 1} \\ \mathbf{2x + 3} \\ \underline{\mathbf{4x^2 + 6x}} \\ \mathbf{4x^2 + 8x + 3} \end{array}$$

■

I'm beginning to look forward to algebra.

Now that you know how multiplication of whole numbers works using expanded notation we will not use it in the remainder of this section.

Example 7 **Multiply the following whole numbers using short notation and the concept of regrouping and carrying.**

$$36 \times 257$$

Step 1: Write the numbers in vertical form.

$$\begin{array}{r} 257 \\ \times\ \underline{36} \end{array}$$

Step 2: Multiply 6 times 7.

$$\begin{array}{r} 257 \\ \times\ \underline{\mathbf{36}} \\ \mathbf{42} \end{array}$$

Step 3: Regroup and carry.

$$\begin{array}{r} \mathbf{4}\ \\ 257 \\ \times\ \underline{36} \\ 2 \end{array}$$

Step 4: Multiply 6 times 5 and then add the 4 which was carried.

$6 \times 5 = 30$

$30 + 4 = 34$

$$\begin{array}{r} \mathbf{4} \\ 2\mathbf{5}7 \\ \times \ 3\mathbf{6} \\ \hline \mathbf{34}2 \end{array}$$

Step 5: Regroup and carry.

$$\begin{array}{r} 34 \\ 257 \\ \times \ 36 \\ \hline 42 \end{array}$$

Step 6: Multiply 6 times 2 and then add the 3 which was carried.

$6 \times 2 = 12$

$12 + 3 = 15$

$$\begin{array}{r} \mathbf{3}4 \\ \mathbf{2}57 \\ \times \ 3\mathbf{6} \\ \hline \mathbf{15}42 \end{array}$$

Step 7: Multiply 3 times 7 and then place the product in the appropriate column.

$$\begin{array}{r} 25\mathbf{7} \\ \times \ \mathbf{3}6 \\ \hline 1542 \\ \mathbf{21} \end{array}$$

Step 8: Regroup and carry.

$$\begin{array}{r} \mathbf{2} \\ 257 \\ \times \ 36 \\ \hline 1542 \\ 1 \end{array}$$

Step 9: Multiply 3 times 5 and then add the 2 which was carried. Then place the result in the appropriate column.

$3 \times 5 = 15$

$15 + 2 = 17$

$$\begin{array}{r} 2\mathbf{5}7 \\ \times \ \mathbf{3}6 \\ \hline 1542 \\ \mathbf{17}1 \end{array}$$

A N S W E R

Step 10: Regroup and carry.

$$\begin{array}{r} \mathbf{1} \\ 257 \\ \times \quad 36 \\ \hline 1542 \\ 71 \end{array}$$

Step 11: Multiply 3 times 2 and then add the 1 which was carried.

$$\begin{array}{r} \mathbf{1} \\ \mathbf{2}57 \\ \times \quad \mathbf{3}6 \\ \hline 1542 \\ \mathbf{7}71 \end{array}$$

Step 12: Add the columns.

$$\begin{array}{r} 257 \\ \times \quad 36 \\ \hline \mathbf{1542} \\ \mathbf{771} \\ \hline \mathbf{9252} \end{array}$$

■

24. ______________

Practice 24: Multiply the following whole numbers using short notation.

25×355

25. ______________

Practice 25: Multiply the following whole numbers using short notation.

34×248

26. ______________

Practice 26: Multiply the following whole numbers using short notation.

26×307

Practice 27: Multiply the following whole numbers using short notation.

30×226

A N S W E R

27. ____________

Practice 28: Multiply the following whole numbers using short notation.

30×300

28. ____________

EXERCISE 1.4

Multiply the following whole numbers in each problem using expanded notation. Check your answer by multiplying in short notation.

A

1. 8×10

2. 6×13

3. 7×16

4. 8×19

5. 5×15

6. 9×12

7. 14×16

8. 30×20

9. 22×22

10. 17×20

11. 18×13

12. 19×17

13. 24×25

14. 33×26

15. 36×63

16. 73×91

17. 57×67

18. 44×77

19. 97×95

20. 39×39

Multiply the following whole numbers in each problem using short notation.

B

21. 45×550

22. 46×504

23. 70×650

24. 47×632

25. 57×905

26. 86×700

27. 87×367

28. 44×668

29. 78×845

30. 57×941

31. 70×356

32. 70×800

33. 82×686

34. 64×587

35. 88×999

C

36. 145×255

37. 104×208

38. 440×563

39. 252×341

40. 405×463

41. 225×625

42. 576 × 586

43. 641 × 499

44. 700 × 679

45. 555 × 777

46. 637 × 985

47. 787 × 900

48. 820 × 6056

49. 483 × 5422

50. 245 × 8860

APPLICATIONS

Work the following problems using short notation. When an asterisk appears by a problem, you may want to use your calculator.

1. The area of a rectangle is found by multiplying the length of the rectangle by the width of the rectangle. Find the area of the following rectangle.

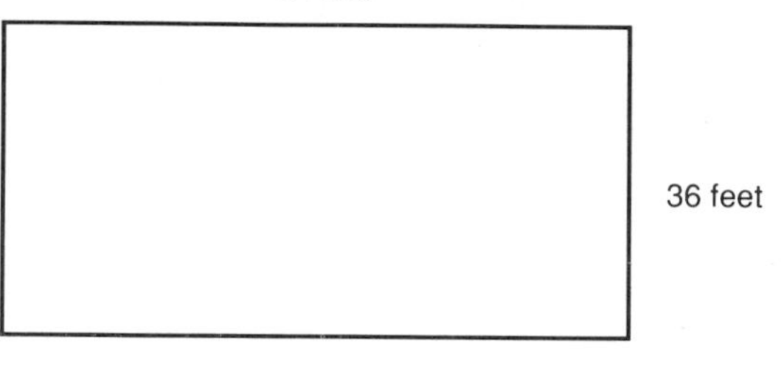
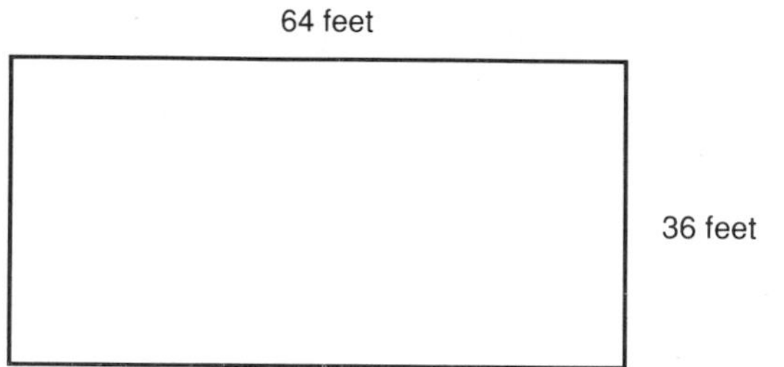

2. The area of a house is usually measured in square feet (ft^2) and the price of the house is usually based on the square feet in the house. Find the area of the following houses.

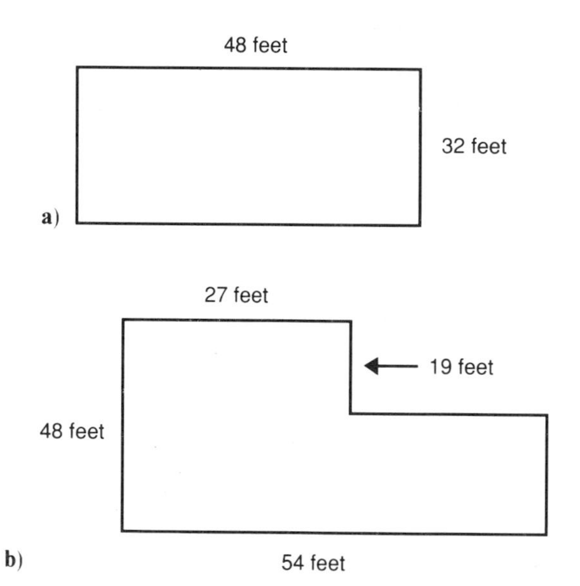

3. Ellie has been working 6 hours a day for the past 23 days at $5 an hour. How much pay will Ellie receive?

4. Emanual makes $253 per month on his part-time job while attending college. How much does he make in one year?

***5.** Mr. Perkins sold his 427 acre farm for $845 an acre. What is the total amount that Mr. Perkins received for his farm?

1.5 DIVIDING WHOLE NUMBERS

In Section 1.4, the multiplication table was used to multiply whole numbers. Since division is the inverse of multiplication, we can use the multiplication table to divide whole numbers by using it in reverse.

Definition of Division. If a, b, and c represent whole numbers and if b is not equal to 0, then $a \div b = c$ if and only if $b \times c = a$.

The division

$$a \div b = ?$$

becomes the multiplication

$$b \times ? = a$$

For example,

$$8 \div 2 = ?$$

is changed to

$$2 \times ? = 8$$

Now, looking back at the multiplication table

$$2 \times 4 = 8$$

thus,

$$8 \div 2 = 4$$

Division problems will be worked using expanded notation. As you will find out, it will take more time and effort to work division problems in expanded notation; however, the rewards will be well worth the investment. You should have a better understanding of the division process by using expanded notation. This will prove to be valuable when studying algebra. Though it may seem difficult at times, we are building an important foundation for your future work in mathematics.

Example 1 **Divide the following whole numbers using expanded notation. Check your answer by dividing in short notation.**

$$46 \div 2$$

To make the work easier, and perhaps more fun, let's think of $46 \div 2$ as \$46 to be divided between two people. This money idea will be used in several of the following examples.

Step 1: Write the numbers in division form using expanded notation.

$$2\overline{)46} = 2\overline{)4(10) + 6(1)}$$

Now, in terms of money the example means that we have four \$10 bills and six \$1 bills to be divided between two people.

Step 2: Divide the four \$10 bills between the two people.

$$\begin{array}{r} \mathbf{2} \\ \mathbf{2}\overline{)\mathbf{4}6} \end{array} = \begin{array}{l} \quad\mathbf{2(10)} \\ \mathbf{2}\overline{)\mathbf{4(10)} + 6(1)} \end{array}$$

Each person gets two \$10 bills.

Step 3: Multiply 2 times 2(10) and then place the result in the proper column.

$$\begin{array}{r} \mathbf{2} \\ \mathbf{2}\overline{)46} \\ \mathbf{4} \end{array} = \begin{array}{l} \quad\mathbf{2(10)} \\ \mathbf{2}\overline{)4(10) + 6(1)} \\ \quad\mathbf{4(10)} \end{array}$$

Step 4: Subtract 4(10) from 4(10).

$$\begin{array}{r} 2 \\ 2\overline{)46} \\ \underline{\mathbf{4}} \\ \mathbf{0} \end{array} = \begin{array}{l} \quad 2(10) \\ 2\overline{)\mathbf{4(10)} + 6(1)} \\ \quad \underline{\mathbf{4(10)}} \\ \quad \mathbf{0(10)} \end{array}$$

Step 5: Since there are no groups of tens remaining, bring 6(1) down—that is, place 6(1) in position to be divided.

$$\begin{array}{r} 2 \\ 2\overline{)46} \\ \underline{4} \\ \mathbf{6} \end{array} = \begin{array}{r} 2(10) \qquad\quad \\ 2\overline{)4(10) + 6(1)} \\ \underline{4(10)} \qquad\quad \\ \mathbf{6(1)} \end{array}$$

Step 6: Divide the six \$1 bills between the two people.

$$\begin{array}{r} 23 \\ \mathbf{2}\overline{)46} \\ \underline{4} \\ \mathbf{6} \end{array} = \begin{array}{r} 2(10) + \mathbf{3(1)} \\ \mathbf{2}\overline{)4(10) + 6(1)} \\ \underline{4(10)} \qquad\quad \\ \mathbf{6(1)} \end{array}$$

Each person gets three \$1 bills.

Step 7: Multiply 2 times 3(1) and then place the result in the proper column.

$$\begin{array}{r} 23 \\ \mathbf{2}\overline{)46} \\ \underline{4} \\ 6 \\ \mathbf{6} \end{array} = \begin{array}{r} 2(10) + \mathbf{3(1)} \\ \mathbf{2}\overline{)4(10) + 6(1)} \\ \underline{4(10)} \qquad\quad \\ 6(1) \\ \mathbf{6(1)} \end{array}$$

Step 8: Subtract 6(1) from 6(1).

$$\begin{array}{r} 23 \\ 2\overline{)46} \\ \underline{4} \\ \mathbf{6} \\ \underline{\mathbf{6}} \\ \mathbf{0} \end{array} = \begin{array}{r} 2(10) + 3(1) \\ 2\overline{)4(10) + 6(1)} \\ \underline{4(10)} \qquad\quad \\ \mathbf{6(1)} \\ \underline{\mathbf{6(1)}} \\ \mathbf{0(1)} \end{array}$$

Thus, each person receives two \$10 bills and three \$1 bills, that is, \$23. ■

Example 2 **Divide the following whole numbers using expanded notation. Check your answer by dividing in short notation.**

$$72 \div 3$$

Step 1: Write the numbers in division form using expanded notation.

$$3\overline{)72} = 3\overline{)7(10) + 2(1)}$$

Now, in terms of money, the example means that we have seven \$10 bills and two \$1 bills to be divided among three people.

Step 2: Divide the seven \$10 bills among the three people.

$$\begin{array}{r} \mathbf{2} \\ \mathbf{3}\overline{)\,72} \end{array} = \begin{array}{l} \quad\;\mathbf{2(10)} \\ \mathbf{3}\overline{)\,\mathbf{7(10)} + 2(1)} \end{array}$$

Since 3 times 2(10) is 6(10), we know that we can give each person two \$10 bills. Let's find out if it would be possible to give each person three \$10 bills. Now, 3 times 3(10) is 9(10); thus, we would need nine \$10 bills in order to give each person three \$10 bills. Since we only have 7(10) or seven \$10 bills, we can only give each person two \$10 bills.

Step 3: Multiply 3 times 2(10) and then place the result in the proper column.

$$\begin{array}{r} \mathbf{2} \\ \mathbf{3}\overline{)\,72} \\ \mathbf{6} \end{array} = \begin{array}{l} \quad\;\mathbf{2(10)} \\ \mathbf{3}\overline{)\,7(10) + 2(1)} \\ \quad\;\mathbf{6(10)} \end{array}$$

Step 4: Subtract 6(10) from 7(10).

$$\begin{array}{r} 2 \\ 3\overline{)\,72} \\ \underline{\mathbf{6}} \\ \mathbf{1} \end{array} = \begin{array}{l} \quad\;2(10) \\ 3\overline{)\,\mathbf{7(10)} + 2(1)} \\ \quad\;\underline{\mathbf{6(10)}\qquad\quad} \\ \quad\;\mathbf{1(10)} \end{array}$$

There is 1(10) or one \$10 bill remaining.

Who gets the one \$10 bill that remains?

Well, Charlie, it would not be fair to give any one person the \$10 bill. The only fair way is to change the \$10 bill into ten \$1 bills and continue to divide.

Step 5: Regroup 1(10) into 10(1).

$$\begin{array}{r} 2 \\ 3\overline{)\,72} \\ \underline{6} \\ 1 \end{array} = \begin{array}{l} \quad\;2(10) \\ 3\overline{)\,7(10) + 2(1)} \\ \quad\;\underline{6(10)\qquad\quad} \\ \quad\;\cancel{1(10)} \\ \quad\;\mathbf{10(1)} \end{array}$$

Step 6: Bring 2(1) down.

$$\begin{array}{r} 2 \\ 3\overline{)\,72} \\ \underline{6} \\ 12 \end{array} = \begin{array}{l} \quad\;2(10) \\ 3\overline{)\,7(10) + 2(1)} \\ \quad\;\underline{6(10)\qquad\quad} \\ \quad\;\cancel{1(10)} \\ \quad\;10(1) + \mathbf{2(1)} \end{array}$$

A N S W E R

Step 7: Add 10(1) and 2(1).

$$\begin{array}{r r r l c r} & 2 & & 2(10) & & \\ 3) & \overline{72} & = 3) & \overline{7(10)} & \overline{+} & \overline{2(1)} \\ & \underline{6} & & \underline{6(10)} & & \\ & \mathbf{12} & & \cancel{1(10)} & & \\ & & & \cancel{10(1)} & + & \cancel{2(1)} \\ & & & & & \mathbf{12(1)} \end{array}$$

Step 8: Divide the twelve \$1 bills among the three people.

$$\begin{array}{r r r l c r} & 24 & & 2(10) & + & \mathbf{4(1)} \\ 3) & \overline{72} & = 3) & \overline{7(10)} & \overline{+} & \overline{2(1)} \\ & \underline{6} & & \underline{6(10)} & & \\ & \mathbf{12} & & \cancel{1(10)} & & \\ & & & \cancel{10(1)} & + & \cancel{2(1)} \\ & & & & & \mathbf{12(1)} \end{array}$$

Each person gets four \$1 bills.

Step 9: Multiply 3 times 4(1) and then place the result in the proper column.

$$\begin{array}{r r r l c r} & 24 & & 2(10) & + & \mathbf{4(1)} \\ 3) & \overline{72} & = 3) & \overline{7(10)} & \overline{+} & \overline{2(1)} \\ & \underline{6} & & \underline{6(10)} & & \\ & 12 & & \cancel{1(10)} & & \\ & \mathbf{12} & & \cancel{10(1)} & + & \cancel{2(1)} \\ & & & & & 12(1) \\ & & & & & \mathbf{12(1)} \end{array}$$

Step 10: Subtract 12(1) from 12(1).

$$\begin{array}{r r r l c r} & 24 & & 2(10) & + & 4(1) \\ 3) & \overline{72} & = 3) & \overline{7(10)} & \overline{+} & \overline{2(1)} \\ & \underline{6} & & \underline{6(10)} & & \\ & \mathbf{12} & & \cancel{1(10)} & & \\ & \underline{\mathbf{12}} & & \cancel{10(1)} & + & \cancel{2(1)} \\ & \mathbf{0} & & & & \mathbf{12(1)} \\ & & & & & \underline{\mathbf{12(1)}} \\ & & & & & \mathbf{0(1)} \end{array}$$

Thus, each person receives two \$10 bills and four \$1 bills, that is, \$24. ■

1. ____________

Practice 1: Divide the following whole numbers using expanded notation. Check your answer by dividing in short notation.

$$96 \div 3$$

Practice 2: Divide the following whole numbers using expanded notation. Check your answer by dividing in short notation.

$$75 \div 5$$

A N S W E R

2. ____________

Practice 3: Divide the following whole numbers using expanded notation. Check your answer by dividing in short notation.

$$72 \div 3$$

3. ____________

Practice 4: Divide the following whole numbers using expanded notation. Check your answer by dividing in short notation.

$$161 \div 7$$

4. ____________

A N S W E R

5. ____________________

Practice 5: Divide the following whole numbers using expanded notation. Check your answer by dividing in short notation.

$$264 \div 8$$

Example 3 **Divide the following whole numbers using *partial* expanded notation. Check your answer by dividing in short notation.**

$$240 \div 15$$

Step 1: Write the numbers in division form using *partial* expanded notation; that is, we will not write the divisor 15 in expanded notation.

$$15\overline{)240} = 15\overline{)2(10^2) + 4(10) + 0(1)}$$

Now, in terms of money, the example means that we have two \$100 bills ($10^2 = 10(10) = 100$), four \$10 bills, and zero \$1 bills to be divided among 15 people.

Step 2: Divide the two \$100 bills among the 15 people.

It is not possible to divide two \$100 bills among 15 people since there are not enough \$100 bills to go around.

That's correct, Charlie. We must change the \$100 bills into \$10 bills.

Step 3: Regroup by changing the \$100 bills into \$10 bills.

$$2(10^2) = 2[10(10)] = 20(10)$$

Thus, two \$100 bills = twenty \$10 bills

$$15\overline{)240} = 15\overline{)\cancel{2(10^2)} + 4(10) + 0(1)}$$
$$\mathbf{20(10)}$$

Step 4: Bring 4(10) down.

$$\begin{array}{rcl}
15\overline{)240} & = & 15\overline{)\cancel{2(10^2)} + 4(10) + 0(1)} \\
 & & \quad 20(10) + \mathbf{4(10)}
\end{array}$$

Step 5: Add 20(10) and 4(10).

$$\begin{array}{rcl}
15\overline{)240} & = & 15\overline{)\cancel{2(10^2)} + \quad 4(10) + 0(1)} \\
 & & \quad \cancel{\mathbf{20(10)}} + \cancel{\mathbf{4(10)}} \\
 & & \qquad\qquad \mathbf{24(10)}
\end{array}$$

Step 6: Divide the twenty-four \$10 bills among the 15 people.

$$\begin{array}{rcl}
1 & & \qquad\qquad \mathbf{1(10)} \\
\mathbf{15}\overline{)\mathbf{240}} & = & \mathbf{15}\overline{)\cancel{2(10^2)} + \quad 4(10) + 0(1)} \\
 & & \quad \cancel{20(10)} + \cancel{4(10)} \\
 & & \qquad\qquad \mathbf{24(10)}
\end{array}$$

Each person gets one \$10 bill.

Step 7: Multiply 15 times 1(10) and then place the result in the proper column.

$$\begin{array}{rcl}
\mathbf{1} & & \qquad\qquad \mathbf{1(10)} \\
\mathbf{15}\overline{)240} & = & \mathbf{15}\overline{)\cancel{2(10^2)} + \quad 4(10) + 0(1)} \\
\mathbf{15} & & \quad \cancel{20(10)} + \cancel{4(10)} \\
 & & \qquad\qquad 24(10) \\
 & & \qquad\qquad \mathbf{15(10)}
\end{array}$$

Step 8: Subtract 15(10) from 24(10).

$$\begin{array}{rcl}
1 & & \qquad\qquad 1(10) \\
15\overline{)\mathbf{240}} & = & 15\overline{)\cancel{2(10^2)} + \quad 4(10) + 0(1)} \\
\underline{\mathbf{15}} & & \quad \cancel{20(10)} + \cancel{4(10)} \\
\mathbf{9} & & \qquad\qquad \mathbf{24(10)} \\
 & & \qquad\qquad \underline{\mathbf{15(10)}} \\
 & & \qquad\qquad \mathbf{9(10)}
\end{array}$$

There are 9(10) or nine \$10 bills remaining.

Step 9: Regroup by changing the nine remaining \$10 bills into \$1 bills.

$$\begin{array}{rcl}
1 & & \qquad\qquad 1(10) \\
15\overline{)240} & = & 15\overline{)\cancel{2(10^2)} + \quad 4(10) + 0(1)} \\
\underline{15} & & \quad \cancel{20(10)} + \cancel{4(10)} \\
9 & & \qquad\qquad 24(10) \\
 & & \qquad\qquad \underline{15(10)} \\
 & & \qquad\qquad \cancel{9(10)} \\
 & & \qquad\qquad \mathbf{90(1)}
\end{array}$$

Step 10: Bring 0(1) down.

$$
\begin{array}{r}
1 \\
15\overline{)240} \\
\underline{15} \\
\mathbf{90}
\end{array}
=
\begin{array}{rrr}
 & 1(10) & \\
15\overline{)\cancel{2(10^2)} +} & \overline{4(10) +} & \overline{0(1)} \\
\cancel{20(10)} + & \cancel{4(10)} & \\
 & 24(10) & \\
 & \underline{15(10)} & \\
 & \cancel{9(10)} & \\
 & 90(1) + & \mathbf{0(1)}
\end{array}
$$

Step 11: Add 90(1) and 0(1).

$$
\begin{array}{r}
1 \\
15\overline{)240} \\
\underline{15} \\
\mathbf{90}
\end{array}
=
\begin{array}{rrr}
 & 1(10) & \\
15\overline{)\cancel{2(10^2)} +} & \overline{4(10) +} & \overline{0(1)} \\
\cancel{20(10)} + & \cancel{4(10)} & \\
 & 24(10) & \\
 & \underline{15(10)} & \\
 & \cancel{9(10)} & \\
 & \cancel{\mathbf{90(1)}} + & \cancel{\mathbf{0(1)}} \\
 & & \mathbf{90(1)}
\end{array}
$$

Step 12: Divide the ninety \$1 bills among the 15 people.

$$
\begin{array}{r}
\mathbf{16} \\
\mathbf{15}\overline{)240} \\
\underline{15} \\
\mathbf{90}
\end{array}
=
\begin{array}{rrr}
 & 1(10) + & \mathbf{6(1)} \\
\mathbf{15}\overline{)\cancel{2(10^2)} +} & \overline{4(10) +} & \overline{0(1)} \\
\cancel{20(10)} + & \cancel{4(10)} & \\
 & 24(10) & \\
 & \underline{15(10)} & \\
 & \cancel{9(10)} & \\
 & \cancel{90(1)} + & \cancel{0(1)} \\
 & & \mathbf{90(1)}
\end{array}
$$

Each person receives six \$1 bills.

Step 13: Multiply 15 times 6(1) and then place the result in the proper column.

$$
\begin{array}{r}
\mathbf{16} \\
\mathbf{15}\overline{)240} \\
\underline{15} \\
90 \\
\mathbf{90}
\end{array}
=
\begin{array}{rrr}
 & 1(10) + & \mathbf{6(1)} \\
\mathbf{15}\overline{)\cancel{2(10^2)} +} & \overline{4(10) +} & \overline{0(1)} \\
\cancel{20(10)} + & \cancel{4(10)} & \\
 & 24(10) & \\
 & \underline{15(10)} & \\
 & \cancel{9(10)} & \\
 & \cancel{90(1)} + & \cancel{0(1)} \\
 & & 90(1) \\
 & & \mathbf{90(1)}
\end{array}
$$

A N S W E R

Step 14: Subtract 90(1) from 90(1).

$$
\begin{array}{r}
16 \\
15\overline{)240} \\
\underline{15} \\
\mathbf{90} \\
\underline{\mathbf{90}} \\
\mathbf{0}
\end{array}
\quad = \quad
\begin{array}{rrr}
 & 1(10) + & 6(1) \\
15\overline{)\cancel{2(10^2)} +} & \overline{4(10) +} & \overline{0(1)} \\
\cancel{20(10)} + & \cancel{4(10)} & \\
 & 24(10) & \\
 & \underline{15(10)} & \\
 & \cancel{9(10)} & \\
 & \cancel{90(1)} + & \cancel{0(1)} \\
 & & \mathbf{90(1)} \\
 & & \underline{\mathbf{90(1)}} \\
 & & \mathbf{0(1)}
\end{array}
$$

Thus, each person receives one \$10 bill and six \$1 bills, that is, \$16. ■

Practice 6: Divide the following whole numbers using partial expanded notation. Check your answer by dividing in short notation.

$$216 \div 12$$

6. ______________________

Practice 7: Divide the following whole numbers using partial expanded notation. Check your answer by dividing in short notation.

$$253 \div 23$$

7. ______________________

A N S W E R

8. ____________________

Practice 8: Divide the following whole numbers using partial expanded notation. Check your answer by dividing in short notation.

$$612 \div 17$$

9. ____________________

Practice 9: Divide the following whole numbers using partial expanded notation. Check your answer by dividing in short notation.

$$828 \div 23$$

10. ____________________

Practice 10: Divide the following whole numbers using partial expanded notation. Check your answer by dividing in short notation.

$$2226 \div 53$$

Example 4 **Divide the following whole numbers using expanded notation. Check your answer by dividing in short notation.**

$$675 \div 21$$

Step 1: Write the numbers in division form using expanded notation.

$$21\overline{)675} = 2(10) + 1\overline{)6(10^2) + 7(10) + 5}$$

Step 2: Divide $6(10^2)$ by $2(10)$.

$$\begin{array}{r} 3 \\ 21\overline{)675} \end{array} = \begin{array}{r} 3(10) \\ 2(10) + 1\overline{)6(10^2) + 7(10) + 5} \end{array}$$

Step 3: Multiply $3(10)$ times $[2(10) + 1]$ and then place the result in the proper columns.

$$\begin{array}{r} 3 \\ 21\overline{)675} \\ 63 \end{array} = \begin{array}{r} 3(10) \\ 2(10) + 1\overline{)6(10^2) + 7(10) + 5} \\ 6(10^2) + 3(10) \quad\; \end{array}$$

Step 4: Subtract $6(10^2) + 3(10)$ from $6(10^2) + 7(10)$.

$$\begin{array}{r} 3 \\ 21\overline{)675} \\ \underline{63} \\ 4 \end{array} = \begin{array}{r} 3(10) \\ 2(10) + 1\overline{)6(10^2) + 7(10) + 5} \\ \underline{6(10^2) + 3(10) \quad\;} \\ 4(10) \quad\; \end{array}$$

Step 5: Bring 5 down.

$$\begin{array}{r} 3 \\ 21\overline{)675} \\ \underline{63} \\ 45 \end{array} = \begin{array}{r} 3(10) \\ 2(10) + 1\overline{)6(10^2) + 7(10) + 5} \\ \underline{6(10^2) + 3(10) \quad\;} \\ 4(10) + 5 \end{array}$$

We will now repeat the above process.

Step 6: Divide $4(10)$ by $2(10)$.

$$\begin{array}{r} 32 \\ 21\overline{)675} \\ \underline{63} \\ 45 \end{array} = \begin{array}{r} 3(10) + 2 \\ 2(10) + 1\overline{)6(10^2) + 7(10) + 5} \\ \underline{6(10^2) + 3(10) \quad\;} \\ 4(10) + 5 \end{array}$$

Step 7: Multiply 2 times $[2(10) + 1]$ and then place the result in the proper columns.

$$\begin{array}{r} 32 \\ 21\overline{)675} \\ \underline{63} \\ 45 \\ 42 \end{array} = \begin{array}{r} 3(10) + 2 \\ 2(10) + 1\overline{)6(10^2) + 7(10) + 5} \\ \underline{6(10^2) + 3(10) \quad\;} \\ 4(10) + 5 \\ 4(10) + 2 \end{array}$$

A N S W E R

Step 8: Subtract.

$$
\begin{array}{r}
32 \\
21\overline{)675} \\
\underline{63} \\
\mathbf{45} \\
\underline{\mathbf{42}} \\
\mathbf{3}
\end{array}
= 2(10) + 1
\begin{array}{r}
3(10) + 2 \\
\overline{)6(10^2) + 7(10) + 5} \\
\underline{6(10^2) + 3(10)} \\
\mathbf{4(10) + 5} \\
\underline{\mathbf{4(10) + 2}} \\
\mathbf{3}
\end{array}
$$

Thus, the answer is $3(10) + 2$ with a remainder of 3 or 32 R3. ■

By working the following practice problems using expanded notation, you should gain a better understanding of arithmetic as well as better preparation for algebra.

11. ______________________

Practice 11: Divide the following whole numbers using expanded notation. Check your answer by dividing in short notation.

$$276 \div 12$$

12. ______________________

Practice 12: Divide the following whole numbers using expanded notation. Check your answer by dividing in short notation.

$$550 \div 25$$

Practice 13: Divide the following whole numbers using expanded notation. Check your answer by dividing in short notation.

$$651 \div 31$$

A N S W E R

13. ______________

Practice 14: Divide the following whole numbers using expanded notation. Check your answer by dividing in short notation.

$$992 \div 31$$

14. ______________

Practice 15: Divide the following whole numbers using expanded notation. Check your answer by dividing in short notation.

$$690 \div 30$$

15. ______________

Now, here's an example that shows how dividing whole numbers in expanded notation can help prepare you for algebra.

Example 5

Arithmetic

$$483 \div 23$$

Step 1: Write the numbers in division form using expanded notation.

$$2(10) + 3\overline{)\,4(10^2) + 8(10) + 3}$$

Step 2: Divide $4(10^2)$ by $2(10)$.

$$\begin{array}{r} \mathbf{2(10)} \\ \mathbf{2(10)} + 3\overline{)\,\mathbf{4(10^2)} + 8(10) + 3} \end{array}$$

Step 3: Multiply $2(10)$ times $[2(10) + 3]$ and then place the result in the proper columns.

$$\begin{array}{r} \mathbf{2(10)} \\ \mathbf{2(10) + 3}\overline{)\,4(10^2) + 8(10) + 3} \\ \mathbf{4(10^2) + 6(10)} \quad\;\; \end{array}$$

Step 4: Subtract.

$$\begin{array}{r} 2(10) \\ 2(10) + 3\overline{)\,\mathbf{4(10^2)} + \mathbf{8(10)} + 3} \\ \underline{\mathbf{4(10^2)} + \mathbf{6(10)}} \quad\;\; \\ \mathbf{2(10)} \quad\;\; \end{array}$$

Step 5: Bring 3 down.

$$\begin{array}{r} 2(10) \\ 2(10) + 3\overline{)\,4(10^2) + 8(10) + 3} \\ \underline{4(10^2) + 6(10)} \quad\;\; \\ 2(10) + \mathbf{3} \end{array}$$

Step 6: Divide $2(10)$ by $2(10)$.

$$\begin{array}{r} 2(10) + \mathbf{1} \\ \mathbf{2(10)} + 3\overline{)\,4(10^2) + 8(10) + 3} \\ \underline{4(10^2) + 6(10)} \quad\;\; \\ \mathbf{2(10)} + 3 \end{array}$$

Algebra

$$(4x^2 + 8x + 3) \div (2x + 3)$$

Step 1: Write in division form.

$$2x + 3\overline{)\,4x^2 + 8x + 3}$$

Step 2: Divide $4x^2$ by $2x$.

$$\begin{array}{r} \mathbf{2x} \\ \mathbf{2x} + 3\overline{)\,\mathbf{4x^2} + 8x + 3} \end{array}$$

Step 3: Multiply $2x$ times $(2x + 3)$ and then place the result in the proper columns.

$$\begin{array}{r} \mathbf{2x} \\ \mathbf{2x + 3}\overline{)\,4x^2 + 8x + 3} \\ \mathbf{4x^2 + 6x} \quad\;\; \end{array}$$

Step 4: Subtract.

$$\begin{array}{r} 2x \\ 2x + 3\overline{)\,\mathbf{4x^2} + \mathbf{8x} + 3} \\ \underline{\mathbf{4x^2} + \mathbf{6x}} \quad\;\; \\ \mathbf{2x} \quad\;\; \end{array}$$

Step 5: Bring 3 down.

$$\begin{array}{r} 2x \\ 2x + 3\overline{)\,4x^2 + 8x + 3} \\ \underline{4x^2 + 6x} \quad\;\; \\ 2x + \mathbf{3} \end{array}$$

Step 6: Divide $2x$ by $2x$.

$$\begin{array}{r} 2x + \mathbf{1} \\ \mathbf{2x} + 3\overline{)\,4x^2 + 8x + 3} \\ \underline{4x^2 + 6x} \quad\;\; \\ \mathbf{2x} + 3 \end{array}$$

Step 7: Multiply 1 times $[2(10) + 3]$.

$$\begin{array}{rl} & 2(10) + \mathbf{1} \\ \mathbf{2(10) + 3} \big) & \overline{4(10^2) + 8(10) + 3} \\ & \underline{4(10^2) + 6(10)} \\ & 2(10) + 3 \\ & \mathbf{2(10) + 3} \end{array}$$

Step 8: Subtract.

$$\begin{array}{rl} & 2(10) + 1 \\ 2(10) + 3 \big) & \overline{4(10^2) + 8(10) + 3} \\ & \underline{4(10^2) + 6(10)} \\ & \mathbf{2(10) + 3} \\ & \underline{\mathbf{2(10) + 3}} \\ & \mathbf{0} \end{array}$$

Step 7: Multiply 1 times $[2x + 3]$.

$$\begin{array}{rl} & 2x + \mathbf{1} \\ \mathbf{2x + 3} \big) & \overline{4x^2 + 8x + 3} \\ & \underline{4x^2 + 6x} \\ & 2x + 3 \\ & \mathbf{2x + 3} \end{array}$$

Step 8: Subtract.

$$\begin{array}{rl} & 2x + 1 \\ 2x + 3 \big) & \overline{4x^2 + 8x + 3} \\ & \underline{4x^2 + 6x} \\ & \mathbf{2x + 3} \\ & \underline{\mathbf{2x + 3}} \\ & \mathbf{0} \end{array}$$

■

Once again, you should see the similarity between algebra and arithmetic. Thus, we will not use expanded notation in the remainder of this section.

Example 6 **Divide the following whole numbers using short notation.**

$$629 \div 27$$

Step 1: Write the numbers in division form.

$$27 \overline{\big) 629}$$

Step 2: Divide 6 by 2.

$$\begin{array}{rl} & \mathbf{3} \\ 27 \big) & \overline{\mathbf{6}29} \end{array}$$

Step 3: Multiply 3 times 27 and then place the result in the proper columns.

$$\begin{array}{rl} & 3 \\ \mathbf{27} \big) & \overline{629} \\ & \mathbf{81} \end{array}$$

Step 4: Since 81 is larger than 62, replace 3 with 2.

$$\begin{array}{rl} & \mathbf{2} \\ 27 \big) & \overline{629} \end{array}$$

Step 5: Multiply 2 times 27 and then place the result in the proper columns.

$$\begin{array}{rl} & 2 \\ \mathbf{27} \big) & \overline{629} \\ & \mathbf{54} \end{array}$$

Step 6: Subtract 54 from 62.

$$\begin{array}{r} 2 \\ 27\overline{)\mathbf{629}} \\ \underline{\mathbf{54}} \\ \mathbf{8} \end{array}$$

Step 7: Bring 9 down.

$$\begin{array}{r} 2 \\ 27\overline{)629} \\ \underline{54} \\ 8\mathbf{9} \end{array}$$

Step 8: Divide 8 by 2.

$$\begin{array}{r} 2\mathbf{4} \\ 2\mathbf{7}\overline{)629} \\ \underline{54} \\ \mathbf{8}9 \end{array}$$

Step 9: Multiply 4 times 27 and then place the result in the proper columns.

$$\begin{array}{r} 2\mathbf{4} \\ \mathbf{27}\overline{)629} \\ \underline{54} \\ 89 \\ \mathbf{108} \end{array}$$

Once again the product is too large so we replace 4 with 3.

Step 10: Replace 4 with 3.

$$\begin{array}{r} 2\mathbf{3} \\ 27\overline{)629} \\ \underline{54} \\ 89 \end{array}$$

Step 11: Multiply 3 times 27 and then place the result in the proper columns.

$$\begin{array}{r} 2\mathbf{3} \\ \mathbf{27}\overline{)629} \\ \underline{54} \\ 89 \\ \mathbf{81} \end{array}$$

Step 12: Subtract.

$$\begin{array}{r} 23 \\ 27\overline{)629} \\ \underline{54} \\ \mathbf{89} \\ \underline{\mathbf{81}} \\ \mathbf{8} \end{array}$$

Thus, the answer is 23 R8. ■

A N S W E R

Practice 16: Divide the following whole numbers using short notation.

$611 \div 13$

16. ____________________

Practice 17: Divide the following whole numbers using short notation.

$2494 \div 43$

17. ____________________

Practice 18: Divide the following whole numbers using short notation.

$635 \div 37$

18. ____________________

Practice 19: Divide the following whole numbers using short notation.

$986 \div 34$

19. ____________________

Practice 20: Divide the following whole numbers using short notation.

$1988 \div 38$

20. ____________________

EXERCISE 1.5

Divide the following whole numbers in each problem using expanded notation. Check your answer by dividing in short notation.

A

1. $28 \div 2$

2. $21 \div 3$

3. $72 \div 8$

4. $81 \div 9$

5. $48 \div 8$

6. $47 \div 6$

7. $87 \div 4$

8. $65 \div 9$

9. $85 \div 5$

10. $91 \div 7$

11. $126 \div 6$

12. $147 \div 7$

13. $126 \div 3$

14. $112 \div 8$

15. $132 \div 6$

16. $288 \div 8$

17. $513 \div 9$

18. $483 \div 7$

19. $474 \div 6$

20. $711 \div 9$

B

21. $42 \div 14$

22. $52 \div 13$

23. $63 \div 21$

24. $460 \div 23$

25. $651 \div 31$

26. $682 \div 62$

27. $992 \div 32$

28. $572 \div 11$

29. $860 \div 20$

30. $384 \div 12$

31. $672 \div 32$

32. $861 \div 21$

Divide the following whole numbers in each problem using short notation.

C

33. $677 \div 48$

34. $963 \div 80$

35. $546 \div 42$

36. $722 \div 19$

37. $848 \div 106$

38. $861 \div 123$

39. $664 \div 161$

40. $1748 \div 46$

41. $1624 \div 56$

42. $6205 \div 85$

43. $3999 \div 43$

44. $1596 \div 19$

45. $7560 \div 35$

46. $6804 \div 21$

47. $7904 \div 32$

48. $9287 \div 251$

49. $6976 \div 218$

50. $10{,}101 \div 37$

51. $34{,}522 \div 421$

52. $36{,}024 \div 632$

APPLICATIONS

Work the following problems using short notation. When an asterisk appears by a problem, you may want to use your calculator.

1. If George received $115 for working 23 hours, how much did he receive per hour?

2. If Nathan made $1164 last year, how much did he make per month?

3. A 3-liter bottle of cola costs 96 cents and a 2-liter bottle costs 68 cents. How much does each bottle cost per liter? Which is the better buy?

4. Twelve people want to purchase plants to place on their block for a beautification project. If the cost of the plants is $1152, how much would each person need to contribute?

***5.** Mr. Arthur purchased 682 acres of land for $486,948. How much did Mr. Arthur pay per acre for the land?

1.6 STRATEGIES FOR SOLVING APPLICATIONS

Here are some suggestions for solving word problems using whole numbers.

1. Read through the word problem one time to get an idea of what it is about.
2. Read through the problem a second time and pick out the key words. Examples:
   Find the sum tells you to add.
   Find the difference tells you to subtract.
3. If you know the *cost per item* and the problem asks for a total price for *so many items,* then you multiply.

Example 1 **If paper costs \$2 a pack and a case of paper contains 24 packs, how much does a case of paper cost?**

Solution: $$2 \times 24 = 48$$

Thus, a case of papers costs \$48. ■

4. If you know the *total amount* and *how many items* there are and the problem asks the price per item, then you *divide.*

Example 2 **If a case of paper costs \$48 and contains 24 packs of paper, how much does each pack of paper cost?**

Solution: $$48 \div 24 = 2$$

Thus, each pack of paper costs \$2. ■

5. To find an *average,* you add all the whole numbers then divide the sum by however many numbers there are.

Example 3 **Find the average of 8, 10, 4, and 6.**

Step 1: Add the numbers.

$$8 + 10 + 4 + 6 = 28$$

Step 2: Divide the sum by however many numbers there are.

$$28 \div 4 = 7$$

Thus, the average is 7. ■

APPLICATION EXERCISES

Work the following problems using short notation. When an asterisk appears by a problem, you may want to use your calculator.

*1. Mr. Smith kept the mileage records shown in the table. How many miles did he drive from January through June?

Month	Mileage
January	823
February	467
March	503
April	782
May	568
June	489

*2. Badcock County has the following items budgeted: highways, $285,347; salaries, $89,587; maintenance, $182,568. What is the county's *total* budget for these three items?

*3. On a construction job, Mr. Foster hauled four truckloads of gravel that weighed as follows: 13,674 pounds, 14,869 pounds, 12,032 pounds, and 13,684 pounds. What was the total weight of the four truckloads of gravel?

4. Bay Minette Elementary School has five first grade classes with the following enrollments: 26, 34, 29, 33, 27. How many first graders attend the school?

5. Vincent Lambeth's car payment each month is $307. How much does he pay the first 6 months of the year?

6. Ms. Bryars had a checking account with a balance of $1807. She wrote checks for a total of $1398. What was her balance after she wrote the checks?

7. The Ralston Company's bid on a construction job was $683,000 and the bid made by Stephens Company was $667,000. Which company had the lower bid? How much lower?

8. If the sum of two numbers is 857 and one of the numbers is 498, what is the other number?

9. The sticker price of a new car is $12,378. The dealer gave Don $2850 for his old car on a trade-in. How much did Don pay for the new car after the trade-in was deducted?

10. An empty truck weighed 4873 kilograms. When loaded, the same truck weighed 9702 kilograms. What was the weight of the load?

11. Mr. and Mrs. Sample pay $378 for rent each month. How much rent will they pay in a year?

12. If you drove at a rate of 52 miles per hour for 8 hours, how many miles did you travel?

*13. The county school district bought 16 new cars at a price of $9550 per car. What was the *total* cost of the new cars?

14. John's car gets 29 highway miles to a gallon of gasoline. How many miles can the car travel on 14 gallons of gasoline?

*15. The average salary of a schoolteacher in Badcock County is $23,704, and there are 585 teachers presently employed by the county. What will be the total cost for teacher salaries?

16. In a math course, April made the following scores on her tests: 78, 92, 83, 98, 69, and 84. What was her average test score for the math course?

*17. If you deposited $530, $720, $1520, $840, and $550 in a checking account during the last five months, what was the average deposit you made?

*18. Hick's TV Store bought 24 television sets of the same model for $8,688. How much did Mr. Hicks pay for each television set?

*19. Bill will earn $30,368 this year. If he is to receive 52 equal pay checks, how much will each check be?

*20. A salesman sold five coats at the following prices: $975, $712, $876, $1128, and $814. What was the average price of the coats?

USEFUL INFORMATION

1. The decimal number system uses groups of tens to form numbers. When a number is written showing its groups of tens, the number is said to be written in **expanded notation.** For example,

$$526$$

written in **expanded notation** is

$$5(10^2) + 2(10) + 6$$

2. The **addition table** allows you to add any two whole numbers 0 through 9.
3. **Addition property of zero.** If the letter a is a symbol representing any whole number, then $a + 0 = a$.
4. **Commutative property of addition.** If a and b represent whole numbers, then $a + b = b + a$.
5. **Associative property of addition.** If a, b, and c represent whole numbers, then $(a + b) + c = a + (b + c)$.
6. **Basic rule for addition.** Only like things can be added. That is, groups of ones can be added to groups of ones, groups of tens to groups of tens, chairs to chairs, and x's to x's.
7. In an addition problem, when the sum in any column is more than 9, you must **regroup and carry.**
8. **Subtraction** is the inverse of addition, which means that the addition table can be used to subtract whole numbers.
9. **Definition of subtraction.** If a, b, and c represent whole numbers, then $a - b = c$ if and only if $b + c = a$.
10. **Basic rule for subtraction.** Only like things can be subtracted. That is, groups of ones can be subtracted from groups of ones, groups of tens from groups of tens, chairs from chairs, and x's from x's.
11. In a subtraction problem, if a larger digit is under a smaller digit in any column, you must **borrow and regroup.**
12. When a number is added to itself several times, such as $5 + 5 + 5 + 5$ it is often written using the shorter notation 4×5 which is called **multiplication.** To help you multiply (perform these special additions), use the multiplication table.
13. When a number is multiplied by itself several times, it can be written using exponents. For example, $5 \times 5 \times 5 \times 5 = 5^4$. The number 5 is called the **base** and the number 4 is called the **exponent.**
14. **Multiplication property of one.** If the letter a is a symbol representing any whole number, then $1 \times a = a$.
15. **Multiplication property of zero.** If the letter a is a symbol representing any whole number, then $0 \times a = 0$.
16. **Commutative property of multiplication.** If a and b represent whole numbers, then $a \times b = b \times a$.
17. **Associative property of multiplication.** If a, b, and c represent whole numbers, then $(a \times b) \times c = a \times (b \times c)$.
18. **Distributive property of multiplication over addition.** If a, b, and c represent whole numbers, then $a \times (b + c) = (a \times b) + (a \times c)$.
19. **Division** is the inverse of multiplication; therefore, the multiplication table can be used to divide whole numbers.
20. **Definition of division.** If a, b, and c represent whole numbers, and if b is not equal to 0, then $a \div b = c$ if and only if $b \times c = a$.
21. **Division property of one.** If the letter a is a symbol representing any whole number, then $a \div 1 = a$.
22. **Any nonzero whole number divided by itself is 1.** If the letter a is a symbol representing a nonzero whole number, then $a \div a = 1$.
23. **Division by zero.** If the letter a is a symbol representing a whole number, then $a \div 0$ is *not defined* (we cannot divide by 0).
24. **Zero divided by any nonzero whole number is zero.** If the letter a is a symbol representing a nonzero whole number, then $0 \div a = 0$.
25. **Important!** Try to understand how whole number arithmetic is similar to algebra. Review: Example 3, Section 1.2; Example 2, Section 1.3; Example 6, Section 1.4; and Example 5, Section 1.5.

REVIEW PROBLEMS

Section 1.1

Write the following whole numbers in expanded notation.

A

1. 17
2. 50
3. 47
4. 93
5. 64
6. 83
7. 71
8. 99
9. 48
10. 80

B

11. 372
12. 583
13. 700
14. 903
15. 222
16. 507
17. 839
18. 711
19. 101
20. 588

C

21. 4734

22. 4200

23. 8037

24. 7007

25. 5002

26. 4355

27. 1000

28. 3333

29. 53,030

30. 45,005

Section 1.2

Add the following whole numbers in each problem using expanded notation. Check your answers by adding in short notation.

A

1. 15 + 12

2. 42 + 33

3. 38 + 19

4. 29 + 52

5. 51 + 78

6. 83 + 61

7. 88 + 77

8. 48 + 84

9. 96 + 74

10. 97 + 86

B

11. 392 + 65

12. 487 + 20

13. 353 + 68

14. 786 + 242

15. 969 + 153

16. 657 + 875

17. 749 + 382

18. 676 + 839

19. 907 + 303

20. 666 + 444

C

21. 5312 + 207

22. 3976 + 589

23. 2798 + 3587

24. 6803 + 7708

25. 8798 + 6864

Section 1.3

Subtract the following whole numbers in each problem using expanded notation. Check your answers by subtracting in short notation.

A

1. 15 − 7

2. 37 − 6

3. 52 − 9

4. 64 − 14

5. 57 − 35

6. 53 − 26

7. 92 − 86

8. 77 − 68

9. 80 − 47

10. 91 − 79

B

11. 173 − 46

12. 598 − 67

13. 234 − 27

14. 656 − 97

15. 857 − 680

16. 922 − 436

17. 804 − 297

18. 901 − 899

19. 703 − 490

20. 600 − 599

C

21. 2356 − 344

22. 7690 − 4591

23. 7932 − 3573

24. 6006 − 5326

25. 3000 − 2999

Section 1.4

Multiply the following whole numbers in each problem using expanded notation. Check your answers by multiplying in short notation.

A

1. 6 × 13

2. 7 × 11

3. 4 × 22

4. 7 × 15

5. 13 × 20

6. 27 × 32

7. 29 × 35

8. 14 × 34

9. 38×66

10. 25×62

Multiply the following whole numbers in each problem using short notation.

B

11. 66×100

12. 48×973

13. 67×504

14. 43×438

15. 67×603

16. 55×222

17. 49×471

18. 69×417

19. 93×824

20. 80×800

C

21. 103×401

22. 220×384

23. 384×275

24. 802×807

25. 900×900

26. 643×1283

27. 488×3007

28. 608×3666

29. 444×7777

30. 800×6000

Section 1.5

Divide the following whole numbers in each problem using expanded notation. Check your answers by dividing in short notation.

A

1. $28 \div 4$

2. $25 \div 5$

3. $56 \div 7$

4. $88 \div 8$

5. $24 \div 7$

6. $138 \div 6$

7. $224 \div 8$

8. $245 \div 7$

9. $522 \div 9$

10. $528 \div 6$

B

11. 72 ÷ 24

12. 105 ÷ 21

13. 66 ÷ 11

14. 286 ÷ 13

15. 363 ÷ 33

16. 231 ÷ 21

17. 891 ÷ 11

18. 792 ÷ 36

19. 682 ÷ 31

20. 624 ÷ 52

Divide the following whole numbers in each problem using short notation.

C

21. 2997 ÷ 37

22. 799 ÷ 47

23. 1536 ÷ 64

24. 1216 ÷ 304

25. 590 ÷ 116

26. 963 ÷ 107

27. 5814 ÷ 51

28. 6206 ÷ 29

29. 7314 ÷ 159

30. 48,034 ÷ 329

ANSWERS TO PRACTICE PROBLEMS

Section 1.1

1. 5 stacks + 6 chairs or 5(10) + 6(1)
3. $3(10^2) + 8(10) + 2$
5. $6(10^2) + 0(10) + 5$
7. $4(10^2) + 0(10) + 0$
9. $2(10^3) + 5(10^2) + 7(10) + 3$
11. $7(10^3) + 0(10^2) + 5(10) + 0$

2. 1 truck + 5 stacks + 4 chairs or $1(10^2) + 5(10) + 4(1)$
4. 238
6. 803
8. 620
10. 3007
12. 8000

Section 1.2

1. $5(10^2) + 8(10) + 9 = 589$
3. $8(10^2) + 7(10) + 9 = 879$
5. $9(10^2) + 5(10) + 8 = 958$
7. $1(10^3) + 4(10^2) + 0(10) + 3 = 1403$

2. $8(10^2) + 9(10) + 9 = 899$
4. $8(10^2) + 7(10) + 1 = 871$
6. $1(10^3) + 4(10^2) + 4(10) + 5 = 1445$
8. $6(10^2) + 7(10) + 5 = 675$

Section 1.3

1. $1(10^2) + 2(10) + 5 = 125$
3. $2(10^2) + 1(10) + 2 = 212$
5. $2(10^2) + 5(10) + 2 = 252$
7. $2(10^2) + 0(10) + 8 = 208$
9. $2(10^2) + 2(10) + 4 = 224$

2. $2(10^2) + 2(10) + 2 = 222$
4. $2(10^2) + 1(10) + 7 = 217$
6. $3(10^2) + 6(10) + 7 = 367$
8. $2(10^2) + 1(10) + 7 = 217$

Section 1.4

1. 6×2
3. $9 + 9 + 9 + 9 + 9$
5. 3×4
7. $2 + 2 + 2 + 2 + 2 + 2 + 2$
9. 7×7
11. $5 + 5 + 5$
13. $6(10) + 4 = 64$
15. $8(10) + 8 = 88$
17. $2(10^2) + 3(10) + 8 = 238$
19. $2(10^3) + 9(10^2) + 8(10) + 2 = 2982$
21. $1(10^3) + 3(10^2) + 7(10) + 6 = 1376$
23. $1(10^3) + 2(10^2) + 9(10) + 0 = 1290$
25. 8432
27. 6780

2. 2^6
4. $9 \times 9 \times 9 \times 9 \times 9$
6. 4^3
8. $2 \times 2 \times 2 \times 2 \times 2 \times 2 \times 2$
10. 7^7
12. $5 \times 5 \times 5$
14. $6(10) + 3 = 63$
16. $1(10^2) + 5(10) + 5 = 155$
18. $5(10^2) + 2(10) + 2 - 522$
20. $1(10^3) + 2(10^2) + 7(10) + 2 = 1272$
22. $4(10^3) + 5(10^2) + 5(10) + 0 = 4550$
24. 8875
26. 7982
28. 9000

Section 1.5

1. $3(10) + 2 = 32$
3. $2(10) + 4 = 24$
5. $3(10) + 3 = 33$
7. $1(10) + 1 = 11$
9. $3(10) + 6 = 36$
11. $2(10) + 3 = 23$
13. $2(10) + 1 = 21$
15. $2(10) + 3 = 23$
17. 58
19. 29

2. $1(10) + 5 = 15$
4. $2(10) + 3 = 23$
6. $1(10) + 8 = 18$
8. $3(10) + 6 = 36$
10. $4(10) + 2 = 42$
12. $2(10) + 2 = 22$
14. $3(10) + 2 = 32$
16. 47
18. 17 R6
20. 52 R12

C H A P T E R 2

Signed Numbers

The objective of this chapter is:

1. To be able to add, subtract, multiply, and divide signed numbers.

In Chapter 1, you worked with zero and numbers greater than zero (positive numbers). Now, we will introduce negative numbers. Negative numbers are less than zero and a negative sign $(-)$ is used to indicate these numbers. For example, the temperature is -5 degrees means that the temperature is 5 degrees below zero. You may also discover negative numbers when you balance your checkbook; for example, you are informed by the bank that you have written checks totaling \$110.25 when you had a starting balance of \$95. You have overdrawn \$15.25; that is, the balance in your checking account is $-\$15.25$.

Every positive number has a corresponding negative number. Examples:

$$\frac{2}{3}, -\frac{2}{3}$$

$$7, -7$$

$$1.75, -1.75$$

$$82, -82$$

$$9\frac{3}{4}, -9\frac{3}{4}$$

$$52.83, -52.83$$

These positive numbers and their corresponding negatives are examples of what we call opposites. For any nonzero number a,

$$a + (-a) = 0.$$

The number $-a$ is called the **opposite** of a. For example,

the opposite of 3 is -3

and the opposite of -5 is $-(-5)$

So, the opposite of a number is just the number with a negative sign.

That's correct, Charlie. The only time students seem to get confused about the opposite of a number is when the number is negative. It is easy to see that the opposite of 7 is -7; however, students sometimes have trouble seeing that the opposite of -8 is $-(-8)$.

What kind of number is $-(-8)$?

Well, Charlie, in the next section we will show that $-(-8) = 8$.

So, the opposite of -8 is really just 8. It looks like the opposite of a negative number is its positive.

That's correct, Charlie. The name opposite comes from the fact that 8 and -8 are the same distance from zero in opposite directions.

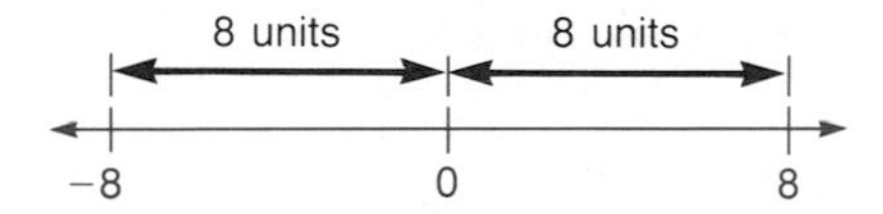

Since every whole number except zero has a corresponding negative number (opposite), we have the following numbers called **integers.**

$$\ldots -5, -4, -3, -2, -1, 0, 1, 2, 3, 4, 5 \ldots$$

The three dots on each end mean that the integers continue indefinitely. The integers

$$-1, -2, -3, -4, -5, -6, -7, -8 \ldots$$

are called **negative integers.** The integers

$$1, 2, 3, 4, 5, 6, 7, 8 \ldots$$

are called **positive integers.** The integer 0 is neither positive nor negative.

A **rational number** is a number that can be written in the form $\frac{a}{b}$, where a and b are integers and $b \neq 0$.

Are rational numbers just fractions?

Well, Charlie, read the definition of a rational number again. Notice that it states "a rational number is a number that can be written in the *form* $\frac{a}{b}$, where a and b are integers and $b \neq 0$." Now, a whole number can be written as a fraction by placing the whole number over 1 ($a = \frac{a}{1}$). This is also true for integers. A mixed number can be written as an improper fraction, and a decimal number can also be written as a fraction. Thus, Charlie, fractions, mixed numbers, integers, and decimal numbers are rational numbers.

Are there any numbers that are not rational?

Yes, Charlie. For example, you may need to measure the length of the diagonal of a square which has sides 3 units long.

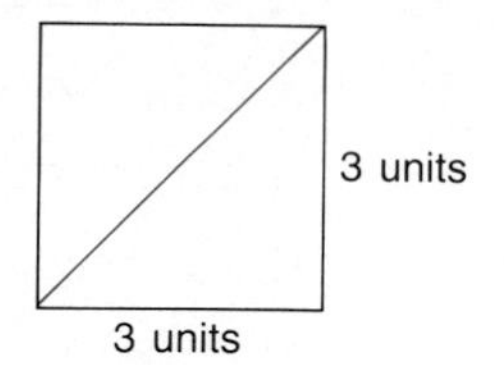

Now, there is no rational number which equals the length of the diagonal.

How do we know?

That's a good question, Charlie. We cannot answer your question at this time; however, you will be able to answer it later when we study algebra.

That sounds interesting. I'm glad that I am taking algebra.

Another example of a length that is not equal to a rational number is the circumference of a circle with diameter 2 units.

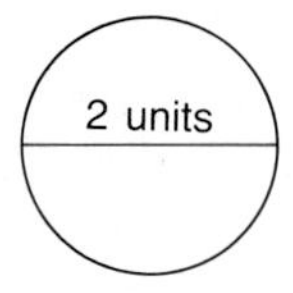

The distance around this circle is 2π (2 pi) units, and 2π is not a rational number. Again, you will have to wait until later to find out why 2π is not a rational number.

The **real numbers** consist of zero and numbers which measure the length of any line segment along with their negatives. Thus, the real numbers are the rational numbers as well as numbers which are not rational (irrational numbers like 2π).

Are all numbers real numbers?

No, Charlie. If you continue to study mathematics, you will learn about some numbers that are not real numbers.

Now back to negative numbers. When writing negative numbers, we must include the negative sign. A nonzero number together with a positive (+) or negative (−) sign is called a **signed number.**

2.1 ADDING SIGNED NUMBERS

We will use the idea of money to help you understand signed numbers. In terms of money, we will think of a negative number (a number with a negative sign) as money we owe and a positive number (a number with a positive sign or no sign) as money we have. For example, −50 tells us we owe $50 and +75 or 75 tells us we have $75. This concept of relating signed numbers to money should make the addition of signed numbers easier to understand.

Rule 1: To add two positive numbers, simply add the two numbers.

Note: When a nonzero number is written without a sign, the number is understood to be positive and assumed to have a positive sign. For example, $8 = +8$.

Example 1 **Add the following numbers.**

$$35 + 45$$

Solution: Since the two numbers are positive simply add the two numbers.

$$\begin{array}{r} 35 \\ +\ 45 \\ \hline 80 \end{array}$$

In terms of money, we have $35 and $45 for a total of $80. ■

Now, let's look at some examples to help prepare you for the next rule.

Example 2 **Add the following numbers.**

$$-35 + 45$$

Solution: In terms of money, −35 means we owe $35 and 45 means we have $45. Now we determine net worth—that is, how much money we owe or have. We take the $45 and pay the $35 we owe. This leaves us with $10. Thus,

$$-35 + 45 = 10$$

■

A N S W E R

Example 3 **Add the following numbers.**

$$35 + (-45)$$

Solution: In terms of money, 35 means we have \$35 and -45 means we owe \$45. Now, we determine our net worth by taking the \$35 and paying on the \$45 we owe. This leaves us owing \$10. Thus,

$$35 + (-45) = -10$$

■

I can really relate to these negative numbers. It seems like I always owe more money than I have.

Work the following practice problems thinking in terms of money.

1. ______________

Practice 1: Add the following numbers.

$$32 + 21$$

2. ______________

Practice 2: Add the following numbers.

$$32 + (-21)$$

3. ______________

Practice 3: Add the following numbers.

$$-32 + 21$$

4. ______________

Practice 4: Add the following numbers.

$$63 + (-78)$$

5. ______________

Practice 5: Add the following numbers.

$$-47 + 54$$

> **Rule 2: To add a positive and negative number,** subtract the smaller number from the larger number (without regard to sign) and place the sign of the larger number in the answer.

Example 4 **Add the following numbers.**

$$4.85 + (-1.52)$$

Step 1: Subtract the smaller number from the larger number (without regard to sign).

$$\begin{array}{r} 4.85 \\ -\ 1.52 \\ \hline 3.33 \end{array}$$

Step 2: The larger number (without regard to sign) is 4.85. Thus, 3.33 has the same sign as 4.85. Therefore,

$$4.85 + (-1.52) = 3.33$$

Even though we now have Rule 2, you may want to continue to think in terms of money when solving this type problem. In fact, let's do Example 4 thinking in terms of money.

The example states that we have \$4.85 and we owe \$1.52. Now, we determine our net worth by taking the \$4.85 and paying the \$1.52 we owe. This leaves us with \$3.33 Thus,

$$4.85 + (-1.52) = 3.33$$ ■

Example 5 **Add the following numbers.**

$$-45 + 15$$

Step 1: Subtract the smaller number from the larger number (without regard to sign).

$$\begin{array}{r} 45 \\ 15 \\ \hline 30 \end{array}$$

Step 2: The larger number (without regard to sign) is 45. Thus, 30 has the same sign as 45. Therefore,

$$-45 + 15 = -30$$

In terms of money, the example states we owe \$45 and we have \$15. Now, we determine our net worth by paying the \$15 on the \$45 we owe. This leaves us owing \$30. Thus,

$$-45 + 15 = -30$$ ■

So, when adding two numbers with unlike signs, we ignore the signs and subtract the smaller number from the larger number and give the difference the sign of the larger number.

A N S W E R

That's correct, Charlie. Let's try some practice problems.

6. ____________________

Practice 6: Add the following numbers.

$$17 + (-8)$$

7. ____________________

Practice 7: Add the following numbers.

$$-24 + 37$$

8. ____________________

Practice 8: Add the following numbers.

$$44 + (-72)$$

9. ____________________

Practice 9: Add the following numbers.

$$-93 + 41$$

10. ____________________

Practice 10: Add the following numbers.

$$83 + (-83)$$

11. ____________________

Practice 11: Add the following numbers.

$$-6.34 + 3.13$$

12. ____________________

Practice 12: Add the following numbers.

$$32.5 + (-21.2)$$

Now, let's look at an example in which we add two negative numbers.

A N S W E R

Example 6 **Add the following numbers.**

$$-15 + (-25)$$

Solution: In terms of money, -15 means we owe \$15 and -25 means we owe \$25. Now, we determine our net worth by putting the two debts together into one. Add the two amounts owed: $15 + 25 = 40$. We owe a total of \$40; thus, our net worth is $-\$40$. Therefore,

$$-15 + (-25) = -40$$ ■

Work the following practice problems thinking in terms of money.

Practice 13: Add the following numbers.

$$-34 + (-23)$$

13. ____________

Practice 14: Add the following numbers.

$$-46 + (-62)$$

14. ____________

> **Rule 3: To add two negative numbers,** ignore the signs and add the numbers giving the sum a negative sign.

Now, let's look at an example in which we use Rule 3.

Example 7 **Add the following numbers.**

$$-41.3 + (-32.5)$$

Step 1: Add the numbers without their signs.

$$\begin{array}{r} 41.3 \\ +\ 32.5 \\ \hline 73.8 \end{array}$$

Step 2: Give the sum a negative sign. Thus,

$$-41.3 + (-32.5) = -73.8$$ ■

A N S W E R

So, to add two numbers with negative signs, we just ignore the signs and add the two numbers giving the sum a negative sign.

That's correct, Charlie. Let's try some practice problems.

15. ____________

Practice 15: Add the following numbers.

$$-83 + (-27)$$

16. ____________

Practice 16: Add the following numbers.

$$-66 + (-63)$$

17. ____________

Practice 17: Add the following numbers.

$$-7.4 + (-12.6)$$

18. ____________

Practice 18: Add the following numbers.

$$-34.67 + (-43.89)$$

EXERCISE 2.1

Add the numbers in each problem.

A

1. $4 + (-5)$

2. $7 + (-9)$

3. $-13 + (-13)$

4. $-23 + (-1)$

5. $-8 + 9$

6. $-7 + 13$

7. $34 + (-11)$

8. $27 + (-19)$

9. $0 + (-9)$

10. $27 + (-28)$

11. $-37 + 32$

12. $-76 + 26$

13. $-38 + 73$

14. $-44 + 56$

15. $-35 + (-23)$

16. $-69 + (-73)$

17. $-43 + (-26)$

18. $-52 + (-78)$

19. $78 + (-103)$

20. $155 + (-210)$

B

21. $7 + (-5) + (-3)$

22. $6 + (-3) + (-12)$

23. $5.55 + (-.5)$

24. $7.6 + (-4.2)$

25. $-8.3 + 5.1$

26. $7.8 + (-6.4)$

27. $8.9 + (-9.1)$

28. $42.7 + (-26)$

29. $-3.34 + (-2.4)$

30. $-23.44 + (-7.88)$

APPLICATIONS

Write your answer using signed numbers.

1. The temperature in Fairbanks, Alaska, is 15 degrees below zero. Write the temperature.

2. Malfunction, Incorporated, lost \$3.5 million last year. Write the loss.

3. Computer sales for this month were down 12% from last month. Write this decline.

4. The price of a particular stock dropped $7\frac{3}{8}$ points last week. Write this drop.

5. If you buy a stereo system for \$834 and later sell it for \$545, what is your loss?

6. The following is a statement of John's profits and losses on his lawn service the last three months. What was John's total profit or loss for the last three months?

May	June	July
\$32	−\$47	\$28

7. Frank's newspaper service received 42 new subscriptions (+42) and 47 cancellations (−47) last year. What was the net change in the number of subscriptions?

8. Sue earned \$637 last month (+\$637) and spent \$426 on school expenses (−\$426). What was her net gain or loss for the month?

2.2 SUBTRACTING SIGNED NUMBERS

When subtracting two signed numbers, we change the subtraction to addition using the following three rules.

> Let a and b be nonzero numbers.
> **Rule 1:** $-(-a) = a$
> **Rule 2:** $a - b = a + (-b)$
> **Rule 3:** $a - (-b) = a + b$

You mean to subtract signed numbers, we change subtraction to addition?

That's correct, Charlie. We will use Rule 2 or Rule 3 above to change the subtraction to addition. Then we will add using the rules discussed in Section 2.1. Let's look at why the above rules are true.

Rule 1: If a is a nonzero number, $-(-a) = a$.

Let's first illustrate the rule using the number line. Now a and $-a$ are opposites; that is, a and $-a$ are both a units from zero in opposite directions.

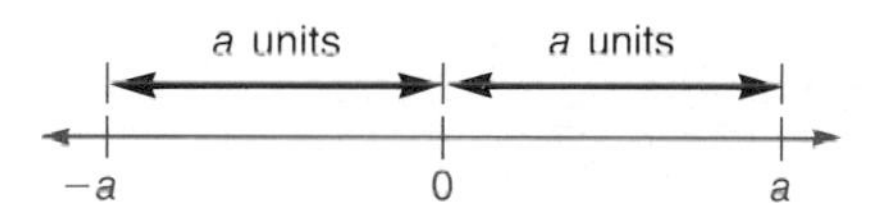

The opposite of $-a$ is $-(-a)$. Now, $-(-a)$ must be the same number of units from zero as $-a$ in the opposite direction. Looking at the number line we see that this places $-(-a)$ at the same position as a.

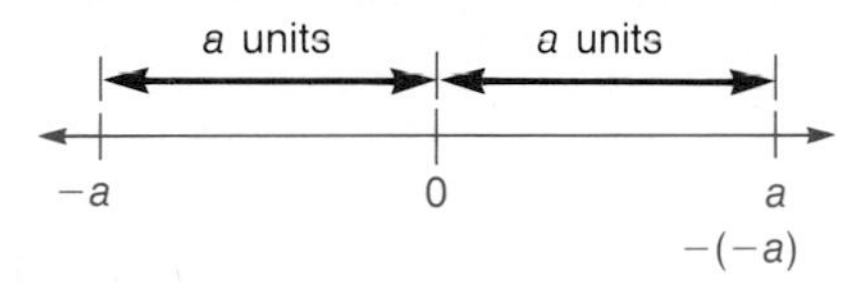

Thus, $-(-a)$ must equal a.

Next, we will prove the rule algebraically.

1. Now, $-a + a = a + (-a)$ — **Commutative property of addition.**
   $= 0$ — **The sum of a number and its opposite is zero.**
2. and $-a + [-(-a)] = 0$ — **The sum of a number and its opposite is zero.**
3. Since there is only one number which added to $-a$ equals 0,

$$-(-a) = a$$

How did you know that $-(-a)$ *was the opposite of* $-a$ *in Step 2?*

Well, Charlie, the opposite of $-a$ is $-a$ with a negative sign, $-(-a)$.

How did you know that there was only one number which added to $-a$ *equals 0 in Step 3?*

Well, Charlie, there is only one correct answer to an arithmetic operation in our number system.

Rule 2: If a and b are nonzero numbers, $a - b = a + (-b)$.

Now, by the definition of subtraction, $a - b$ is the number which added to b equals a. So,

$$(a - b) + b = a$$ **Definition of subtraction.**

Also, $$[a + (-b)] + b = a + [(-b) + b]$$ **Associative property of addition.**

$$= a + 0$$ **The sum of a number and its opposite is zero.**

$$= a$$ **Additive property of zero.**

That is, $$[a + (-b)] + b = a$$

Since there is only one number which added to b equals a,

$$a - b = a + (-b)$$

Thus, subtracting b from a is the same as adding the opposite of b to a.

Rule 3: If a and b are nonzero numbers, $a - (-b) = a + b$.

Now, $$a - (-b) = a + [-(-b)]$$ **Rule 2:** $a - b = a + (-b)$
$$= a + b$$ **Rule 1:** $-(-a) = a$

Thus, $a - (-b) = a + b$.
So, subtracting is the same as adding the opposite.
Now, let's look at some examples in which we use these three rules.

Example 1 **Subtract the following numbers.**

$$5 - 8.2$$

Step 1: Change the subtraction to addition.

$$5 - 8.2 = 5 + (-8.2)$$ **Rule 2:** $a - b = a + (-b)$

ANSWER

Step 2: Add using the rules of Section 2.1.

$$5 + (-8.2) = -3.2$$

Thus, $5 - 8.2 = -3.2$. ■

Example 2 **Subtract the following numbers.**

$$3 - (-7)$$

Step 1: Change the subtraction to addition.

$$3 - (-7) = 3 + 7 \qquad \textbf{Rule 3: } a - (-b) = a + b$$

Step 2: Add using the rules of Section 2.1.

$$3 + 7 = 10$$

Thus, $3 - (-7) = 10$. ■

Example 3 **Subtract the following numbers.**

$$-(-4.6) - 8.7$$

Step 1: Change the subtraction to addition.

$$-(-4.6) - 8.7 = -(-4.6) + (-8.7) \qquad \textbf{Rule 2: } a - b = a + (-b)$$

Step 2: Rewrite $-(-4.6)$ as 4.6.

$$-(-4.6) + (-8.7) = 4.6 + (-8.7) \qquad \textbf{Rule 1: } -(-a) = a$$

Step 3: Add using the rules of Section 2.1.

$$4.6 + (-8.7) = -4.1$$

Thus, $-(-4.6) - 8.7 = -4.1$. ■

So, to subtract signed numbers, we change the sign of the number to be subtracted and add, using the rules for addition of signed numbers.

That's true, Charlie. Let's do some practice problems.

Practice 1: Subtract the following numbers.

$$34 - 27$$

1. ______________

A N S W E R

2. ____________________

Practice 2: Subtract the following numbers.

$$23 - 38$$

3. ____________________

Practice 3: Subtract the following numbers.

$$-47 - 58$$

4. ____________________

Practice 4: Subtract the following numbers.

$$69 - (-30)$$

5. ____________________

Practice 5: Subtract the following numbers.

$$-(-42) - 73$$

6. ____________________

Practice 6: Subtract the following numbers.

$$-(-23) - (-17)$$

7. ____________________

Practice 7: Subtract the following numbers.

$$-(-93) - 93$$

8. ____________________

Practice 8: Subtract the following numbers.

$$-9 - 6 - (-10)$$

Practice 9: Subtract the following numbers.

$$-8.2 - 5.8$$

A N S W E R

9. ________________

Practice 10: Subtract the following numbers.

$$-7.38 - (-3.21)$$

10. ________________

EXERCISE 2.2

Subtract the numbers in each problem.

A

1. $17 - 3$

2. $18 - 9$

3. $4 - 32$

4. $9 - 23$

5. $-9 - 7$

6. $-6 - 8$

7. $-(-7) - 8$

8. $-(-22) - 17$

9. $23 - (-15)$

10. $26 - (-42)$

11. $-(-5) - (-5)$

12. $47 - (-23)$

13. $-22 - 18$

14. $-(-23) - 37$

15. $-52 - (-48)$

16. $88 - 72$

17. $-43 - (-43)$

18. $-(-79) - 97$

19. $48 - (-27)$

20. $-99 - (-99)$

B

21. $-8 - 3 - 9$

22. $6 - (-9) - 4$

23. $-(-25) - 9 - (-22)$

24. $-8.7 - 3.3$

25. $6.8 - (-3.2)$

26. $-9.6 - 4.7$

27. $37.4 - 9.32$

28. $-8.2 - 3.4$

29. $-4.37 - (-2.7)$

30. $-(-8) - 4.55$

31. $-(-7.2) - (-3.3)$

APPLICATIONS

Write your answer using signed numbers.

1. Paul buys and sells real estate. When Paul purchases property, he has the property appraised and later subtracts the appraisal cost from his profit or loss. On a recent transaction, Paul lost \$3455 (−\$3455) and his appraisal cost was \$525. What was Paul's total loss?

2. When Mary buys and sells stock, she subtracts the brokerage fee from her profit or loss. Mary purchased some stock last year for \$936 and sold it this year for \$638; her brokerage fee was \$65. What was Mary's loss? What was Mary's total loss after the brokerage fee was subtracted?

2.3 MULTIPLYING SIGNED NUMBERS

When multiplying two signed numbers, we use the following three rules.

Rule 1: The product of two positive numbers is a positive number.
Rule 2: The product of a positive and a negative number is a negative number.
Rule 3: The product of two negative numbers is a positive number.

Let's look at why Rules 2 and 3 are true.

Rule 2: If a and b are positive numbers, $a \times (-b) = -(a \times b)$.

Now, $(a \times b) + [a \times (-b)] = a \times [b + (-b)]$ **Distributive property.**

$= a \times 0$ **The sum of a number and its opposite is zero.**

$= 0$ **Multiplication property of zero.**

That is, $(a \times b) + [a \times (-b)] = 0.$

Also, $(a \times b) + [-(a \times b)] = 0$ **The sum of a number and its opposite is zero.**

Since there is only one number which added to $a \times b$ equals 0,

$$a \times (-b) = -(a \times b)$$

Thus, a positive number a times a negative number $-b$ is a negative number $[-(a \times b)]$. Since multiplication is commutative $a \times (-b) = (-b) \times a$ and we know that a negative number $-b$ times a positive number a is also a negative number.

Rule 3: If a and b are positive numbers, $(-a) \times (-b) = a \times b$.

Now,

$-(a \times b) + [(-a) \times (-b)] = [a \times (-b)] + [(-a) \times (-b)]$

Rule 2: $-(a \times b) = a \times (-b)$

$= [(-b) \times a] + [(-b) \times (-a)]$

Commutative property of multiplication.

$= (-b) \times [a + (-a)]$

Distributive property.

$$= (-b) \times 0$$ The sum of a number and its opposite is zero.

$$= 0$$ Multiplication property of zero.

That is, $-(a \times b) + [(-a) \times (-b)] = 0.$

Also, $-(a \times b) + (a \times b) = 0$ The sum of a number and its opposite is zero.

Since there is only one number which added to $-(a \times b)$ equals 0,

$$(-a) \times (-b) = a \times b$$

Thus, a negative number $-a$ times a negative number $-b$ is a positive number $(a \times b)$.

Now, let's look at some examples in which we use these three rules.

Example 1 **Multiply the following numbers.**

$$5 \times 6$$

Step 1: Ignore the signs and multiply.

$$5 \times 6 = 30$$

Step 2: According to Rule 1, the product of two positive numbers is a positive number. Thus,

$$5 \times 6 = 30$$ ■

Example 2 **Multiply the following numbers.**

$$2.4 \times (-10)$$

Step 1: Ignore the signs and multiply.

$$2.4 \times 10 = 24$$

Step 2: According to Rule 2, the product of a positive and a negative number is a negative number. Thus,

$$2.4 \times (-10) = -24$$ ■

Example 3 **Multiply the following numbers.**

$$-6 \times (-1.5)$$

Step 1: Ignore the signs and multiply.

$$6 \times 1.5 = 9$$

Step 2: According to Rule 3, the product of two negative numbers is a positive number. Thus,

$$-6 \times (-1.5) = 9$$ ■

ANSWER

So, when multiplying two signed numbers, we just ignore the signs and multiply the two numbers. Then we use Rule 1, 2, or 3 to determine the sign in the answer.

That's correct, Charlie. Let's try some practice problems.

Practice 1: Multiply the following numbers.

$$5 \times 9$$

1. ________________

Practice 2: Multiply the following numbers.

$$-7 \times 15$$

2. ________________

Practice 3: Multiply the following numbers.

$$-4 \times (-9)$$

3. ________________

Practice 4: Multiply the following numbers.

$$12 \times (-8)$$

4. ________________

A N S W E R

5. ______________

Practice 5: Multiply the following numbers.

$$-13 \times 2.4$$

6. ______________

Practice 6: Multiply the following numbers.

$$-3.2 \times (-2.5)$$

7. ______________

Practice 7: Multiply the following numbers.

$$.6 \times (-5.2)$$

What happens when we multiply more than two signed numbers?

Well, Charlie, we can only multiply two numbers at a time. Thus, we use Rules 1, 2, and 3 as many times as necessary. Also, we can put Rules 1, 2, and 3 together to form a fourth rule which may be easier to use.

Rule 4: To multiply more than two signed numbers:

1. Ignore the signs and multiply the numbers.
2. Count the negative signs in the original problem.
3. If the number of negative signs is even (0, 2, 4, 6, . . .) then give the product a positive sign or no sign. Otherwise, give the product a negative sign.

A N S W E R

Example 4 **Multiply the following numbers.**

$$-2 \times (-3) \times (-5) \times (-4)$$

Step 1: Ignore the signs and multiply.

$$2 \times 3 \times 5 \times 4 = 120$$

Step 2: Count the negative signs in the original problem. There are four negative signs.

Step 3: Since 4 is an even number, give the product a positive sign or no sign. Thus,

$$-2 \times (-3) \times (-5) \times (-4) = 120$$ ■

Example 5 **Multiply the following numbers.**

$$-(-3) \times (-2) \times (-3) \times (-5) \times 2$$

Step 1: Ignore the signs and multiply.

$$3 \times 2 \times 3 \times 5 \times 2 = 180$$

Step 2: Count the negative signs in the original problem. There are five negative signs.

Step 3: Since 5 is not an even number, give the product a negative sign. Thus,

$$-(-3) \times (-2) \times (-3) \times (-5) \times 2 = -180$$ ■

So, when we multiply signed numbers, we just ignore the signs and multiply the numbers. Then we count the negative signs in the original problem. If the number of negative signs is even, we give the product a positive sign or no sign; otherwise, we give the product a negative sign.

That's correct, Charlie. Let's try some practice problems.

Practice 8: Multiply the following numbers.

$$-(-7) \times (-5) \times 2$$

8. ______________

ANSWER

9. ________

Practice 9: Multiply the following numbers.

$$2 \times (-2) \times (-5) \times (-4) \times (-1)$$

10. ________

Practice 10: Multiply the following numbers.

$$7 \times (-12) \times 3$$

11. ________

Practice 11: Multiply the following numbers.

$$-3 \times (-6) \times 5 \times (-2)$$

12. ________

Practice 12: Multiply the following numbers.

$$4 \times (-8) \times (-2) \times 5 \times (-11)$$

In the following practice problems do not actually multiply; just determine whether the product is positive, negative, or zero.

13. ________

Practice 13: Determine whether the answer to the following problem is positive, negative, or zero.

$$-32 \times (-15) \times [-(-23)] \times (-45) \times (-12) \times (-7)$$

A N S W E R

Practice 14: Determine whether the answer to the following problem is positive, negative, or zero.

$$-1.75 \times (-2.3) \times (-1.3) \times (-5.67) \times (-2.34) \times 5.83 \times (-6.72)$$

14. ______________

Practice 15: Determine whether the answer to the following problem is positive, negative, or zero.

$$-\frac{1}{2} \times \frac{2}{3} \times \left(-\frac{5}{8}\right) \times \left(-\frac{3}{5}\right) \times \frac{1}{2}$$

15. ______________

Practice 16: Determine whether the answer to the following problem is positive, negative, or zero.

$$-1 \times (-2) \times (-1) \times (-3) \times (-5) \times (-6) \times (-2) \times (-9)$$

16. ______________

Practice 17: Determine whether the answer to the following problem is positive, negative, or zero.

$$23 \times (-21) \times (-2.5) \times 3.75 \times 0 \times (-2.7) \times 5.6$$

17. ______________

Well, Charlie, how did you do on the practice problems?

I had a little trouble with Practice 17. My answer was negative; however, the answer at the end of the chapter is zero.

Well, Charlie, I can understand your trouble with Practice 17. If you simply count the negative signs, you get a negative answer; however, the zero in the problem makes the answer zero.

EXERCISE 2.3

Multiply the numbers in each problem.

A

1. 7×8

2. -6×3

3. $25 \times (-2)$

4. $-9 \times (-3)$

5. -6×4

6. $7 \times (-8)$

7. $-8 \times (-12)$

8. $5 \times (-15)$

9. $-21 \times (-8)$

10. $-25 \times (-8)$

B

11. $-(-4) \times -5$

12. $5 \times (-2) \times (-7)$

13. $(-5) \times (-11) \times (-4)$

14. $(-3) \times (-3) \times (-6)$

15. $3 \times (-2) \times (-4) \times 2$

16. $(-4) \times [-(-2)] \times (\quad 3) \times (-2)$

17. $3 \times (-5) \times 6 \times 4 \times (-2)$

18. $3 \times 2 \times (-2) \times 5 \times (-4)$

19. $(-5) \times (-5) \times 2 \times (-2) \times 4$

20. $(-2) \times (-2) \times (-2) \times (-2)$

21. $-(-7) \times [-(-8)]$

22. $(-2) \times (-3.7)$

23. $3.1 \times (-2.8)$

24. $-3.2 \times [-(-1.3)]$

25. -25.5×4

26. $-5.2 \times (-3.5)$

27. $6.1 \times (-15.15)$

28. $2.5 \times (-3.7)$

29. $-(-2.2) \times 4.61$

30. $-9.2 \times (-1.5)$

31. $-(-5.1) \times (-6)$

32. $3.4 \times (-7.2)$

APPLICATIONS

Write your answers using signed numbers.

1. The average change in the stock market the past five days was -17 points a day. What was the net change in the stock market the past five days?

2. Frank's records show that each of his hogs had a weight change of -1.5 pounds a day during a particularly cold 12-day period. What was the net weight change for a hog during this 12-day period? If hogs were selling for \$1.15 a pound, what was Frank's loss per hog during this 12-day period?

2.4 DIVIDING SIGNED NUMBERS

In order to divide two signed numbers, we will use the following three rules.

> **Rule 1:** When two positive numbers are divided, the answer (quotient) is a positive number.
> **Rule 2:** When a positive number and a negative number are divided, the answer (quotient) is a negative number.
> **Rule 3:** When two negative numbers are divided, the answer (quotient) is a positive number.

Let's look at why Rules 2 and 3 are true.

Rule 2: If a and b are positive numbers, $\frac{-a}{b} = \frac{a}{-b} = -\frac{a}{b}$.

Now, $\frac{a}{b} + \frac{-a}{b} = \frac{a + (-a)}{b}$ **Addition of fractions.**

$= \frac{0}{b}$ **The sum of a number and its opposite is zero.**

$= 0$ **Zero divided by any nonzero number is zero.**

That is, $\frac{a}{b} + \frac{-a}{b} = 0$.

Now, $\frac{a}{b} + \frac{a}{-b} = \frac{a \times (-b)}{b \times (-b)} + \frac{a \times b}{(-b) \times b}$ **Getting like denominators.**

$= \frac{[a \times (-b)] + (a \times b)}{b \times (-b)}$ **Addition of fractions.**

$= \frac{a \times [(-b) + b]}{b \times (-b)}$ **Distributive property.**

$= \frac{a \times 0}{b \times (-b)}$ **The sum of a number and its opposite is zero.**

$= \frac{0}{b \times (-b)}$ **Multiplication property of zero.**

$= 0$ **Zero divided by any nonzero number is zero.**

That is, $\frac{a}{b} + \frac{a}{(-b)} = 0$.

Also, $\frac{a}{b} + \left(-\frac{a}{b}\right) = 0$ The sum of a number and its opposite is zero.

So we have

$$\frac{a}{b} + \frac{-a}{b} = 0$$

$$\frac{a}{b} + \frac{a}{-b} = 0$$

and $$\frac{a}{b} + \left(-\frac{a}{b}\right) = 0$$

Since there is only one number which added to $\frac{a}{b}$ equals 0,

$$\frac{-a}{b} = \frac{a}{-b} = -\frac{a}{b}$$

Thus, a negative number $-a$ divided by a positive number b is a negative number $-\frac{a}{b}$; a positive number a divided by a negative number $-b$ is a negative number $-\frac{a}{b}$.

Rule 3: If a and b are positive numbers, $\frac{-a}{-b} = \frac{a}{b}$.

Now,

$$-\frac{a}{b} + \frac{-a}{-b} = \frac{a}{-b} + \frac{-a}{-b}$$ Rule 2: $-\frac{a}{b} = \frac{a}{-b}$

$$= \frac{a + (-a)}{-b}$$ Addition of fractions.

$$= \frac{0}{(-b)}$$ The sum of a number and its opposite is zero.

$$= 0$$ Zero divided by any nonzero number is zero.

That is, $-\frac{a}{b} + \frac{-a}{-b} = 0$.

Also, $-\frac{a}{b} + \frac{a}{b} = 0$ The sum of a number and its opposite is zero.

Since there is only one number which added to $-\frac{a}{b}$ equals 0,

$$\frac{-a}{-b} = \frac{a}{b}$$

Thus, a negative number $-a$ divided by a negative number $-b$ is a positive number $\frac{a}{b}$.

Now, let's look at some examples in which we use these three rules.

Example 1 **Divide the following numbers.**

$$1.8 \div 9$$

Step 1: Ignore the signs and divide.

$$1.8 \div 9 = .2$$

A N S W E R

Step 2: According to Rule 1, when two positive numbers are divided, the answer is a positive number. Thus,

$$1.8 \div 9 = .2$$

■

Example 2 **Divide the following numbers.**

$$-25 \div 5$$

Step 1: Ignore the signs and divide.

$$25 \div 5 = 5$$

Step 2: According to Rule 2, when a positive and a negative number are divided, the answer is a negative number. Thus,

$$-25 \div 5 = -5$$

■

Example 3 **Divide the following numbers.**

$$-4.2 \div (-6)$$

Step 1: Ignore the signs and divide.

$$4.2 \div 6 = .7$$

Step 2: According to Rule 3, when two negative numbers are divided, the answer is a positive number. Thus,

$$-4.2 \div (-6) = .7$$

■

So, when dividing two signed numbers, we just ignore the signs and divide the two numbers. Then we use Rule 1, 2, or 3 to determine the sign in the answer.

That's correct, Charlie. Let's try some practice problems.

Practice 1: Divide the following numbers.

$$72 \div 6$$

1. ____________

A N S W E R

2. ____________

Practice 2: Divide the following numbers.

$$42 \div (-7)$$

3. ____________

Practice 3: Divide the following numbers.

$$-60 \div 5$$

4. ____________

Practice 4: Divide the following numbers.

$$(-6.3) \div (-9)$$

5. ____________

Practice 5: Divide the following numbers.

$$12 \div (-1.5)$$

6. ____________

Practice 6: Divide the following numbers.

$$(-16.25) \div (-2.5)$$

Practice 7: Divide the following numbers.

$$(-4.98) \div 3.32$$

A N S W E R

7. ____________

EXERCISE 2.4

Divide the numbers in each problem.

A

1. $24 \div (-4)$

2. $(-63) \div (-9)$

3. $(\ \ 72) \div (-8)$

4. $-56 \div 7$

5. $42 \div (-3)$

6. $-65 \div 5$

7. $(-98) \div (-7)$

8. $(-143) \div (-11)$

9. $(-189) \div (-7)$

10. $(-128) \div (-8)$

11. $87 \div (-3)$

12. $-195 \div 15$

B

13. $-5.7 \div 1.9$

14. $(-2.6) \div 4$

15. $(-11) \div (-2.5)$

16. $8.5 \div (-3.4)$

17. $-6.96 \div (-1.2)$

18. $-4.9 \div 3.5$

19. $(-6.72) \div (-2.1)$

20. $0 \div (-4.9)$

APPLICATIONS

Write your answers using signed numbers.

1. The change in the stock market the last six months was -78 points. What was the average monthly change during this six-month period?

2. The daily low temperatures in Fairbanks, Alaska, last week were -14, -11, 6, 9, -21, -18, and -7. What was the average low temperature in Fairbanks last week?

2.5 MORE SIGNED NUMBERS

In this section, we will perform arithmetic operations on fractions and mixed numbers containing negative signs. Since the fraction $\frac{a}{b}$ is actually a (the numerator) divided by b (the denominator), we will use the rules for dividing signed numbers when working with fractions. Let's review Rules 2 and 3 from Section 2.4 using fractional notation.

Rule 2: $\frac{-a}{b} = \frac{a}{-b} = -\frac{a}{b}$, where a and b are positive numbers.
Rule 3: $\frac{-a}{-b} = \frac{a}{b}$, where a and b are positive numbers.

When adding or subtracting fractions involving negative signs, we will use the above rules to write each fraction in the form

$$\frac{-a}{b} \text{ or } \frac{a}{b}$$

Why do we want to write each fraction in the form $\frac{-a}{b}$ or $\frac{a}{b}$?

Well, Charlie, we want to have the negative sign, if any, in the numerator of the fraction.

How does it help having the negative sign in the numerator of the fraction?

When the negative sign is in the numerator of the fraction, all of the signed number arithmetic can be done with integers. This simplifies the signed number arithmetic. Let's look at some examples.

A N S W E R

Example 1 **Add the following fractions.**

$$\frac{-7}{12} + \frac{5}{12}$$

Solution: Since the only negative sign is in the numerator, and the denominators are the same, we simply add the numerators using the rules for adding signed numbers from Section 2.1, and reduce.

$$\frac{-7}{12} + \frac{5}{12} = \frac{-7+5}{12} \quad \text{Combine fractions.}$$

$$= \frac{-2}{12} \quad \text{Simplify the numerator.}$$

$$= -\frac{2}{12} \quad \frac{-a}{b} = -\frac{a}{b}$$

$$= -\frac{1}{6} \quad \text{Reduce.}$$

Thus, $\frac{-7}{12} + \frac{5}{12} = -\frac{1}{6}$. ■

We added these fractions the same way that we have always added fractions.

That's true, Charlie. The only difference is we used the rules for signed numbers when adding the numerators. Let's try some practice problems.

1. ____________________

Practice 1: Add the following fractions.

$$\frac{-3}{5} + \frac{4}{5}$$

2. ____________________

Practice 2: Add the following fractions.

$$\frac{3}{8} + \frac{-5}{8}$$

Practice 3: Add the following fractions.

$$\frac{-3}{8} + \frac{-5}{8}$$

A N S W E R

3. ______________

Practice 4: Add the following fractions.

$$\frac{7}{18} + \frac{-5}{18}$$

4. ______________

Example 2 **Add the following fractions.**

$$-\frac{7}{15} + \frac{4}{15}$$

Solution: Rewrite $-\frac{7}{15}$ in the form $\frac{-a}{b}$ and add.

$$-\frac{7}{15} + \frac{4}{15} = \frac{-7}{15} + \frac{4}{15} \qquad -\frac{a}{b} = \frac{-a}{b}$$

$$= \frac{-7+4}{15} \qquad \text{Combine fractions.}$$

$$= \frac{-3}{15} \qquad \text{Simplify the numerator.}$$

$$= -\frac{3}{15} \qquad \frac{-a}{b} = -\frac{a}{b}$$

$$= -\frac{1}{5} \qquad \text{Reduce.}$$

Thus, $-\frac{7}{15} + \frac{4}{15} = -\frac{1}{5}$. ■

Why did you rewrite $-\frac{7}{15}$ as $\frac{-7}{15}$?

Because we were then able to perform all of the signed number arithmetic with integers.

A N S W E R

5. ____________

Practice 5: Add the following fractions.

$$-\frac{5}{7} + \frac{6}{7}$$

6. ____________

Practice 6: Add the following fractions.

$$\frac{5}{16} + \left(-\frac{9}{16}\right)$$

7. ____________

Practice 7: Add the following fractions.

$$-\frac{11}{14} + \left(-\frac{1}{14}\right)$$

8. ____________

Practice 8: Add the following fractions.

$$-\frac{9}{20} + \left(-\frac{7}{20}\right)$$

Example 3 **Add the following fractions.**

$$-\frac{3}{10} + \frac{2}{5}$$

Step 1: Rewrite $-\frac{3}{10}$ in the form $\frac{-a}{b}$.

$$-\frac{3}{10} + \frac{2}{5} = \frac{-3}{10} + \frac{2}{5} \qquad -\frac{a}{b} = \frac{-a}{b}$$

Step 2: Find the least common denominator.

$$\frac{-3}{10} + \frac{2}{5} = \frac{-3}{2 \times 5} + \frac{2}{5}$$ **Completely factor the denominators.**

$$= \frac{-3}{2 \times 5} + \frac{\mathbf{2 \times 2}}{\mathbf{2 \times 5}}$$ **Get a common denominator, $\frac{a}{b} = \frac{a \times c}{b \times c}$.**

$$= \frac{-3}{10} + \frac{4}{10}$$ **Perform the multiplications.**

ANSWER

Step 3: Add the fractions.

$$\frac{-3}{10} + \frac{4}{10} = \frac{-3+4}{10} \quad \text{Combine fractions.}$$

$$= \frac{1}{10} \quad \text{Simplify the numerator.}$$

Thus, $-\frac{3}{10} + \frac{2}{5} = \frac{1}{10}$. ■

Once we rewrite $-\frac{3}{10}$ as $\frac{-3}{10}$, we just find the least common denominator and add the two fractions, just as we have always done.

That's true, Charlie. In fact, to find the least common denominator, you don't have to completely factor the denominators. If you can find the least common denominator without completely factoring, that is okay.

Practice 9: Add the following fractions.

$$-\frac{3}{5} + \frac{7}{15}$$

9. ______

Practice 10: Add the following fractions.

$$\frac{7}{8} + \frac{-5}{12}$$

10. ______

Practice 11: Add the following fractions.

$$\frac{-2}{9} + \frac{-5}{12}$$

11. ______

Practice 12: Add the following fractions.

$$\frac{5}{18} + \left(-\frac{11}{24}\right)$$

12. ______

Now, let's look at some examples of subtracting fractions containing negative signs.

Example 4 **Subtract the following fractions.**

$$\frac{-5}{16} - \frac{3}{16}$$

Solution: Since the only negative sign is in the first numerator, and the denominators are the same, we simply subtract the numerators using the rules for subtracting signed numbers from Section 2.2 and reduce.

$$\frac{-5}{16} - \frac{3}{16} = \frac{-5 - 3}{16} \qquad \text{Combine fractions.}$$

$$= \frac{-5 + (-3)}{16} \qquad a - b = a + (-b)$$

$$= \frac{-8}{16} \qquad \text{Simplify the numerator.}$$

$$= -\frac{8}{16} \qquad \frac{-a}{b} = -\frac{a}{b}$$

$$= -\frac{1}{2} \qquad \text{Reduce.}$$

Thus, $\frac{-5}{16} - \frac{3}{16} = -\frac{1}{2}$. ■

When you started to find the solution, you said the only negative sign was in the first numerator. Isn't the sign in front of $\frac{3}{16}$ a negative sign?

No, Charlie. The sign in front of $\frac{3}{16}$ is the subtraction symbol.

Example 5 **Subtract the following fractions.**

$$\frac{-7}{24} - \frac{-5}{24}$$

Solution: Since the only negative signs are in the numerators, and the denominators are the same, we simply subtract the numerators using the rules for subtracting signed numbers from Section 2.2 and reduce.

$$\frac{-7}{24} - \frac{-5}{24} = \frac{-7 - (-5)}{24} \qquad \text{Combine fractions.}$$

$$= \frac{-7 + 5}{24} \qquad a - (-b) = a + b$$

ANSWER

$$= \frac{-2}{24} \qquad \textbf{Simplify the numerator.}$$

$$= -\frac{2}{24} \qquad \frac{-a}{b} = -\frac{a}{b}$$

$$= -\frac{1}{12} \qquad \textbf{Reduce.}$$

Thus, $\frac{-7}{24} - \frac{-5}{24} = -\frac{1}{12}$. ■

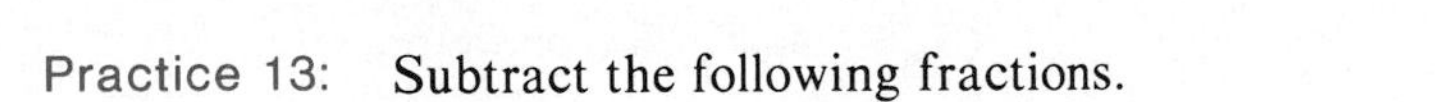

Practice 13: Subtract the following fractions.

$$\frac{-3}{8} - \frac{5}{8}$$

13. ____________________

Practice 14: Subtract the following fractions.

$$\frac{-7}{10} - \frac{-3}{10}$$

14. ____________________

Practice 15: Subtract the following fractions.

$$\frac{-8}{9} - \frac{-5}{9}$$

15. ____________________

Practice 16: Subtract the following fractions.

$$\frac{7}{12} - \frac{-5}{12}$$

16. ____________________

Example 6 **Subtract the following fractions.**

$$-\frac{5}{16} - \frac{7}{16}$$

A N S W E R

Solution: Rewrite $-\frac{5}{16}$ in the form $\frac{-a}{b}$ and subtract.

$$-\frac{5}{16} - \frac{7}{16} = \frac{-5}{16} - \frac{7}{16} \qquad -\frac{a}{b} = \frac{-a}{b}$$

$$= \frac{-5 - 7}{16} \qquad \textbf{Combine fractions.}$$

$$= \frac{-5 + (-7)}{16} \qquad a - b = a + (-b)$$

$$= \frac{-12}{16} \qquad \textbf{Simplify the numerator.}$$

$$= -\frac{12}{16} \qquad \frac{-a}{b} = -\frac{a}{b}$$

$$= -\frac{3}{4} \qquad \textbf{Reduce.}$$

Thus, $-\frac{5}{16} - \frac{7}{16} = -\frac{3}{4}$. ■

Example 7 **Subtract the following fractions.**

$$\frac{-3}{24} - \left(-\frac{11}{24}\right)$$

Solution: Rewrite $-\frac{11}{24}$ in the form $\frac{-a}{b}$ and subtract.

$$\frac{-3}{24} - \left(-\frac{11}{24}\right) = \frac{-3}{24} - \frac{-11}{24} \qquad -\frac{a}{b} = \frac{-a}{b}$$

$$= \frac{-3 - (-11)}{24} \qquad \textbf{Combine fractions.}$$

$$= \frac{-3 + 11}{24} \qquad a - (-b) = a + b$$

$$= \frac{8}{24} \qquad \textbf{Simplify the numerator.}$$

$$= \frac{1}{3} \qquad \textbf{Reduce.}$$

Thus, $\frac{-3}{24} - (-\frac{11}{24}) = \frac{1}{3}$. ■

17. ______________________

Practice 17: Subtract the following fractions.

$$-\frac{7}{12} - \frac{-5}{12}$$

18. ______________________

Practice 18: Subtract the following fractions.

$$-\frac{5}{9} - \left(-\frac{2}{9}\right)$$

Practice 19: Subtract the following fractions.

$$-\frac{4}{15} - \frac{8}{15}$$

ANSWER

19. ______

Practice 20: Subtract the following fractions.

$$\frac{7}{24} - \left(-\frac{11}{24}\right)$$

20. ______

Example 8 **Subtract the following fractions.**

$$-\frac{5}{8} - \frac{-11}{12}$$

Step 1: Rewrite $-\frac{5}{8}$ in the form $\frac{-a}{b}$.

$$-\frac{5}{8} - \frac{-11}{12} = \frac{-5}{8} - \frac{-11}{12} \qquad -\frac{a}{b} = \frac{-a}{b}$$

Step 2: Find the least common denominator.

$$\frac{-5}{8} - \frac{-11}{12} = \frac{-5}{2 \times 2 \times 2} - \frac{-11}{2 \times 2 \times 3}$$ **Completely factor the denominators.**

$$= \frac{3(-5)}{2 \times 2 \times 2 \times 3} - \frac{2(-11)}{2 \times 2 \times 2 \times 3}$$ **Get a common denominator,** $\frac{a}{b} = \frac{a \times c}{b \times c}$.

$$= \frac{-15}{24} - \frac{-22}{24}$$ **Perform the multiplications.**

Step 3: Subtract the fractions.

$$\frac{-15}{24} - \frac{-22}{24} = \frac{-15 - (-22)}{24}$$ **Combine fractions.**

$$= \frac{-15 + 22}{24}$$ $a - (-b) = a + b$

$$= \frac{7}{24}$$ **Simplify the numerator.**

Thus, $-\frac{5}{8} - \frac{-11}{12} = \frac{7}{24}$. ■

A N S W E R

In Step 2, I was able to find the least common denominator without completely factoring the denominators. Was that okay?

Yes, Charlie.

21. ____________

Practice 21: Subtract the following fractions.

$$-\frac{7}{22} - \frac{-10}{11}$$

22. ____________

Practice 22: Subtract the following fractions.

$$-\frac{9}{14} - \left(-\frac{8}{21}\right)$$

23. ____________

Practice 23: Subtract the following fractions.

$$\frac{-11}{20} - \frac{9}{15}$$

24. ____________

Practice 24: Subtract the following fractions.

$$-\frac{3}{4} - \left(-\frac{5}{6}\right)$$

ANSWER

Now, let's look at the addition and subtraction of signed mixed numbers. When adding and subtracting signed mixed numbers, we will use the rules from Sections 2.1 and 2.2.

Example 9 **Add the following mixed numbers.**

$$2\frac{7}{16} + \left(-5\frac{3}{16}\right)$$

Step 1: Subtract the smaller mixed number from the larger mixed number (without regard to sign). Now, $5\frac{3}{16} = 4 + 1 + \frac{3}{16} = 4 + \frac{16}{16} + \frac{3}{16} = 4 + \frac{19}{16} = 4\frac{19}{16}$.

$$\begin{array}{r} 4\frac{19}{16} \\ \not{5}\frac{3}{16} \\ -\ 2\frac{7}{16} \\ \hline 2\frac{12}{16} \end{array}$$

Step 2: The larger mixed number (without regard to sign) is $5\frac{3}{16}$. Thus, $2\frac{12}{16}$ has the same sign as $5\frac{3}{16}$. Therefore,

$$2\frac{7}{16} + \left(-5\frac{3}{16}\right) = -2\frac{12}{16}$$ Rule 2, Section 2.1.

Step 3: Reduce the fractional part.

$$-2\frac{12}{16} = -2\frac{3}{4}$$

Thus, $2\frac{7}{16} + (-5\frac{3}{16}) = -2\frac{3}{4}$. ■

Practice 25: Add the following mixed numbers.

$$8\frac{3}{4} + \left(-5\frac{1}{4}\right)$$

25. ____________________

A N S W E R

Practice 26: Add the following mixed numbers.

$$7\frac{5}{8} + \left(-9\frac{3}{8}\right)$$

26. ____________________

27. ____________________

Practice 27: Add the following mixed numbers.

$$1\frac{5}{16} + \left(-6\frac{7}{16}\right)$$

28. ____________________

Practice 28: Add the following mixed numbers.

$$-12\frac{11}{12} + 16\frac{7}{12}$$

Example 10 **Add the following mixed numbers.**

$$-4\frac{3}{8} + \left(-2\frac{7}{8}\right)$$

Step 1: Add the mixed numbers without their sign.

a. Add the fractions.

$$\begin{array}{r} 4\mathbf{\frac{3}{8}} \\ +\ 2\mathbf{\frac{7}{8}} \\ \hline \mathbf{\frac{10}{8}} \end{array}$$

ANSWER

b. Regroup the sum and carry (change the improper fraction $\frac{10}{8}$ to the mixed number $1\frac{2}{8}$ and carry the 1).

$$\begin{array}{r} 1 \\ 4\frac{3}{8} \\ +\ 2\frac{7}{8} \\ \hline \frac{2}{8} \end{array}$$

c. Add the whole numbers.

$$\begin{array}{r} \mathbf{1} \\ \mathbf{4}\frac{3}{8} \\ +\ \mathbf{2}\frac{7}{8} \\ \hline \mathbf{7}\frac{2}{8} \end{array}$$

Step 2: Give the sum a negative sign.

$$-4\frac{3}{8} + \left(-2\frac{7}{8}\right) = -7\frac{2}{8}$$ **Rule 3, Section 2.1.**

Step 3: Reduce the fractional part.

$$-7\frac{2}{8} = -7\frac{1}{4}$$

Thus, $-4\frac{3}{8} + (-2\frac{7}{8}) = -7\frac{1}{4}$. ■

Practice 29: Add the following mixed numbers.

$$-5\frac{3}{4} + \left(-6\frac{1}{4}\right)$$

29. ____________

Practice 30: Add the following mixed numbers.

$$-3\frac{2}{5} + \left(-5\frac{4}{5}\right)$$

30. ____________

A N S W E R

31. ____________________

Practice 31: Add the following mixed numbers.

$$-8\frac{5}{16} + \left(-9\frac{15}{16}\right)$$

32. ____________________

Practice 32: Add the following mixed numbers.

$$-12\frac{11}{24} + \left(-8\frac{17}{24}\right)$$

Example 11 **Add the following mixed numbers.**

$$6\frac{7}{10} + \left(-8\frac{4}{15}\right)$$

Step 1: Find the least common denominator of the two fractional parts.

$$\frac{7}{10} \quad \frac{4}{15}$$

$$\frac{7}{2 \times 5} \quad \frac{4}{3 \times 5}$$ **Completely factor the denominators.**

$$\frac{3 \times 7}{2 \times \mathbf{3} \times 5} \quad \frac{\mathbf{2} \times 4}{\mathbf{2} \times 3 \times 5}$$ **Get a common denominator,** $\frac{a}{b} = \frac{a \times c}{b \times c}$.

$$\frac{21}{30} \quad \frac{8}{30}$$ **Perform the multiplications.**

Thus, $6\frac{7}{10} + (-8\frac{4}{15}) = 6\frac{21}{30} + (-8\frac{8}{30})$.

Step 2: Subtract the smaller mixed number from the larger mixed number (without regard to sign). Now, $8\frac{8}{30} = 7 + 1 + \frac{8}{30} = 7 + \frac{30}{30} + \frac{8}{30} = 7 + \frac{38}{30} = 7\frac{38}{30}$.

$$\begin{array}{r} 7\frac{38}{30} \\ \not{8}\frac{8}{30} \\ -\ 6\frac{21}{30} \\ \hline 1\frac{17}{30} \end{array}$$

A N S W E R

Step 3: The larger mixed number without regard to sign is $8\frac{8}{30}$. Thus, $1\frac{17}{30}$ has the same sign as $8\frac{8}{30}$. Hence,

$$6\frac{21}{30} + \left(-8\frac{8}{30}\right) = -1\frac{17}{30}$$ **Rule 2, Section 2.1.**

Therefore, $$6\frac{7}{10} + \left(-8\frac{4}{15}\right) = -1\frac{17}{30}$$ ■

Practice 33: Add the following mixed numbers.

$$3\frac{1}{4} + \left(-6\frac{1}{2}\right)$$

33. ______________

Practice 34: Add the following mixed numbers.

$$-8\frac{2}{5} + \left(-5\frac{3}{10}\right)$$

34. ______________

Practice 35: Add the following mixed numbers.

$$-18\frac{7}{12} + \left(-9\frac{11}{16}\right)$$

35. ______________

A N S W E R

36. ______________

Practice 36: Add the following mixed numbers.

$$-22\frac{7}{10} + 8\frac{5}{6}$$

Example 12 **Subtract the following mixed numbers.**

$$-9\frac{3}{8} - \left(-6\frac{5}{12}\right)$$

Step 1: Change the subtraction to addition.

$$-9\frac{3}{8} - \left(-6\frac{5}{12}\right) = -9\frac{3}{8} + 6\frac{5}{12} \qquad a - (-b) = a + b$$

Step 2: Find the least common denominator of the two fractional parts.

$$\frac{3}{8} \qquad \frac{5}{12}$$

$$\frac{3}{2 \times 2 \times 2} \qquad \frac{5}{2 \times 2 \times 3}$$ **Completely factor the denominators.**

$$\frac{3 \times \mathbf{3}}{2 \times 2 \times 2 \times \mathbf{3}} \qquad \frac{\mathbf{2} \times 5}{\mathbf{2} \times 2 \times 2 \times 3}$$ **Get a common denominator, $\frac{a}{b} = \frac{a \times c}{b \times c}$.**

$$\frac{9}{24} \qquad \frac{10}{24}$$ **Perform the multiplications.**

Thus, $-9\frac{3}{8} + 6\frac{5}{12} = -9\frac{9}{24} + 6\frac{10}{24}$.

Step 3: Subtract the smaller mixed number from the larger mixed number (without regard to sign). Now, $9\frac{9}{24} = 8 + 1 + \frac{9}{24} = 8 + \frac{24}{24} + \frac{9}{24} = 8 + \frac{33}{24} = 8\frac{33}{24}$.

$$\begin{array}{r} 8\frac{33}{24} \\ \not{9}\frac{9}{24} \\ -\ 6\frac{10}{24} \\ \hline 2\frac{23}{24} \end{array}$$

A N S W E R

Step 4: The larger mixed number (without regard to sign) is $9\frac{9}{24}$. Thus, $2\frac{23}{24}$ has the same sign as $9\frac{9}{24}$. Hence,

$$-9\frac{9}{24} + 6\frac{10}{24} = -2\frac{23}{24}$$ **Rule 2, Section 2.1.**

Therefore, $-9\frac{3}{8} - (-6\frac{5}{12}) = -2\frac{23}{24}$. ■

Practice 37: Subtract the following mixed numbers.

$$5\frac{2}{3} - \left(-2\frac{1}{6}\right)$$

37. ______________________

Practice 38: Subtract the following mixed numbers.

$$-15\frac{3}{8} - 5\frac{1}{6}$$

38. ______________________

Practice 39: Subtract the following mixed numbers.

$$-23\frac{7}{12} - \left(-18\frac{7}{8}\right)$$

39. ______________________

A combination of the rules from Sections 2.3 and 2.4 gives us the following rule for multiplying or dividing fractions containing negative signs.

To multiply or divide fractions containing negative signs:
1. Ignore the signs and multiply or divide the fractions.
2. Count the negative signs in the original problem.
3. If the number of negative signs is even (0, 2, 4, 6, . . .), give the product or quotient a positive sign or no sign. Otherwise, give the product or quotient a negative sign.

Let's look at an example in which we multiply two fractions containing negative signs.

Example 13 **Multiply the following fractions.**

$$\frac{-3}{15} \times \frac{5}{-12}$$

Step 1: Ignore the signs and multiply.

$$\frac{3}{15} \times \frac{5}{12} = \frac{3 \times 5}{15 \times 12}$$

Definition of multiplication of fractions: $\frac{a}{b} \times \frac{c}{d} = \frac{a \times c}{b \times d}$.

$$= \frac{\cancel{3} \times \cancel{5}}{\cancel{3} \times \cancel{5} \times 12}$$

Factor 15.

$$= \frac{1}{12}$$

Reduce.

Step 2: Count the negative signs in the original problem. There are two negative signs.

Step 3: Since there are an even number of negative signs, give the product a positive sign or no sign. Thus,

$$\frac{-3}{15} \times \frac{5}{-12} = \frac{1}{12}$$

■

Now, let's look at an example in which we divide two fractions containing negative signs.

Example 14 **Divide the following fractions.**

$$\frac{-8}{25} \div \frac{-16}{-35}$$

ANSWER

Step 1: Ignore the signs and divide.

$$\frac{8}{25} \div \frac{16}{35} = \frac{8}{25} \times \frac{35}{16}$$

Definition of division of fractions: $\frac{a}{b} \div \frac{c}{d} = \frac{a}{b} \times \frac{d}{c}$.

$$= \frac{8 \times 35}{25 \times 16}$$

Definition of multiplication of fractions: $\frac{a}{b} \times \frac{c}{d} = \frac{a \times c}{b \times d}$.

$$= \frac{\cancel{8} \times (\cancel{5} \times 7)}{(\cancel{5} \times 5) \times (2 \times \cancel{8})}$$

Factor.

$$= \frac{7}{5 \times 2}$$

Reduce.

$$= \frac{7}{10}$$

Perform the multiplications.

Step 2: Count the negative signs in the original problem. There are three negative signs.

Step 3: Since there are not an even number of negative signs, give the quotient a negative sign. Thus,

$$\frac{-8}{25} \div \frac{-16}{-35} = -\frac{7}{10}$$

■

So, you just multiply or divide the fractions and then count the negative signs in the original problem to find the sign of the answer.

That's right, Charlie.

Practice 40: Multiply the following fractions.

$$\frac{-9}{-16} \times \frac{-6}{27}$$

40. ______________

A N S W E R

41. ____________________

Practice 41: Multiply the following fractions.

$$-\frac{5}{18} \times \frac{-24}{25}$$

42. ____________________

Practice 42: Divide the following fractions.

$$\frac{6}{-21} \div \frac{-9}{28}$$

43. ____________________

Practice 43: Divide the following fractions.

$$-\frac{11}{39} \div \frac{-22}{-26}$$

Example 15 **Multiply the following mixed numbers.**

$$-5\frac{1}{3} \times 6\frac{3}{4}$$

Step 1: Ignore the signs and multiply.

$$5\frac{1}{3} \times 6\frac{3}{4} = \frac{16}{3} \times \frac{27}{4}$$ Write the mixed numbers as improper fractions.

$$= \frac{16 \times 27}{3 \times 4}$$ Definition of multiplication of fractions: $\frac{a}{b} \times \frac{c}{d} = \frac{a \times c}{b \times d}$.

$$= \frac{4 \times \cancel{4} \times \cancel{3} \times 9}{\cancel{3} \times \cancel{4}}$$ Factor.

$$= 4 \times 9$$ Reduce.

$$= 36$$ Perform the multiplications.

Step 2: Count the negative signs in the original problem. There is one negative sign.

Step 3: Since there is not an even number of negative signs, give the product a negative sign. Thus,

$$-5\frac{1}{3} \times 6\frac{3}{4} = -36$$ ■

Example 16 **Divide the following mixed numbers.**

$$-3\frac{1}{8} \div \left(-2\frac{3}{16}\right)$$

Step 1: Ignore the signs and divide.

$$3\frac{1}{8} \div 2\frac{3}{16} = \frac{25}{8} \div \frac{35}{16}$$ Write the mixed numbers as improper fractions.

$$= \frac{25}{8} \times \frac{16}{35}$$ Definition of division of fractions: $\frac{a}{b} \div \frac{c}{d} = \frac{a}{b} \times \frac{d}{c}$.

$$= \frac{25 \times 16}{8 \times 35}$$ Definition of multiplication of fractions. $\frac{a}{b} \times \frac{c}{d} = \frac{a \times c}{b \times d}$.

$$= \frac{5 \times \cancel{5} \times \cancel{8} \times 2}{\cancel{8} \times \cancel{5} \times 7}$$ Factor.

$$= \frac{5 \times 2}{7}$$ Reduce.

$$= \frac{10}{7}$$ Perform the multiplication.

$$= 1\frac{3}{7}$$ Write the improper fraction as a mixed number.

A N S W E R

Step 2: Count the negative signs in the original problem. There are two negative signs.

Step 3: Since there are an even number of negative signs, give the quotient a positive sign or no sign. Thus,

$$-3\frac{1}{8} \div \left(-2\frac{3}{16}\right) = 1\frac{3}{7}$$

■

44. ______________

Practice 44: Multiply the following mixed numbers.

$$2\frac{2}{5} \times \left(-1\frac{7}{8}\right)$$

45. ______________

Practice 45: Multiply the following mixed numbers.

$$-1\frac{5}{9} \times \left(-5\frac{1}{4}\right)$$

46. ______________

Practice 46: Divide the following mixed numbers.

$$3\frac{3}{5} \div \left(-2\frac{1}{4}\right)$$

Practice 47: Divide the following mixed numbers.

$$-8\frac{1}{3} \div \left(-5\frac{5}{6}\right)$$

A N S W E R

47. ____________

Sometimes we need to simplify expressions that contain more than one arithmetic operation. Let's look at some examples.

Example 17 **Simplify the following expression.**

$$\frac{-2(-5+2)}{-8-7}$$

Step 1: Simplify the numerator.

$$\frac{-2(-5+2)}{-8-7} = \frac{-2(-3)}{-8-7}$$ **Simplify within the parentheses.**

$$= \frac{6}{-8-7}$$ **Perform the multiplication.**

Step 2: Simplify the denominator.

$$\frac{6}{-8-7} = \frac{6}{-8+(-7)}$$ $a - b = a + (-b)$

$$= \frac{6}{-15}$$ **Perform the addition.**

$$= -\frac{6}{15}$$ $\frac{a}{-b} = -\frac{a}{b}$

Step 3: Reduce the fraction.

$$-\frac{6}{15} = -\frac{2}{5}$$

Thus, $\frac{-2(-5+2)}{-8-7} = -\frac{2}{5}$. ■

A N S W E R

Example 18 **Simplify the following expression.**

$$\frac{5(-1) - 8(-5)}{7(2 - 3)}$$

Step 1: Simplify the numerator.

$$\frac{5(-1) - 8(-5)}{7(2 - 3)} = \frac{-5 - (-40)}{7(2 - 3)}$$ **Perform the multiplications.**

$$= \frac{-5 + 40}{7(2 - 3)}$$ $a - (-b) = a + b$

$$= \frac{35}{7(2 - 3)}$$ **Perform the addition.**

Step 2: Simplify the denominator.

$$\frac{35}{7(2 - 3)} = \frac{35}{7[2 + (-3)]}$$ $a - b = a + (-b)$

$$= \frac{35}{7(-1)}$$ **Perform the addition.**

$$= \frac{35}{-7}$$ **Perform the multiplication.**

$$= -\frac{35}{7}$$ $\frac{a}{-b} = -\frac{a}{b}$

Step 3: Reduce the fraction.

$$-\frac{35}{7} = -5$$

Thus, $\frac{5(-1) - 8(-5)}{7(2 - 3)} = -5$. ■

I noticed in Examples 17 and 18 you first simplified the numerator, then simplified the denominator.

That's correct, Charlie. It would be a good method for you to use when working the practice problems.

48. ____________________

Practice 48: Simplify the following expression.

$$\frac{3(-5 + 8)}{4(-6)}$$

Practice 49: Simplify the following expression.

$$\frac{-4(-9) + 8}{2(-6 + 8)}$$

A N S W E R

49. ____________

Practice 50: Simplify the following expression.

$$\frac{-2(6) + 3(-6)}{(-3)(-12 + 10)}$$

50. ____________

Practice 51: Simplify the following expression.

$$\frac{-2[-5 + 3(-6)]}{(-6)(-2) - (-11)}$$

51. ____________

EXERCISE 2.5

PART 1

Add or subtract the fractions in each of the following problems.

A

1. $\frac{1}{2} + \left(-\frac{1}{2}\right)$

2. $-\frac{3}{4} + \frac{1}{4}$

3. $\frac{-7}{8} + \frac{3}{8}$

4. $\frac{5}{16} - \left(-\frac{3}{16}\right)$

5. $-\frac{3}{8} - \frac{5}{8}$

6. $-\frac{7}{12} - \left(-\frac{5}{12}\right)$

7. $-\frac{25}{36} + \left(-\frac{5}{36}\right)$

8. $-\frac{7}{32} - \left(-\frac{9}{32}\right)$

9. $\frac{-14}{35} + \frac{9}{35}$

10. $\frac{11}{42} + \left(-\frac{5}{42}\right)$

B

11. $\frac{1}{2} + \frac{-1}{4}$

12. $-\frac{5}{8} + \frac{5}{6}$

13. $\frac{2}{3} + \left(-\frac{1}{6}\right)$

14. $\frac{1}{4} - \frac{3}{8}$

15. $\frac{8}{15} - \frac{7}{20}$

16. $-\frac{3}{4} + \left(-\frac{1}{8}\right)$

17. $-\frac{3}{10} - \frac{-7}{10}$

18. $-\frac{5}{12} + \left(-\frac{5}{9}\right)$

19. $\frac{-9}{14} - \frac{-5}{21}$

20. $-\frac{5}{33} - \left(-\frac{15}{22}\right)$

PART 2

Multiply the fractions in each of the following problems.

A

1. $-\frac{1}{2} \times \frac{2}{3}$

2. $\frac{3}{4} \times \frac{-2}{9}$

3. $-\frac{5}{8} \times \frac{-3}{10}$

4. $\frac{-3}{7} \times \frac{5}{9}$

5. $\frac{7}{12} \times \frac{-3}{7}$

6. $\frac{-7}{11} \times \frac{5}{-14}$

7. $-\frac{3}{5} \times \left(-\frac{5}{9}\right)$

8. $\frac{8}{-9} \times \frac{-6}{7}$

9. $\frac{11}{15} \times \frac{10}{-11}$

10. $\frac{-3}{25} \times \frac{-15}{18}$

B

11. $\frac{-2}{5} \times \frac{-7}{-10}$

12. $-\frac{(-3)}{8} \times \frac{12}{21}$

13. $-\frac{14}{25} \times \frac{-5}{7}$

14. $-\frac{3}{-49} \times \frac{14}{15}$

15. $-\frac{18}{35} \times \frac{-14}{-36}$

PART 3

Divide the fractions in each of the following problems.

A

1. $\frac{-1}{2} \div \frac{3}{4}$

2. $\frac{2}{3} \div \left(-\frac{5}{6}\right)$

3. $-\frac{1}{4} \div \left(-\frac{3}{8}\right)$

4. $\frac{-3}{5} \div \frac{4}{5}$

5. $\frac{7}{8} \div \left(-\frac{1}{4}\right)$

6. $\frac{-1}{9} \div \frac{-5}{12}$

7. $\frac{5}{14} \div \frac{-5}{21}$

8. $\frac{7}{15} \div \left(-\frac{21}{25}\right)$

9. $\frac{6}{-35} \div \frac{-9}{14}$

10. $\frac{8}{15} \div \frac{-16}{25}$

B

11. $\frac{-2}{-3} \div \left(-\frac{4}{5}\right)$

12. $-\frac{(-3)}{8} \div \frac{9}{20}$

13. $-\frac{6}{25} \div \frac{8}{-15}$

14. $-\frac{11}{-21} \div \frac{-3}{-28}$

15. $-\frac{-13}{-45} \div \frac{39}{40}$

PART 4

Add or subtract the mixed numbers in each of the following problems.

A

1. $1\frac{1}{2} - 2\frac{3}{4}$

2. $-2\frac{1}{3} + 1\frac{1}{4}$

3. $5\frac{1}{8} - 2\frac{3}{8}$

4. $3\frac{5}{6} - 5\frac{1}{9}$

5. $-9\frac{1}{10} + 3\frac{2}{5}$

6. $8\frac{1}{5} + \left(-3\frac{1}{4}\right)$

7. $12\frac{3}{4} - 15\frac{7}{12}$

8. $-5\frac{3}{16} - 8\frac{5}{16}$

9. $-15\frac{7}{15} - \left(-12\frac{11}{15}\right)$

10. $-21\frac{9}{14} - \left(-7\frac{19}{21}\right)$

PART 5

Multiply or divide the mixed numbers in each of the following problems.

A

1. $-3\frac{1}{2} \times 5\frac{1}{2}$

2. $2\frac{1}{3} \times \left(-1\frac{1}{2}\right)$

3. $-5\frac{1}{5} \times \left(-2\frac{1}{2}\right)$

4. $-1\frac{1}{8} \times \left(-2\frac{2}{9}\right)$

5. $4\frac{2}{5} \div \left(-2\frac{1}{5}\right)$

6. $-5\frac{5}{9} \div 4\frac{1}{6}$

7. $-6\frac{2}{5} \div \left(-1\frac{1}{15}\right)$

8. $-7\frac{1}{7} \div \left(-5\frac{5}{14}\right)$

9. $9\frac{3}{5} \div \left(-1\frac{29}{35}\right)$

10. $-12\frac{2}{9} \times \left(-1\frac{40}{77}\right)$

PART 6

Simplify each of the following expressions.

A

1. $\dfrac{2(7-3)}{8}$

2. $\dfrac{-2(6-7)}{12}$

3. $\dfrac{-5(2)+8(3)}{-7}$

4. $\dfrac{3(-3)-8(-2)}{3(-7)}$

5. $\dfrac{5[-3+(-2)]}{(-5)(-6)}$

6. $\dfrac{-4(-3)+(-5)4}{-6+2}$

7. $\dfrac{-3-12}{-12-18}$

8. $\dfrac{-4[3-(-6)]}{2(-5+2)}$

9. $\dfrac{(-3)(-2)-(-3)(-5)}{3(-6)}$

10. $\dfrac{-2[(-3)7+8(-5)]}{-12-(-18)}$

APPLICATION EXERCISES

Write your answers using signed numbers.

1. The daily low temperatures during one week were recorded as follows:

$$-9, 3, 0, -8, 2, 1, 4$$

Find the average daily low temperature for the week.

2. The average temperature throughout the earth's stratosphere is -70 degrees Fahrenheit. The average temperature on the earth's surface is 57 degrees Fahrenheit. Find the difference between these average temperatures.

3. The temperature dropped 10 degrees below the previous temperature of -5 degrees. Find the new temperature.

4. Jason has $17; Steve is $14 in debt. Find the difference in their net worth.

5. One company made a profit of $86,000, while another company lost $39,000. Find the difference.

6. One reading of a dial was 6.93; a second reading was -2.45. How much drop was there?

USEFUL INFORMATION

1. A nonzero number together with a positive (+) or negative (−) sign is called a **signed number.**
2. If a nonzero number does not have a sign in front of it, then the number is understood to be positive and assumed to have a positive sign.
3. **To add two positive numbers,** simply add the two numbers.
4. **To add a positive and negative number,** subtract the smaller from the larger (without regard to sign) and place the sign of the larger number in the answer.
5. **To add two negative numbers,** ignore the signs and add the numbers giving the sum a negative sign.
6. The sum of a number and its opposite is zero.

   $$a + (-a) = 0$$

7. In order **to subtract two signed numbers,** change the subtraction problem to an addition problem using the following rules.
   Rule 1. $-(-a) = a$
   Rule 2. $a - b = a + (-b)$
   Rule 3. $a - (-b) = a + b$
8. **To multiply two positive numbers,** simply multiply the two numbers.
9. **To multiply a positive and a negative number,** ignore the signs and multiply the two numbers giving the product a negative sign.
10. **To multiply two negative numbers,** ignore the signs and multiply the two numbers giving the product a positive sign.
11. **To divide two positive numbers,** simply divide the two numbers.
12. **To divide a positive and a negative number,** ignore the signs and divide the two numbers giving the quotient a negative sign.
13. **To divide two negative numbers,** ignore the signs and divide the two numbers, giving the quotient a positive sign.
14. To summarize, when multiplying or dividing signed numbers:
    1. Ignore the signs and multiply or divide the numbers.
    2. Count the negative signs in the original problem.
    3. If the number of negative signs is even (0, 2, 4, 6, . . .), give the product or quotient a positive sign or no sign. Otherwise, give the product or quotient a negative sign.

REVIEW PROBLEMS

The answer for each problem in the back of the textbook for Part 1 is only one of many possible answers and it should not be used to check your answer. The purpose of giving you an answer is to help you solve that problem with which you are having trouble. Use the rules of signed number operations to check your answers.

PART 1

1. Find two positive numbers whose sum is a positive number.

2. Find two negative numbers whose sum is a negative number.

3. Find a positive number and a negative number whose sum is a positive number.

4. Find a positive number and a negative number whose sum is a negative number.

5. Find two positive numbers whose difference is a positive number.

6. Find two negative numbers whose difference is a positive number.

7. Find two positive numbers whose difference is a negative number.

8. Find two negative numbers whose difference is a negative number.

9. Find a positive number and a negative number whose difference is a positive number.

10. Find a negative number and a positive number whose difference is a negative number.

11. Find two positive numbers whose product is a positive number.

12. Find two negative numbers whose product is a positive number.

13. Find a positive number and a negative number whose product is a negative number.

14. Find two positive numbers whose quotient is a positive number.

15. Find two negative numbers whose quotient is a positive number.

16. Find a positive number and a negative number whose quotient is a negative number.

PART 2

Section 2.1

Add the numbers in each problem.

A

1. $-4 + 5$

2. $14 + (-6)$

3. $18 + (-14)$

4. $-12 + 7$

5. $-3 + (-5)$

6. $-10 + (-32)$

7. $-205 + 110$

8. $-14 + (-66)$

9. $62 + (-115)$

10. $-32 + (-25)$

B

11. $-5 + 3 + (-2)$

12. $6 + (-9) + (-12)$

13. $-2 + (-1) + (-8)$

14. $7 + (-11) + (-21)$

15. $3.8 + (-1.8)$

16. $-5.6 + 2.3$

17. $-12.4 + (-3.4)$

18. $-34.1 + 16.5$

Section 2.2

Subtract the numbers in each problem.

A

1. $7 - (-1)$

2. $-4 - 6$

3. $-5 - (-12)$

4. $10 - (-4)$

5. $-12 - 3$

6. $-(-4) + 6$

7. $-(-9) - 3$

8. $-(-15) - (-4)$

9. $17 - (-6)$

10. $4 - 3 - 6$

B

11. $-11 - 2 - (-14)$

12. $7.9 - 1.3$

13. $-3.4 - 1.8$

14. $-2.5 - 3.7$

15. $-5.18 - 4.6$

Section 2.3

Multiply the numbers in each problem.

1. $(-6) \times 2$

2. $(-3) \times (-4)$

3. $7 \times (-8)$

4. -8×6

5. $(-7) \times (-7)$

6. $-12 \times (-6)$

7. $(-30) \times (-3)$

8. $25 \times (-6)$

9. -22×4

10. $33 \times (-5)$

B

11. $-(-2) \times 7$

12. $-(-5) \times (-6)$

13. $(-4) \times (-7) \times (-1)$

14. $(-3) \times 2 \times (-5)$

15. $(-20) \times (-15)$

16. $(-2) \times (-1) \times (-4) \times (-3)$

17. $-3 \times (-1) \times (-5) \times 7$

18. $-2 \times 3 \times (-4) \times (-3) \times (-4)$

19. $(-1.4) \times 16$

20. $1.4 \times (-1.8)$

21. $(-3.4) \times 2.5$

22. $(-12.15) \times (-10.4)$

23. $(-2.6) \times (1.7)$

24. $(-3) \times (-4.17)$

25. $(-21.2) \times (4.25)$

Section 2.4

Divide the numbers in each problem.

A

1. $-15 \div 5$

2. $(-20) \div (-4)$

3. $(-18) \div 2$

4. $(-48) \div (-6)$

5. $-51 \div (-17)$

6. $(-45) \div 9$

7. $63 \div (-9)$

8. $(-65) \div (-13)$

9. $(-69) \div (-23)$

10. $121 \div (-11)$

B

11. $-(-18) \div 6$

12. $-(-45) \div (-3)$

13. $15 \div (-15)$

14. $0 \div (-1)$

15. $12.8 \div (-4)$

16. $31.5 \div (-15)$

17. $-(-12) \div (1.6)$

18. $(-8.37) \div (-3.1)$

19. $(-2.24) \div (.14)$

20. $9.378 \div (-5.21)$

Section 2.5

PART 1

Add or subtract the fractions in each of the following problems.

1. $\frac{13}{20} + \left(\frac{-3}{20}\right)$

2. $\frac{-5}{9} + \frac{2}{9}$

3. $\frac{-5}{21} - \frac{2}{21}$

4. $\frac{5}{18} - \frac{7}{18}$

5. $\frac{3}{16} - \left(\frac{-5}{16}\right)$

6. $\frac{-5}{20} + \frac{7}{30}$

7. $\frac{3}{15} - \frac{2}{10}$

8. $\frac{-7}{18} - \left(\frac{-2}{27}\right)$

9. $\frac{-11}{27} + \left(\frac{-5}{9}\right)$

10. $-\left(\frac{-2}{19}\right) - \frac{3}{38}$

PART 2

Multiply or divide the fractions in each of the following problems.

1. $-\frac{3}{5} \times \frac{2}{3}$

2. $-\frac{7}{20} \times \left(-\frac{5}{21}\right)$

3. $-\frac{5}{9} \div \left(\frac{-5}{18}\right)$

4. $-\frac{15}{19} \div \frac{5}{38}$

5. $\frac{-11}{12} \times \left(\frac{-18}{22}\right)$

6. $\frac{-5}{-11} \times \frac{-3}{5}$

7. $-\frac{19}{27} \div \frac{-19}{36}$

8. $\frac{-15}{49} \div \frac{-25}{42}$

9. $\frac{7}{22} \times \left(-\frac{11}{35}\right)$

10. $\frac{-22}{36} \div \frac{-11}{54}$

PART 3

Add or subtract the mixed numbers in each of the following problems.

1. $5\frac{3}{8} - 7\frac{3}{4}$

2. $-8\frac{2}{5} + 10\frac{1}{3}$

3. $-12\frac{3}{10} + 5\frac{3}{5}$

4. $-5\frac{7}{8} + \left(-3\frac{5}{8}\right)$

5. $-6\frac{1}{2} - \left(-5\frac{7}{12}\right)$

6. $-\left(-15\frac{1}{6}\right) - 2\frac{1}{3}$

7. $10\frac{5}{16} + \left(-12\frac{1}{4}\right)$

8. $7\frac{11}{20} - \left(-8\frac{5}{10}\right)$

9. $-21\frac{9}{11} + 25\frac{3}{22}$

10. $-24\frac{5}{14} - \left(-13\frac{11}{21}\right)$

PART 4

Multiply or divide the mixed numbers in each of the following problems.

1. $-3\frac{2}{3} \times 2\frac{6}{11}$

2. $-9\frac{3}{7} \times \left(-4\frac{5}{11}\right)$

3. $-5\frac{13}{15} \div 2\frac{16}{25}$

4. $2\frac{4}{7} \div \left(-3\frac{3}{14}\right)$

5. $-3\frac{3}{5} \times \left(-6\frac{2}{3}\right)$

6. $-3\frac{3}{8} \times \left(-2\frac{2}{3}\right)$

7. $-8\frac{1}{10} \times 1\frac{17}{18}$

8. $1\frac{7}{11} \div \left(-2\frac{1}{22}\right)$

9. $-3\frac{3}{49} \div \left(-2\frac{6}{7}\right)$

10. $-9\frac{3}{8} \times \left(-5\frac{13}{15}\right)$

PART 5

Simplify each of the following expressions.

1. $\frac{-3(2-5)}{18}$

2. $\frac{-5(7+3)}{(-5)5}$

3. $\frac{-3-5}{(-4)(-6)}$

4. $\frac{-3[-2+(-3)]}{-10-5}$

5. $\frac{-6(-5)+(-3)4}{3(-2-4)}$

6. $\frac{(-2)3-(5)(2)}{2(-4)}$

7. $\frac{(-3)(-6)}{5(-6+4)}$

8. $\frac{[6+(-3)]2}{-5+7}$

9. $\frac{-8+(4-3)}{-5-(-7)}$

10. $\frac{3[-5-(-5)]}{8(3-9)}$

ANSWERS TO PRACTICE PROBLEMS

Section 2.1

1. 53
2. 11
3. -11
4. -15
5. 7
6. 9
7. 13
8. -28
9. -52
10. 0
11. -3.21
12. 11.3
13. -57
14. -108
15. -110
16. -129
17. -20
18. -78.56

Section 2.2

1. 7
2. -15
3. -105
4. 99
5. -31
6. 40
7. 0
8. -5
9. -14
10. -4.17

Section 2.3

1. 45
2. -105
3. 36
4. -96
5. -31.2
6. 8
7. -3.12
8. -70
9. 80
10. -252
11. -180
12. -3520
13. negative
14. positive
15. negative
16. positive
17. zero

Section 2.4

1. 12
2. -6
3. -12
4. .7
5. -8
6. 6.5
7. -1.5

Section 2.5

1. $\frac{1}{5}$
2. $-\frac{1}{4}$
3. -1
4. $\frac{1}{9}$
5. $\frac{1}{7}$
6. $-\frac{1}{4}$
7. $-\frac{6}{7}$
8. $-\frac{4}{5}$
9. $-\frac{2}{15}$
10. $\frac{11}{24}$
11. $-\frac{23}{36}$
12. $-\frac{13}{72}$
13. -1
14. $-\frac{2}{5}$
15. $-\frac{1}{3}$
16. 1
17. $-\frac{1}{6}$
18. $-\frac{1}{3}$
19. $-\frac{4}{5}$
20. $\frac{3}{4}$

21. $\frac{13}{22}$ **22.** $-\frac{11}{42}$ **23.** $-\frac{23}{20}$ **24.** $\frac{1}{12}$

25. $3\frac{1}{2}$ **26.** $-1\frac{3}{4}$ **27.** $-5\frac{1}{8}$ **28.** $3\frac{2}{3}$

29. -12 **30.** $-9\frac{1}{5}$ **31.** $-18\frac{1}{4}$ **32.** $-21\frac{1}{6}$

33. $-3\frac{1}{4}$ **34.** $-13\frac{7}{10}$ **35.** $-28\frac{13}{48}$ **36.** $-13\frac{13}{15}$

37. $7\frac{5}{6}$ **38.** $-20\frac{13}{24}$ **39.** $-4\frac{17}{24}$ **40.** $-\frac{1}{8}$

41. $\frac{4}{15}$ **42.** $\frac{8}{9}$ **43.** $-\frac{1}{3}$ **44.** $-4\frac{1}{2}$

45. $8\frac{1}{6}$ **46.** $-1\frac{3}{5}$ **47.** $1\frac{3}{7}$ **48.** $-\frac{3}{8}$

49. 11 **50.** -5 **51.** 2

C H A P T E R 3

Polynomials

The objective of this chapter is:

1. **To be able to add, subtract, multiply, and divide polynomials.**

In Chapter 1 when arithmetic procedures were presented, algebra was frequently compared to arithmetic. In this chapter and later chapters when algebraic procedures are presented, algebra will also be compared to arithmetic. We will make constant use of the fact that algebra is a generalization of arithmetic and obeys the rules of arithmetic. We will begin our study of algebra with polynomials.

You should find your work with expanded notation in Chapter 1 helpful when working with polynomials. This is a good time to review Chapter 1, reading carefully how expanded notation was used.

Definition: Any combination of letters and numbers related by addition, subtraction, multiplication, or division is an **algebraic expression.**

Example 1 **The following are algebraic expressions.**

$$3a + 7b$$

$$5x^2 + 2x - 3$$

$$7ab$$

$$\frac{6x}{y}$$

$$5x^{2/3} + 3x^{1/3} + 2$$

■

Definition: An algebraic expression consisting of one or more distinct parts connected by plus or minus signs is called an **algebraic sum.** Each distinct part, *together with its sign,* is a **term** of the expression.

In the algebraic expression

$$5x^2 - 2x + 3$$

the terms are

$$5x^2$$

$$-2x$$

and $$3$$

Thus, $5x^2 - 2x + 3$ consists of three terms.

The numerical part, 5, of the term, $5x^2$, is called the numerical **coefficient** of the letter part, x^2. Also, in the term $5x^2$, 2 is called the **exponent** of the **base** x.

Example 2 **In the algebraic expression**

$$12x^2 - 3x - 8$$

the terms are

$$12x^2$$

$$-3x$$

and $$-8$$

■

Thus, $12x^2 - 3x - 8$ consists of three terms.

In the term $-3x$, -3 is the coefficient of x. In the term $12x^2$, 2 is the exponent of x and 12 is the coefficient of x^2.

Definition: A **polynomial** is an algebraic sum consisting of one or more terms in which the exponents of the letters are whole numbers and no letter occurs in a denominator.

Example 3 **The following are polynomials.**

$$3x^2 - 5x + 1$$

$$2x^3 - 6$$

$$5a - 8b$$

$$6y + 7$$

■

Definition: If a polynomial consists of just *one* term, it is a **monomial.**

Example 4 **The following are monomials.**

$$5a$$

$$3x^2$$

$$-7$$

$$2x$$

■

Definition: A polynomial consisting of *two* terms is a **binomial.**

Example 5 **The following are binomials.**

$$2x - 3$$

$$4a - 5b$$

$$4x^2 + 8$$

■

Let's look at how a polynomial is similar to a whole number. The polynomial $4x^2 + 2x + 3$ has no apparent relationship to the whole number 423. However, in expanded notation the whole number

$$423 = 4(10^2) + 2(10) + 3$$

and the polynomial

$$4x^2 + 2x + 3$$

appear to be similar. The only difference is that in the first case groups of tens are used and in the second case groups of x's are used, where x represents some number.

Thus, it appears that working with polynomials will be similar to working with whole numbers in expanded notation. In fact, this is one of the main reasons for using expanded notation when working with whole numbers. One of the goals of this textbook is to show that Webster was indeed correct when he defined algebra as a generalization of arithmetic. Hopefully, as you work through this textbook you will see that algebra is not a totally new topic but simply an extension of arithmetic. Anytime you have difficulty working with polynomials, ask the question, How is it done in arithmetic? Then use the same procedure with polynomials.

Notice that the exponents of x decrease from left to right in $4x^2 + 2x + 3$. When a polynomial is written this way, it is in **standard form.** That is, a polynomial in standard form is written the same way an arithmetic expression is written when using expanded notation.

3.1 ADDING POLYNOMIALS

Polynomials are added the same way that arithmetic expressions are added. Let's start with the simplest polynomial, one with only one term, a monomial.

To add monomials, use the arithmetic rule that only like things can be added (apples can be added to apples and oranges to oranges, but apples cannot be added to oranges). Now, let's use this rule with some a's, b's, x's, and y's.

Example 1

$$5a + 3a = 8a$$

$$6y + 7y = 13y$$

$$5x + 8x = 13x$$ ■

Example 2

$$5a + 3b = 5a + 3b$$

$$6x + 7y = 6x + 7y$$ ■

Wait a minute! In Example 2, you did not add.

That's right, Charlie. Things that are not alike cannot be added. We can only indicate the addition.

Basic Rule for Algebraic Addition. Only like things can be added.

Two monomials (terms) that differ only in their numerical parts are said to be **like monomials (terms).**

ANSWER

Example 3 **The following are like monomials (terms).**

$$2xy \text{ and } 3xy$$
$$5x^2y \text{ and } -4x^2y$$
$$3mn^3 \text{ and } 15mn^3$$
$$2ab \text{ and } -3ab$$

■

Example 4 **The following are *not* like monomials (terms).**

$$8x^2y \text{ and } 5xy$$
$$6x^2y^2 \text{ and } -8x^2y$$
$$17m^2n^3 \text{ and } -10mn^3$$

■

Practice 1: Which pairs of terms are like terms?

a. $5x$ and $7x$
b. $3xy$ and $7y$
c. $-7x^3$ and $9x^3$
d. $14mn^2$ and $5m^2n$
e. $17m^2n^2$ and $-5m^2n^2$

1. ______________

When there are more than two monomials (terms) to add, group the like monomials (terms) and add.

Example 5 **Collect like terms and add.**

$$3a + 2b + 5a + 2a + 3b$$

Solution: $3a + 2b + 5a + 2a + 3b = (3a + 5a + 2a) + (2b + 3b)$
$= 10a + 5b$

Do not add terms that are not alike ($10a$ and $5b$ cannot be added). ■

Example 6 **Collect like terms and add (remember the rules of signed numbers).**

$$3xy + (-5x) + 3x + (-7xy) + 2xy + 6x$$

Solution:

$$3xy + (-5x) + 3x + (-7xy) + 2xy + 6x =$$
$$[3xy + (-7xy) + 2xy] + (-5x + 3x + 6x) =$$
$$(-4xy + 2xy) + (-2x + 6x) =$$
$$-2xy + 4x$$

■

Practice 2: Collect like terms and add.

$$7a + 3a$$

2. ______________

ANSWER

3. ____________

Practice 3: Collect like terms and add.

$$2a + (-5a) + 7a$$

4. ____________

Practice 4: Collect like terms and add.

$$3x + 5y + 4x + 1y$$

5. ____________

Practice 5: Collect like terms and add.

$$-2x + (-3y) + (-5x) + 4y$$

EXERCISE 3.1.1

Collect like terms and add.

A

1. $2x + 7x$

2. $5a + 7a$

3. $6y + 8y$

4. $12b + 15b$

5. $14x + 6x$

6. $3a + 2b + 7a$

7. $5x + 7x + 2y$

8. $7c + 3c + 5b + 4b$

9. $2a + 3b + 8a + 5b$

10. $5x + 12y + 15x + 8y$

B

11. $5a + (-7a)$

12. $-12x + 15x$

13. $-8y + (-12y)$

14. $7b + 8b + (-12b)$

15. $-7a + 8b + (-8a) + 2b$

16. $-3x + 5y + 7x + 8y + (-2x)$

17. $12xy + 8x + (-15xy)$

18. $5ab + 3c + 7ab + 2ab + (-12c)$

19. $8x^2 + 3x + 2x^2 + (-6x)$

20. $-15x^2 + 9x + 12x^2 + 5 + 3x$

C

21. $7x^2 + 5x + (-13x^2) + 7 + (-8x) + (-24)$

22. $5a^3b + 7a^2 + (-8a^3b) + 5a + 3a^3b + (-4a^2)$

23. $15xy^2 + 20xy + 4xy^2 + (-16xy^2) + (-18xy) + 23x$

24. $3abc^2 + 2abc + 5ab + (-7ab) + 8abc + (-26abc^2)$

Now, let's consider polynomials with more than one term, such as $2x^2 + 3x + 2$, $5x + 7$, $4x^3 - 5x^2 + 7x - 2$, $4y - 5$, and $5y^2 + 3$. Adding polynomials such as these is similar to adding whole numbers written in expanded notation. Let's look at some addition examples to check this similarity.

Example 7

Algebra	**Arithmetic**
$(2x + 3) + (5x + 4)$	$23 + 54$
Step 1: Write the polynomials in vertical form.	*Step 1:* Write the numbers in vertical form using expanded notation.
$\begin{array}{r} 2x + 3 \\ +\ \underline{5x + 4} \end{array}$	$\begin{array}{rcr} 23 & = & 2(10) + 3 \\ +\ \underline{54} & = & +\ \underline{5(10) + 4} \end{array}$
There are two additions. Starting on the right:	There are two additions. Starting on the right:
$\begin{array}{r} 3 \\ +\ \underline{4} \end{array}$ **Addition 1**	$\begin{array}{r} 3 \\ +\ \underline{4} \end{array}$ **Addition 1**
$\begin{array}{r} 2x \\ +\ \underline{5x} \end{array}$ **Addition 2**	$\begin{array}{r} 2(10) \\ +\ \underline{5(10)} \end{array}$ **Addition 2**
Step 2: Add the groups of ones.	*Step 2:* Add the groups of ones.
$\begin{array}{r} 2x + \mathbf{3} \\ +\ \underline{5x + \mathbf{4}} \\ \mathbf{7} \end{array}$	$\begin{array}{rcr} 23 & = & 2(10) + \mathbf{3} \\ +\ \underline{54} & = & +\ \underline{5(10) + \mathbf{4}} \\ 7 & & \mathbf{7} \end{array}$
Step 3: Add the groups of x's.	*Step 3:* Add the groups of tens.
$\begin{array}{r} \mathbf{2x} + 3 \\ +\ \underline{\mathbf{5x} + 4} \\ \mathbf{7x} + 7 \end{array}$	$\begin{array}{rcr} \mathbf{23} & = & \mathbf{2(10)} + 3 \\ +\ \underline{\mathbf{54}} & = & +\ \underline{\mathbf{5(10)} + 4} \\ \mathbf{77} & & \mathbf{7(10)} + 7 \end{array}$ ■

The algebra problem is worked just like the arithmetic problem.

That's correct, Charlie. In one case, groups of x's are added and in the other case groups of tens are added. Since it makes no difference whether x's or tens are added, these two problems are worked exactly the same. The rules for algebra are basically the same as the rules for arithmetic.

To emphasize the similarity between arithmetic and algebra, work the practice problems in this section using the vertical format.

Practice 6: Add the following polynomials.

$$(3x + 4) + (2x + 1)$$

Practice 7: Add the following polynomials.

$$(x + 5) + (7x + 3)$$

Practice 8: Add the following polynomials.

$$(5x + 6) + (4x + 2)$$

Practice 9: Add the following polynomials.

$$(7x + 1) + (2x + 8)$$

A N S W E R

6. ____________

7. ____________

8. ____________

9. ____________

Let's look at another example.

Example 8

Algebra	**Arithmetic**
$(4x^2 + 2x + 1) + (3x^2 + 5x + 7)$	$421 + 357$
Step 1: Write the polynomials in vertical form.	*Step 1:* Write the numbers in vertical form using expanded notation.
$\begin{array}{r} 4x^2 + 2x + 1 \\ + \underline{3x^2 + 5x + 7} \end{array}$	$\begin{array}{rcr} 421 & = & 4(10^2) + 2(10) + 1 \\ + \underline{357} & = & + \underline{3(10^2) + 5(10) + 7} \end{array}$
There are three additions. Starting on the right:	There are three additions. Starting on the right:
$\begin{array}{r} 1 \\ + \underline{7} \end{array}$ **Addition 1**	$\begin{array}{r} 1 \\ + \underline{7} \end{array}$ **Addition 1**
$\begin{array}{r} 2x \\ + \underline{5x} \end{array}$ **Addition 2**	$\begin{array}{r} 2(10) \\ + \underline{5(10)} \end{array}$ **Addition 2**
$\begin{array}{r} 4x^2 \\ + \underline{3x^2} \end{array}$ **Addition 3**	$\begin{array}{r} 4(10^2) \\ + \underline{3(10^2)} \end{array}$ **Addition 3**

ANSWER

Step 2: Add the groups of ones.

$$\begin{array}{r} 4x^2 + 2x + \mathbf{1} \\ + \; 3x^2 + 5x + \mathbf{7} \\ \hline \mathbf{8} \end{array}$$

Step 3: Add the groups of x's.

$$\begin{array}{r} 4x^2 + \mathbf{2x} + 1 \\ + \; 3x^2 + \mathbf{5x} + 7 \\ \hline \mathbf{7x} + 8 \end{array}$$

Step 4: Add the groups of x^2.

$$\begin{array}{r} \mathbf{4x^2} + 2x + 1 \\ + \; \mathbf{3x^2} + 5x + 7 \\ \hline \mathbf{7x^2} + 7x + 8 \end{array}$$

Step 2: Add the groups of ones.

$$\begin{array}{rcr} 42\mathbf{1} & = & 4(10) + 2(10^2) + \mathbf{1} \\ + \; 35\mathbf{7} & = & + \; 3(10) + 5(10^2) + \mathbf{7} \\ \hline \mathbf{8} & & \mathbf{8} \end{array}$$

Step 3: Add the groups of tens.

$$\begin{array}{rcr} 4\mathbf{2}1 & = & 4(10^2) + \mathbf{2(10)} + 1 \\ + \; 3\mathbf{5}7 & = & + \; 3(10^2) + \mathbf{5(10)} + 7 \\ \hline \mathbf{7}8 & & \mathbf{7(10)} + 8 \end{array}$$

Step 4: Add the groups of 10^2.

$$\begin{array}{rcr} \mathbf{4}21 & = & \mathbf{4(10^2)} + 2(10) + 1 \\ + \; \mathbf{3}57 & = & + \; \mathbf{3(10^2)} + 5(10) + 7 \\ \hline \mathbf{7}78 & & \mathbf{7(10^2)} + 7(10) + 8 \end{array}$$ ■

Once again, the two problems are basically worked the same way; groups of x's are added in the algebra problem, whereas groups of tens are added in the arithmetic problem.

Now I see why we used expanded notation in arithmetic. Algebra may not be as hard as I thought.

That's right, Charlie. Algebra is not difficult. It is just an extension of arithmetic. Let's do some practice problems.

10. ____________

Practice 10: Add the following polynomials.

$$(5x^2 + 2x + 7) + (x^2 + 4x + 1)$$

11. ____________

Practice 11: Add the following polynomials.

$$(3x^2 + 5x + 2) + (2x^2 + 3x + 5)$$

Practice 12: Add the following polynomials.

$$(5y^2 + 2y + 6) + (3y^2 + 4y + 2)$$

A N S W E R

12. ____________________

Practice 13: Add the following polynomials.

$$(2x^2 + 7x + 3) + (4x^2 + 2x + 6)$$

13. ____________________

In arithmetic addition, we regroup and carry when a sum is greater than 9. This is one area where algebra may actually be easier than arithmetic since *numbers are not regrouped or carried in algebraic addition.* In algebra, the groups are groups of x's; thus, it is not known how many it takes to make a group. In arithmetic, $6 + 8 = 14 = 1(10) + 4$ since arithmetic uses groups of tens. In algebra, $6 + 8 = 14$ but, we cannot regroup the 14 into $1x + 4$ since x may not be ten.

How many chairs do I put in a stack? *x chairs.*

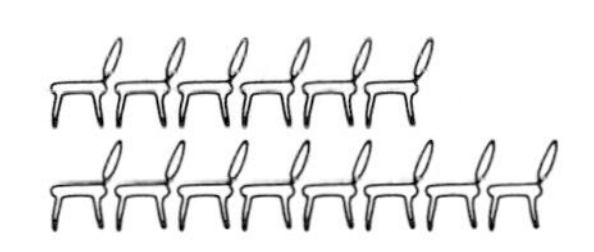

But how many is x?

We don't know how many, Charlie.

Let's look at another example in which algebraic addition is compared to arithmetic addition. You might be convinced as you read through this example that algebraic addition is sometimes easier than arithmetic addition.

Example 9

Algebra

$(5x^2 + 7x + 8) + (3x^2 + 8x + 6)$

Step 1: Write the polynomials in vertical form.

$$\begin{array}{r} 5x^2 + 7x + 8 \\ +\ \underline{3x^2 + 8x + 6} \end{array}$$

There are three additions to be performed.

Step 2: Add the groups of ones.

$$\begin{array}{r} 5x^2 + 7x + \mathbf{8} \\ +\ \underline{3x^2 + 8x + \mathbf{6}} \\ \mathbf{14} \end{array}$$

Step 3: Add the groups of x's.

$$\begin{array}{r} 5x^2 + \mathbf{7x} + 8 \\ +\ \underline{3x^2 + \mathbf{8x} + 6} \\ \mathbf{15x} + 14 \end{array}$$

Step 4: Add the groups of x^2.

$$\begin{array}{r} \mathbf{5x^2} + 7x + 8 \\ +\ \underline{\mathbf{3x^2} + 8x + 6} \\ \mathbf{8x^2} + 15x + 14 \end{array}$$

Arithmetic

$578 + 386$

Step 1: Write the numbers in vertical form using expanded notation.

$$\begin{array}{rcr} 578 & = & 5(10^2) + 7(10) + 8 \\ +\ \underline{386} & = & +\ \underline{3(10^2) + 8(10) + 6} \end{array}$$

There are three additions to be performed.

Step 2: Add the groups of ones.

$$\begin{array}{rcr} 578 & = & 5(10^2) + 7(10) + \mathbf{8} \\ +\ \underline{386} & = & +\ \underline{3(10^2) + 8(10) + \mathbf{6}} \\ \mathbf{14} & & \mathbf{14} \end{array}$$

Step 3: Regroup and carry.

$$\begin{array}{rcr} \mathbf{1} & & \mathbf{1(10)} \\ 578 & = & 5(10^2) + 7(10) + 8 \\ +\ \underline{386} & = & +\ \underline{3(10^2) + 8(10) + 6} \\ \mathbf{4} & & \mathbf{4} \end{array}$$

Step 4: Add the groups of tens, including the 1(10) that was carried.

$$\begin{array}{rcr} \mathbf{1} & & \mathbf{1(10)} \\ 578 & = & 5(10^2) + \mathbf{7(10)} + 8 \\ +\ \underline{386} & = & +\ \underline{3(10^2) + \mathbf{8(10)} + 6} \\ \mathbf{164} & & \mathbf{16(10)} + 4 \end{array}$$

Step 5: Regroup and carry.

$$\begin{array}{rcr} \mathbf{1}1 & & \mathbf{1(10^2)} \quad 1(10) \\ 578 & = & 5(10^2) + 7(10) + 8 \\ +\ \underline{386} & = & +\ \underline{3(10^2) + 8(10) + 6} \\ \mathbf{6}4 & & \mathbf{6(10)} + 4 \end{array}$$

Step 6: Add the groups of 10^2, including the $1(10^2)$ that was carried.

$$\begin{array}{rcr} \mathbf{1}1 & & \mathbf{1(10^2)} \quad 1(10) \\ \mathbf{5}78 & = & \mathbf{5(10^2)} + 7(10) + 8 \\ +\ \underline{\mathbf{3}86} & = & +\ \underline{\mathbf{3(10^2)} + 8(10) + 6} \\ \mathbf{9}64 & & \mathbf{9(10^2)} + 6(10) + 4 \end{array}$$ ■

A N S W E R

The algebra was less work.

That's right, Charlie. We did not have to regroup or carry in the algebraic addition since it was not known how many made an x. However, the procedure was still basically the same as in the arithmetic addition.

I like this algebra. It's nice not having to carry numbers.

That's good Charlie. Let's work some practice problems.

Practice 14: Add the following polynomials.

$$(4x^2 + 7x + 3) + (3x^2 + 6x + 5)$$

14. ____________

Practice 15: Add the following polynomials.

$$(8a^2 + 5a + 7) + (4a^2 + 7a + 4)$$

15. ____________

Practice 16: Add the following polynomials.

$$(3x^2 + 8x + 5) + (9x^2 + 7x + 8)$$

16. ____________

Practice 17: Add the following polynomials.

$$(7y^2 + 9y + 4) + (8y^2 + 6y + 7)$$

17. ____________

A N S W E R

It is common practice in algebra to leave out zero terms. For example, $x^2 + 0x + 9$ is usually written $x^2 + 9$. Since only like things are added, simply line up the polynomials vertically before adding so that like things are in the same column. Let's look at an example.

Example 10

Algebra	**Arithmetic**
$(5x^2 + 2) + (2x^2 + 7x + 3)$	$502 + 273$
Step 1: Write the polynomials in vertical form. (Be sure that like things are in the same column.)	*Step 1:* Write the numbers in vertical form using expanded notation.
$\begin{array}{r} 5x^2 + 2 \\ + \underline{2x^2 + 7x + 3} \end{array}$	$\begin{array}{rl} 502 = & 5(10^2) + 0(10) + 2 \\ + \underline{273} = & + \underline{2(10^2) + 7(10) + 3} \end{array}$
There are three additions.	There are three additions.
Step 2: Add the groups of ones.	*Step 2:* Add the groups of ones.
$\begin{array}{r} 5x^2 + \mathbf{2} \\ + \underline{2x^2 + 7x + \mathbf{3}} \\ \mathbf{5} \end{array}$	$\begin{array}{rr} 502 = & 5(10^2) + 0(10) + \mathbf{2} \\ + \underline{273} = & + \underline{2(10^2) + 7(10) + \mathbf{3}} \\ \mathbf{5} & \mathbf{5} \end{array}$
Step 3: Add the groups of x's.	*Step 3:* Add the groups of tens.
$\begin{array}{r} 5x^2 + 2 \\ + \underline{2x^2 + \mathbf{7x} + 3} \\ \mathbf{7x} + 5 \end{array}$	$\begin{array}{rr} 502 = & 5(10^2) + \mathbf{0(10)} + 2 \\ + \underline{273} = & + \underline{2(10^2) + \mathbf{7(10)} + 3} \\ 75 & \mathbf{7(10)} + 5 \end{array}$
Step 4: Add the groups of x^2.	*Step 4:* Add the groups of 10^2.
$\begin{array}{r} \mathbf{5x^2} + 2 \\ + \underline{\mathbf{2x^2} + 7x + 3} \\ \mathbf{7x^2} + 7x + 5 \end{array}$	$\begin{array}{rr} 502 = & \mathbf{5(10^2)} + 0(10) + 2 \\ + \underline{273} = & + \underline{\mathbf{2(10^2)} + 7(10) + 3} \\ 775 & \mathbf{7(10^2)} + 7(10) + 5 \end{array}$ ■

Let's try some practice problems.

18. ______________

Practice 18: Add the following polynomials.

$$(3x^2 + 8) + (7x^2 + 3x + 1)$$

Practice 19: Add the following polynomials.

$$(2x^2 + 3x + 7) + (5x^2 + 7x)$$

A N S W E R

19. ______________

Practice 20: Add the following polynomials.

$$(7x^2 + 3x + 5) + (5x + 4)$$

20. ______________

Next we look at an example in which we have to use the rules of signed number operations. Now is a good time to read the Point of Interest at the end of this section.

Example 11 **Add the following polynomials.**

$$(3x^2 - 5x + 7) + (7x^2 + 2x - 12)$$

Since this example contains negative signs, it is more difficult to translate this example into an arithmetic problem. However, the procedure is still to add like things using the rules for signed numbers.

Step 1: Write the polynomials in vertical form.

$$\begin{array}{r} 3x^2 - 5x + 7 \\ + \; 7x^2 + 2x - 12 \\ \hline \end{array}$$

There are three additions. Starting on the right of the polynomials:

$$\begin{array}{r} 7 \\ -12 \\ \hline \end{array} \quad \textbf{Addition 1} \qquad \begin{array}{r} -5x \\ 2x \\ \hline \end{array} \quad \textbf{Addition 2} \qquad \begin{array}{r} 3x^2 \\ 7x^2 \\ \hline \end{array} \quad \textbf{Addition 3}$$

Since there is no regrouping or carrying in algebra, these three additions can be worked separately.

Step 2: Add the groups of ones. (Remember your rules for adding signed numbers.)

$$7 + (-12) = -5$$

Thus,

$$\begin{array}{r} 3x^2 - 5x + \mathbf{7} \\ + \; 7x^2 + 2x - \mathbf{12} \\ \hline - \; \mathbf{5} \end{array}$$

Step 3: Add the groups of x's.

$$-5x + 2x = -3x$$

Thus,

$$\begin{array}{r} 3x^2 - \mathbf{5x} + 7 \\ + \; 7x^2 + \mathbf{2x} - 12 \\ \hline - \; \mathbf{3x} - 5 \end{array}$$

A N S W E R

Step 4: Add the groups of x^2.

$$3x^2 + 7x^2 = 10x^2$$

Thus,

$$\begin{array}{r} 3x^2 - 5x + 7 \\ + \ 7x^2 + 2x - 12 \\ \hline 10x^2 - 3x - 5 \end{array}$$

■

When adding polynomials containing negative signs, I will have to remember to include the sign when adding.

That's correct, Charlie. Adding polynomials often involves the addition of signed numbers. However, we do not have to regroup or carry when adding polynomials. Let's try some practice problems.

21. ______________

Practice 21: Add the following polynomials.

$$(8x^2 - 5x + 3) + (5x^2 + 7x - 8)$$

22. ______________

Practice 22: Add the following polynomials.

$$(4y^2 + 7y - 3) + (8y^2 + 5y - 11)$$

23. ______________

Practice 23: Add the following polynomials.

$$(3x^2 - 8x - 2) + (4x^2 - 3x - 1)$$

Practice 24: Add the following polynomials.

$$(9b^2 + 3b - 7) + (6b^2 - 3b + 8)$$

A N S W E R

24. ____________

Example 12 **Add the following polynomials.**

$$(5x^2 - 3x + 4) + (7 - 3x - 8x^2)$$

Step 1: Write the polynomials in vertical form. The first polynomial is in standard form—that is, the exponents of x decrease from left to right. However, it is necessary to rewrite the second polynomial in standard form.

$$\begin{array}{r} 5x^2 - 3x + 4 \\ + \; -8x^2 - 3x + 7 \\ \hline \end{array}$$

Now, like terms are in the same column. There are three additions to be performed.

$\begin{array}{r} 4 \\ 7 \\ \hline \end{array}$	**Addition 1**	$\begin{array}{r} -3x \\ -3x \\ \hline \end{array}$	**Addition 2**	$\begin{array}{r} 5x^2 \\ -8x^2 \\ \hline \end{array}$	**Addition 3**

Step 2: Add the groups of ones.

$4 + 7 = 11$

Thus,

$$\begin{array}{r} 5x^2 - 3x + \mathbf{4} \\ + \; -8x^2 - 3x + \mathbf{7} \\ \hline \mathbf{11} \end{array}$$

Step 3: Add the groups of x's.

$-3x + (-3x) = -6x$

Thus,

$$\begin{array}{r} 5x^2 \mathbf{- 3x} + 4 \\ + \; -8x^2 \mathbf{- 3x} + 7 \\ \hline \mathbf{- 6x} + 11 \end{array}$$

Step 4: Add the groups of x^2.

$5x^2 + (-8x^2) = -3x^2$

Thus,

$$\begin{array}{r} \mathbf{5x^2} - 3x + 4 \\ + \; \mathbf{-8x^2} - 3x + 7 \\ \hline \mathbf{-3x^2} - 6x + 11 \end{array}$$

■

Now, let's try some practice problems in which you first write the polynomials in standard form.

A N S W E R

25. ____________________

Practice 25: Add the following polynomials.

$$(-3x + 7x^2 + 4) + (5x^2 + 8x - 6)$$

26. ____________________

Practice 26: Add the following polynomials.

$$(2x^2 - 7x + 8) + (8 - 3x + 5x^2)$$

27. ____________________

Practice 27: Add the following polynomials.

$$(4x - 3x^2 + 8) + (4 - 7x - 6x^2)$$

Point of Interest

Let's translate $3x^2 - 5x + 1$ into an arithmetic expression by replacing the x's with tens. Now

$$3x^2 - 5x + 1$$

translates into

$$\begin{aligned} 3(10^2) - 5(10) + 1 &= \\ 2(10^2) + 10(10) - 5(10) + 1 &= \\ 2(10^2) + 5(10) + 1 &= 251 \end{aligned}$$

In arithmetic, the terms are combined so that all the terms are either positive or negative. Let's look at another example.

$$-3x^2 + 5x - 1$$

translates into

$$\begin{aligned} -3(10^2) + 5(10) - 1 &= \\ -2(10^2) - 10(10) + 5(10) - 1 &= \\ -2(10^2) - 5(10) - 1 &= \\ -[2(10^2) + 5(10) + 1] &= -251 \end{aligned}$$

In order to accomplish this, it was necessary to borrow and regroup as is often done in arithmetic. However, in algebra it is not possible to borrow or regroup since it is not known how many it takes to make an x.

EXERCISE 3.1.2

Simplify each of the following.

A

1. $2x + 5x$

2. $4x + (-2x) + 1$

3. $5x + 3 + 2x$

4. $3y + 2x + 4y$

5. $(4y + 2) + (2y + 3)$

6. $(2x + 4) + (x + 7)$

7. $(3x + 2) + (4x + 1)$

8. $(8x + 3) + (7x - 2)$

9. $(5x - 8) + (3x + 2)$

10. $(12x + 5) + (3x - 4)$

B

11. $(7x^2 + 2x + 5) + (x^2 + 5x + 2)$

12. $(2x^2 + 3x + 6) + (5x^2 + 4x + 1)$

13. $(2x^2 + 3x + 5) + (6x^2 + 2x - 7)$

14. $(7x^2 + 8x - 3) + (5x^2 + 9x + 7)$

15. $(5x^2 - x - 1) + (4x^2 - x + 1)$

16. $(2a^2 + a + 1) + (a^2 - 7a - 8)$

17. $(7x^2 + 5x - 8) + (6x^2 + 8x + 3)$

18. $(10x^2 + 15x + 12) + (8x^2 - 10x - 20)$

19. $(7x^2 + 9x - 24) + (23x^2 - 11x - 6)$

20. $(15x^2 - 13x - 8) + (7x^2 - 5x + 12)$

C

21. $(y^2 + y + 1) + (2y^2 + 3 + 2y)$

22. $(x^2 + 2x) + (4x^2 + 6x)$

23. $(-4y^2 + 1) + (2y^2 - 3)$

24. $(-5x^2 - 13) + (8x^2 + 3x - 2)$

25. $(8x^2 + 16) + (-5x - 7)$

26. $(5x - 7 + 2x^2) + (3x^2 + 15 - 9x)$

27. $(-8x + 5x^2 - 9) + (5x + 12)$

3.2 SUBTRACTING POLYNOMIALS

There are two considerations when subtracting polynomials. First, subtracting polynomials (algebraic expressions) is basically the same as subtracting arithmetic expressions. Second, the rules for signed numbers are used when subtracting polynomials. Once again, let's start with the simplest polynomial, a monomial.

To subtract monomials, use the rules of signed numbers and the arithmetic rule that only like things can be added or subtracted.

Example 1

$$3a - 7a = 3a + (-7a) = -4a$$

$$-6x - (-8x) = -6x + 8x = 2x$$

■

Example 2

$$3a - 7b = 3a - 7b$$

$$6x - 5y = 6x - 5y$$

■

Wait a minute! In Example 2, you did not subtract.

That's right, Charlie. Things that are not alike cannot be subtracted, we can only indicate the subtraction. When there are more than two monomials to add or subtract, group the like monomials and combine them.

Example 3 **Combine like terms.**

$$2a - 3b - (-5a) + 7b$$

Solution:
$$\begin{aligned} 2a - 3b - (-5a) + 7b &= [2a - (-5a)] + (-3b + 7b) \\ &= (2a + 5a) + (-3b + 7b) \\ &= 7a + 4b \end{aligned}$$

■

Example 4 **Combine like terms.**

$$3x - 5x - (-7x) + 5y - 8x - 7y$$

Solution:

$$\begin{aligned} &3x - 5x - (-7x) + 5y - 8x - 7y \\ &= [3x - 5x - (-7x) - 8x] + (5y - 7y) \\ &= [3x + (-5x) + 7x + (-8x)] + (5y - 7y) \\ &= [-2x + 7x + (-8x)] + (-2y) \\ &= [5x + (-8x)] + (-2y) \\ &= -3x + (-2y) \\ &= -3x - 2y \end{aligned}$$

■

Practice 1: Combine like terms.

$$6a - 2a$$

1. ______________

Practice 2: Combine like terms.

$$4a - 7a$$

2. ______________

Practice 3: Combine like terms.

$$-5x - 8x$$

3. ______________

A N S W E R

4. ____________________

Practice 4: Combine like terms.

$$3x - 4y + 2x - 8y$$

5. ____________________

Practice 5: Combine like terms.

$$-2x - (-3y) + 5x + 2y$$

EXERCISE 3.2.1

Combine like terms.

A

1. $7x - 2x$

2. $3a - 5a$

3. $5a - 8a + 13a$

4. $7b - (-2b)$

5. $3x - (-7x)$

6. $-2y - 5y + 12y$

7. $2a - 5b - 8a - 6b$

8. $-2x - 5y - 7x + 8y$

9. $3y - 7y + 2y - 8x - 5x + 9y$

10. $-6ab + 7b - 8ab - 12b - 3ab$

B

11. $-7x - (-8x) - 5x$

12. $-2xy + 3y - 7x - (-3xy)$

13. $5x^2 - 2xy + 7xy - 18x^2$

14. $3y^2 - (-7y) + 23y^2 - 14y - 3y$

15. $-15abc + (-8ab) - 12bc + 15ab - 18abc + 15ab$

Now, let's consider subtracting polynomials containing two or more terms. The same rules used in arithmetic will also apply to polynomials. Let's look at some subtraction examples to check this similarity.

Example 5

Algebra	Arithmetic
$(5x + 7) - (2x + 3)$	$57 - 23$
Step 1: Write the polynomials in vertical form.	*Step 1:* Write the numbers in vertical form using expanded notation.
$\begin{array}{r} 5x + 7 \\ - \underline{2x + 3} \end{array}$	$\begin{array}{rcr} 57 & = & 5(10) + 7 \\ - \underline{23} & = & - \underline{2(10) + 3} \end{array}$
There are two subtractions.	There are two subtractions.
$\begin{array}{r} 7 \\ - \underline{3} \end{array}$ **Subtraction 1**	$\begin{array}{r} 7 \\ - \underline{3} \end{array}$ **Subtraction 1**
$\begin{array}{r} 5x \\ - \underline{2x} \end{array}$ **Subtraction 2**	$\begin{array}{r} 5(10) \\ - \underline{2(10)} \end{array}$ **Subtraction 2**
Step 2: Subtract the groups of ones.	*Step 2:* Subtract the groups of ones.
$\begin{array}{r} 5x + \mathbf{7} \\ - \underline{2x + \mathbf{3}} \\ \mathbf{4} \end{array}$	$\begin{array}{rcr} 57 & = & 5(10) + \mathbf{7} \\ - \underline{23} & = & - \underline{2(10) + \mathbf{3}} \\ \mathbf{4} & & \mathbf{4} \end{array}$
Step 3: Subtract the groups of x's.	*Step 3:* Subtract the groups of tens.
$\begin{array}{r} \mathbf{5x} + 7 \\ - \underline{\mathbf{2x} + 3} \\ \mathbf{3x} + 4 \end{array}$	$\begin{array}{rcr} 57 & = & \mathbf{5(10)} + 7 \\ - \underline{23} & = & - \underline{\mathbf{2(10)} + 3} \\ 34 & & \mathbf{3(10)} + 4 \end{array}$ ■

The two problems are basically worked the same way, except that in the first case groups of x's are subtracted and in the second case groups of tens are subtracted. It makes no difference whether x's or tens are subtracted; the procedure is the same.

ANSWER

6. ______

7. ______

8. ______

To emphasize the similarity between arithmetic and algebra, work the practice problems in this section using the vertical format.

Practice 6: Subtract the following polynomials.

$$(8x + 7) - (3x + 2)$$

Practice 7: Subtract the following polynomials.

$$(9x + 5) - (4x + 3)$$

Practice 8: Subtract the following polynomials.

$$(7x + 8) - (5x + 4)$$

Let's look at another example.

Example 6

Algebra	Arithmetic
$(5x^2 + 7x + 8) - (3x^2 + x + 7)$	$578 - 317$
Step 1: Write the polynomials in vertical form.	*Step 1:* Write the numbers in vertical form using expanded notation.
$\begin{array}{r} 5x^2 + 7x + 8 \\ - \underline{3x^2 + x + 7} \end{array}$	$\begin{array}{rcr} 578 & = & 5(10^2) + 7(10) + 8 \\ -\underline{317} & = & -\ \underline{3(10^2) + 1(10) + 7} \end{array}$
There are three subtractions.	There are three subtractions.
$\begin{array}{r} 8 \\ -\underline{7} \end{array}$ **Subtraction 1**	$\begin{array}{r} 8 \\ -\underline{7} \end{array}$ **Subtraction 1**
$\begin{array}{r} 7x \\ -\underline{x} \end{array}$ **Subtraction 2**	$\begin{array}{r} 7(10) \\ -\underline{1(10)} \end{array}$ **Subtraction 2**
$\begin{array}{r} 5x^2 \\ -\underline{3x^2} \end{array}$ **Subtraction 3**	$\begin{array}{r} 5(10^2) \\ -\underline{3(10^2)} \end{array}$ **Subtraction 3**

A N S W E R

Step 2: Subtract the groups of ones.

$$\begin{array}{r} 5x^2 + 7x + \mathbf{8} \\ -\ 3x^2 + x + \mathbf{7} \\ \hline \mathbf{1} \end{array}$$

Step 3: Subtract the groups of x's.

$$\begin{array}{r} 5x^2 + \mathbf{7x} + 8 \\ -\ 3x^2 + \mathbf{x} + 7 \\ \hline \mathbf{6x} + 1 \end{array}$$

Step 4: Subtract the groups of x^2.

$$\begin{array}{r} \mathbf{5x^2} + 7x + 8 \\ -\ \mathbf{3x^2} + x + 7 \\ \hline \mathbf{2x^2} + 6x + 1 \end{array}$$

Step 2: Subtract the groups of ones.

$$\begin{array}{rcr} 578 & = & 5(10^2) + 7(10) + \mathbf{8} \\ -\ 317 & = & -\ 3(10^2) + 1(10) + \mathbf{7} \\ \hline & & \mathbf{1} \end{array}$$

Step 3: Subtract the groups of tens.

$$\begin{array}{rcr} 578 & = & 5(10^2) + \mathbf{7(10)} + 8 \\ -\ 317 & = & -\ 3(10^2) + \mathbf{1(10)} + 7 \\ \hline 61 & & \mathbf{6(10)} + 1 \end{array}$$

Step 4: Subtract the groups of 10^2.

$$\begin{array}{rcr} 578 & = & \mathbf{5(10^2)} + 7(10) + 8 \\ -\ 317 & = & -\ \mathbf{3(10^2)} + 1(10) + 7 \\ \hline 261 & & \mathbf{2(10^2)} + 6(10) + 1 \end{array}$$ ■

I can see the similarity between algebra and arithmetic. I believe I will do okay in algebra.

Good, Charlie. Let's try some practice problems.

Practice 9: Subtract the following polynomials.

$$(8x^2 + 7x + 4) - (5x^2 + 3x + 2)$$

9. ____________

Practice 10: Subtract the following polynomials.

$$(6x^2 + 4x + 5) - (2x^2 + 3x + 4)$$

10. ____________

Practice 11: Subtract the following polynomials.

$$(7x^2 + 9x + 2) - (4x^2 + 5x + 1)$$

11. ____________

Just as numbers cannot be regrouped or carried in algebra, they also *cannot be regrouped or borrowed in algebra.* This often makes it necessary to use the rules of signed number operations when subtracting polynomials. Let's look at an example.

Example 7

Algebra

$(8x^2 + 2x + 4) - (5x^2 + 4x + 7)$

Step 1: Write the polynomials in vertical form.

$$\begin{array}{r} 8x^2 + 2x + 4 \\ -\ \underline{5x^2 + 4x + 7} \end{array}$$

Step 2: Subtract the groups of ones using the rules for signed numbers.

$$\begin{aligned} 4 - 7 &= 4 + (-7) \\ &= -3 \end{aligned}$$

Thus,

$$\begin{array}{r} 8x^2 + 2x + \mathbf{4} \\ -\ \underline{5x^2 + 4x + \mathbf{7}} \\ -\ \mathbf{3} \end{array}$$

Step 3: Subtract the groups of x's using the rules for signed numbers.

$$\begin{aligned} 2x - 4x &= 2x + (-4x) \\ &= -2x \end{aligned}$$

Thus,

$$\begin{array}{r} 8x^2 + \mathbf{2x} + 4 \\ -\ \underline{5x^2 + \mathbf{4x} + 7} \\ -\ \mathbf{2x} - 3 \end{array}$$

Arithmetic

$824 - 547$

Step 1: Write the numbers in vertical form using expanded notation.

$$\begin{array}{rcl} 824 & = & 8(10^2) + 2(10) + 4 \\ -\ \underline{547} & = & -\ \underline{5(10^2) + 4(10) + 7} \end{array}$$

Step 2: Borrow and regroup.

$$\begin{array}{rcl} 1\ 14 & & 1(10) \quad 14 \\ 8\,\not{2}\,\not{4} & = & 8(10^2) + \not{2}\not{(10)} + \not{4} \\ -\ \underline{5\ 4\ 7} & = & -\ \underline{5(10^2) + 4(10) + 7} \end{array}$$

Step 3: Subtract the groups of ones.

$$\begin{array}{rcl} 1\ \mathbf{14} & & 1(10) \quad \mathbf{14} \\ 8\,\not{2}\,\not{4} & = & 8(10^2) + \not{2}\not{(10)} + \not{4} \\ -\ \underline{5\ 4\ \mathbf{7}} & = & -\ \underline{5(10^2) + 4(10) + \mathbf{7}} \\ \mathbf{7} & & +\ \mathbf{7} \end{array}$$

Step 4: Borrow and regroup.

$$\begin{array}{rcl} 11 & & 11(10) \\ 7\,\not{1}\,14 & & 7(10^2) \quad \not{1}\not{(10)} \quad 14 \\ \not{8}\,\not{2}\,\not{4} & = & \not{8}\not{(10^2)} + \not{2}\not{(10)} + \not{4} \\ -\ \underline{5\ 4\ 7} & = & -\ \underline{5(10^2) + 4(10) + 7} \\ 7 & & +\ 7 \end{array}$$

Step 5: Subtract the groups of tens.

$$\begin{array}{rcl} \mathbf{11} & & \mathbf{11(10)} \\ 7\,\not{1}\,14 & & 7(10^2) \quad \not{1}\not{(10)} \quad 14 \\ 8\,\not{2}\,4 & = & \not{8}\not{(10^2)} + \not{2}\not{(10)} + 4 \\ -\ \underline{5\ \mathbf{4}\ 7} & = & -\ \underline{5(10^2) + \mathbf{4(10)} + 7} \\ 7\ 7 & & +\ \mathbf{7(10)} + 7 \end{array}$$

ANSWER

Step 4: Subtract the groups of x^2.

$$8x^2 - 5x^2 = 3x^2$$

Thus,

$$\begin{array}{r} 8x^2 + 2x + 4 \\ -\ 5x^2 + 4x + 7 \\ \hline 3x^2 - 2x - 3 \end{array}$$

Step 6: Subtract the groups of 10^2.

$$\begin{array}{r} 11 \\ 7\,\cancel{1}\,14 \\ \cancel{8\,2\,4} \\ -\ 5\,4\,7 \\ \hline 2\,7\,7 \end{array} \quad = \quad \begin{array}{r} 7(10^2) \quad \overset{11(10)}{\cancel{1(10)}} \quad 14 \\ \cancel{8(10^2)} + \cancel{2(10)} + \cancel{4} \\ -\ 5(10^2) + 4(10) + 7 \\ \hline 2(10^2) + 7(10) + 7 \end{array}$$ ■

So, we don't have to borrow and regroup in algebra.

That's right, Charlie. There is no borrowing or regrouping in algebra since it is not known how many it takes to make an x. That's why signed numbers were discussed before subtraction of polynomials. However, subtraction of polynomials is still basically the same as arithmetic subtraction.

I think I like algebra better than arithmetic. I believe it is easier to work with signed numbers than it is to borrow and regroup.

That's good, Charlie. Let's try some practice problems.

Practice 12: Subtract the following polynomials.

$$(8x^2 + 3x + 2) - (5x^2 + 7x + 8)$$

12. ____________

Practice 13: Subtract the following polynomials.

$$(5y^2 + 2y + 6) - (2y^2 + 6y + 3)$$

13. ____________

ANSWER

14. ____________

Practice 14: Subtract the following polynomials.

$$(2x^2 + 7x + 3) - (8x^2 + 3x + 5)$$

15. ____________

Practice 15: Subtract the following polynomials.

$$(4a^2 + 2a + 6) - (7a^2 + 5a + 9)$$

Now, let's look at an example in which the polynomials contain negative signs.

Example 8 **Subtract the following polynomials.**

$$(3x^2 - 7x - 2) - (5x^2 - 3x + 5)$$

Solution: Write the polynomials in vertical form.

$$\begin{array}{r} 3x^2 - 7x - 2 \\ - \underline{5x^2 - 3x + 5} \end{array}$$

There are three subtractions to be performed. Remember the rules for subtracting signed numbers (Section 2.2). *Since there is no borrowing or regrouping in algebra, the three subtractions can be worked separately.*

$$\begin{array}{r} -2 \\ - \underline{5} \\ -7 \end{array} \quad \text{or} \quad \begin{aligned} -2 - 5 &= -2 + (-5) \\ &= -7 \end{aligned}$$ **Subtraction 1**

$$\begin{array}{r} -7x \\ - \underline{-3x} \\ -4x \end{array} \quad \text{or} \quad \begin{aligned} -7x - (-3x) &= -7x + 3x \\ &= -4x \end{aligned}$$ **Subtraction 2**

$$\begin{array}{r} 3x^2 \\ - \underline{5x^2} \\ -2x^2 \end{array} \quad \text{or} \quad \begin{aligned} 3x^2 - 5x^2 &= 3x^2 + (-5x^2) \\ &= -2x^2 \end{aligned}$$ **Subtraction 3**

Thus,

$$\begin{array}{r} 3x^2 - 7x - 2 \\ - \underline{5x^2 - 3x + 5} \\ -2x^2 - 4x - 7 \end{array}$$

■

The examples let us know that subtraction of polynomials, like the subtraction of whole numbers, often consists of several simpler subtractions. In algebra, the simpler subtractions can be worked separately since there is no borrowing or regrouping; however, signed numbers are used.

Practice 16: Subtract the following polynomials.

$$(5x^2 - 4x + 2) - (3x^2 + 2x + 5)$$

ANSWER

16. ______

Practice 17: Subtract the following polynomials.

$$(4b^2 + 8b - 5) - (7b^2 - 4b + 3)$$

17. ______

Practice 18: Subtract the following polynomials.

$$(10x^2 - 7x - 3) - (2x^2 - 5x - 4)$$

18. ______

Practice 19: Subtract the following polynomials.

$$(15x^2 - 6x + 5) - (9x^2 + 8x - 6)$$

19. ______

Example 9 **Subtract the following polynomials.**

$$(3x - 5 - 8x^2) - (12x^2 - 3x - 8)$$

Solution: Write the polynomials in vertical form putting the first polynomial in standard form.

$$\begin{array}{r} -8x^2 + 3x - 5 \\ -\quad 12x^2 - 3x - 8 \\ \hline \end{array}$$

There are three subtractions to be performed. Since there is no borrowing or regrouping in algebra, the three subtractions can be worked separately.

ANSWER

$$\begin{array}{r} -5 \\ -\ \underline{-8} \\ 3 \end{array} \quad \text{or} \quad \begin{aligned} -5-(-8) &= -5+8 \\ &= 3 \end{aligned} \qquad \textbf{Subtraction 1}$$

$$\begin{array}{r} 3x \\ -\ \underline{-3x} \\ 6x \end{array} \quad \text{or} \quad \begin{aligned} 3x-(-3x) &= 3x+3x \\ &= 6x \end{aligned} \qquad \textbf{Subtraction 2}$$

$$\begin{array}{r} -8x^2 \\ -\ \underline{12x^2} \\ -20x^2 \end{array} \quad \text{or} \quad \begin{aligned} -8x^2-12x^2 &= -8x^2+(-12x^2) \\ &= -20x^2 \end{aligned} \qquad \textbf{Subtraction 3}$$

Thus,

$$\begin{array}{r} -8x^2+3x-5 \\ -\ \underline{12x^2-3x-8} \\ -20x^2+6x+3 \end{array}$$

■

Now, let's try some practice problems in which you first write the polynomials in standard form.

20. ______________

Practice 20: Subtract the following polynomials.

$$(7x-8+5x^2)-(3x^2-2x+3)$$

21. ______________

Practice 21: Subtract the following polynomials.

$$(8y^2-5y+7)-(3y^2-8-3y)$$

22. ______________

Practice 22: Subtract the following polynomials.

$$(5x^2+3)-(7x^2-2x-5)$$

A N S W E R

Practice 23: Subtract the following polynomials.

$$(8 - 5x^2 + 6x) - (7 - 2x^2)$$

23. ______

Example 10 **Subtract the following polynomials.**

$$(-12a + 3b - 15c) - (-3a - 6b + 7c)$$

Solution: Write the polynomials in vertical form.

$$\begin{array}{r} -12a + 3b - 15c \\ - \quad -3a - 6b + \ \ 7c \\ \hline \end{array}$$

There are three subtractions.

$$\begin{array}{r} -15c \\ - \quad 7c \\ \hline -22c \end{array} \quad \text{or} \quad \begin{aligned} -15c - 7c &= -15c + (-7c) \\ &= -22c \end{aligned} \quad \textbf{Subtraction 1}$$

$$\begin{array}{r} 3b \\ - \ -6b \\ \hline 9b \end{array} \quad \text{or} \quad \begin{aligned} 3b - (-6b) &= 3b + 6b \\ &= 9b \end{aligned} \quad \textbf{Subtraction 2}$$

$$\begin{array}{r} -12a \\ - \ - \ 3a \\ \hline - \ 9a \end{array} \quad \text{or} \quad \begin{aligned} -12a - (-3a) &= -12a + 3a \\ &= -9a \end{aligned} \quad \textbf{Subtraction 3}$$

Thus,

$$\begin{array}{r} -12a + 3b - 15c \\ - \quad -3a - 6b + \ \ 7c \\ \hline -9a + 9b - 22c \end{array}$$

■

Practice 24: Subtract the following polynomials.

$$(-8a + 3b - 2c) - (3a - 2b + 5c)$$

24. ______

Practice 25: Subtract the following polynomials.

$$(3a - 2c + 5b) - (5b - 4a + 3c)$$

25. ______

EXERCISE 3.2.2

Simplify each of the following.

A

1. $5a - 2a$
2. $3x - 6x$
3. $-2a - (-2a)$
4. $3x + 2y - x - y$
5. $(4x + 7) - (3x + 2)$
6. $(5x + 9) - (2x + 7)$
7. $(8x + 6) - (7x + 1)$
8. $(9x + 4) - (3x + 2)$
9. $(12x + 15) - (7x + 10)$
10. $(15x + 9) - (8x + 6)$

B

11. $(5x^2 + 7x + 8) - (2x^2 + 3x + 6)$
12. $(8x^2 + 4x + 5) - (6x^2 + x + 3)$
13. $(3x^2 + 9x + 7) - (3x^2 + 8x + 2)$
14. $(9x^2 + 5x + 9) - (6x^2 + 2x + 7)$
15. $(5x^2 + 6x + 3) - (2x^2 + 3x + 1)$
16. $(3x^2 + 2x + 1) - (x^2 + 4x + 5)$
17. $(6x^2 + 5x + 3) - (4x^2 + 8x + 7)$
18. $(9x^2 + 2x + 5) - (3x^2 + 12x + 11)$

19. $(12x^2 + 7x + 9) - (15x^2 + 3x + 15)$

20. $(2x^2 - 5x + 7) - (x^2 - 2x - 3)$

21. $(4x^2 + 7x - 3) - (2x^2 + 3x - 5)$

22. $(10x^2 - 11x + 7) - (6x^2 + 5x - 8)$

23. $(15x^2 + 12x + 9) - (11x^2 - 10x - 3)$

24. $(4x^2 - 7x + 15) - (-3x^2 + 5x - 7)$

25. $(6x^2 - 2x + 1) - (-2x^2 - x - 1)$

26. $(x^2 - 2x + 1) - (x^2 + 2x + 1)$

27. $(8a - 7b + 3c) - (5a + 8b - 2c)$

28. $(-6a + 3b + 5c) - (2a - 6b + 4c)$

29. $(5a - 2b - 7c) - (-3a + 4b - 5c)$

30. $(-3a + 4b - 2c) - (-2a - 5b + 8c)$

C

31. $(2y^2 - 7) - (-y^2 + y)$

32. $(-3x + 2x^2) - (2x^2 - 3x)$

33. $(-3x + 2 - x^2) - (3 + 2x - 3x^2)$

34. $(a^2 + 2ab + b^2) - (ab - b^2)$

35. $(5a - 3c + 8b) - (5c - 7a - 3b)$

3.3 MULTIPLYING POLYNOMIALS

This section will concentrate on the similarity between multiplication of polynomials (algebraic expressions) and multiplication of arithmetic expressions. In Chapter 5, where exponents will be discussed, additional multiplication of polynomials will be explained.

A multiplication symbol is sometimes not used in algebraic multiplication. Examples:

2×3 is written $2(3)$

$a \times b$ is written ab

$4 \times x$ is written $4x$

$5 \times (3x)$ is written $5(3x)$

Let's review some of the properties of multiplication from Chapter 1. Remember that the letters in a polynomial represent numbers; thus, the following rules also apply when working with polynomials.

Commutative Property of Multiplication: If a and b represent real numbers, then $ab = ba$.

Example 1

$$2x \times 3x = 3x \times 2x$$

$$(7x)5 = 5(7x)$$

$$5(2x + 3) = (2x + 3)5$$

$$(7x - 8)(3x + 2) = (3x + 2)(7x - 8)$$

■

Associative Property of Multiplication: If a, b, and c represent real numbers, then $(ab)c = a(bc)$.

Example 2

$$2 \times (3x) = (2 \times 3)x$$

$$(5x) \times x = 5(x \times x)$$

■

Distributive Property of Multiplication over Addition: If a, b, and c represent real numbers, then $a(b + c) = ab + ac$.

Example 3

$$2 \times (2x + 3) = 2 \times (2x) + 2 \times 3$$

$$5(7x + 8) = 5(7x) + 5(8)$$

$$8x(2x + 7) = 8x(2x) + 8x(7)$$

$$(5x + 3)(2x + 9) = (5x + 3)(2x) + (5x + 3)9$$

■

Let's look at some examples in which more than one property is used.

Example 4

$$\begin{aligned} (5x)3 &= 3(5x) && \text{Commutative property.} \\ &= (3 \times 5)x && \text{Associative property.} \\ &= 15x \end{aligned}$$ ■

Example 5

$$\begin{aligned} (6x)(5x) &= 6[x(5x)] && \text{Associative property.} \\ &= 6[(5x)x] && \text{Commutative property.} \\ &= 6\{5[x(x)]\} && \text{Associative property.} \\ &= (6 \times 5)[x(x)] && \text{Associative property.} \\ &= 30x^2 \end{aligned}$$ ■

That was a lot of work just to get the numbers and letters together.

That's true, Charlie. We just wanted you to see how the commutative and associative properties allow us to multiply the numbers and group the letters. In future problems, we will skip many of the steps when rearranging the terms. For example,

$$\begin{aligned} (3x)(8x) &= (3 \times 8)[x(x)] \\ &= 24x^2 \end{aligned}$$

Let's look again at the similarity between whole numbers written in expanded notation and polynomials. The number 5872, in expanded notation, is written

$$5(10^3) + 8(10^2) + 7(10) + 2$$

where 10^2 represents ten groups of tens

$$10^2 = 10(10) = 100,$$

and 10^3 represents ten groups of ten groups of tens

$$10^3 = 10(10^2) = 10[10(10)] = 1000$$

The number 5872 written in expanded notation is similar to the polynomial

$$5x^3 + 8x^2 + 7x + 2$$

In fact, the only difference is that the groups of tens have been replaced with groups of x's. Now, x^2 represents x groups of x

$$x^2 = x(x),$$

and x^3 represents x groups of x groups of x

$$x^3 = x(x^2) = x[x(x)]$$

Now, let's look at some examples that show the similarity between algebraic multiplication and arithmetic multiplication.

ANSWER

Example 6

Algebra	Arithmetic
$2 \times (3x + 4)$	2×34
Step 1: Write the polynomials in vertical form.	*Step 1:* Write the numbers in vertical form using expanded notation.
$\begin{array}{r} 3x + 4 \\ \times\ 2 \\ \hline \end{array}$	$\begin{array}{rcr} 34 & = & 3(10) + 4 \\ \times\ 2 & = & \times\ 2 \\ \hline \end{array}$
Use the distributive property and multiply.	Use the distributive property and multiply.
Step 2: Multiply 2 times 4.	*Step 2:* Multiply 2 times 4.
$\begin{array}{r} 3x + \mathbf{4} \\ \mathbf{2} \\ \hline \mathbf{8} \end{array}$	$\begin{array}{rcr} 34 & = & 3(10) + \mathbf{4} \\ \mathbf{2} & = & \mathbf{2} \\ \hline \mathbf{8} & & \mathbf{8} \end{array}$
Step 3: Multiply 2 times $3x$.	*Step 3:* Multiply 2 times 3(10).
$\begin{array}{r} \mathbf{3x} + 4 \\ \mathbf{2} \\ \hline \mathbf{6x} + 8 \end{array}$	$\begin{array}{rcr} 34 & = & \mathbf{3(10)} + 4 \\ \mathbf{2} & = & \mathbf{2} \\ \hline 68 & & \mathbf{6(10)} + 8 \end{array}$ ■

Now, let's work Example 6 using a horizontal format to illustrate the use of the distributive property.

$$\begin{aligned} 2 \times (3x + 4) &= 2 \times (3x) + 2 \times 4 && \textbf{Distributive property.} \\ &= (2 \times 3)x + 2 \times 4 && \textbf{Associative property.} \\ &= 6x + 8 && \textbf{Perform the multiplications.} \end{aligned}$$

To emphasize the similarity between arithmetic and algebra, work the practice problems in this section using the vertical format.

1. ____________________

Practice 1: Multiply the following polynomials.

$$3 \times (2x + 1)$$

2. ____________________

Practice 2: Multiply the following polynomials.

$$2 \times (3x + 2)$$

Practice 3: Multiply the following polynomials.

$$3 \times (2x + 3)$$

A N S W E R

3. ______________

Example 7

Algebra

$$2 \times (2x^2 + 4x + 3)$$

Step 1: Write the polynomials in vertical form.

$$\begin{array}{r} 2x^2 + 4x + 3 \\ \times\ 2 \\ \hline \end{array}$$

Use the distributive property and multiply.

Step 2: Multiply 2 times 3.

$$\begin{array}{r} 2x^2 + 4x + \mathbf{3} \\ \times\ \mathbf{2} \\ \hline \mathbf{6} \end{array}$$

Step 3: Multiply 2 times $4x$.

$$\begin{array}{r} 2x^2 + \mathbf{4x} + 3 \\ \times\ \mathbf{2} \\ \hline \mathbf{8x} + 6 \end{array}$$

Step 4: Multiply 2 times $2x^2$.

$$\begin{array}{r} \mathbf{2x^2} + 4x + 3 \\ \times\ \mathbf{2} \\ \hline \mathbf{4x^2} + 8x + 6 \end{array}$$

Arithmetic

$$2 \times 243$$

Step 1: Write the numbers in vertical form using expanded notation.

$$\begin{array}{rcr} 243 & = & 2(10^2) + 4(10) + 3 \\ \times\ \ 2 & = & \times\ 2 \\ \hline \end{array}$$

Use the distributive property and multiply.

Step 2: Multiply 2 times 3.

$$\begin{array}{rcr} 243 & = & 2(10^2) + 4(10) + \mathbf{3} \\ \times\ \ \mathbf{2} & = & \times\ \mathbf{2} \\ \hline \mathbf{6} & & \mathbf{6} \end{array}$$

Step 3: Multiply 2 times 4(10).

$$\begin{array}{rcr} 243 & = & 2(10^2) + \mathbf{4(10)} + 3 \\ \times\ \ \mathbf{2} & = & \times\ \mathbf{2} \\ \hline \mathbf{8}6 & & \mathbf{8(10)} + 6 \end{array}$$

Step 4: Multiply 2 times $2(10^2)$.

$$\begin{array}{rcr} 243 & = & \mathbf{2(10^2)} + 4(10) + 3 \\ \times\ \ \mathbf{2} & = & \times\ \mathbf{2} \\ \hline \mathbf{4}86 & & \mathbf{4(10^2)} + 8(10) + 6 \end{array}$$ ■

Practice 4: Multiply the following polynomials.

$$3 \times (3x^2 + 2x + 1)$$

4. ______________

A N S W E R

5. ______________

Practice 5: Multiply the following polynomials.

$$2 \times (2x^2 + 3x + 2)$$

6. ______________

Practice 6: Multiply the following polynomials.

$$4 \times (2x^2 + x + 2)$$

Well, Charlie, how are you doing so far?

I think that I'm doing okay.

In that case, let's look at some algebra examples that are slightly different from arithmetic.

Example 8

Algebra	**Arithmetic**
$5 \times (3x^2 + 2x + 4)$	5×324
Step 1: Write the polynomials in vertical form.	*Step 1:* Write the numbers in vertical form using expanded notation.
$\begin{array}{r} 3x^2 + 2x + 4 \\ \times 5 \\ \hline \end{array}$	$\begin{array}{rcr} 324 & = & 3(10^2) + 2(10) + 4 \\ \times \ \ 5 & = & \times 5 \\ \hline \end{array}$
Use the distributive property and multiply.	Use the distributive property and multiply.
Step 2: Multiply 5 times 4.	*Step 2:* Multiply 5 times 4.
$\begin{array}{r} 3x^2 + 2x + \mathbf{4} \\ \times \mathbf{5} \\ \hline \mathbf{20} \end{array}$	$\begin{array}{r} 3(10^2) + 2(10) + \mathbf{4} \\ \times \mathbf{5} \\ \hline \mathbf{20} \end{array}$

Step 3: Regroup and carry.

$$\begin{array}{r} \mathbf{2(10)} \\ 3(10^2) + 2(10) + 4 \\ \times\ 5 \\ \hline \mathbf{0} \end{array}$$

Step 3: Multiply 5 times $2x$.

$$\begin{array}{r} 3x^2 + \mathbf{2x} + 4 \\ \times\ \mathbf{5} \\ \hline \mathbf{10x} + 20 \end{array}$$

Step 4: Multiply 5 times 2(10) and add the groups of tens that were carried.

$$\begin{array}{r} \mathbf{2(10)} \\ 3(10^2) + \mathbf{2(10)} + 4 \\ \times\ \mathbf{5} \\ \hline \mathbf{12(10)} + 0 \end{array}$$

Step 5: Regroup and carry.

$$\begin{array}{r} \mathbf{1(10^2)} \\ 3(10^2) + 2(10) + 4 \\ \times\ 5 \\ \hline \mathbf{2(10)} + 0 \end{array}$$

Step 4: Multiply 5 times $3x^2$.

$$\begin{array}{r} \mathbf{3x^2} + 2x + 4 \\ \times\ \mathbf{5} \\ \hline \mathbf{15x^2} + 10x + 20 \end{array}$$

Step 6: Multiply 5 times $3(10^2)$ and add the groups of 10^2 that were carried.

$$\begin{array}{r} \mathbf{1(10^2)} \\ \mathbf{3(10^2)} + 2(10) + 4 \\ \times\ \mathbf{5} \\ \hline \mathbf{16(10^2)} + 2(10) + 0 \end{array}$$

Step 7: Regroup and carry.

$$\begin{array}{r} 3(10^2) + 2(10) + 4 \\ \times\ 5 \\ \hline \mathbf{1(10^3)} + \mathbf{6(10^2)} + 2(10) + 0 \end{array}$$ ■

Notice that the two products differ since there is no regrouping or carrying in algebra.

Good! Carrying numbers can be exhausting.

Let's try some practice problems.

A N S W E R

7. ____________

Practice 7: Multiply the following polynomials.

$$5 \times (2x^2 + 7x + 3)$$

8. ____________

Practice 8: Multiply the following polynomials.

$$4 \times (7x^2 + 6x + 8)$$

9. ____________

Practice 9: Multiply the following polynomials.

$$6 \times (3x^2 + 4x + 5)$$

Now, let's look at an example involving signed numbers.

Example 9 **Multiply the following polynomials.**

$$3 \times (2x^2 - 5x - 6)$$

Step 1: Write the polynomials in vertical form.

$$\begin{array}{r} 2x^2 - 5x - 6 \\ \times\ 3 \\ \hline \end{array}$$

Step 2: Multiply 3 times -6.

$$\begin{array}{r} 2x^2 - 5x - \mathbf{6} \\ \times\ \mathbf{3} \\ \hline -\ \mathbf{18} \end{array}$$

Step 3: Multiply 3 times $-5x$.

$$\begin{array}{r} 2x^2 - \mathbf{5x} - 6 \\ \times\ \mathbf{3} \\ \hline -\ \mathbf{15x} - 18 \end{array}$$

Step 4: Multiply 3 times $2x^2$.

$$\begin{array}{r} \mathbf{2x^2} - 5x - 6 \\ \times\ \mathbf{3} \\ \hline \mathbf{6x^2} - 15x - 18 \end{array}$$

■

ANSWER

Even though multiplying polynomials sometimes involves signed numbers, the procedure is still basically the same as in arithmetic. Let's work some practice problems.

Practice 10: Multiply the following polynomials.

$$7 \times (2x^2 + 5x - 3)$$

10. ____________

Practice 11: Multiply the following polynomials.

$$3 \times (5y^2 - 4y + 8)$$

11. ____________

Practice 12: Multiply the following polynomials.

$$7 \times (3x^2 - 6x - 9)$$

12. ____________

Practice 13: Multiply the following polynomials.

$$-4 \times (5x^2 - 3x + 4)$$

13. ____________

Now, let's look at multiplying two binomials.

Example 10

Algebra	**Arithmetic**
$(3x + 2) \times (3x + 1)$	32×31
Step 1: Write the binomials in vertical form.	*Step 1:* Write the numbers in vertical form using expanded notation.
$\begin{array}{r} 3x + 1 \\ \times\ \underline{3x + 2} \end{array}$	$\begin{array}{rcr} 31 & = & 3(10) + 1 \\ \times\ \underline{32} & = & \times\ \underline{3(10) + 2} \end{array}$

Step 2: Multiply 2 times 1.

$$\begin{array}{r} 3x + \mathbf{1} \\ \times\ \underline{3x + \mathbf{2}} \\ \mathbf{2} \end{array}$$

Step 3: Multiply 2 times $3x$.

$$\begin{array}{r} \mathbf{3x} + 1 \\ \times\ \underline{3x + \mathbf{2}} \\ \mathbf{6x} + 2 \end{array}$$

Step 4: Multiply $3x$ times 1 and place the result in the appropriate column. Remember, only like things can be added.

$$\begin{array}{rr} & 3x + \mathbf{1} \\ \times & \underline{\mathbf{3x} + 2} \\ & 6x + 2 \\ & \mathbf{3x} \quad\ \ \end{array}$$

Step 5: Multiply $3x$ times $3x$ and place the result in the appropriate column. Remember, only like things can be added.

$$\begin{aligned}(3x) \times (3x) &= (3 \times 3)[x(x)] \\ &= 9x^2\end{aligned}$$

$$\begin{array}{rrr} & \mathbf{3x} & + 1 \\ \times & \mathbf{3x} & + 2 \\ \hline & 6x & + 2 \\ \mathbf{9x^2} & + 3x & \end{array}$$

Step 6: Add the columns.

$$\begin{array}{rrr} & 3x & + 1 \\ \times & 3x & + 2 \\ \hline & \mathbf{6x} & \mathbf{+ 2} \\ \mathbf{9x^2} & \mathbf{+ 3x} & \\ \hline \mathbf{9x^2} & \mathbf{+ 9x} & \mathbf{+ 2} \end{array}$$

Step 2: Multiply 2 times 1.

$$\begin{array}{rcr} 31 & = & 3(10) + \mathbf{1} \\ \times\ \underline{3\mathbf{2}} & = & \times\ \underline{3(10) + \mathbf{2}} \\ \mathbf{2} & & \mathbf{2} \end{array}$$

Step 3: Multiply 2 times 3(10).

$$\begin{array}{rcr} \mathbf{3}1 & = & \mathbf{3(10)} + 1 \\ \times\ \underline{3\mathbf{2}} & = & \times\ \underline{3(10) + \mathbf{2}} \\ \mathbf{6}2 & & \mathbf{6(10)} + 2 \end{array}$$

Step 4: Multiply 3(10) times 1 and place the result in the appropriate column. Remember, only like things can be added.

$$\begin{array}{rcr} 3\mathbf{1} & = & 3(10) + \mathbf{1} \\ \times\ \underline{\mathbf{3}2} & = & \times\ \underline{\mathbf{3(10)} + 2} \\ 62 & & 6(10) + 2 \\ \mathbf{3}\ \ & & \mathbf{3(10)} \quad\ \ \end{array}$$

Step 5: Multiply 3(10) times 3(10) and place the result in the appropriate column. Remember, only like things can be added.

$$\begin{aligned}3(10) \times 3(10) &= (3 \times 3)[10(10)] \\ &= 9(10^2)\end{aligned}$$

$$\begin{array}{rcrrr} 31 & = & & \mathbf{3(10)} & + 1 \\ \times\ \underline{32} & = & \times & \mathbf{3(10)} & + 2 \\ 62 & & & 6(10) & + 2 \\ \mathbf{93}\ \ & & \mathbf{9(10^2)} & + 3(10) & \end{array}$$

Step 6: Add the columns.

$$\begin{array}{rcrrr} 31 & = & & 3(10) & + 1 \\ \times\ \underline{32} & = & \times & 3(10) & + 2 \\ \mathbf{62} & & & \mathbf{6(10)} & \mathbf{+ 2} \\ \underline{\mathbf{93}}\ \ & & \mathbf{9(10^2)} & \mathbf{+ 3(10)} & \\ \mathbf{992} & & \mathbf{9(10^2)} & \mathbf{+ 9(10)} & \mathbf{+ 2} \end{array}$$ ■

I think I can multiply binomials since I already know how to multiply two-digit numbers.

A N S W E R

That's good, Charlie. Now, let's work Example 5 using a horizontal format to illustrate the use of the distributive property.

$(3x + 2) \times (3x + 1)$	
$= (3x + 2) \times (3x) + (3x + 2) \times 1$	**Distributive property.**
$= (3x) \times (3x + 2) + 1 \times (3x + 2)$	**Commutative property.**
$= (3x) \times (3x) + (3x) \times 2 + 1 \times (3x) + 1 \times 2$	**Distributive property.**
$= (3 \times 3)(x \times x) + (2 \times 3)x + (1 \times 3)x + 1 \times 2$	**Commutative and associative properties.**
$= 9x^2 + 6x + 3x + 2$	**Perform the multiplications.**
$= 9x^2 + 9x + 2$	**Combine like terms.**

To emphasize the similarity between arithmetic and algebra, work the practice problems in this section using the vertical format.

Practice 14: Multiply the following binomials.

$$(x + 3) \times (2x + 1)$$

14. ____________

Practice 15: Multiply the following binomials.

$$(2x + 1) \times (2x + 2)$$

15. ____________

Now, let's look at some examples where it is necessary to regroup and carry in the corresponding arithmetic problem. The answers will be different; however, the procedure will be the same.

Example 11

Algebra	**Arithmetic**
$(5x + 2) \times (3x + 4)$	52×34
Step 1: Write the binomials in vertical form.	*Step 1:* Write the numbers in vertical form using expanded notation.
$\begin{array}{r} 3x + 4 \\ \times\ \underline{5x + 2} \end{array}$	$\begin{array}{rcr} 34 & = & 3(10) + 4 \\ \times\ \underline{52} & = & \times\ \underline{5(10) + 2} \end{array}$

Step 2: Multiply 2 times 4.

$$\begin{array}{r} 3x + \mathbf{4} \\ \times\ \underline{5x + \mathbf{2}} \\ \mathbf{8} \end{array}$$

Step 3: Multiply 2 times $3x$.

$$\begin{array}{r} \mathbf{3x} + 4 \\ \times\ \underline{5x + \mathbf{2}} \\ \mathbf{6x} + 8 \end{array}$$

Step 4: Multiply $5x$ times 4 and place the result in the appropriate column.

$$\begin{aligned} (5x) \times 4 &= (5 \times 4)\,x \\ &= 20x \end{aligned}$$

$$\begin{array}{rr} & 3x + \mathbf{4} \\ \times & \underline{\mathbf{5x} + 2} \\ & 6x + 8 \\ \mathbf{20x} & \end{array}$$

Step 5: Multiply $5x$ times $3x$ and place the result in the appropriate column.

$$\begin{aligned} 5x \times 3x &= (5 \times 3)[x(x)] \\ &= 15x^2 \end{aligned}$$

$$\begin{array}{rr} & \mathbf{3x} + 4 \\ \times & \underline{\mathbf{5x} + 2} \\ & 6x + 8 \\ \mathbf{15x^2} + 20x & \end{array}$$

Step 2: Multiply 2 times 4.

$$\begin{array}{rcr} 3\mathbf{4} & = & 3(10) + \mathbf{4} \\ \times\ \underline{5\mathbf{2}} & = & \times\ \underline{5(10) + \mathbf{2}} \\ \mathbf{8} & & \mathbf{8} \end{array}$$

Step 3: Multiply 2 times 3(10).

$$\begin{array}{rcr} \mathbf{3}4 & = & \mathbf{3(10)} + 4 \\ \times\ \underline{5\mathbf{2}} & = & \times\ \underline{5(10) + \mathbf{2}} \\ \mathbf{6}8 & & \mathbf{6(10)} + 8 \end{array}$$

Step 4: Multiply 5(10) times 4 and place the result in the appropriate column.

$$\begin{aligned} 5(10) \times 4 &= (5 \times 4)(10) \\ &= 20(10) \end{aligned}$$

$$\begin{array}{rcr} 3\mathbf{4} & = & 3(10) + \mathbf{4} \\ \times\ \underline{\mathbf{5}2} & = \times & \underline{\mathbf{5(10)} + 2} \\ 68 & & 6(10) + 8 \\ \mathbf{20} & & \mathbf{20(10)} \end{array}$$

Step 5: Regroup and carry.

$$\begin{array}{rcr} \mathbf{2} & & \mathbf{2(10^2)} \\ 34 & = & 3(10) + 4 \\ \times\ \underline{52} & = & \times\ \underline{5(10) + 2} \\ 68 & & 6(10) + 8 \\ \mathbf{0} & & \mathbf{0(10)} \end{array}$$

Step 6: Multiply 5(10) times 3(10), add the $2(10^2)$ that was carried and place the result in the appropriate column.

$$\begin{aligned} 5(10) \times 3(10) &= (5 \times 3)[10(10)] \\ &= 15(10^2) \end{aligned}$$

$$\begin{array}{rcr} \mathbf{2} & & \mathbf{2(10^2)} \\ \mathbf{3}4 & = & \mathbf{3(10)} + 4 \\ \times\ \underline{\mathbf{5}2} & = & \times\ \underline{\mathbf{5(10)} + 2} \\ 68 & & 6(10) + 8 \\ \mathbf{17}0 & & \mathbf{17(10^2)} + 0(10) \end{array}$$

Step 7: Regroup and carry.

$$\begin{array}{rcr} 34 & = & 3(10) + 4 \\ \times\ \underline{52} & = & \times\ \underline{5(10) + 2} \\ 68 & & 6(10) + 8 \\ 170 & & \mathbf{1(10^3)} + \mathbf{7(10^2)} + 0(10) \end{array}$$

A N S W E R

Step 6: Add the columns.

$$\begin{array}{r} 3x + 4 \\ \times\ 5x + 2 \\ \hline 6x + 8 \\ 15x^2 + 20x \\ \hline 15x^2 + 26x + 8 \end{array}$$

Step 8: Add the columns.

$$\begin{array}{rr} 34 = & 3(10) + 4 \\ \times\ 52 = & \times\ 5(10) + 2 \\ \hline 68 & 6(10) + 8 \\ 170 & 1(10^3) + 7(10^2) + 0(10) \\ \hline 1768 & 1(10^3) + 7(10^2) + 6(10) + 8 \end{array}$$ ■

As you can see, the answers are different since we do not regroup or carry in algebra; however, the procedure is the same. Let's try some practice problems.

Practice 16: Multiply the following binomials.

$$(3x + 5) \times (2x + 4)$$

16. ____________

Practice 17: Multiply the following binomials.

$$(4a + 1) \times (3a + 6)$$

17. ____________

Practice 18: Multiply the following binomials.

$$(2y + 7) \times (3y + 2)$$

18. ____________

Practice 19: Multiply the following binomials.

$$(5x + 6) \times (8x + 7)$$

19. ____________

Well, Charlie, how did you do on the practice problems?

Fine! The things I learned in arithmetic seem to apply to algebra.

That's good, Charlie. Now, let's look at some examples involving negative signs.

Example 12 **Multiply the following binomials.**

$$(3x - 2) \times (2x - 5)$$

Step 1: Write the binomials in vertical form.

$$\begin{array}{rr} & 3x - 2 \\ \times & \underline{2x - 5} \end{array}$$

Step 2: Multiply -5 times -2.

$$\begin{array}{rr} & 3x - \mathbf{2} \\ \times & \underline{2x - \mathbf{5}} \\ & \mathbf{+\ 10} \end{array}$$

Step 3: Multiply -5 times $3x$.

$$\begin{array}{rr} & \mathbf{3x} - 2 \\ \times & \underline{2x - \mathbf{5}} \\ & \mathbf{-\ 15x} + 10 \end{array}$$

Step 4: Multiply $2x$ times -2 and place the result in the appropriate column.

$$\begin{array}{rr} & 3x - \mathbf{2} \\ \times & \underline{\mathbf{2x} - 5} \\ & -\ 15x + 10 \\ & \mathbf{-\ 4x} \quad\quad \end{array}$$

Step 5: Multiply $2x$ times $3x$ and place the result in the appropriate column.

$$\begin{array}{rrr} & & \mathbf{3x} - 2 \\ & \times & \underline{\mathbf{2x} - 5} \\ & & -\ 15x + 10 \\ \mathbf{6x^2} & - & 4x \quad\quad \end{array}$$

Step 6: Add the columns.

$$\begin{array}{rrr} & & 3x - 2 \\ & \times & \underline{2x - 5} \\ & & \mathbf{-\ 15x + 10} \\ \underline{\mathbf{6x^2}} & \underline{-} & \underline{\mathbf{4x} \quad\quad} \\ \mathbf{6x^2} & - & \mathbf{19x + 10} \end{array}$$

■

The above example shows that even when signed numbers are used, the procedure for multiplying algebraic expressions (polynomials) is still the same as the procedure for multiplying arithmetic expressions. Now, let's work some practice problems.

ANSWER

Practice 20: Multiply the following binomials.

$$(5x - 1) \times (3x + 7)$$

20. ______________

Practice 21: Multiply the following binomials.

$$(3b - 6) \times (5b - 4)$$

21. ______________

Practice 22: Multiply the following binomials.

$$(2x - 8) \times (6x - 7)$$

22. ______________

Practice 23: Multiply the following binomials.

$$(5y + 9) \times (2y - 6)$$

23. ______________

Now, let's look at an example multiplying a binomial and a trinomial (a polynomial with three terms).

Example 13

Algebra	Arithmetic
$(2x + 1) \times (2x^2 + 3x + 2)$	21×232
Step 1: Write the polynomials in vertical form. $\begin{array}{r} 2x^2 + 3x + 2 \\ \times\ 2x + 1 \\ \hline \end{array}$	*Step 1:* Write the numbers in vertical form using expanded notation. $\begin{array}{rcr} 232 & = & 2(10^2) + 3(10) + 2 \\ \times\ 21 & = & \times\ 2(10) + 1 \\ \hline \end{array}$
Step 2: Multiply 1 times 2. $\begin{array}{r} 2x^2 + 3x + \mathbf{2} \\ \times\ 2x + \mathbf{1} \\ \hline \mathbf{2} \end{array}$	*Step 2:* Multiply 1 times 2. $\begin{array}{rcr} 232 & = & 2(10^2) + 3(10) + \mathbf{2} \\ \times\ \mathbf{21} & = & \times\ 2(10) + \mathbf{1} \\ \hline \mathbf{2} & & \mathbf{2} \end{array}$
Step 3: Multiply 1 times $3x$. $\begin{array}{r} 2x^2 + \mathbf{3x} + 2 \\ \times\ 2x + \mathbf{1} \\ \hline \mathbf{3x} + 2 \end{array}$	*Step 3:* Multiply 1 times 3(10). $\begin{array}{rcr} 232 & = & 2(10^2) + \mathbf{3(10)} + 2 \\ \times\ \mathbf{21} & = & \times\ 2(10) + \mathbf{1} \\ \hline \mathbf{3}2 & & \mathbf{3(10)} + 2 \end{array}$
Step 4: Multiply 1 times $2x^2$. $\begin{array}{r} \mathbf{2x^2} + 3x + 2 \\ \times\ 2x + \mathbf{1} \\ \hline \mathbf{2x^2} + 3x + 2 \end{array}$	*Step 4:* Multiply 1 times $2(10^2)$. $\begin{array}{rcr} 232 & = & \mathbf{2(10^2)} + 3(10) + 2 \\ \times\ \mathbf{21} & = & \times\ 2(10) + \mathbf{1} \\ \hline 232 & & \mathbf{2(10^2)} + 3(10) + 2 \end{array}$
Step 5: Multiply $2x$ times 2 and place the result in the appropriate column. $\begin{array}{r} 2x^2 + 3x + \mathbf{2} \\ \times\ \mathbf{2x} + 1 \\ \hline 2x^2 + 3x + 2 \\ \mathbf{4x} \end{array}$	*Step 5:* Multiply 2(10) times 2 and place the result in the appropriate column. $\begin{array}{rcr} 23\mathbf{2} & = & 2(10^2) + 3(10) + \mathbf{2} \\ \times\ \mathbf{2}1 & = & \times\ \mathbf{2(10)} + 1 \\ \hline 232 & & 2(10^2) + 3(10) + 2 \\ \mathbf{4} & & \mathbf{4(10)} \end{array}$
Step 6: Multiply $2x$ times $3x$ and place the result in the appropriate column. $\begin{array}{r} 2x^2 + \mathbf{3x} + 2 \\ \times\ \mathbf{2x} + 1 \\ \hline 2x^2 + 3x + 2 \\ \mathbf{6x^2} + 4x \end{array}$	*Step 6:* Multiply 2(10) times 3(10) and place the result in the appropriate column. $\begin{array}{rcr} 2\mathbf{3}2 & = & 2(10^2) + \mathbf{3(10)} + 2 \\ \times\ \mathbf{2}1 & = & \times\ \mathbf{2(10)} + 1 \\ \hline 232 & & 2(10^2) + 3(10) + 2 \\ \mathbf{6}4 & & \mathbf{6(10^2)} + 4(10) \end{array}$

A N S W E R

Step 7: Multiply $2x$ times $2x^2$ and place the result in the appropriate column.

$$(2x) \times (2x^2) = (2 \times 2)[x(x^2)] = 4x^3$$

$$\begin{array}{r} 2x^2 + 3x + 2 \\ \times\ 2x + 1 \\ \hline 2x^2 + 3x + 2 \\ 4x^3 + 6x^2 + 4x \end{array}$$

Step 8: Add the columns.

$$\begin{array}{r} 2x^2 + 3x + 2 \\ \times\ 2x + 1 \\ \hline 2x^2 + 3x + 2 \\ 4x^3 + 6x^2 + 4x \\ \hline 4x^3 + 8x^2 + 7x + 2 \end{array}$$

Step 7: Multiply $2(10)$ times $2(10^2)$ and place the result in the appropriate column.

$$2(10) \times 2(10^2) = (2 \times 2)[10(10^2)] = 4(10^3)$$

$$\begin{array}{rcr} 232 & = & 2(10^2) + 3(10) + 2 \\ \times\ 21 & = & \times\ 2(10) + 1 \\ \hline 232 & & 2(10^2) + 3(10) + 2 \\ 464 & & 4(10^3) + 6(10^2) + 4(10) \end{array}$$

Step 8: Add the columns.

$$\begin{array}{rcr} 232 & = & 2(10^2) + 3(10) + 2 \\ \times\ 21 & = & \times\ 2(10) + 1 \\ \hline 232 & & 2(10^2) + 3(10) + 2 \\ 464 & & 4(10^3) + 6(10^2) + 4(10) \\ \hline 4872 & & 4(10^3) + 8(10^2) + 7(10) + 2 \end{array}$$ ■

The similarity between algebra and arithmetic is amazing. I always thought they were totally different.

Well, Charlie, now that you see the similarity between them, we will not continue to work the arithmetic examples. Let's work some practice problems.

Practice 24: Multiply the following polynomials.

$$(x + 2) \times (x^2 + 2x + 3)$$

24. ________________

Practice 25: Multiply the following polynomials.

$$(3x + 2) \times (2x^2 + 5x + 7)$$

25. ________________

A N S W E R

26. ____________________

Practice 26: Multiply the following polynomials.

$$(2x + 5) \times (4x^2 + 2x + 6)$$

27. ____________________

Practice 27: Multiply the following polynomials.

$$(5x + 1) \times (6x^2 + 7x + 8)$$

Example 14 **Multiply the following polynomials.**

$$(2x + 3) \times (7x + 4x^2 + 6)$$

Step 1: Write the second polynomial in standard form.

$$(2x + 3) \times (4x^2 + 7x + 6)$$

Step 2: Write the polynomials in vertical form.

$$\begin{array}{r} 4x^2 + 7x + 6 \\ \times\ 2x + 3 \\ \hline \end{array}$$

Step 3: Multiply 3 times 6.

$$\begin{array}{r} 4x^2 + 7x + \mathbf{6} \\ \times\ 2x + \mathbf{3} \\ \hline \mathbf{18} \end{array}$$

Step 4: Multiply 3 times $7x$.

$$\begin{array}{r} 4x^2 + \mathbf{7x} + 6 \\ \times\ 2x + \mathbf{3} \\ \hline \mathbf{21x} + 18 \end{array}$$

A N S W E R

Step 5: Multiply 3 times $4x^2$.

$$\begin{array}{r} \mathbf{4x^2} + 7x + 6 \\ \times\ 2x + \mathbf{3} \\ \hline \mathbf{12x^2} + 21x + 18 \end{array}$$

Step 6: Multiply $2x$ times 6 and place the result in the appropriate column.

$$\begin{array}{r} 4x^2 + 7x + \mathbf{6} \\ \times\ \mathbf{2x} + 3 \\ \hline 12x^2 + 21x + 18 \\ \mathbf{12x} \end{array}$$

Step 7: Multiply $2x$ times $7x$ and place the result in the appropriate column.

$$\begin{array}{r} 4x^2 + \mathbf{7x} + 6 \\ \times\ \mathbf{2x} + 3 \\ \hline 12x^2 + 21x + 18 \\ \mathbf{14x^2} + 12x \end{array}$$

Step 8: Multiply $2x$ times $4x^2$ and place the result in the appropriate column.

$$\begin{array}{r} \mathbf{4x^2} + 7x + 6 \\ \times\ \mathbf{2x} + 3 \\ \hline 12x^2 + 21x + 18 \\ \mathbf{8x^3} + 14x^2 + 12x \end{array}$$

Step 9: Add the columns.

$$\begin{array}{r} 4x^2 + 7x + 6 \\ \times\ 2x + 3 \\ \hline \mathbf{12x^2} + \mathbf{21x} + \mathbf{18} \\ \mathbf{8x^3} + \mathbf{14x^2} + \mathbf{12x} \\ \hline \mathbf{8x^3} + \mathbf{26x^2} + \mathbf{33x} + \mathbf{18} \end{array}$$

■

Let's try some practice problems in which you need to write the polynomials in standard form before multiplying.

Practice 28: Multiply the following polynomials.

$$(3x + 2) \times (5x + 4 + 2x^2)$$

28. ____________________

ANSWER

Practice 29: Multiply the following polynomials.

$$(5x + 3) \times (4x^2 + 8 + 2x)$$

29. ____________

Now, let's multiply a binomial and a trinomial involving signed numbers.

Example 15 **Multiply the following polynomials.**

$$(5x - 7) \times (4x^2 - 6x - 8)$$

Step 1: Write the polynomials in vertical form.

$$\begin{array}{r} 4x^2 - 6x - 8 \\ \times\ 5x - 7 \\ \hline \end{array}$$

Step 2: Multiply -7 times -8.

$$\begin{array}{r} 4x^2 - 6x - \mathbf{8} \\ \times\ 5x - \mathbf{7} \\ \hline \mathbf{+\ 56} \end{array}$$

Step 3: Multiply -7 times $-6x$.

$$\begin{array}{r} 4x^2 - \mathbf{6x} - 8 \\ \times\ 5x - \mathbf{7} \\ \hline +\ \mathbf{42x} + 56 \end{array}$$

Step 4: Multiply -7 times $4x^2$.

$$\begin{array}{r} \mathbf{4x^2} - 6x - 8 \\ \times\ 5x - \mathbf{7} \\ \hline \mathbf{-\ 28x^2} + 42x + 56 \end{array}$$

Step 5: Multiply $5x$ times -8 and place the result in the appropriate column.

$$\begin{array}{r} 4x^2 - 6x - \mathbf{8} \\ \times\ \mathbf{5x} - 7 \\ \hline -\ 28x^2 + 42x + 56 \\ \mathbf{-\ 40x} \end{array}$$

ANSWER

Step 6: Multiply $5x$ times $-6x$ and place the result in the appropriate column.

$$\begin{array}{rrrr} & 4x^2 - & 6x - & 8 \\ & \times & 5x - & 7 \\ \hline & -28x^2 + & 42x + & 56 \\ & -30x^2 - & 40x & \end{array}$$

Step 7: Multiply $5x$ times $4x^2$ and place the result in the appropriate column.

$$\begin{array}{rrrr} & 4x^2 - & 6x - & 8 \\ & \times & 5x - & 7 \\ \hline & -28x^2 + & 42x + & 56 \\ 20x^3 - & 30x^2 - & 40x & \end{array}$$

Step 8: Add the columns.

$$\begin{array}{rrrr} & 4x^2 - & 6x - & 8 \\ & \times & 5x - & 7 \\ \hline & -28x^2 + & 42x + & 56 \\ 20x^3 - & 30x^2 - & 40x & \\ \hline 20x^3 - & 58x^2 + & 2x + & 56 \end{array}$$

■

Now, let's do some practice problems.

Practice 30: Multiply the following polynomials.

$$(2x - 3) \times (x^2 + 3x + 2)$$

30. ____________

Practice 31: Multiply the following polynomials.

$$(x + 5) \times (3x^2 - 5x - 3)$$

31. ____________

A N S W E R

32. ______________________

Practice 32: Multiply the following polynomials.

$$(3y - 4) \times (2y^2 - 5y + 4)$$

33. ______________________

Practice 33: Multiply the following polynomials.

$$(7a - 2) \times (3a^2 - 4a - 8)$$

34. ______________________

Practice 34: Multiply the following polynomials.

$$(4x - 2) \times (3 - 2x + 5x^2)$$

35. ______________________

Practice 35: Multiply the following polynomials.

$$(5x - 3) \times (2x^2 + 8)$$

Well, Charlie, how did you do on the practice problems?

The practice problems were tough. However, anytime I had trouble, I simply did what I would have done in arithmetic.

Good, Charlie. Now, let's look at the multiplication of two slightly different binomials.

Example 16 **Multiply the following binomials.**

$$(7a - 3b) \times (5a + 2b)$$

Step 1: Write the binomials in vertical form.

$$\begin{array}{r} 5a + 2b \\ \times\ \underline{7a - 3b} \end{array}$$

Step 2: Multiply $-3b$ times $2b$.

$$\begin{array}{r} 5a + \mathbf{2b} \\ \times\ \underline{7a - \mathbf{3b}} \\ -\ \mathbf{6b^2} \end{array}$$

Step 3: Multiply $-3b$ times $5a$. (Remember that ab is the same as ba according to the commutative property of multiplication.)

$$\begin{array}{rrr} & \mathbf{5a} & +\ 2b \\ \times & \underline{7a} & \underline{-\ \mathbf{3b}} \\ & -\ \mathbf{15ab} & -\ 6b^2 \end{array}$$

Step 4: Multiply $7a$ times $2b$ and place the result in the appropriate column.

$$\begin{array}{rrr} & 5a & +\ \mathbf{2b} \\ \times & \underline{\mathbf{7a}} & \underline{-\ 3b} \\ & -\ 15ab & -\ 6b^2 \\ & +\ \mathbf{14ab} & \end{array}$$

Step 5: Multiply $7a$ times $5a$ and place the result in the appropriate column.

$$\begin{array}{rrrr} & & \mathbf{5a} & +\ 2b \\ & \times & \underline{\mathbf{7a}} & \underline{-\ 3b} \\ & & -\ 15ab & -\ 6b^2 \\ \mathbf{35a^2} & +\ 14ab & & \end{array}$$

Step 6: Add the columns.

$$\begin{array}{rrrr} & & 5a & +\ 2b \\ & \times & \underline{7a} & \underline{-\ 3b} \\ & & -\ \mathbf{15ab} & -\ \mathbf{6b^2} \\ \underline{\mathbf{35a^2}} & \underline{+\ \mathbf{14ab}} & & \\ \mathbf{35a^2} & -\ \mathbf{1ab} & & -\ \mathbf{6b^2} \end{array}$$

■

A N S W E R

Let's work some practice problems.

36. ____________________

Practice 36: Multiply the following binomials.

$$(3a + b) \times (2a + 4b)$$

37. ____________________

Practice 37: Multiply the following binomials.

$$(2a - 5b) \times (6a - 3b)$$

38. ____________________

Practice 38: Multiply the following binomials.

$$(5x + 3y) \times (5x - 3y)$$

EXERCISE 3.3

Multiply the following polynomials.

A

1. $2 \times (3x + 1)$

2. $3 \times (4x - 2)$

3. $7 \times (5a - 8)$

4. $-5 \times (3x + 7)$

5. $-4 \times (6b - 3)$

6. $2 \times (2x^2 + 4x + 3)$

7. $5 \times (8x^2 - 4x + 2)$

8. $6 \times (3y^2 + 2y - 5)$

9. $12 \times (5x^2 - x + 4)$

10. $-15 \times (4x^2 - 6x - 2)$

B

11. $-7 \times (-5x^2 + 6x - 1)$

12. $(2x + 3) \times (3x + 1)$

13. $(6y + 1) \times (2y - 3)$

14. $(2x - 5) \times (4x - 2)$

15. $(9x - 5) \times (3x + 7)$

16. $(3a - 2) \times (7a + 8)$

17. $(7x - 3) \times (9x - 2)$

18. $(3x + 10) \times (2x - 5)$

19. $(12b - 1) \times (5b - 3)$

20. $(15x - 3) \times (5x - 12)$

21. $(3a - b) \times (2a + b)$

22. $(5a + 7b) \times (3a + 2b)$

23. $(4a - 3b) \times (5a - 2b)$

24. $(6a - 3b) \times (2a + 3b)$

25. $(7x + 3y) \times (7x - 5y)$

C

26. $(x + 3) \times (3x^2 + 2x + 1)$

27. $(3x + 5) \times (2x^2 + x + 4)$

28. $(2x - 1) \times (x^2 + 2x + 1)$

29. $(3x - 4) \times (x^2 - 3x + 5)$

30. $(9y - 2) \times (y^2 - y - 1)$

31. $(5x + 1) \times (8x^2 - 3x + 9)$

32. $(-x^2 - x - 3) \times (-2x + 3)$

33. $(6b - 8) \times (3b + 5b^2 - 7)$

34. $(-2x - 5) \times (7 - 3x^2 - 2x)$

35. $(3x^2 + 2x + 3) \times (x^2 - 2x - 1)$

3.4 DIVIDING POLYNOMIALS

This section will concentrate on the similarity between division of polynomials (algebraic expressions) and division of arithmetic expressions. In Chapter 5, where exponents will be discussed, additional division of polynomials will be explained. Let's look at some examples that show the similarity between algebraic division and arithmetic division.

Example 1

Algebra	Arithmetic
$(6x + 9) \div 3$	$69 \div 3$
Step 1: Write the polynomials in division form.	*Step 1:* Write the numbers in division form using expanded notation.
$3\overline{)6x + 9}$	$3\overline{)69} = 3\overline{)6(10) + 9}$
Step 2: Divide $6x$ by 3.	*Step 2:* Divide $6(10)$ by 3.
$\begin{array}{r} \mathbf{2x} \\ 3\overline{)\mathbf{6x} + 9} \end{array}$	$\begin{array}{rcr} \mathbf{2} & & \mathbf{2(10)} \\ 3\overline{)\mathbf{69}} & = & 3\overline{)\mathbf{6(10)} + 9} \end{array}$
Step 3: Multiply $2x$ times 3 and place the result in the proper column.	*Step 3:* Multiply $2(10)$ times 3 and place the result in the proper column.
$\begin{array}{r} \mathbf{2x} \\ \mathbf{3}\overline{)6x + 9} \\ \mathbf{6x} \end{array}$	$\begin{array}{rcr} \mathbf{2} & & \mathbf{2(10)} \\ \mathbf{3}\overline{)69} & = & \mathbf{3}\overline{)6(10) + 9} \\ \mathbf{6} & & \mathbf{6(10)} \end{array}$
Step 4: Subtract $6x$ from $6x$.	*Step 4:* Subtract $6(10)$ from $6(10)$
$\begin{array}{r} 2x \\ 3\overline{)\mathbf{6x} + 9} \\ \underline{\mathbf{6x} } \\ \mathbf{0} \end{array}$	$\begin{array}{rcr} 2 & & 2(10) \\ 3\overline{)\mathbf{69}} & = & 3\overline{)\mathbf{6(10)} + 9} \\ \underline{\mathbf{6}} & & \underline{\mathbf{6(10)} } \\ \mathbf{0} & & \mathbf{0} \end{array}$
Step 5: Bring 9 down.	*Step 5:* Bring 9 down.
$\begin{array}{r} 2x \\ 3\overline{)6x + 9} \\ \underline{6x } \\ \mathbf{9} \end{array}$	$\begin{array}{rcr} 2 & & 2(10) \\ 3\overline{)69} & = & 3\overline{)6(10) + 9} \\ \underline{6} & & \underline{6(10) } \\ \mathbf{9} & & \mathbf{9} \end{array}$
Step 6: Divide 9 by 3.	*Step 6:* Divide 9 by 3.
$\begin{array}{r} 2x + \mathbf{3} \\ \mathbf{3}\overline{)6x + 9} \\ \underline{6x } \\ \mathbf{9} \end{array}$	$\begin{array}{rcr} 2\mathbf{3} & & 2(10) + \mathbf{3} \\ \mathbf{3}\overline{)69} & = & \mathbf{3}\overline{)6(10) + 9} \\ \underline{6} & & \underline{6(10) } \\ \mathbf{9} & & \mathbf{9} \end{array}$

ANSWER

Step 7: Multiply 3 times 3 and place the result in the proper column.

$$\begin{array}{r} 2x + \mathbf{3} \\ \mathbf{3}\overline{)6x + 9} \\ \underline{6x } \\ 9 \\ \mathbf{9} \end{array}$$

Step 7: Multiply 3 times 3 and place the result in the proper column.

$$\begin{array}{r} 2\mathbf{3} \\ \mathbf{3}\overline{)69} \\ \underline{6} \\ 9 \\ \mathbf{9} \end{array} = \begin{array}{r} 2(10) + \mathbf{3} \\ \mathbf{3}\overline{)6(10) + 9} \\ \underline{6(10)} \\ 9 \\ \mathbf{9} \end{array}$$

Step 8: Subtract 9 from 9.

$$\begin{array}{r} 2x + 3 \\ 3\overline{)6x + 9} \\ \underline{6x } \\ \mathbf{9} \\ \underline{\mathbf{9}} \\ \mathbf{0} \end{array}$$

Step 8: Subtract 9 from 9.

$$\begin{array}{r} 23 \\ 3\overline{)69} \\ \underline{6} \\ \mathbf{9} \\ \underline{\mathbf{9}} \\ \mathbf{0} \end{array} = \begin{array}{r} 2(10) + 3 \\ 3\overline{)6(10) + 9} \\ \underline{6(10)} \\ \mathbf{9} \\ \underline{\mathbf{9}} \\ \mathbf{0} \end{array}$$ ■

I'm still amazed at the similarity between algebra and arithmetic. I think it will help me with division of polynomials.

That's right, Charlie. If you do have trouble, just remember the rules for dividing whole numbers and apply them to the division of polynomials. Also, since division is the inverse of multiplication ($a \div b = c$ if and only if $b \times c = a$), you can check your answer by multiplying. Let's check the algebra part of Example 1 by multiplying 3 times $2x + 3$.

$$\begin{array}{r} 2x + 3 \\ \underline{\times\ 3} \\ 6x + 9 \end{array}$$

Since 3 times $2x + 3$ equals $6x + 9$, Example 1 is correct. Now, let's try some practice problems.

Practice 1: Divide the following polynomials.

$$(3x + 6) \div 3$$

1. ____________________

A N S W E R

2. ____________________

Practice 2: Divide the following polynomials.

$$(6x + 4) \div 2$$

3. ____________________

Practice 3: Divide the following polynomials.

$$(8x + 6) \div 2$$

4. ____________________

Practice 4: Divide the following polynomials.

$$(6x^2 + 3x + 9) \div 3$$

5. ____________________

Practice 5: Divide the following polynomials.

$$(8x^2 + 4x + 6) \div 2$$

Now, let's look at another example.

Example 2

Algebra

$$(6x^2 + 7x + 2) \div (2x + 1)$$

Step 1: Write the polynomials in division form.

$$2x + 1\overline{)6x^2 + 7x + 2}$$

Step 2: Divide $6x^2$ by $2x$.

$$\begin{array}{r} 3x \\ 2x + 1\overline{)6x^2 + 7x + 2} \end{array}$$

Step 3: Multiply $3x$ times $2x + 1$ and place the results in the proper columns.

$$\begin{array}{r} 3x \\ 2x + 1\overline{)6x^2 + 7x + 2} \\ 6x^2 + 3x \end{array}$$

Step 4: Subtract $6x^2 + 3x$ from $6x^2 + 7x$.

$$\begin{array}{r} 3x \\ 2x + 1\overline{)6x^2 + 7x + 2} \\ \underline{6x^2 + 3x } \\ 4x \end{array}$$

Step 5: Bring 2 down.

$$\begin{array}{r} 3x \\ 2x + 1\overline{)6x^2 + 7x + 2} \\ \underline{6x^2 + 3x } \\ 4x + 2 \end{array}$$

Step 6: Divide $4x$ by $2x$.

$$\begin{array}{r} 3x + 2 \\ 2x + 1\overline{)6x^2 + 7x + 2} \\ \underline{6x^2 + 3x } \\ 4x + 2 \end{array}$$

Arithmetic

$$672 \div 21$$

Step 1: Write the numbers in division form using expanded notation.

$$21\overline{)672} = 2(10) + 1\overline{)6(10^2) + 7(10) + 2}$$

Step 2: Divide $6(10^2)$ by $2(10)$.

$$\begin{array}{r} 3 \\ 21\overline{)672} \end{array} = \begin{array}{r} 3(10) \\ 2(10) + 1\overline{)6(10^2) + 7(10) + 2} \end{array}$$

Step 3: Multiply $3(10)$ times $2(10) + 1$ and place the results in the proper columns.

$$\begin{array}{r} 3 \\ 21\overline{)672} \\ 63 \end{array} = \begin{array}{r} 3(10) \\ 2(10) + 1\overline{)6(10^2) + 7(10) + 2} \\ 6(10^2) + 3(10) \end{array}$$

Step 4: Subtract $6(10^2) + 3(10)$ from $6(10^2) + 7(10)$.

$$\begin{array}{r} 3 \\ 21\overline{)672} \\ \underline{63} \\ 4 \end{array} = \begin{array}{r} 3(10) \\ 2(10) + 1\overline{)6(10^2) + 7(10) + 2} \\ \underline{6(10^2) + 3(10) } \\ 4(10) \end{array}$$

Step 5: Bring 2 down.

$$\begin{array}{r} 3 \\ 21\overline{)672} \\ \underline{63} \\ 42 \end{array} = \begin{array}{r} 3(10) \\ 2(10) + 1\overline{)6(10^2) + 7(10) + 2} \\ \underline{6(10^2) + 3(10) } \\ 4(10) + 2 \end{array}$$

Step 6: Divide $4(10)$ by $2(10)$.

$$\begin{array}{r} 32 \\ 21\overline{)672} \\ \underline{63} \\ 42 \end{array} = \begin{array}{r} 3(10) + 2 \\ 2(10) + 1\overline{)6(10^2) + 7(10) + 2} \\ \underline{6(10^2) + 3(10) } \\ 4(10) + 2 \end{array}$$

ANSWER

Step 7: Multiply 2 times $2x + 1$ and place the results in the proper columns.	*Step 7:* Multiply 2 times $2(10) + 1$ and place the results in the proper columns.

$$\begin{array}{r} 3x + \mathbf{2} \\ \mathbf{2x + 1}\overline{)\,6x^2 + 7x + 2} \\ \underline{6x^2 + 3x} \\ 4x + 2 \\ \mathbf{4x + 2} \end{array} \qquad \begin{array}{r} 32 \\ \mathbf{21}\overline{)\,672} \\ \underline{63} \\ 42 \\ \mathbf{42} \end{array} = \begin{array}{r} 3(10) + \mathbf{2} \\ \mathbf{2(10) + 1}\overline{)\,6(10^2) + 7(10) + 2} \\ \underline{6(10^2) + 3(10)} \\ 4(10) + 2 \\ \mathbf{4(10) + 2} \end{array}$$

Step 8: Subtract $4x + 2$ from $4x + 2$.	*Step 8:* Subtract $4(10) + 2$ from $4(10) + 2$.

$$\begin{array}{r} 3x + 2 \\ 2x + 1\overline{)\,6x^2 + 7x + 2} \\ \underline{6x^2 + 3x} \\ \mathbf{4x + 2} \\ \underline{\mathbf{4x + 2}} \\ \mathbf{0} \end{array} \qquad \begin{array}{r} 32 \\ 21\overline{)\,672} \\ \underline{63} \\ \mathbf{42} \\ \underline{\mathbf{42}} \\ \mathbf{0} \end{array} = \begin{array}{r} 3(10) + 2 \\ 2(10) + 1\overline{)\,6(10^2) + 7(10) + 2} \\ \underline{6(10^2) + 3(10)} \\ \mathbf{4(10) + 2} \\ \underline{\mathbf{4(10) + 2}} \\ \mathbf{0} \end{array}$$ ■

Let's check the algebra part of Example 2.

$$\begin{array}{r} 3x + 2 \\ \times\ 2x + 1 \\ \hline 3x + 2 \\ 6x^2 + 4x \\ \hline 6x^2 + 7x + 2 \end{array}$$

So, Example 2 is correct.

Well, Charlie, do you understand everything so far?

Example 2 was harder; however, the algebra part seemed to work the same as the arithmetic part. I guess I'm ready for some practice problems.

6. ____________

Practice 6: Divide the following polynomials.

$$(2x^2 + 5x + 3) \div (2x + 3)$$

Practice 7: Divide the following polynomials.

$$(6x^2 + 8x + 2) \div (3x + 1)$$

A N S W E R

7. ____________

Practice 8: Divide the following polynomials.

$$(4x^2 + 9x + 2) \div (x + 2)$$

8. ____________

Let's look at an example where there is a nonzero remainder.

Example 3

Algebra	**Arithmetic**
$(6x^2 + 7x + 5) \div (3x + 2)$	$675 \div 32$
Step 1: Write the polynomials in division form.	*Step 1:* Write the numbers in division form.
$3x + 2\overline{)6x^2 + 7x + 5}$	$32\overline{)675} = 3(10) + 2\overline{)6(10^2) + 7(10) + 5}$
Step 2: Divide $6x^2$ by $3x$.	*Step 2:* Divide $6(10^2)$ by $3(10)$.
$\begin{array}{r} \mathbf{2x} \\ 3x + 2\overline{)\mathbf{6x^2} + 7x + 5} \end{array}$	$\begin{array}{r} \mathbf{2} \\ 32\overline{)675} \end{array} = \begin{array}{r} \mathbf{2(10)} \\ \mathbf{3(10)} + 2\overline{)\mathbf{6(10^2)} + 7(10) + 5} \end{array}$
Step 3: Multiply $2x$ times $3x + 2$ and place the results in the proper columns.	*Step 3:* Multiply $2(10)$ times $3(10) + 2$ and place the results in the proper columns.
$\begin{array}{r} \mathbf{2x} \\ \mathbf{3x + 2}\overline{)6x^2 + 7x + 5} \\ \mathbf{6x^2 + 4x} \end{array}$	$\begin{array}{r} \mathbf{2} \\ \mathbf{32}\overline{)675} \\ \mathbf{64} \end{array} = \begin{array}{r} \mathbf{2(10)} \\ \mathbf{3(10) + 2}\overline{)6(10^2) + 7(10) + 5} \\ \mathbf{6(10^2) + 4(10)} \end{array}$

Step 4: Subtract $6x^2 + 4x$ from $6x^2 + 7x$.

$$\begin{array}{r} 2x \\ 3x+2\overline{)\,6x^2+7x+5} \\ \underline{6x^2+4x} \\ 3x \end{array}$$

Step 5: Bring 5 down.

$$\begin{array}{r} 2x \\ 3x+2\overline{)\,6x^2+7x+5} \\ \underline{6x^2+4x} \\ 3x+5 \end{array}$$

Step 6: Divide $3x$ by $3x$.

$$\begin{array}{r} 2x+1 \\ 3x+2\overline{)\,6x^2+7x+5} \\ \underline{6x^2+4x} \\ 3x+5 \end{array}$$

Step 7: Multiply 1 times $3x + 2$ and place the results in the proper columns.

$$\begin{array}{r} 2x+1 \\ 3x+2\overline{)\,6x^2+7x+5} \\ \underline{6x^2+4x} \\ 3x+5 \\ 3x+2 \end{array}$$

Step 8: Subtract $3x + 2$ from $3x + 5$.

$$\begin{array}{r} 2x+1 \\ 3x+2\overline{)\,6x^2+7x+5} \\ \underline{6x^2+4x} \\ 3x+5 \\ \underline{3x+2} \\ 3 \end{array}$$

Step 9: Write the answer.

$$2x + 1 + \frac{3}{3x + 2}$$

Step 4: Subtract $6(10^2) + 4(10)$ from $6(10^2) + 7(10)$.

$$\begin{array}{r} 2 \\ 32\overline{)\,675} \\ \underline{64} \\ 3 \end{array} = \begin{array}{r} 2(10) \\ 3(10)+2\overline{)\,6(10^2)+7(10)+5} \\ \underline{6(10^2)+4(10)} \\ 3(10) \end{array}$$

Step 5: Bring 5 down.

$$\begin{array}{r} 2 \\ 32\overline{)\,675} \\ \underline{64} \\ 35 \end{array} = \begin{array}{r} 2(10) \\ 3(10)+2\overline{)\,6(10^2)+7(10)+5} \\ \underline{6(10^2)+4(10)} \\ 3(10)+5 \end{array}$$

Step 6: Divide $3(10)$ by $3(10)$.

$$\begin{array}{r} 21 \\ 32\overline{)\,675} \\ \underline{64} \\ 35 \end{array} = \begin{array}{r} 2(10)+1 \\ 3(10)+2\overline{)\,6(10^2)+7(10)+5} \\ \underline{6(10^2)+4(10)} \\ 3(10)+5 \end{array}$$

Step 7: Multiply 1 times $3(10) + 2$ and place the results in the proper columns.

$$\begin{array}{r} 21 \\ 32\overline{)\,675} \\ \underline{64} \\ 35 \\ 32 \end{array} = \begin{array}{r} 2(10)+1 \\ 3(10)+2\overline{)\,6(10^2)+7(10)+5} \\ \underline{6(10^2)+4(10)} \\ 3(10)+5 \\ 3(10)+2 \end{array}$$

Step 8: Subtract $3(10) + 2$ from $3(10) + 5$.

$$\begin{array}{r} 21 \\ 32\overline{)\,675} \\ \underline{64} \\ 35 \\ \underline{32} \\ 3 \end{array} = \begin{array}{r} 2(10)+1 \\ 3(10)+2\overline{)\,6(10^2)+7(10)+5} \\ \underline{6(10^2)+4(10)} \\ 3(10)+5 \\ \underline{3(10)+2} \\ 3 \end{array}$$

Step 9: Write the answer.

$$21\frac{3}{32} = 2(10) + 1 + \frac{3}{3(10) + 2}$$

■

Let's check the algebra part of Example 3. First multiply $3x + 2$ times $2x + 1$.

$$\begin{array}{r} 2x+1 \\ \underline{\times\ 3x+2} \\ 4x+2 \\ \underline{6x^2+3x} \\ 6x^2+7x+2 \end{array}$$

Now, add the remainder.

A N S W E R

$$\begin{array}{r} 6x^2 + 7x + 2 \\ + 3 \\ \hline 6x^2 + 7x + 5 \end{array}$$

So, Example 3 is correct. The remainder from the division of two polynomials is treated the same as the remainder from the division of two whole numbers.

Let's work some practice problems that have a nonzero remainder.

Practice 9: Divide the following polynomials.

$$(2x^2 + 7x + 7) \div (2x + 3)$$

9. ________________

Practice 10: Divide the following polynomials.

$$(2y^2 + 9y + 6) \div (2y + 1)$$

10. ________________

Practice 11: Divide the following polynomials.

$$(6x^2 + 7x + 7) \div (3x + 2)$$

11. ________________

Well, Charlie, how did you do on the practice problems?

I think I did okay. I just worked the problems in a way similar to the way I would have worked the division of two whole numbers.

Good! In the examples that follow, we will not work the corresponding arithmetic problem.

Example 4 **Divide the following polynomials.**

$$(-x + 15x^2 - 6) \div (3x - 2)$$

Step 1: Before dividing, rewrite $-x + 15x^2 - 6$ in standard form. (When dividing polynomials, it is essential that they be in standard form.)

In division of polynomials, it is important for every term to be in its proper place.

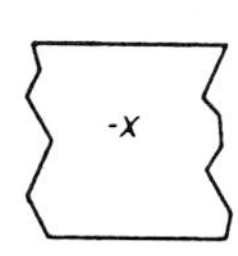

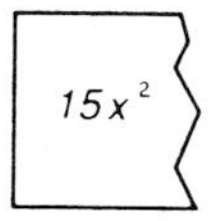

That's better!

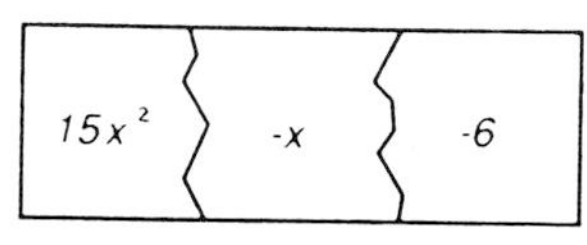

Now, divide $15x^2 - x - 6$ by $3x - 2$.

Step 2: Write the polynomials in division form.

$$3x - 2\overline{)\,15x^2 - x - 6}$$

Step 3: Divide $15x^2$ by $3x$.

$$\begin{array}{r} \mathbf{5x} \\ \mathbf{3x - 2)\overline{\,15x^2 - \quad x - 6}} \end{array}$$

Step 4: Multiply $5x$ times $3x - 2$ and place the results in the proper columns.

$$\begin{array}{rl} & \quad\;\; \mathbf{5x} \\ \mathbf{3x - 2} & \overline{)\,15x^2 - \quad x - 6} \\ & \;\; \mathbf{15x^2 - 10x} \end{array}$$

Step 5: Subtract $15x^2 - 10x$ from $15x^2 - x$. (You may want to review Section 3.2 on subtraction of polynomials.)

Now,

$$\begin{array}{r} 15x^2 - x \\ -\ \underline{15x^2 - 10x} \end{array}$$

can be broken into two separate subtractions since there is no regrouping or borrowing in algebra.

$$\begin{array}{r} -x \\ -\ \underline{-10x} \\ 9x \end{array} \quad \text{or} \quad \begin{aligned} -x - (-10x) &= -x + 10x \\ &= 9x \end{aligned} \qquad \textbf{Subtraction 1}$$

$$\begin{array}{r} 15x^2 \\ \underline{-15x^2} \\ 0 \end{array} \qquad \textbf{Subtraction 2}$$

Thus,

$$\begin{array}{r} 5x \phantom{{}-6} \\ 3x - 2 \overline{)\,\mathbf{15x^2} - x - 6} \\ \underline{\mathbf{15x^2} - \mathbf{10x}} \phantom{{}-6} \\ \mathbf{9x} \phantom{{}-6} \end{array}$$

Step 6: Bring −6 down.

$$\begin{array}{r} 5x \phantom{{}-6} \\ 3x - 2 \overline{)\,15x^2 - x - 6} \\ \underline{15x^2 - 10x \phantom{{}-6}} \\ 9x \mathbf{{}-6} \end{array}$$

Step 7: Divide $9x$ by $3x$.

$$\begin{array}{r} 5x + \mathbf{3} \\ \mathbf{3x} - 2 \overline{)\,15x^2 - x - 6} \\ \underline{15x^2 - 10x \phantom{{}-6}} \\ \mathbf{9x} - 6 \end{array}$$

Step 8: Multiply 3 times $3x - 2$ and place the results in the proper columns.

$$\begin{array}{r} 5x + \mathbf{3} \\ \mathbf{3x - 2} \overline{)\,15x^2 - x - 6} \\ \underline{15x^2 - 10x \phantom{{}-6}} \\ 9x - 6 \\ \mathbf{9x - 6} \end{array}$$

Step 9: Subtract $9x - 6$ from $9x - 6$.

$$\begin{array}{r} 5x + 3 \\ 3x - 2 \overline{)\,15x^2 - x - 6} \\ \underline{15x^2 - 10x \phantom{{}-6}} \\ \mathbf{9x - 6} \\ \underline{\mathbf{9x - 6}} \\ \mathbf{0} \end{array}$$

■

A N S W E R

Now, let's do some practice problems.

12. ____________________

Practice 12: Divide the following polynomials.

$$(2x^2 - 4x - 30) \div (2x + 6)$$

13. ____________________

Practice 13: Divide the following polynomials.

$$(15x^2 - 23x + 6) \div (3x - 1)$$

14. ____________________

Practice 14: Divide the following polynomials.

$$(8x^2 - 10x - 7) \div (4x - 7)$$

15. ____________________

Practice 15: Divide the following polynomials.

$$(x + 6x^2 - 15) \div (2x - 3)$$

Practice 16: Divide the following polynomials.

$$(2a - 24 + 15a^2) \div (3a + 4)$$

A N S W E R

16. ____________

Practice 17: Divide the following polynomials.

$$(12x^2 - 6 - x) \div (4x - 3)$$

17. ____________

Well, Charlie, how did you do on the practice problems?

I did okay. However, if I had not reviewed Section 3.2, I would have had trouble with some of the subtractions.

Okay, Charlie. Let's look at an example where one of the polynomials has a missing term.

Example 5 **Divide the following polynomials.**

$$(8x^2 + 7) \div (4x - 6)$$

<u>*Step 1:*</u> Before dividing, rewrite $8x^2 + 7$ in standard form inserting any missing terms.

It is difficult to divide polynomials when a term is missing.

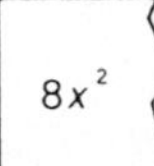

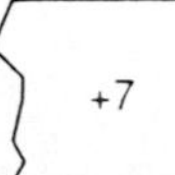

That's much better!

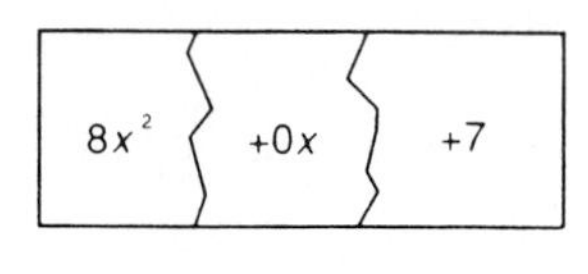

Now, divide $8x^2 + 0x + 7$ by $4x - 6$.

Step 2: Write the polynomials in division form.

$$4x - 6 \overline{)\, 8x^2 + 0x + 7}$$

Step 3: Divide $8x^2$ by $4x$.

$$\begin{array}{r} \mathbf{2x} \\ \mathbf{4x} - 6 \overline{)\, \mathbf{8x^2} + 0x + 7} \end{array}$$

Step 4: Multiply $2x$ times $4x - 6$ and place the results in the proper columns.

$$\begin{array}{r} \mathbf{2x} \\ \mathbf{4x - 6} \overline{)\, 8x^2 + \quad 0x + 7} \\ \mathbf{8x^2 - 12x} \quad\quad \end{array}$$

Step 5: Subtract $8x^2 - 12x$ from $8x^2 + 0x$. Now,

$$\begin{array}{r} 8x^2 + \quad 0x \\ -\ \underline{8x^2 - 12x} \end{array}$$

can be broken into two separate subtractions.

$$\begin{array}{r} 0x \\ -\ \underline{-12x} \\ 12x \end{array} \quad \text{or} \quad \begin{aligned} 0x - (-12x) &= 0x + 12x \\ &= 12x \end{aligned}$$

Subtraction 1

$$\begin{array}{r} 8x^2 \\ -\ \underline{8x^2} \\ 0 \end{array}$$

Subtraction 2

Thus,

$$\begin{array}{r} 2x \quad\quad \\ 4x - 6 \overline{)\, \mathbf{8x^2} + \quad \mathbf{0x} + 7} \\ \underline{\mathbf{8x^2 - 12x}} \quad\quad \\ \mathbf{12x} \quad\quad \end{array}$$

Step 6: Bring 7 down.

$$\begin{array}{r} 2x \quad\quad \\ 4x - 6 \overline{)\, 8x^2 + \quad 0x + 7} \\ \underline{8x^2 - 12x} \quad\quad \\ 12x + \mathbf{7} \end{array}$$

A N S W E R

Step 7: Divide $12x$ by $4x$.

$$\begin{array}{rl} & 2x + \mathbf{3} \\ \mathbf{4x} - 6 & \overline{)\,8x^2 + 0x + 7} \\ & \underline{8x^2 - 12x} \\ & \mathbf{12x} + 7 \end{array}$$

Step 8: Multiply 3 times $4x - 6$ and place the results in the proper columns.

$$\begin{array}{rl} & 2x + \mathbf{3} \\ \mathbf{4x} - \mathbf{6} & \overline{)\,8x^2 + 0x + 7} \\ & \underline{8x^2 - 12x} \\ & 12x + 7 \\ & \mathbf{12x - 18} \end{array}$$

Step 9: Subtract $12x - 18$ from $12x + 7$.

$$\begin{array}{rl} & 2x + 3 \\ 4x - 6 & \overline{)\,8x^2 + 0x + 7} \\ & \underline{8x^2 - 12x} \\ & \mathbf{12x + 7} \\ & \underline{\mathbf{12x - 18}} \\ & \mathbf{25} \end{array}$$

Remember, $7 - (-18) = 7 + 18 = 25$.

Step 10: Write the answer.

$$(8x^2 + 7) \div (4x - 6) = 2x + 3 + \frac{25}{4x - 6}$$ ■

Now, let's do some practice problems with missing terms.

Practice 18: Divide the following polynomials.

$$(4x^2 - 9) \div (2x + 3)$$

18. ______________________

A N S W E R

19. ____________________

Practice 19: Divide the following polynomials.

$$(12x^2 - 12) \div (4x - 4)$$

20. ____________________

Practice 20: Divide the following polynomials.

$$(8x^2 - 50) \div (4x - 10)$$

EXERCISE 3.4

Divide the following polynomials.

A

1. $(8x + 4) \div 4$

2. $(7x + 14) \div 7$

3. $(12a + 8) \div 4$

4. $(50a - 25) \div 5$

5. $(3x^2 - 9x + 27) \div 3$

6. $(x^2 + 5x + 6) \div (x + 2)$

7. $(x^2 + 8x + 7) \div (x + 7)$

8. $(35x^2 + 31x + 6) \div (5x + 3)$

9. $(6x^2 + 17x + 5) \div (3x + 1)$

10. $(8x^2 + 30x + 18) \div (4x + 3)$

B

11. $(6x^2 - 16x - 6) \div (3x + 1)$

12. $(15x^2 - 21x + 6) \div (5x - 2)$

13. $(6x^2 - 23x + 20) \div (2x - 5)$

14. $(3a^2 + 13a - 30) \div (a + 6)$

15. $(6x^2 - 19x + 15) \div (2x - 3)$

16. $(35x^2 + 6x - 8) \div (7x + 4)$

17. $(20x^2 - 51x + 28) \div (5x - 4)$

18. $(16x^2 + 42x - 18) \div (2x + 6)$

19. $(21x^2 - 13x + 2) \div (3x - 1)$

20. $(24x^2 - 28x - 40) \div (6x + 5)$

C

21. $(7a + 3a^2 - 6) \div (a + 4)$

22. $(-17x + 6x^2 + 5) \div (3x - 1)$

23. $(x^2 - 9) \div (x - 3)$

24. $(y^2 - 9) \div (y + 4)$

25. $(-3y - 28 + 6y^2) \div (2y - 7)$

26. $(3x - 6 + 63x^2) \div (3 + 9x)$

27. $(-35x^2 + 26x + 48) \div (8 - 5x)$

28. $(2x^2 - x) \div (x + 1)$

3.5 STRATEGIES FOR SOLVING APPLICATIONS

To be able to change an English phrase into an algebraic expression is a very important skill that will help you set up equations when working word problems in later chapters.

To change an English phrase into an algebraic expression, you need to know what the key words mean in the English phrase. For example,

A number increased by 1 means $x + 1$

5 less than a number means $x - 5$

The product of a number and 3 means $3x$

The quotient of a number and 4 means $\frac{x}{4}$

2 times a number decreased by 6 means $2x - 6$

The following two examples are slightly more difficult.

Example 1 **How many inches are in x feet?**

Solution: There are 12 inches in 1 foot. Thus, there are $12x$ inches in x feet. ■

Example 2 **If a dress cost x dollars and it was discounted 25%, what amount do you save? What is the selling price?**

Solution: Since the dress cost x dollars, to find the discount amount (savings) you would take 25% of x—that is, $.25x$. The selling price equals the original price minus the discount—that is, $x - .25x$. ■

APPLICATION EXERCISES

Change the following English phrases into algebraic expressions.

1. 5 added to a number
2. The sum of 15 and a number
3. 11 less than a number
4. The difference between 12 and a number
5. A number increased by 3
6. A number decreased by 9
7. The product of a number and 8
8. 3 times a number increased by 7
9. The quotient of a number and -2
10. 10 times a number decreased by 4
11. 16 decreased by two times a number
12. 5 less 3 times a number

Write an algebraic expression for each of the following.

13. If a gallon of unleaded gasoline costs x cents and the price goes up 5 cents a gallon, what is the new price?

14. The length of a rectangle is 5 feet longer than its width. If w is the width, what is the length?

15. How many feet are in x yards?

16. Daniel scored x points in 4 basketball games. What was his average score?

17. A train travels at an average rate of x miles per hour. A second train averages 7 miles faster per hour than the first train. What is the second train's average rate?

18. Susan rode a bike x miles in 5 hours. What was the average speed that Susan traveled?

19. If a jacket cost c dollars, and it was discounted 20%, what amount do you save?

20. What is the new selling price of the jacket in Exercise 19?

21. What is the difference between the square of a number and 4?

USEFUL INFORMATION

1. Any combination of letters and numbers related by addition, subtraction, multiplication, or division is an **algebraic expression.**
2. An algebraic expression consisting of one or more distinct parts connected by plus or minus signs is called an **algebraic sum.** Each distinct part, *together with its sign,* is a **term** of the expression.
3. A **polynomial** is an algebraic sum consisting of one or more terms in which the exponents of the letters are whole numbers and no letter occurs in a denominator.
4. If a polynomial consists of just *one* term, it is a **monomial.**
5. A polynomial consisting of *two* terms is a **binomial.**
6. A polynomial is in **standard form** when the exponents of the variables decrease from left to right. It is the same way an arithmetic expression is written when using expanded notation. For example, $5x^2 + 3x + 2$ is in standard form, while $3x + 2 + 5x^2$ is *not* in standard form.
7. **Basic rule for algebraic addition and subtraction:** Only like things can be added or subtracted.
8. Use the rules for signed numbers when adding, subtracting, multiplying, or dividing polynomials.
   a. **To add two numbers with like signs,** ignore the signs and add the two numbers giving the sum the common sign. *Example:* $-5 + (-7) = -12$
   b. **To add two numbers with unlike signs,** ignore the signs and find the difference of the two numbers giving the difference the sign of the larger number. *Example:* $8 + (-12) = -4$
   c. **To subtract two numbers,** change the subtraction to additon and apply the addition rules. *Example:* $8 - (-12) = 8 + 12 = 20$ also $5 - 7 = 5 + (-7) = -2$
   d. **To multiply two numbers with like signs,** ignore the signs and multiply the two numbers giving the product a positive sign. *Example:* $(-5) \times (-3) = +15$
   e. **To multiply two numbers with unlike signs,** ignore the signs and multiply the two numbers giving the product a negative sign. *Example:* $7 \times (-3) = -21$
   f. **To divide two numbers with like signs,** ignore the signs and divide the two numbers giving the quotient a positive sign. *Example:* $(-12) \div (-3) = 4$
   g. **To divide two numbers with unlike signs,** ignore the signs and divide the two numbers giving the quotient a negative sign. *Example:* $(-15) \div 5 = -3$
9. Neither regrouping, borrowing, nor carrying is possible when adding, subtracting, multiplying, or dividing polynomials.
10. When adding or subtracting polynomials vertically, like terms must be in the same column.
11. Polynomials should be in *standard form* before adding, subtracting, multiplying, or dividing.
12. The commutative, associative, and distributive properties hold in algebra.

Commutative property: $a + b = b + a$ and $a \times b = b \times a$
Associative property: $a + (b + c) = (a + b) + c$ and $a \times (b \times c) = (a \times b) \times c$
Distributive property: $a \times (b + c) = (a \times b) + (a \times c)$

13. $x^1 = x$, $x^2 = x(x)$, $x^3 = x(x^2) = x[x(x)]$
14. Before dividing polynomials, it is helpful to insert any missing terms. Example: $6x^2 + 7$ divided by $3x - 6$ should be rewritten as $6x^2 + 0x + 7$ divided by $3x - 6$.

REVIEW PROBLEMS

Section 3.1

A

Simplify the following polynomials.

1. $-8xy + 3xy + 2x + (-8x)$
2. $-4a + 3b + (-6a) + (-7b)$
3. $-5x + 2x + (-7y) + (-10x) + 9y$
4. $(2x + 3) + (5x + 1)$
5. $(3x - 12) + (7x - 8)$
6. $(9x + 15) + (7x - 12)$
7. $(15x - 18) + (13x - 14)$
8. $(22y + 21) + (8y - 11)$
9. $(5z - 17) + (4z - 13)$
10. $(8a + 23) + (7a + 17)$

B

11. $(z^2 + 8z + 5) + (3z^2 + z + 1)$
12. $(3x^2 + 2x + 7) + (4x^2 + 3x + 1)$
13. $(4y^2 - 6y + 3) + (y^2 - 2y + 1)$
14. $(6x^2 - 4x + 2) + (9x^2 - 6x + 3)$
15. $(4y^2 - 3y + 2) + (5y^2 + 12y - 4)$
16. $(5a^2 - 14a - 9) + (15a^2 - 6a - 6)$

17. $(12x^2 - 23x + 14) + (8x^2 + 13x - 19)$

18. $(8 - 6x^2 + 3x) + (7x^2 - 8x - 15)$

19. $(7x^2 + 3x + 6) + (-3x^2 - 6)$

20. $(6x^2 - 1) + (7x^2 + 6x - 2)$

Section 3.2

A

Simplify the following polynomials.

1. $12x + 3y - 5x - 2y$

2. $8xy - (-7x) - 15xy + 8x$

3. $21a - 17ab - (-8a) - 5a + 12ab$

4. $-7x^2y - (-12xy) + 15x^2y - 15xy + 5xy$

5. $(4a + 9) - (3a + 6)$

6. $(2x + 5) - (x + 3)$

7. $(10a - 3) - (2a + 19)$

8. $(12x + 7) - (8x - 8)$

9. $(9y - 8) - (16y - 9)$

10. $(22x - 17) - (18x + 12)$

B

11. $(8x^2 + 7x + 3) - (6x^2 + 4x + 1)$

12. $(7x^2 + 5x + 9) - (2x^2 + x + 7)$

13. $(5x^2 + 9x + 7) - (x^2 + 8x + 4)$

14. $(6x^2 + 12x + 8) - (3x^2 + 6x + 5)$

15. $(6a^2 + 7a + 6) - (2a^2 + 9a + 1)$

16. $(9x^2 + 7x + 3) - (4x^2 - 3x + 9)$

17. $(3x^2 - 8x + 5) - (2x^2 - 5x + 8)$

18. $(4x^2 + 2x - 12) - (8x^2 - 15x + 17)$

19. $(21y^2 - 18y + 7) - (15y^2 + 12y - 18)$

20. $(2a^2 + 24a - 13) - (12a^2 - 6a + 7)$

C

21. $(6x^2 + 2x) - (-3x^2 - 7x + 8)$

22. $(7x^2 - 2x + 6) - (7x^2 - 4)$

23. $(5y - 8y^2 - 17) - (8y^2 + 6y - 15)$

24. $(7 - 12y^2 - 3y) - (2y - 15 - 7y^2)$

25. $(9x^5 + x^3 - 2x^2 + 4) - (2x^5 + x^4 - 4x^3 - 3x^2)$

Section 3.3

Multiply the following polynomials.

A

1. $8(x - 1)$

2. $5(7x + 3)$

3. $-5(2a^2 - 7a + 3)$

4. $8(3y^2 - 7y - 6)$

5. $(x + 3)(2x + 3)$

6. $(a - 1)(a - 1)$

7. $(x + 5)(x + 6)$

8. $(3a + 5)(2a - 1)$

9. $(4b - 5)(3b + 6)$

10. $(5x + 4)(3x - 2)$

B

11. $(2x + 1)(3x^2 + 2x + 3)$

12. $(3x + 5)(4x^2 + 6x + 4)$

13. $(5x + 7)(6x^2 + 3x + 2)$

14. $(7x - 2)(2x^2 - 3x + 2)$

15. $(3y - 4)(4y^2 + 5y - 6)$

16. $(7a + 3)(6a^2 - 8a - 7)$

17. $(2x - 8)(5x^2 - 8x - 4)$

18. $(8y - 2)(5y^2 + 2y - 3)$

19. $(6b - 6)(3b^2 + 8b - 2)$

20. $(12x + 2)(3x^2 - 5x - 2)$

C

21. $(3x - 9)(7 - 8x^2 - 5x)$

22. $(5y - 7)(5y^2 + 7)$

23. $(1 - 2x + x^2)(2x^2 + 1)$

24. $(2x^2 + 3x - 4)(2x^2 - x + 3)$

Section 3.4

Divide the following polynomials.

A

1. $(9x + 3) \div 3$

2. $(6y + 3) \div 3$

3. $(60a - 24) \div 12$

4. $(21y^2 + 7y + 14) \div 7$

5. $(a^2 + 5a + 4) \div (a + 4)$

6. $(y^2 + 6y + 9) \div (y + 3)$

7. $(x^2 + 8x + 15) \div (x + 3)$

8. $(4x^2 + 16x + 7) \div (2x + 1)$

9. $(10x^2 + 41x + 40) \div (2x + 5)$

10. $(18x^2 + 27x + 7) \div (3x + 1)$

B

11. $(b^2 - 7b + 10) \div (b - 5)$

12. $(6x^2 + x - 15) \div (2x - 3)$

13. $(9x^2 - 9x + 2) \div (3x - 1)$

14. $(56x^2 - 3x - 20) \div (7x + 4)$

15. $(24x^2 + 16x - 30) \div (6x - 5)$

16. $(6x^2 + 3x - 63) \div (3x - 9)$

17. $(49x^2 + 7x - 2) \div (7x + 2)$

18. $(72y^2 - 21y - 15) \div (8y - 5)$

19. $(8x^2 + 8x - 5) \div (2x + 3)$

20. $(10a^2 - 9a + 3) \div (5a - 7)$

C

21. $(2a^2 + 11a - 4) \div (a + 5)$

22. $(6x^2 - 48 - 2x) \div (3x + 8)$

23. $(5 - 8x + 21x^2) \div (7x + 2)$

24. $(49x^2 - 100) \div (7x + 10)$

25. $(-1 + 8x^2) \div (2x + 1)$

ANSWERS TO PRACTICE PROBLEMS

Section 3.1

1. a,c,e
2. $10a$
3. $4a$
4. $7x + 6y$
5. $-7x + y$
6. $5x + 5$
7. $8x + 8$
8. $9x + 8$
9. $9x + 9$
10. $6x^2 + 6x + 8$
11. $5x^2 + 8x + 7$
12. $8y^2 + 6y + 8$
13. $6x^2 + 9x + 9$
14. $7x^2 + 13x + 8$
15. $12a^2 + 12a + 11$
16. $12x^2 + 15x + 13$
17. $15y^2 + 15y + 11$
18. $10x^2 + 3x + 9$
19. $7x^2 + 10x + 7$
20. $7x^2 + 8x + 9$
21. $13x^2 + 2x - 5$
22. $12y^2 + 12y - 14$
23. $7x^2 - 11x - 3$
24. $15b^2 + 1$
25. $12x^2 + 5x - 2$
26. $7x^2 - 10x + 16$
27. $-9x^2 - 3x + 12$

Section 3.2

1. $4a$
2. $-3a$
3. $-13x$
4. $5x - 12y$
5. $3x + 5y$
6. $5x + 5$
7. $5x + 2$
8. $2x + 4$
9. $3x^2 + 4x + 2$
10. $4x^2 + x + 1$
11. $3x^2 + 4x + 1$
12. $3x^2 - 4x - 6$
13. $3y^2 - 4y + 3$
14. $-6x^2 + 4x - 2$
15. $-3a^2 - 3a - 3$
16. $2x^2 - 6x - 3$
17. $-3b^2 + 12b - 8$
18. $8x^2 - 2x + 1$
19. $6x^2 - 14x + 11$
20. $2x^2 + 9x - 11$
21. $5y^2 - 2y + 15$
22. $-2x^2 + 2x + 8$
23. $-3x^2 + 6x + 1$
24. $-11a + 5b - 7c$
25. $7a - 5c$

Section 3.3

1. $6x + 3$
2. $6x + 4$
3. $6x + 9$
4. $9x^2 + 6x + 3$
5. $4x^2 + 6x + 4$
6. $8x^2 + 4x + 8$
7. $10x^2 + 35x + 15$
8. $28x^2 + 24x + 32$
9. $18x^2 + 24x + 30$
10. $14x^2 + 35x - 21$
11. $15y^2 - 12y + 24$
12. $21x^2 - 42x - 63$
13. $-20x^2 + 12x - 16$
14. $2x^2 + 7x + 3$
15. $4x^2 + 6x + 2$
16. $6x^2 + 22x + 20$
17. $12a^2 + 27a + 6$
18. $6y^2 + 25y + 14$
19. $40x^2 + 83x + 42$
20. $15x^2 + 32x - 7$
21. $15b^2 - 42b + 24$
22. $12x^2 - 62x + 56$
23. $10y^2 - 12y - 54$
24. $x^3 + 4x^2 + 7x + 6$
25. $6x^3 + 19x^2 + 31x + 14$
26. $8x^3 + 24x^2 + 22x + 30$
27. $30x^3 + 41x^2 + 47x + 8$
28. $6x^3 + 19x^2 + 22x + 8$
29. $20x^3 + 22x^2 + 46x + 24$
30. $2x^3 + 3x^2 - 5x - 6$
31. $3x^3 + 10x^2 - 28x - 15$
32. $6y^3 - 23y^2 + 32y - 16$
33. $21a^3 - 34a^2 - 48a + 16$
34. $20x^3 - 18x^2 + 16x - 6$
35. $10x^3 - 6x^2 + 40x - 24$
36. $6a^2 + 14ab + 4b^2$
37. $12a^2 - 36ab + 15b^2$
38. $25x^2 - 9y^2$

Section 3.4

1. $x + 2$
2. $3x + 2$
3. $4x + 3$
4. $2x^2 + x + 3$
5. $4x^2 + 2x + 3$
6. $x + 1$
7. $2x + 2$
8. $4x + 1$
9. $x + 2 + \dfrac{1}{2x + 3}$
10. $y + 4 + \dfrac{2}{2y + 1}$
11. $2x + 1 + \dfrac{5}{3x + 2}$
12. $x - 5$
13. $5x - 6$
14. $2x + 1$
15. $3x + 5$
16. $5a - 6$
17. $3x + 2$
18. $2x - 3$
19. $3x + 3$
20. $2x + 5$

C H A P T E R 4

Linear Equations

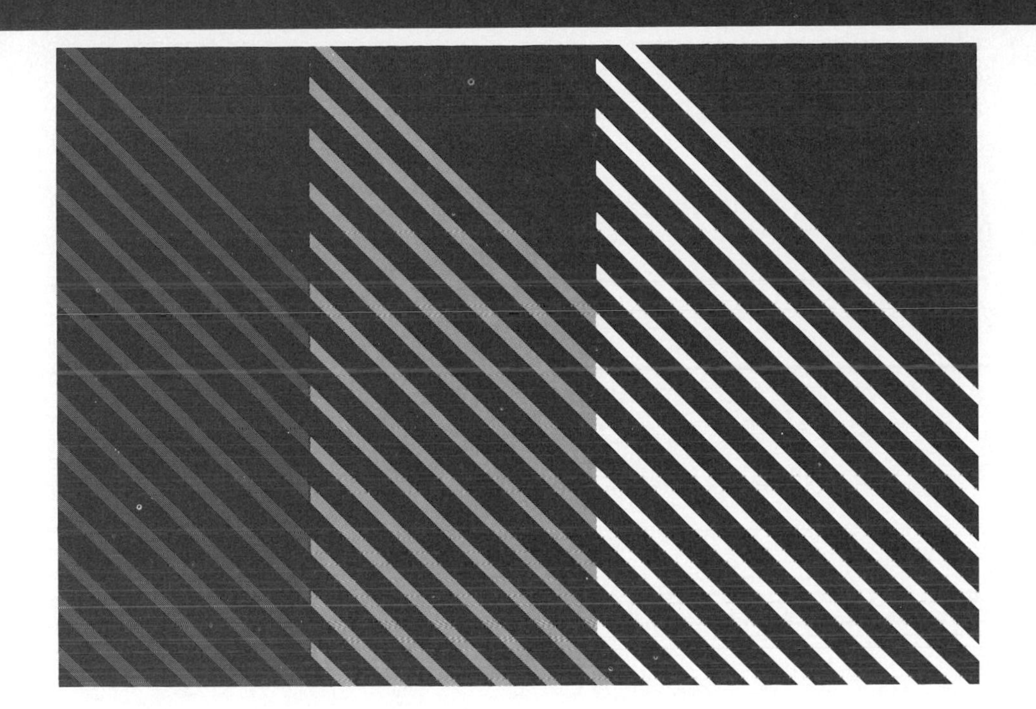

The objective of this chapter is:

1. To be able to solve linear equations in one variable.

An **equation** is a statement that two mathematical expressions are equal. The following are examples of equations.

$$x = 5$$
$$a + 3 = 7$$
$$2y - 5 = 9$$
$$3x^2 + 6x - 5 = 0$$
$$3(p^2 - 5) + 7 = 2p^2 + 4(p - 6)$$

A **linear equation** in one variable is an equation that can be written in the form $ax + b = 0$, where a and b are real numbers with $a \neq 0$, and x is a variable.

Example 1 **The following are linear equations in one variable.**

$$4x + 7 = 0$$
$$8x - 5 = 0$$
$$5y + 15 = 0$$
$$2b + 8 = 6$$
$$2t = 7$$
$$9 = 5 - 3a$$

■

How can $5y + 15 = 0$, $2b + 8 = 6$, $2t = 7$, *and* $9 = 5 - 3a$ *be written in the form* $ax + b = 0$ *when there is no* x?

That's a good point, Charlie. However, it makes no difference whether the variable is x or some other letter such as y, b, t, or a. Thus, $5y + 15 = 0$ is in the form $ax + b = 0$. Also, $2b + 8 = 6$, $2t = 7$, and $9 = 5 - 3a$ can be written in the form $ax + b = 0$ as you will find out later in the chapter.

Example 2 **The following are *not* linear equations.**

$$x^2 = 4$$
$$3a^2 - 5a + 6 = 0$$
$$-3x^3 = 4x^2 - 9$$

■

The expression on each side of the equal sign of an equation is called a **side** of the equation. For example, in the equation $2x + 3 = 5$, $2x + 3$ is the left side of the equation and 5 is the right side of the equation.

A N S W E R

> A **solution** or **root** of an equation is a number replacement for the variable(s) which makes the equation a true statement—that is, it makes the left side equal to the right.

Example 3

3 is the solution of $2x = 6$ since $2(3) = 6$

$\frac{1}{2}$ is the solution of $2t + 3 = 4$ since $2\left(\frac{1}{2}\right) + 3 = 4$

1.6 is the solution of $5y - 6 = 2$ since $5(1.6) - 6 = 2$ ■

To **solve an equation,** you find all its solutions. In the case of linear equations in one variable (the equations we will study in this chapter), there is only one solution.

Example 4 **Determine if the following number is the solution of the given equation.**

Is 3 the solution of $2x + 5 = 11$?

Solution: Replace x with 3.

$$2x + 5 = 11$$
$$2(\mathbf{3}) + 5 \stackrel{?}{=} 11$$
$$6 + 5 \stackrel{?}{=} 11$$
$$11 = 11$$

Since 11 equals 11, 3 is the solution of $2x + 5 = 11$. ■

Example 5 **Determine if the following number is the solution of the given equation.**

Is 4 the solution of $5y + 3 = 15$?

Solution: Replace y with 4.

$$5y + 3 = 15$$
$$5(\mathbf{4}) + 3 \stackrel{?}{=} 15$$
$$20 + 3 \stackrel{?}{=} 15$$
$$23 \neq 15$$

Since 23 does not equal 15, 4 is *not* the solution of $5y + 3 = 15$. ■

Practice 1: Determine if the following number is the solution of the given equation.

Is 1 the solution of $x + 3 = 4$?

1. ____________________

A N S W E R

2. ______________

Practice 2: Determine if the following number is the solution of the given equation.

Is 3 the solution of $t - 1 = 5$?

3. ______________

Practice 3: Determine if the following number is the solution of the given equation.

Is 6.5 the solution of $2x - 8 = 5$?

4. ______________

Practice 4: Determine if the following number is the solution of the given equation.

Is 5 the solution of $3a + 2 = 2a + 7$?

5. ______________

Practice 5: Determine if the following number is the solution of the given equation.

Is $\frac{-2}{5}$ the solution of $5x + 12 = 10$?

6. ______________

Practice 6: Determine if the following number is the solution of the given equation.

Is -3 the solution of $\frac{x}{3} + 8 = 2$?

7. ______________

Practice 7: Determine if the following number is the solution of the given equation.

Is $\frac{1}{4}$ the solution of $2x + \frac{5}{2} = 3$?

EXERCISE 4.0

Determine if the following numbers are solutions of the given equations.

A

1. Is 1 the solution of $x - 5 = -4$?

2. Is 7 the solution of $x + 9 = 14$?

3. Is 3 the solution of $5a - 4 = 11$?

4. Is 10 the solution of $-4x + 20 = -20$?

5. Is 3 the solution of $-2x + 2 = 8$?

6. Is -2 the solution of $-5p - 8 = 4$?

7. Is -1 the solution of $2x - 3 = 4x - 1$?

8. Is 2 the solution of $3(x - 3) = 2(x - 1)$?

9. Is -4 the solution of $2(b + 1) = 3b - 2$?

10. Is -5 the solution of $2(3x - 5) = 7x - 5$?

B

11. Is 1.2 the solution of $5x - 4 = 2$?

12. Is $\frac{1}{2}$ the solution of $8x - 4 = 0$?

13. Is $\frac{3}{5}$ the solution of $3m - 1 = \frac{1}{5}$?

14. Is 3.8 the solution of $5x - 14 = 2x - 2.6$?

15. Is $\frac{3}{2}$ the solution of $3x - \frac{1}{2} = 4x - 2$?

16. Is 2.7 the solution of $2.5w - 2 = 3w - 3.3$?

17. Is $\frac{17}{8}$ the solution of $4x - \frac{3}{2} = \frac{2}{3}x$?

18. Is 5.28 the solution of $3.5x = 7x - 18.48$?

4.1 SOLVING LINEAR EQUATIONS OF THE FORM $x + a = b$

Two equations are **equivalent** if they have exactly the same solution(s).

Example 1 **The following equations are equivalent.**

a. $x + 1 = 6$ and $x = 5$ are equivalent.
Since 5 is the only solution of $x = 5$ and 5 is also the only solution of $x + 1 = 6$, the two equations have exactly the same solution.

b. $x + 3 = 5$ and $x + 1 = 3$ are equivalent.
Since 2 is the only solution of $x + 3 = 5$ and 2 is also the only solution of $x + 1 = 3$, the two equations have exactly the same solution. ■

To solve a linear equation we find the equivalent linear equation $x = d$. The number d is the solution of the original equation. In order to do this we need to be able to add or subtract the *same* quantity on both sides of an equation.

> **Addition and Subtraction Property of Equality:** You can change an equation into an equivalent one by adding or subtracting the *same* quantity on both sides of the equation. That is,
>
> if $a = b$ then $a + c = b + c$
> if $a = b$ then $a - c = b - c$

For example, if

$$12 = 12$$

then

$$12 + 3 = 12 + 3$$

or

$$15 = 15$$

Also,

$$12 - 3 = 12 - 3$$

or

$$9 = 9$$

Example 2 **Solve the following linear equation and check the answer.**

$$x + 3 = 7$$

Solution: Subtract 3 from both sides of the equation in order to isolate x; that is, in order to get x by itself on one side of the equation.

$x + 3 = 7$	Original equation.
$x + 3 - \mathbf{3} = 7 - \mathbf{3}$	Subtract 3 from both sides: Subtraction property of equality.
$x + 0 = 4$	Perform the subtractions.
$x = 4$	The addition property of zero.

How did you know to subtract 3 from both sides of the equation?

Well, Charlie, we had $x + 3$ on the left side of the equation and we wanted just x. In order to isolate x, we had to remove the 3 by subtracting 3 from both sides.

Check: Replace x with 4 in the original equation.

$x + 3 = 7$	Original equation.
$\mathbf{4} + 3 \stackrel{?}{=} 7$	Substitute 4 for x.
$7 = 7$	Perform the addition.

Thus, 4 is the solution. ■

A N S W E R

Now, let's work some practice problems.

Do I have to give the reason for each step when solving a linear equation?

No, Charlie. When solving linear equations, you will not need to give the reasons. The reasons are given in the examples for your understanding.

1. ____________

Practice 1: Solve the following linear equation and check your answer.

$$x + 5 = 9$$

2. ____________

Practice 2: Solve the following linear equation and check your answer.

$$y + 8 = 15$$

3. ____________

Practice 3: Solve the following linear equation and check your answer.

$$x + 23 = 10$$

4. ____________

Practice 4: Solve the following linear equation and check your answer.

$$a + 35 = 15$$

A N S W E R

Example 3 **Solve the following linear equation and check the answer.**

$$y - 2 = -5$$

Solution: Add 2 to both sides of the equation in order to isolate y; that is, in order to get y by itself on one side of the equation.

$y - 2 = -5$	**Original equation.**
$y - 2 + \mathbf{2} = -5 + \mathbf{2}$	**Add 2 to both sides: Addition property of equality.**
$y + 0 = -3$	**Perform the additions.**
$y = -3$	**Addition property of zero.**

So, we added 2 to both sides of the equation in order to isolate y.

That's correct, Charlie.

Check: Replace y with -3 in the original equation.

$y - 2 = -5$	**Original equation.**
$\mathbf{-3} - 2 \stackrel{?}{=} -5$	**Substitute -3 for y.**
$-3 + (-2) \stackrel{?}{=} -5$	$a - b = a + (-b)$
$-5 = -5$	**Perform the addition.**

Thus, -3 is the solution. ■

Practice 5: Solve the following linear equation and check your answer.

$$x - 5 = 3$$

5. ______________

Practice 6: Solve the following linear equation and check your answer.

$$y - 3 = -7$$

6. ______________

A N S W E R

7. ____________

Practice 7: Solve the following linear equation and check your answer.

$$w - 12 = -4$$

8. ____________

Practice 8: Solve the following linear equation and check your answer.

$$x - 15 = -25$$

Example 4 **Solve the following linear equation and check the answer.**

$$-7 = x - 3$$

Solution: Add 3 to both sides of the equation in order to isolate x.

	$-7 = x - 3$	**Original equation.**
	$-7 + \mathbf{3} = x - 3 + \mathbf{3}$	**Add 3 to both sides: Addition property of equality.**
	$-4 = x + 0$	**Perform the additions.**
	$-4 = x$	**Addition property of zero.**
or	$x = -4$	**Symmetric property of equality: if $a = b$, then $b = a$.**

Check: Replace x with -4 in the original equation.

$-7 = x - 3$	**Original equation.**
$-7 \stackrel{?}{=} \mathbf{-4} - 3$	**Substitute -4 for x.**
$-7 \stackrel{?}{=} -4 + (-3)$	$a - b = a + (-b)$
$-7 = -7$	**Perform the addition.**

Thus, -4 is the solution. ■

9. ____________

Practice 9: Solve the following linear equation and check your answer.

$$8 = x + 5$$

Practice 10: Solve the following linear equation and check your answer.

$$-15 = p - 12$$

A N S W E R

10. ____________

Practice 11: Solve the following linear equation and check your answer.

$$21 + x = 11$$

11. ____________

Practice 12: Solve the following linear equation and check your answer.

$$-3 + x = -14$$

12. ____________

Now, let's look at an example involving fractions.

Example 5 **Solve the following linear equation and check the answer.**

$$x - \frac{5}{2} = \frac{7}{8}$$

Solution: Add $\frac{5}{2}$ to both sides of the equation in order to isolate x.

$x - \frac{5}{2} = \frac{7}{8}$	**Original equation.**
$x - \frac{5}{2} + \mathbf{\frac{5}{2}} = \frac{7}{8} + \mathbf{\frac{5}{2}}$	**Add $\frac{5}{2}$ to both sides: Addition property of equality.**
$x + 0 = \frac{7}{8} + \frac{5}{2}$	**Perform the addition on the left side.**
$x = \frac{7}{8} + \frac{5}{2}$	**Addition property of zero.**
$x = \frac{7}{8} + \frac{20}{8}$	**Get the least common denominator.**
$x = \frac{27}{8}$ or $3\frac{3}{8}$	**Perform the addition on the right side.**

A N S W E R

This time we had to add $\frac{5}{2}$ to both sides of the equation in order to isolate x.

That's correct, Charlie.

Check: Replace x with $\frac{27}{8}$ in the original equation.

$$x - \frac{5}{2} = \frac{7}{8}$$ **Original equation.**

$$\mathbf{\frac{27}{8}} - \frac{5}{2} \stackrel{?}{=} \frac{7}{8}$$ **Substitute $\frac{27}{8}$ for x.**

$$\frac{27}{8} - \frac{20}{8} \stackrel{?}{=} \frac{7}{8}$$ **Get a common denominator.**

$$\frac{7}{8} = \frac{7}{8}$$ **Perform the subtraction.**

Thus, $\frac{27}{8}$ or $3\frac{3}{8}$ is the solution. ■

13. ______________________

Practice 13: Solve the following linear equation and check your answer.

$$x - \frac{2}{5} = \frac{8}{5}$$

14. ______________________

Practice 14: Solve the following linear equation and check your answer.

$$t - 7.5 = -8.5$$

15. ______________________

Practice 15: Solve the following linear equation and check your answer.

$$\frac{7}{6} = x + \frac{5}{3}$$

Practice 16: Solve the following linear equation and check your answer.

$$a - 12.75 = 6.30$$

A N S W E R

16. ______________

Practice 17: Solve the following linear equation and check your answer.

$$x + 5\frac{3}{8} = -4\frac{5}{12}$$

17. ______________

EXERCISE 4.1

Solve the following linear equations and check your answers.

A

1. $x + 1 = 3$

2. $x - 2 = 4$

3. $a - 5 = -1$

4. $b + 4 = 4$

5. $7 = n - 9$

6. $-20 = x + 20$

7. $a - 7 = 12$

8. $y + 11 = -5$

9. $-9 = w - 6$

10. $-11 = p + 3$

11. $x + 24 = 14$

12. $y - 25 = 30$

B

13. $5 + x = 8$

14. $10 = t - 14$

15. $21 = -6 + t$

16. $13 = -17 + w$

17. $x - 7.25 = 8.25$

18. $3.4 + y = 8.2$

19. $x - 6 = 2\frac{1}{2}$

20. $a + \frac{1}{2} = 9$

21. $6 = t - 4\frac{1}{3}$

22. $-31 = p + 14$

23. $1.2 + x = 4.9$

24. $x - 3\frac{1}{3} = 2$

25. $-\frac{1}{5} = m - 21$

26. $\frac{1}{2} + x = -\frac{7}{2}$

4.2 SOLVING LINEAR EQUATIONS OF THE FORM $ax = b$

In order to solve a linear equation of the form $ax = b$, where $a \neq 0$, we will need a second rule.

> **Multiplication and Division Property of Equality:** You can change an equation to an equivalent one by multiplying or dividing both sides of the equation by the *same* nonzero quantity. That is,
>
> if $a = b$, then $ca = cb$
>
> if $a = b$, then $\frac{a}{c} = \frac{b}{c}$, where $c \neq 0$.

ANSWER

Let's look at an example in which we use the division property of equality.

Example 1 **Solve the following linear equation and check the answer.**

$$2x = 6$$

Solution: Divide both sides of the equation by 2 in order to isolate x; that is, in order to get x by itself on one side of the equation.

$2x = 6$	**Original equation.**
$\frac{2x}{\mathbf{2}} = \frac{6}{\mathbf{2}}$	**Divide both sides by 2: Division property of equality.**
$x = 3$	**Perform the divisions.**

Why did you divide both sides of the equation by 2?

Well, Charlie, we had $2x$ on the left side of the equation and we wanted x. In order to isolate x, we divided by 2.

Check: Replace x with 3 in the original equation.

$2x = 6$	**Original equation.**
$2(\mathbf{3}) \stackrel{?}{=} 6$	**Substitute 3 for x.**
$6 = 6$	**Perform the multiplication.**

Thus, 3 is the solution. ■

Note: When isolating the variable we divide both sides of the equation by the number in front of the variable.

1. ____________________

Practice 1: Solve the following linear equation and check your answer.

$$2x = 8$$

2. ____________________

Practice 2: Solve the following linear equation and check your answer.

$$5y = 20$$

Practice 3: Solve the following linear equation and check your answer.

$$36 = 4x$$

A N S W E R

3. ______

Practice 4: Solve the following linear equation and check your answer.

$$7b = 28$$

4. ______

Example 2 **Solve the following linear equation and check the answer.**

$$15 = -5y$$

Solution: Divide both sides of the equation by -5 in order to isolate y.

	$15 = -5y$	**Original equation.**
	$\frac{15}{-5} = \frac{-5y}{-5}$	**Divide both sides by −5: Division property of equality.**
	$-3 = y$	**Perform the divisions.**
or	$y = -3$	**Symmetric property of equality: if $a = b$ then $b = a$.**

So, we divided both sides of the equation by -5 *in order to isolate* y.

That's correct, Charlie.

Check: Replace y with -3 in the original equation.

$-5y = 15$	**Original equation.**
$-5(\mathbf{-3}) \stackrel{?}{=} 15$	**Substitute −3 for y.**
$15 = 15$	**Perform the multiplication.**

Thus, -3 is the solution. ■

Practice 5: Solve the following linear equation and check your answer.

$$-3x = 12$$

5. ______

A N S W E R

6. ____________________

Practice 6: Solve the following linear equation and check your answer.

$$5a = -30$$

7. ____________________

Practice 7: Solve the following linear equation and check your answer.

$$-36 = -9x$$

8. ____________________

Practice 8: Solve the following linear equation and check your answer.

$$-15t = -45$$

9. ____________________

Practice 9: Solve the following linear equation and check your answer.

$$-x = 12$$

Well, Charlie, how did you do on the practice problems?

I did not understand Practice 9. In the equation $-x = 12$ *there was no number to divide by.*

That's a good point, Charlie. Let's take a look at Practice 9:

$-x = 12$	**Original equation.**
$(-1)x = 12$	$-a = (-1)a$
$\dfrac{-1x}{-1} = \dfrac{12}{-1}$	**Divide both sides by -1: Division property of equality.**
$x = -12$	**Perform the divisions.**

Therefore, Charlie, you needed to divide both sides by -1.

You could also have used the multiplication property of equality. That is, multiply both sides of the equation by -1.

$-x = 12$	**Original equation.**
$(-1)(-x) = (-1)12$	**Multiply both sides by -1: Multiplication property of equality.**
$x = -12$	**Perform the multiplications.**

Now, let's look at an example containing a fraction.

Example 3 **Solve the following linear equation and check the answer.**

$$4t = \frac{-12}{5}$$

Solution: Divide both sides of the equation by 4 in order to isolate t.

$4t = \frac{-12}{5}$	**Original equation.**
$\frac{4t}{\mathbf{4}} = \frac{\frac{-12}{5}}{\mathbf{4}}$	**Divide both sides by 4: Division property of equality.**
$t = \frac{\frac{-12}{5}}{\frac{4}{1}}$	**Perform the division on the left side and change 4 to $\frac{4}{1}$ on the right side.**
$t = \frac{-12}{5} \times \frac{1}{4}$	**Definition of division of fractions: $\frac{a}{b} \div \frac{c}{d} = \frac{a}{b} \times \frac{d}{c}$.**
$t = \frac{-12 \times 1}{5 \times 4}$	**Definition of multiplication of fractions: $\frac{a}{b} \times \frac{c}{d} = \frac{a \times c}{b \times d}$.**
$t = \frac{-3 \times 4 \times 1}{5 \times 4}$	**Factor -12.**
$t = \frac{-3 \times 1}{5}$	**Reduce.**
$t = \frac{-3}{5}$	**Perform the multiplication.**

Check: Replace t with $\frac{-3}{5}$ in the original equation.

$4t = \frac{-12}{5}$	**Original equation.**
$4\left(\frac{\mathbf{-3}}{\mathbf{5}}\right) \stackrel{?}{=} \frac{-12}{5}$	**Substitute $\frac{-3}{5}$ for t.**
$\left(\frac{4}{1}\right)\left(\frac{-3}{5}\right) \stackrel{?}{=} \frac{-12}{5}$	**Write 4 as $\frac{4}{1}$.**
$\frac{4 \times (-3)}{1 \times 5} \stackrel{?}{=} \frac{-12}{5}$	**Definition of multiplication of fractions: $\frac{a}{b} \times \frac{c}{d} = \frac{a \times c}{b \times d}$.**
$\frac{-12}{5} = \frac{-12}{5}$	**Perform the multiplications.**

Thus, $\frac{-3}{5}$ is the solution. ■

A N S W E R

Practice 10: Solve the following linear equation and check your answer.

$$6x = \frac{3}{4}$$

10. ____________________

11. ____________________

Practice 11: Solve the following linear equation and check your answer.

$$\frac{2}{3} = 8w$$

12. ____________________

Practice 12: Solve the following linear equation and check your answer.

$$-5x = \frac{15}{16}$$

13. ____________________

Practice 13: Solve the following linear equation and check your answer.

$$\frac{-5}{9} = -5a$$

A N S W E R

Example 4 **Solve the following linear equation and check the answer.**

$$28.5 = -3.8b$$

Solution: Divide both sides of the equation by -3.8 in order to isolate b.

$$28.5 = -3.8b \qquad \textbf{Original equation.}$$

$$\frac{28.5}{\mathbf{-3.8}} = \frac{-3.8b}{\mathbf{-3.8}} \qquad \textbf{Divide both sides by } \mathbf{-3.8}\textbf{: Division property of equality.}$$

$$-7.5 = b \qquad \textbf{Perform the divisions.}$$

or

$$b = -7.5 \qquad \textbf{Symmetric property of equality: if } \boldsymbol{a = b} \textbf{ then } \boldsymbol{b = a}.$$

Check: Replace b with -7.5 in the original equation.

$$28.5 = -3.8b \qquad \textbf{Original equation.}$$

$$28.5 \stackrel{?}{=} -3.8(\mathbf{-7.5}) \qquad \textbf{Substitute } \mathbf{-7.5} \textbf{ for } \boldsymbol{b}.$$

$$28.5 = 28.5 \qquad \textbf{Perform the multiplication.}$$

Thus, -7.5 is the solution. ■

Practice 14: Solve the following linear equation and check your answer.

$$-5t = -5.5$$

14. ____________________

Practice 15: Solve the following linear equation and check your answer.

$$105.4 = -8.5y$$

15. ____________________

Practice 16: Solve the following linear equation and check your answer.

$$8x = -6.4$$

16. ____________________

Example 5 **Solve the following linear equation and check the answer.**

$$\frac{x}{4} = 5$$

Solution: Multiply both sides of the equation by 4 in order to isolate x.

$\frac{x}{4} = 5$ **Original equation.**

$\mathbf{4}\left(\frac{x}{4}\right) = \mathbf{4}(5)$ **Multiply both sides by 4: Multiplication property of equality.**

$x = 20$ **Perform the multiplications.**

How did you know to multiply both sides of the equation by 4 in order to isolate x?

Well, Charlie, we wanted the variable, x, by itself. Since x is divided by 4, we multiplied both sides of the equation by 4 in order to remove the 4 and isolate x.

Check: Replace x with 20 in the original equation.

$\frac{x}{4} = 5$ **Original equation.**

$\frac{\mathbf{20}}{4} \stackrel{?}{=} 5$ **Substitute 20 for x.**

$5 = 5$ **Reduce.**

Thus, 20 is the solution. ■

Example 6 **Solve the following linear equation and check the answer.**

$$\frac{1}{5}x = 7$$

Solution: Multiply both sides of the equation by 5 in order to isolate x.

$\frac{1}{5}x = 7$ **Original equation.**

$\mathbf{5}\left(\frac{1}{5}x\right) = \mathbf{5}(7)$ **Multiply both sides by 5: Multiplication property of equality.**

$x = 35$ **Perform the multiplications.**

A N S W E R

How did you know to multiply both sides of the equation by 5 in order to isolate x?

Well, Charlie, we wanted the variable, x, by itself. In this case, we had two choices: We could have multiplied both sides of the equation by 5 or divided both sides by $\frac{1}{5}$. We chose to multiply both sides by 5 since it is easier to multiply by a whole number than to divide by a fraction.

Check: Replace x with 35 in the original equation.

$$\frac{1}{5}x = 7 \qquad \textbf{Original equation.}$$

$$\left(\frac{1}{5}\right)(\mathbf{35}) \stackrel{?}{=} 7 \qquad \textbf{Substitute 35 for } x.$$

$$7 = 7 \qquad \textbf{Perform the multiplication.}$$

Thus, 35 is the solution. ■

Note: The equation $\frac{1}{5}x = 7$ is equivalent to the equation $\frac{x}{5} = 7$, since $\frac{1}{5}x = \frac{x}{5}$.

Practice 17: Solve the following linear equation and check your answer.

$$\frac{1}{7}x = 11$$

17. ______________

Practice 18: Solve the following linear equation and check your answer.

$$-15 = \frac{1}{3}x$$

18. ______________

ANSWER

Practice 19: Solve the following linear equation and check your answer.

$$\frac{x}{3} = \frac{-5}{6}$$

19. ____________________

20. ____________________

Practice 20: Solve the following linear equation and check your answer.

$$\frac{x}{(-10)} = \frac{-3}{5}$$

How would we solve an equation like $\frac{5x}{4} = 10$?

That's a good question, Charlie. Let's solve $\frac{5x}{4} = 10$ as an example.

Example 7 **Solve the following linear equation and check the answer.**

$$\frac{5x}{4} = 10$$

Step 1: Multiply both sides of the equation by 4 in order to isolate $5x$.

$\frac{5x}{4} = 10$	**Original equation.**
$\mathbf{4}\left(\frac{5x}{4}\right) = \mathbf{4}(10)$	**Multiply both sides by 4: Multiplication property of equality.**
$5x = 40$	**Perform the multiplications.**

ANSWER

Step 2: Divide both sides of the resulting equation by 5 in order to isolate x.

$5x = 40$	**Equation from Step 1.**
$\frac{5x}{5} = \frac{40}{5}$	**Divide both sides by 5: Division property of equality.**
$x = 8$	**Perform the divisions.**

Check: Replace x with 8 in the original equation.

$\frac{5x}{4} = 10$	**Original equation.**
$\frac{5(8)}{4} \stackrel{?}{=} 10$	**Substitute 8 for x.**
$\frac{40}{4} \stackrel{?}{=} 10$	**Perform the multiplication.**
$10 = 10$	**Reduce.**

Thus, 8 is the solution. ■

Would we solve $\frac{5}{4}x = 10$ the same way that we solved $\frac{5x}{4} = 10$?

Yes, Charlie. The equation $\frac{5}{4}x = 10$ is equivalent to the equation $\frac{5x}{4} = 10$, since $\frac{5}{4}x = \frac{5x}{4}$.

Practice 21: Solve the following linear equation and check your answer.

$$\frac{3x}{5} = 12$$

21. ____________

Practice 22: Solve the following linear equation and check your answer.

$$\frac{5}{6}x = \frac{-2}{3}$$

22. ____________

A N S W E R

23. ______________

Practice 23: Solve the following linear equation and check your answer.

$$-6 = \frac{-3x}{8}$$

24. ______________

Practice 24: Solve the following linear equation and check your answer.

$$\frac{-2}{3}x = 8$$

EXERCISE 4.2

Solve the following linear equations and check your answers.

A

1. $4x = 20$

2. $3a = 18$

3. $-5x = 30$

4. $-3t = -12$

5. $2a = 32$

6. $-5x = -35$

7. $-30 = -6p$

8. $15 = 3y$

9. $24 = -6x$

10. $-5 = -x$

B

11. $\frac{1}{3}x = 7$

12. $2.4b = -12$

13. $-5x = 0$

14. $6x = 15$

15. $10 = \frac{5}{6}x$

16. $\frac{7}{4} = -\frac{1}{2}x$

17. $-\frac{2}{3}a = -4$

18. $28 = -3.2y$

19. $\frac{t}{7} = 3$

20. $8 = \frac{x}{-3}$

21. $-20 = \frac{5}{7}x$

22. $5w = -9.5$

23. $\frac{x}{8} = \frac{11}{16}$

24. $-6 = -\frac{3}{4}y$

25. $5.2a = -44.2$

26. $18.6 = -1.2b$

27. $\frac{x}{6} = \frac{5}{12}$

28. $-\frac{3}{4} = \frac{a}{12}$

29. $6 = -\frac{4}{5}x$

30. $\frac{5}{2}x = -20$

31. $\frac{4}{3}a = 12$

32. $-\frac{2}{3}x = 10$

C

33. $\frac{2x}{7} = 6$

34. $\frac{5t}{3} = -20$

35. $-12 = \frac{3y}{5}$

36. $21 = \frac{7w}{5}$

37. $-\frac{3a}{8} = 6$

38. $-\frac{2x}{3} = -4$

39. $-12 = -\frac{6x}{5}$

40. $\frac{5}{6} = \frac{2x}{-3}$

41. $\frac{3b}{-4} = -\frac{5}{8}$

42. $\frac{4t}{-5} = \frac{8}{15}$

4.3 SOLVING LINEAR EQUATIONS OF THE FORM $ax + b = c$

To solve a linear equation of the form $ax + b = c$, we will need to use the techniques from both Section 4.1 and 4.2.

Example 1 **Solve the following linear equation and check the answer.**

$$3x + 9 = 4$$

Step 1: Subtract 9 from both sides of the equation in order to isolate $3x$.

$3x + 9 = 4$	**Original equation.**
$3x + 9 - \mathbf{9} = 4 - \mathbf{9}$	**Subtract 9 from both sides: Subtraction property of equality.**
$3x = -5$	**Simplify.**

A N S W E R

Step 2: Divide both sides of the resulting equation by 3 in order to isolate x.

$3x = -5$	**Equation from Step 1.**
$\dfrac{3x}{3} = \dfrac{-5}{3}$	**Divide both sides by 3: Division property of equality.**
$x = \dfrac{-5}{3}$ or $-1\dfrac{2}{3}$	**Perform the divisions.**

Check: Replace x with $\frac{-5}{3}$ in the original equation.

$3x + 9 = 4$	**Original equation.**
$3\left(\dfrac{-5}{3}\right) + 9 \stackrel{?}{=} 4$	**Substitute $\frac{-5}{3}$ for x.**
$-5 + 9 \stackrel{?}{=} 4$	**Perform the multiplication.**
$4 = 4$	**Perform the addition.**

Thus, $\frac{-5}{3}$ or $-1\frac{2}{3}$ is the solution. ■

Why did you subtract 9 from both sides first instead of dividing both sides by 3?

Well, Charlie, it is easier if you get $3x$ by itself before dividing by 3. Let's make a note of this.

Note: When solving equations of the form $ax + b = c$, you should first isolate the x *term* (ax).

Practice 1: Solve the following linear equation and check your answer.

$$2x + 7 = 15$$

1. ____________

Practice 2: Solve the following linear equation and check your answer.

$$5y + 7.5 = 15.25$$

2. ____________

A N S W E R

3. ____________________

4. ____________________

Practice 3: Solve the following linear equation and check your answer.

$$15 = 8x + 31$$

Practice 4: Solve the following linear equation and check your answer.

$$6a + 48 = 16$$

Example 2 **Solve the following linear equation and check the answer.**

$$-.4p - 2 = 8$$

Step 1: Add 2 to both sides of the equation in order to isolate $-.4p$.

$-.4p - 2 = 8$	**Original equation.**
$-.4p - 2 + \mathbf{2} = 8 + \mathbf{2}$	**Add 2 to both sides: Addition property of equality.**
$-.4p = 10$	**Simplify.**

Step 2: Divide both sides of the resulting equation by $-.4$ in order to isolate p.

$-.4p = 10$	**Equation from Step 1.**
$\frac{-.4p}{\mathbf{-.4}} = \frac{10}{\mathbf{-.4}}$	**Divide both sides by $-.4$: Division property of equality.**
$p = -25$	**Perform the divisions.**

Check: Replace p with -25 in the original equation.

$-.4p - 2 = 8$	**Original equation.**
$-.4(\mathbf{-25}) - 2 \stackrel{?}{=} 8$	**Substitute -25 for p.**
$10 - 2 \stackrel{?}{=} 8$	**Perform the multiplication.**
$8 = 8$	**Perform the subtraction.**

Thus, -25 is the solution. ■

Practice 5: Solve the following linear equation and check your answer.

$$5x - 3.5 = 6.5$$

A N S W E R

5. ________________

Practice 6: Solve the following linear equation and check your answer.

$$4.4 - .6x = 35$$

6. ________________

Practice 7: Solve the following linear equation and check your answer.

$$-17 = 4b - 3$$

7. ________________

Practice 8: Solve the following linear equation and check your answer.

$$6 - 8w = -25$$

8. ________________

Example 3 **Solve the following linear equation and check the answer.**

$$3 = \frac{-5x}{8} + 6$$

Step 1: Subtract 6 from both sides of the equation in order to isolate $\frac{-5x}{8}$.

$3 = \frac{-5x}{8} + 6$	**Original equation.**
$3 - \mathbf{6} = \frac{-5x}{8} + 6 - \mathbf{6}$	**Subtract 6 from both sides: Subtraction property of equality.**
$-3 = \frac{-5x}{8}$	**Simplify.**

Step 2: Multiply both sides of the resulting equation by 8 in order to isolate $-5x$.

$-3 = \frac{-5x}{8}$	**Equation from Step 1.**
$\mathbf{8}(-3) = \mathbf{8}\left(\frac{-5x}{8}\right)$	**Multiply both sides by 8: Multiplication property of equality.**
$-24 = -5x$	**Perform the multiplications.**

Step 3: Divide both sides of the resulting equation by -5 in order to isolate x.

$-24 = -5x$	**Equation from Step 2.**
$\frac{-24}{\mathbf{-5}} = \frac{-5x}{\mathbf{-5}}$	**Divide both sides by −5: Division property of equality.**
$4.8 = x$	**Perform the divisions.**
or $x = 4.8$	**Symmetric property of equality: if $a = b$ then $b = a$.**

Check: Replace x with 4.8 in the original equation.

$3 = \frac{-5x}{8} + 6$	**Original equation.**
$3 \stackrel{?}{=} \frac{-5(\mathbf{4.8})}{8} + 6$	**Substitute 4.8 for x.**
$3 \stackrel{?}{=} \frac{-24}{8} + 6$	**Perform the multiplication.**
$3 \stackrel{?}{=} -3 + 6$	**Perform the division.**
$3 = 3$	**Perform the addition.**

Thus, 4.8 is the solution. ■

Practice 9: Solve the following linear equation and check your answer.

$$\frac{1}{5}x + 3 = -2$$

A N S W E R

9. ________________

Practice 10: Solve the following linear equation and check your answer.

$$\frac{-t}{3} - 7 = -2$$

10. ________________

Practice 11: Solve the following linear equation and check your answer.

$$\frac{3}{4}x - 12 = 3$$

11. ________________

Practice 12: Solve the following linear equation and check your answer.

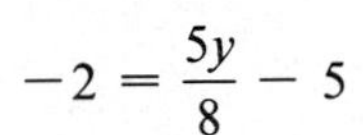

$$-2 = \frac{5y}{8} - 5$$

12. ________________

EXERCISE 4.3

Solve the following linear equations and check your answers.

A

1. $2x + 5 = 19$

2. $5a - 1 = 39$

3. $22 = 3b + 1$

4. $19 = 7x - 2$

5. $3x + 15 = 9$

6. $5y - 8 = 7$

7. $13 = 2x + 25$

8. $17 = 5t - 18$

9. $-3a + 18 = 0$

10. $-4x - 5 = 23$

B

11. $2 - 4x = 18$

12. $4 = 3y - 11$

13. $19 = 12 - x$

14. $11 = 5 + 2x$

15. $5 - 2x = 11$

16. $3b - 15 = 2$

17. $-17 = 8 - 5x$

18. $11 = 7 - 3x$

19. $5y - 18 = -22$

20. $-14 = 5x + 18$

C

21. $\frac{3}{2}x - 5 = 10$

22. $5.8y - 7 = 30.7$

23. $3.6 - 8a = -70$

24. $\frac{5x}{3} + 2 = 8$

25. $0 = \frac{2}{3}x + 8$

26. $15.6 = 5.2w - 26$

27. $-\frac{3x}{2} + 10 = -5$

28. $\frac{3y}{4} + 15 = 0$

29. $17 = \frac{3x}{5} - 19$

30. $7 = \frac{3x}{4} - 2$

4.4 SOLVING LINEAR EQUATIONS USING TRANSPOSITION (Optional)

I left out some steps when solving some of the equations in the last section. Is that okay?

Perhaps, Charlie. Let's solve an equation and find out which steps you left out.

Example 1 **Solve the following linear equation.**

$$x - 5 = 6$$

Solution: Add 5 to both sides of the equation in order to isolate x.

	$x - 5 = 6$	**Original equation.**
1.	$x - 5 + 5 = 6 + 5$	**Add 5 to both sides: Addition property of equality.**
2.	$x + 0 = 6 + 5$	**Perform the addition on the left side.**
3.	$x = 6 + 5$	**Addition property of zero.**
4.	$x = 11$	**Perform the addition on the right side.**

■

Now, which steps did you leave out?

I left out Steps 1 and 2 and just did Steps 3 and 4.

Well, Charlie, what you did was use the transposition rule without realizing it. Let's write the transposition rule.

> **Transposition Rule:** In an equation, you can move any term from one side to the other (transpose) if you also change its sign.

Since the transposition rule can often make our work easier, let's take a look at why it is true.

	$ax + \boxed{b} = c$	**Original equation.**
1.	$ax + b + (-b) = c + (-b)$	**Add $-b$ to both sides: Addition property of equality.**
2.	$ax + 0 = c + (-b)$	**The sum of a number and its opposite is zero.**
3.	$ax = c + \boxed{-b}$	**Addition property of zero.**

As you can see, we started with b on the left side of the equation and after adding $-b$ to both sides, we have $-b$ on the right side of the equation. Thus, the transposition rule allows us to omit Steps 1 and 2 of the addition process.

Say, that's pretty neat. In an equation, you can change sides with a term if you change its sign.

That's correct, Charlie. Let's look at another example.

	$ax - b = c$	**Original equation.**
1.	$ax + \boxed{-b} = c$	$a - b = a + (-b)$
2.	$ax + (-b) + b = c + b$	**Add b to both sides: Addition property of equality.**
3.	$ax + 0 = c + b$	**The sum of a number and its opposite is zero.**
4.	$ax = c + \boxed{b}$	**Addition property of zero.**

This time we had $-b$ on the left side of the equation, and after adding b to both sides of the equation, we have b on the right side of the equation.

The transposition rule is a short cut based on the addition property of equality. However, there is sometimes a danger in using short cuts; we forget the process behind the short cut. That is why the transposition rule was not introduced until after you were familar with using the addition property of equality. Let's use the transposition rule in some examples.

Example 2 **Solve the following linear equation using transposition and check the answer.**

$$x - 7 = 5$$

Solution: Transpose -7 in order to isolate x.

$x - 7 = 5$	**Original equation.**
$x + (-7) = 5$	$a - b = a + (-b)$
$x = 5 + \mathbf{7}$	**Transposition rule.**
$x = 12$	**Perform the addition.**

Check: Replace x with 12 in the original equation.

$x - 7 = 5$	**Original equation.**
$\mathbf{12} - 7 \stackrel{?}{=} 5$	**Substitute 12 for x.**
$5 = 5$	**Perform the subtraction.**

Thus, 12 is the solution. ■

A N S W E R

Can I omit changing $x - 7$ to $x + (-7)$?

Yes, Charlie, if you understand that you are transposing -7.

Example 3 **Solve the following linear equation using transposition and check the answer.**

$$2x + 22 = 12$$

Step 1: Transpose 22 in order to isolate $2x$.

$2x + \mathbf{22} = 12$	**Original equation.**
$2x = 12 + (-\mathbf{22})$	**Transposition rule.**
$2x = -10$	**Perform the addition.**

Step 2: Divide both sides of the resulting equation by 2 in order to isolate x.

$2x = -10$	**Equation from Step 1.**
$\frac{2x}{\mathbf{2}} = \frac{-10}{\mathbf{2}}$	**Divide both sides by 2: Division property of equality.**
$x = -5$	**Perform the divisions.**

Check: Replace x with -5 in the original equation.

$2x + 22 = 12$	**Original equation.**
$2(-\mathbf{5}) + 22 \stackrel{?}{=} 12$	**Substitute −5 for x.**
$-10 + 22 \stackrel{?}{=} 12$	**Perform the multiplication.**
$12 = 12$	**Perform the addition.**

Thus, -5 is the solution. ■

Practice 1: Solve the following linear equation using transposition and check your answer.

$$x + 2 = 27$$

1. ____________________

ANSWER

2. ____________

Practice 2: Solve the following linear equation using transposition and check your answer.

$$3x - 7 = -13$$

3. ____________

Practice 3: Solve the following linear equation using transposition and check your answer.

$$5x - 23 = 7$$

4. ____________

Practice 4: Solve the following linear equation using transposition and check your answer.

$$-6 = 3x + 3$$

EXERCISE 4.4

Solve the following linear equations using transposition and check your answers.

A

1. $x + 5 = 3$

2. $x - 7 = -1$

3. $y + 6 = -3$

4. $y + 12 = 8$

5. $a - 15 = -5$

6. $b + 7 = -13$

7. $2t + 3 = 11$

8. $3x - 12 = 15$

9. $5x - 12 = 23$

10. $7y + 15 = 29$

B

11. $-5a - 3 = 12$

12. $3 - 2b = -5$

13. $7 = 2x - 3$

14. $14 = 6y + 2$

15. $5 - 2x = -7$

16. $9 - 3y = -6$

17. $12 = 8 - 2x$

18. $-3 = 15 - 9x$

19. $-8 = 12 - 4x$

20. $7 = -13 - 5y$

4.5 SOLVING LINEAR EQUATIONS WITH LIKE TERMS ON OPPOSITE SIDES

In order to solve a linear equation with like terms on opposite sides of the equation, we first get the terms that do not contain the variable (the numbers) on one side and the terms containing the variable on the other side. Let's look at an example.

Example 1 **Solve the following linear equation and check the answer.**

$$5x + 3 = 3x - 7$$

Step 1: Subtract 3 from both sides of the equation in order to get the terms that do not contain the variable (the numbers) on one side.

$5x + 3 = 3x - 7$	**Original equation.**
$5x + 3 - \mathbf{3} = 3x - 7 - \mathbf{3}$	**Subtract 3 from both sides.**
$5x = 3x - 10$	**Simplify.**

Step 2: Subtract $3x$ from both sides of the resulting equation in order to isolate the terms containing the variable.

$5x = 3x - 10$	**Equation from Step 1.**
$5x - \mathbf{3x} = 3x - 10 - \mathbf{3x}$	**Subtract $3x$ from both sides.**
$2x = -10$	**Simplify.**

Step 3: Divide both sides of the resulting equation by 2 in order to isolate x.

$2x = -10$	**Equation from Step 2.**
$\frac{2x}{\mathbf{2}} = \frac{-10}{\mathbf{2}}$	**Divide both sides by 2.**
$x = -5$	**Perform the divisions.**

Check: Replace x with -5 in the original equation.

$5x + 3 = 3x - 7$	**Original equation.**
$5(\mathbf{-5}) + 3 \stackrel{?}{=} 3(\mathbf{-5}) - 7$	**Substitute -5 for x.**
$-25 + 3 \stackrel{?}{=} -15 - 7$	**Perform the multiplications.**
$-22 = -22$	**Perform the addition and subtraction.**

Thus, -5 is the solution. ■

Do we have to get the variable on the left side?

No, Charlie. We usually get the variable on the left side; however, if it is easier, we will put the variable on the right side. Let's do an example in which we get the variable on the right side.

Example 2 **Solve the following linear equation and check the answer.**

$$2x - 4 = 7x - 1$$

A N S W E R

Step 1: Add 1 to both sides of the equation in order to get the terms that do not contain the variable (the numbers) on one side.

$2x - 4 = 7x - 1$	**Original equation.**
$2x - 4 + \mathbf{1} = 7x - 1 + \mathbf{1}$	**Add 1 to both sides.**
$2x - 3 = 7x$	**Simplify.**

Step 2: Subtract $2x$ from both sides of the resulting equation in order to isolate the terms containing the variable.

$2x - 3 = 7x$	**Equation from Step 1.**
$2x - 3 - \mathbf{2x} = 7x - \mathbf{2x}$	**Subtract $2x$ from both sides.**
$-3 = 5x$	**Simplify.**

Step 3: Divide both sides of the resulting equation by 5 in order to isolate x.

$-3 = 5x$	**Equation from Step 2.**
$-\frac{3}{\mathbf{5}} = \frac{5x}{\mathbf{5}}$	**Divide both sides by 5.**
$-\frac{3}{5}$ or $-.6 = x$	**Perform the divisions.**

Check: Replace x with $-.6$ in the original equation.

$2x - 4 = 7x + 1$	**Original equation.**
$2(\mathbf{-.6}) - 4 \stackrel{?}{=} 7(\mathbf{-.6}) + 1$	**Substitute $-.6$ for x.**
$-1.2 - 4 \stackrel{?}{=} -4.2 - 1$	**Perform the multiplications.**
$-5.2 = -5.2$	**Perform the subtractions.**

Thus, $-.6$ is the solution. ■

1. ____________________

Practice 1: Solve the following linear equation and check your answer.

$$5x - 2 = 3x + 4$$

Practice 2: Solve the following linear equation and check your answer.

$$8x + 9 = 5x$$

A N S W E R

2. ______________

Practice 3: Solve the following linear equation and check your answer.

$$-3x + 5 = 2x - 15$$

3. ______________

Practice 4: Solve the following linear equation and check your answer.

$$2x + 6 = -6 - 6x$$

4. ______________

Practice 5: Solve the following linear equation and check your answer.

$$-3x - 8 = -5 + 2x$$

5. ______________

EXERCISE 4.5

Solve the following linear equations and check your answers.

1. $2x - 7 = x + 9$

2. $5 - 2b = 11 + 4b$

3. $8x - 1 = 4x + 19$

4. $1.4 + 3m = 6m - 1.6$

5. $7a - 9 = 16a$

6. $5y - 10 = -3y$

7. $.5 - x = 3x + 8.1$

8. $5y - 7 = y - 11$

9. $3x - 1 = 4x - 3$

10. $3a + 1 = a - .8$

11. $1 - x = x - 1$

12. $7t - 1 = 4t - 3$

13. $3x - 5 = 8x + 10$

14. $3x - 6 = 8x + 9$

15. $2b - 9 = 5b$

16. $4 - 2t = 2t - 3$

17. $6x + 3 = 7 - 2x$

18. $4y = 3y - 11$

19. $4a - 10 = 8 - a$

20. $2y + 9 = 5y - 6$

A N S W E R

4.6 REMOVING PARENTHESES

We will use the distributive property when removing parentheses from an expression.

Distributive Property of Multiplication Over Addition. If a, b, and c represent real numbers, then $a(b + c) = ab + ac$.

What is the purpose of parentheses?

Well, Charlie, parentheses are grouping symbols and their purpose is to group things together. In mathematics, we use parentheses much like we might use a box in everyday life. We place things that we want to keep together in a box; in mathematics, we place numbers and symbols that we want to keep together in parentheses.

Why would we want to remove parentheses?

We often need the numbers and symbols found within parentheses and to get them we have to remove the parentheses. Removing parentheses is similar to opening a box; it allows us to use what is inside. Let's look at some examples.

Example 1 **Remove the parentheses and simplify the expression.**

$$5(2x + 3)$$

Solution: Use the distributive property.

$$\begin{aligned} 5(2x + 3) &= 5(2x) + 5(3) && \textbf{Distributive property.} \\ &= 10x + 15 && \textbf{Perform the multiplications.} \end{aligned}$$

■

1. ____________________

Practice 1: Remove the parentheses and simplify the expression.

$$3(4x + 7)$$

Practice 2: Remove the parentheses and simplify the expression.

$$7(2x + 3)$$

A N S W E R

2. ______________

Practice 3: Remove the parentheses and simplify the expression.

$$5(6x + 8)$$

3. ______________

Example 2 **Remove the parentheses and simplify the expression.**

$$-2(3x + 4)$$

Solution: Use the distributive property.

$$\begin{aligned} -2(3x + 4) &= -2(3x) + (-2)(4) && \textbf{Distributive property.} \\ &= -6x + (-8) && \textbf{Perform the multiplications.} \\ &= -6x - 8 && a + (-b) = a - b \end{aligned}$$

■

Practice 4: Remove the parentheses and simplify the expression.

$$-3(4x + 6)$$

4. ______________

Practice 5: Remove the parentheses and simplify the expression.

$$-7(5x + 1)$$

5. ______________

Example 3 **Remove the parentheses and simplify the expression.**

$$-4(2x - 5)$$

Solution: Use the distributive property.

$$\begin{aligned} -4(2x - 5) &= -4[2x + (-5)] && a - b = a + (-b) \\ &= -4(2x) + (-4)(-5) && \textbf{Distributive property.} \\ &= -8x + 20 && \textbf{Perform the multiplications.} \end{aligned}$$

■

A N S W E R

6. ____________________

Practice 6: Remove the parentheses and simplify the expression.

$$2(3x - 7)$$

7. ____________________

Practice 7: Remove the parentheses and simplify the expression.

$$-5(4x - 6)$$

8. ____________________

Practice 8: Remove the parentheses and simplify the expression.

$$-3(-5x - 2)$$

9. ____________________

Practice 9: Remove the parentheses and simplify the expression.

$$5(-7x + 8)$$

10. ____________________

Practice 10: Remove the parentheses and simplify the expression.

$$-6(-3x - 5)$$

Example 4 **Remove the parentheses and simplify the expression.**

$$(-2x + 5)$$

Solution: Insert 1 and use the distributive property.

$$\begin{aligned}(-2x + 5) &= 1(-2x + 5) && a = 1 \times a\\ &= 1(-2x) + 1(5) && \text{Distributive property.}\\ &= -2x + 5 && 1 \times a = a\end{aligned}$$

■

Nothing changed when you removed the parentheses in Example 4.

That's correct, Charlie. In fact, that's the next rule.

> **Rule 1.** When you remove parentheses that have a positive sign or nothing in front of them, you simply omit the parentheses.

Example 5 **Remove the parentheses and simplify the expression.**

$$7x + 5 + (6x - 9)$$

Step 1: Remove the parentheses. Since the parentheses have a positive sign in front of them, simply omit the parentheses.

$$7x + 5 + (6x - 9) = 7x + 5 + 6x - 9 \qquad \textbf{Rule 1.}$$

Step 2: Combine like terms.

$$\begin{aligned} 7x + 5 + 6x - 9 &= 7x + 6x + 5 - 9 && \textbf{Rearrange terms.} \\ &= 13x - 4 && \textbf{Combine like terms.} \end{aligned}$$

Thus, $7x + 5 + (6x - 9) = 13x - 4$. ■

How did you know that you could rearrange the terms in Step 2?

Well, Charlie, rearranging the terms is nothing more than using the commutative and associative properties of addition.

How can you rearrange terms using the commutative and associative properties of addition when you have the subtraction 6x − 9*?*

Well, Charlie, $6x - 9$ is the same as $6x + (-9)$.

A N S W E R

Example 6 **Remove the parentheses and simplify the expression.**

$$2x + (-3x - 5)$$

Step 1: Remove the parentheses around $-3x - 5$. Since the parentheses have a positive sign in front of them, omit the parentheses.

$$2x + (-3x - 5) = 2x + (-3x) - 5$$

Step 2: Combine like terms.

$$2x + (-3x) - 5 = -x - 5$$

Thus, $2x + (-3x - 5) = -x - 5$. ■

In Step 1 you didn't just omit the parentheses around $-3x - 5$, you also placed parentheses around $-3x$. Why?

Well, Charlie, in mathematics we don't write $2x + -3x - 5$; we write $2x + (-3x) - 5$ in order not to have two signs together.

11. 

Practice 11: Remove the parentheses and simplify the expression.

$$8x + (5x + 2)$$

12. ____________________

Practice 12: Remove the parentheses and simplify the expression.

$$7x + 5 + (3x - 7)$$

Practice 13: Remove the parentheses and simplify the expression.

$$-2x + 9 + (-8x + 6)$$

A N S W E R

13. ______________

Practice 14: Remove the parentheses and simplify the expression.

$$9x - 25 + (-38x - 26)$$

14. ______________

Example 7 **Remove the parentheses and simplify the expression.**

$$-(12x - 3)$$

Solution: Insert 1 and use the distributive property.

$-(12x - 3) = -[1(12x - 3)]$	$a = 1 \times a$
$= (-1)(12x - 3)$	$-(ab) = (-a)b$
$= (-1)[12x + (-3)]$	$a - b = a + (-b)$
$= (-1)(12x) + (-1)(-3)$	Distributive property.
$= -12x + 3$	Perform the multiplications. ■

When the parentheses were removed, the sign of each term that was within the parentheses changed.

That's correct, Charlie. This leads us to the next rule.

Rule 2. When you remove parentheses that have a negative sign in front of them, the negative sign changes the sign of *each* term that was within the parentheses.

A N S W E R

Example 8 **Remove the parentheses and simplify the expression.**

$$-4x + 2 - (-2x - 7)$$

Step 1: Remove the parentheses. Since the parentheses have a negative sign in front of them, when we remove the parentheses the negative sign changes the sign of each term that was within the parentheses.

$$-4x + 2 - (-2x - 7) = -4x + 2 + 2x + 7 \quad \text{Rule 2.}$$

Step 2: Combine like terms.

$$\begin{aligned} -4x + 2 + 2x + 7 &= -4x + 2x + 2 + 7 && \text{Rearrange terms.} \\ &= -2x + 9 && \text{Combine like terms.} \end{aligned}$$

Thus, $-4x + 2 - (-2x - 7) = -2x + 9$. ■

Example 9 **Remove the parentheses and simplify the expression.**

$$3x - 5 - (4x - 9)$$

Step 1: Remove the parentheses. Since the parentheses have a negative sign in front of them, when we remove the parentheses the negative sign changes the sign of each term that was within the parentheses.

$$3x - 5 - (4x - 9) = 3x - 5 - 4x + 9 \quad \text{Rule 2.}$$

Step 2: Combine like terms.

$$\begin{aligned} 3x - 5 - 4x + 9 &= 3x - 4x - 5 + 9 && \text{Rearrange terms.} \\ &= -x + 4 && \text{Combine like terms.} \end{aligned}$$

Thus, $3x - 5 - (4x - 9) = -x + 4$. ■

15. ____________

Practice 15: Remove the parentheses and simplify the expression.

$$2x - 13 - (8x - 6)$$

16. ____________

Practice 16: Remove the parentheses and simplify the expression.

$$5x + 6 - (-4x + 7)$$

Practice 17: Remove the parentheses and simplify the expression.

$$-12x + 9 - (-15x - 25)$$

A N S W E R

17. ______

Practice 18: Remove the parentheses and simplify the expression.

$$24x + 7 - (-25x - 32)$$

18. ______

Example 10 **Remove the parentheses and simplify the expression.**

$$14x + 7 + 6(-3x + 5)$$

Step 1: Remove the parentheses around $-3x + 5$.

$$14x + 7 + 6(-3x + 5) = 14x + 7 + 6(-3x) + 6(5)$$ **Distributive property.**

$$= 14x + 7 + (-18x) + 30$$ **Perform the multiplications.**

Step 2: Combine like terms.

$$14x + 7 + (-18x) + 30 = 14x + (-18x) + 7 + 30$$ **Rearrange terms.**

$$= -4x + 37$$ **Combine like terms.**

Thus, $14x + 7 + 6(-3x + 5) = -4x + 37$. ■

Example 11 **Remove the parentheses and simplify the expression.**

$$-7x + 9 - 3(5x - 8)$$

Step 1: Remove the parentheses around $5x - 8$.

$$-7x + 9 - 3(5x - 8) = -7x + 9 + (-3)[5x + (-8)]$$

$a - b = a + (-b)$

$$= -7x + 9 + (-3)(5x) + (-3)(-8)$$

Distributive property.

$$= -7x + 9 + (-15x) + 24$$

Perform the multiplications.

In summary, we just multiplied $-3(5x - 8)$ getting $-15x + 24$.

ANSWER

Step 2: Combine like terms.

$$-7x + 9 + (-15x) + 24 = -7x + (-15x) + 9 + 24 \quad \text{Rearrange terms.}$$
$$= -22x + 33 \quad \text{Combine like terms.}$$

Thus, $-7x + 9 - 3(5x - 8) = -22x + 33$. ■

Do I have to put in each step from Example 11 when removing parentheses?

No, Charlie, you can leave out some of the steps. The extra steps were put in Example 11 for your understanding.

19. ____________

Practice 19: Remove the parentheses and simplify the expression.

$$4x + 3 + 2(-5x - 3)$$

20. ____________

Practice 20: Remove the parentheses and simplify the expression.

$$-7x + 24 - 3(8x - 4)$$

21. ____________

Practice 21: Remove the parentheses and simplify the expression.

$$32x - 17 - 5(-2x - 3)$$

Practice 22: Remove the parentheses and simplify the expression.

$$24x + 12 + 5(-7x + 2)$$

ANSWER

22. ______________

EXERCISE 4.6

Remove the parentheses and simplify each of the following expressions.

A

1. $3(2x + 1)$

2. $3(x + 1)$

3. $5(2a + 3)$

4. $7(3t + 5)$

5. $6(5b - 9)$

6. $4(3x - 5)$

7. $8(-2y - 7)$

8. $6(-5b - 4)$

9. $-2(x + 3)$

10. $-5(2a + 3)$

11. $-(2a + 1)$

12. $-(7y - 3)$

13. $-(-6x - 2)$

14. $-3(-2a - 5)$

15. $-4(-3t + 7)$

16. $-8(10b - 2)$

17. $-1(-3y + 5)$

18. $-4(-12x - 15)$

19. $-3(7 - 3x)$

20. $-6(8 - 5t)$

B

21. $3t + (t - 2)$

22. $x + (5x + 7)$

23. $5x - 7 + (8x - 8)$

24. $3a + 2 + (5a - 7)$

25. $-7b - 3 + (10b + 5)$

26. $2x - (x + 1)$

27. $y - (6 - y)$

28. $5a - 3 - (2a + 1)$

29. $7t + 8 - (12t - 7)$

30. $-5x + 7 - (-9x + 3)$

31. $-3x - 2 - (-5x - 8)$

32. $-12x - 15 - (-10x + 5)$

33. $(x - 2) - (x - 2)$

34. $2b - 3 - (2b + 1)$

35. $(\mathrm{a} + 1) - (a - 1)$

C

36. $3x - 5 + 2(5x + 7)$

37. $3a + 7 + 5(2a + 1)$

38. $7t - 8 + 3(5t - 2)$

39. $8y - 7 + 6(y - 5)$

40. $-3x - 2 + 4(-2x + 5)$

41. $7b - 1 + 8(-3b - 2)$

42. $-12y - 15 + 3(-5y - 4)$

43. $8x - 7 - 3(2x + 5)$

44. $3x + 2 - 5(6x + 1)$

45. $5a - 8 - 7(3a - 1)$

46. $-2t + 5 - 3(2t - 4)$

47. $-8y - 3 - 5(-2y - 1)$

48. $-3w - 12 - 6(-5w + 2)$

49. $5x - 17 - 4(-5x - 3)$

50. $-2(3x - 5) + 5(2x - 1)$

51. $3(-5x - 4) - 2(7x - 1)$

52. $-5(-2x + 3) - 2(5x - 7)$

4.7 SOLVING LINEAR EQUATIONS CONTAINING PARENTHESES

To solve a linear equation containing parentheses, we will use the distributive property and the rules of Section 4.6 to remove the parentheses. We will then use the rules discussed earlier in this chapter to solve the resulting equation.

Example 1 **Solve the following linear equation and check the answer.**

$$2x + 2(5x - 7) = 10$$

Step 1: Remove the parentheses around $5x - 7$.

$$2x + 2(5x - 7) = 10$$
$$2x + 10x - 14 = 10$$

Step 2: Combine like terms.

$$2x + 10x - 14 = 10$$
$$12x - 14 = 10$$

A N S W E R

Step 3: Add 14 to both sides of the resulting equation in order to isolate $12x$.

$$12x - 14 = 10$$
$$12x - 14 + \mathbf{14} = 10 + \mathbf{14}$$
$$12x = 24$$

Step 4: Divide both sides of the resulting equation by 12 in order to isolate x.

$$12x = 24$$
$$\frac{12x}{\mathbf{12}} = \frac{24}{\mathbf{12}}$$
$$x = 2$$

Check: Replace x with 2 in the original equation.

$$2x + 2(5x - 7) = 10$$
$$2(\mathbf{2}) + 2[5(\mathbf{2}) - 7] \stackrel{?}{=} 10$$
$$4 + 2(10 - 7) \stackrel{?}{=} 10$$
$$4 + 2(3) \stackrel{?}{=} 10$$
$$4 + 6 \stackrel{?}{=} 10$$
$$10 = 10$$

Thus, 2 is the solution. ■

Note: To solve an equation containing parentheses, you first remove the parentheses.

Practice 1: Solve the following linear equation and check your answer.

$$5x + 3(2x - 1) = 8$$

1. ____________

Practice 2: Solve the following linear equation and check your answer.

$$-4x + 2 - 2(3x + 2) = 18$$

2. ____________

ANSWER

3. ____________

Practice 3: Solve the following linear equation and check your answer.

$$-2x - 7 - 3(-2x - 5) = 0$$

4. ____________

Practice 4: Solve the following linear equation and check your answer.

$$5x + 3 + 5(-2x - 1) = 23$$

Example 2 **Solve the following linear equation and check your answer.**

$$4x - 6 - 2(3x - 1) = 3(2x - 4)$$

Step 1: Remove the parentheses around $3x - 1$ and $2x - 4$.

$$4x - 6 - 2(3x - 1) = 3(2x - 4)$$

$$4x - 6 - 6x + 2 = 6x - 12$$

Step 2: Combine like terms.

$$4x - 6 - 6x + 2 = 6x - 12$$

$$4x - 6x - 6 + 2 = 6x - 12$$

$$-2x - 4 = 6x - 12$$

Step 3: Add 4 to both sides of the resulting equation in order to get the terms that do not contain the variable (the numbers) on one side.

$$-2x - 4 = 6x - 12$$

$$-2x - 4 + \mathbf{4} = 6x - 12 + \mathbf{4}$$

$$-2x = 6x - 8$$

A N S W E R

Step 4: Subtract $6x$ from both sides of the resulting equation in order to isolate the terms containing the variable.

$$-2x = 6x - 8$$
$$-2x - \mathbf{6x} = 6x - 8 - \mathbf{6x}$$
$$-8x = -8$$

Step 5: Divide both sides of the resulting equation by -8 in order to isolate x.

$$-8x = -8$$
$$\frac{-8x}{\mathbf{-8}} = \frac{-8}{\mathbf{-8}}$$
$$x = 1$$

Check: Replace x with 1 in the original equation.

$$4x - 6 - 2(3x - 1) = 3(2x - 4)$$
$$4(\mathbf{1}) - 6 - 2[3(\mathbf{1}) - 1] \stackrel{?}{=} 3[2(\mathbf{1}) - 4]$$
$$4 - 6 - 2(3 - 1) \stackrel{?}{=} 3(2 - 4)$$
$$4 - 6 - 2(2) \stackrel{?}{=} 3(-2)$$
$$4 - 6 - 4 \stackrel{?}{=} -6$$
$$-6 = -6$$

Thus, 1 is the solution. ■

Practice 5: Solve the following linear equation and check your answer.

$$2(2x - 1) = 3(x + 2)$$

5. ________________

Practice 6: Solve the following linear equation and check your answer.

$$3x - 2(2x - 1) = 5(2x - 4)$$

6. ________________

A N S W E R

7. ____________________

Practice 7: Solve the following linear equation and check your answer.

$$2(-3y - 5) - 5y = y + 2$$

8. ____________________

Practice 8: Solve the following linear equation and check your answer.

$$-5(2n - 3) - 4n = -2(3n - 1) - 3$$

EXERCISE 4.7

Solve the following linear equations and check your answers.

A

1. $3(x + 2) = -3$

2. $-2(3 - y) = -7$

3. $-8 = 4 - 3(y + 1)$

4. $3(2a + 7) = 30$

5. $-(y - 3) = 10$

6. $2(y - 1) = 8$

7. $3(a + 2) - a = 4$

8. $y - (2 + 2y) = 6$

9. $x + 2(1 - x) = 3$

10. $8(2x - 5) = x - 40$

B

11. $4(x - 1) = 5(2x - 8)$

12. $2(x + 1) + 3(x - 1) = 4$

13. $4 - 2(x + 1) = 5(x - 1)$

14. $3(y + 2) - 1 = 5 - (y - 2)$

15. $5 - (y + 2) = 2y + 9$

16. $2y - (5 - y) = y - 9$

17. $x - 1 = 2[2 - (x + 1)]$

18. $y + 1 = 3[2 + (y + 1)]$

19. $1.5(6 - y) = 14 - 2.5(y + 4)$

20. $1.2(2x + 5) = 3.5(2x - 4) - 9.9$

4.8 STRATEGIES FOR SOLVING APPLICATIONS

Applications of algebra often involve setting up the equation in a word problem. In this section you will find some suggestions for setting up the linear equation in a word problem and then solving for the unknown.

1. **Read the word problem** one time to get an idea of what it is about.
2. Read the word problem a second time and ask yourself: What quantity/number am I looking for? Then **let this unknown equal a letter, usually x.** Write down your meaning for x. Now, draw a diagram if possible, and label it.
3. Read through the word problem a third time and **translate the English phrases into mathematical expressions.** The following is a list of some of the key words in this section and what they mean.

   sum means add
   increase means add
   difference means subtract
   minus means subtract
   times means multiply
   product means multiply
   twice a number means $2x$, where x is the number
   quotient means divide
   more than means add
   is means equals
   result means equal

   Sometimes translating a phrase will involve recalling or looking up a formula. For example, if a problem uses the perimeter of a rectangle, you will need to recall or look up the formula for the perimeter of a rectangle.

   Let's translate a phrase; for example, a number increased by 3 is written $x + 3$ since number is unknown and increased means add.

4. After writing the phrases as mathematical expressions, you relate them to each other to **form the equation.**
5. **Solve the equation** for the unknown.
6. Reread the problem and **make sure your answer makes sense in the application.** For example, the length of a rectangle cannot be -8 inches.

Now, let's look at some examples in which the above suggestions are applied.

Example 1 **When 15 is subtracted from the sum of a number and 6 the result is 3. Find the number then check the answer.**

Step 1: Read the problem.

Step 2: Let $x =$ the number (the unknown).

Step 3: Write the mathematical expressions using the key words.

Sum of a number and 6 is written $x + 6$
15 subtracted from the sum of a number and 6 is written $(x + 6) - 15$

Step 4: Write the linear equation. Since result means equal, the equation is

$$(x + 6) - 15 = 3$$

Step 5: Solve the equation for x.

a. Remove the parentheses.

$$(x + 6) - 15 = 3$$
$$x + 6 - 15 = 3$$

b. Combine like terms.

$$x + 6 - 15 = 3$$
$$x - 9 = 3$$

c. Isolate x.

$$x - 9 = 3$$
$$x = 3 + 9$$
$$x = 12$$

Step 6: Now, 12 is a reasonable number for this application.

Check: Replace x with 12 in the original equation.

$$(x + 6) - 15 = 3$$
$$(12 + 6) - 15 \stackrel{?}{=} 3$$
$$18 - 15 \stackrel{?}{=} 3$$
$$3 = 3$$

Thus, 12 is the number. ■

Example 2 **The length of a rectangle is 9 inches longer than its width. The perimeter of the rectangle is 50 inches. Find the width and length.**

Step 1: Read the problem.

Step 2: Let w = width. Then, $w + 9$ = length since the length is 9 inches longer than the width. Draw the rectangle and label it.

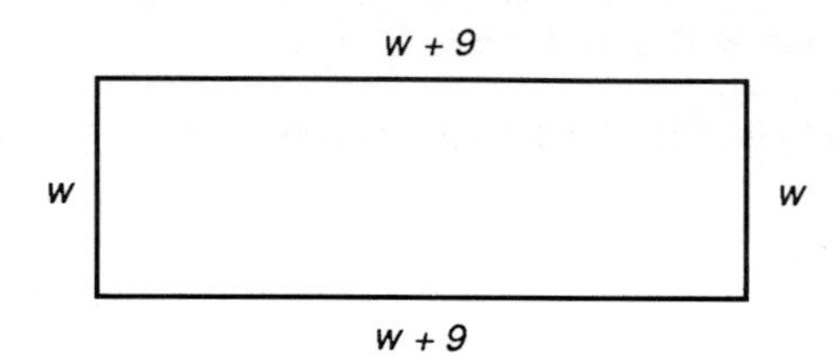

Step 3: Write the mathematical expressions using the key words. Since perimeter means distance around,

$$\text{Perimeter} = w + (w + 9) + w + (w + 9)$$

Step 4: Write the equation. It is also given that the perimeter is 50 inches, so

$$w + (w + 9) + w + (w + 9) = 50$$

since each side of the equation is the perimeter of the same rectangle.

Step 5: Solve the equation for w.

a. Remove the parentheses.

$$w + (w + 9) + w + (w + 9) = 50$$
$$w + w + 9 + w + w + 9 = 50$$

b. Combine like terms.

$$w + w + 9 + w + w + 9 = 50$$
$$4w + 18 = 50$$

c. Isolate $4w$ and combine like terms.

$$4w + 18 = 50$$
$$4w = 50 - 18$$
$$4w = 32$$

d. Divide both sides by 4.

$$4w = 32$$
$$\frac{4w}{4} = \frac{32}{4}$$
$$w = 8$$

Step 6: Now 8 is a reasonable number for the width of this rectangle.

Thus, the width is 8 inches and the length is 8 inches + 9 inches = 17 inches. ■

Example 3 **Nine years ago, Mary was $\frac{1}{2}$ as old as she is now. How old is Mary now?**

Step 1: Read the problem.

Step 2: Let x = Mary's age now.

Step 3: Write the mathematical expressions using the key words.

Mary's age nine years ago was $x - 9$

$\frac{1}{2}$ as old as she is now is written $\frac{1}{2}x$

Step 4: Write the equation. Nine years ago $(x - 9)$ Mary was $\frac{1}{2}$ as old as she is now $(\frac{1}{2}x)$ is written

$$x - 9 = \frac{1}{2}x$$

Step 5: Solve the equation for x.

a. Add 9 to both sides.

$$x - 9 = \frac{1}{2}x$$

$$x = \frac{1}{2}x + 9$$

b. Subtract $\frac{1}{2}x$ from both sides.

$$x = \frac{1}{2}x + 9$$

$$x - \frac{1}{2}x = 9$$

$$\frac{1}{2}x = 9$$

c. Multiply both sides by 2.

$$\frac{1}{2}x = 9$$

$$2\left(\frac{1}{2}x\right) = 2(9)$$

$$x = 18$$

Step 6: Now, 18 is a reasonable number for Mary's age.

Thus, Mary is 18 years old. ■

APPLICATION EXERCISES

Work the following word problems by writing an equation and then solving it.

1. The sum of 7 and a number is equal to 15. What is the number?

2. The sum of two consecutive integers is 21. What are the integers?

3. The difference between a number and 11 is 48. What is the number?

4. Two times the sum of a number and -1 is 10. What is the number?

5. The product of a number and 6 is 15 more than the number. What is the number?

6. The product of 3 and a number is 1 less than 31. Find the number.

7. If twice a number is increased by 4 times the same number, the result is 72. What is the number?

8. The sum of -4 and a number is 9 more than twice the number. Find the number.

9. The width of a rectangle is w inches. The length is 5 inches more than the width. If the perimeter of the rectangle is 70 inches, what are the width and length?

10. The quotient of a number and 2 increased by 20 is the number. Find the number.

11. A student bought two books. One book cost $4.50 more than the other. If the total cost was $10.00, what was the cost of each book?

12. In a seventh grade class of 29 students, there are 7 more girls than boys. How many boys are there? How many girls are there?

13. Rob has a wire 36 inches long. He wants it cut into two pieces so that one piece is 8 inches longer than the other. How long will each piece be?

14. A television store received 35 television sets. There were four times as many color sets as black and white sets. How many sets of each were there?

15. Lynn is 11 years older than Mary. The sum of their ages is 39. How old are Mary and Lynn?

16. Three times a number less 7 is 11. Find the number.

17. Susan's car travels at 55 m.p.h. How long will it take her to travel 330 miles?

18. Steve mixed 21 pounds of sand and gravel. He mixed twice as much sand as gravel. How much of each did he mix?

19. A basketball player made 12 shots in 20 attempts during one game. What was his percentage of shots made in that game?

20. Lisa is 7 years older than Donna. If twice the sum of their ages is 66, how old is each one?

USEFUL INFORMATION

1. An **equation** is a statement that two mathematical expressions are equal.
2. A **linear equation** in one variable is an equation that can be written in the form $ax + b = 0$, where a and b are real numbers with $a \neq 0$, and x is a variable.
3. A **solution** of an equation is a number replacement for the variable(s) which makes the equation a true statement—that is, it makes the left side equal to the right.
4. To **solve an equation,** you find all its solutions. In the case of linear equations in one variable (the equations studied in this chapter), there is only one solution.
5. Two equations are **equivalent** if they have exactly the same solution(s).
6. **Addition and subtraction property of equality:** You can change an equation to an equivalent one by adding or subtracting the *same* quantity on both sides of the equation. That is,

$$\text{if } a = b, \text{ then } a + c = b + c$$
$$\text{if } a = b, \text{ then } a - c = b - c$$

7. **Transposition rule:** In an equation, you can move any term from one side to the other (transpose) if you also change its sign.
8. **Symmetric property of equality:** If $a = b$, then $b = a$.
9. **Multiplication and division property of equality:** You can change an equation to an equivalent one by multiplying or dividing both sides of the equation by the *same* nonzero quantity. That is,

$$\text{if } a = b, \text{ then } ca = cb$$
$$\text{if } a = b, \text{ then } \frac{a}{c} = \frac{b}{c}, \text{ where } c \neq 0.$$

10. The distributive property of multiplication over addition allows us to remove parentheses in an equation.
11. When you remove parentheses that have a positive sign or nothing in front of them, simply omit the parentheses.
12. When you remove parentheses that have a negative sign in front of them, the negative sign changes the sign of *each* term that was within the parentheses.
13. To solve a linear equation containing parentheses, first remove the parentheses.
14. To solve a linear equation in one variable
    a. Remove all grouping symbols (parentheses).
    b. Get all terms with the unknown (usually x) on one side of the equality sign.
    c. Get all other terms on the opposite side of the equality sign. Combine like terms to get an equation of the form $ax = b$.
    d. Divide both sides of the equality by the coefficient of the unknown to get the solution to the equation.

REVIEW PROBLEMS

Section 4.1

Solve the following linear equations and check your answers.

A

1. $x + 3 = 5$
2. $y - 4 = 9$
3. $t + 12 = -2$
4. $x - 7 = -1$

5. $4 = x + 1$

6. $5 = 7 + x$

7. $-4 = t + 3$

8. $-7 = y - 3$

9. $x - 3 = -4$

10. $10 + x = 7$

B

11. $x - 2.5 = 8.3$

12. $a + 8.1 = -5.2$

13. $w - \frac{1}{2} = \frac{3}{2}$

14. $t + \frac{3}{8} = \frac{5}{6}$

15. $y - \frac{3}{5} = -\frac{1}{10}$

Section 4.2

Solve the following linear equations and check your answers.

A

1. $3y = 12$

2. $5b = -10$

3. $-2x = 8$

4. $-4x = -20$

5. $-18 = 3t$

6. $5x = -65$

7. $60 = -12y$

8. $-x = 15$

9. $-x = -15$

10. $-3x = 0$

B

11. $\frac{1}{2}t = 6$

12. $\frac{1}{3}x = -7$

13. $-\frac{2}{3}x = \frac{1}{2}$

14. $-2x = 5$

15. $\frac{2}{5}x = 7$

16. $\frac{5}{6}y = -4$

17. $3.5t = 21.7$

18. $7.2x = -39.6$

19. $-8.1y = -19.44$

20. $-18.2 = 4x$

21. $-93.5 = -5x$

22. $\frac{x}{-2} = -3$

23. $\frac{t}{5} = \frac{3}{5}$

24. $\frac{b}{-6} = 1.2$

25. $-5.2 = \frac{x}{-3}$

C

26. $\frac{3x}{2} = 6$

27. $\frac{5x}{8} = -\frac{1}{2}$

28. $-\frac{6a}{7} = -3$

29. $-1.5 = \frac{5t}{2}$

30. $-\frac{3}{5} = \frac{2x}{-15}$

Section 4.3

Solve the following linear equations and check your answers.

1. $3y - 4 = 8$

2. $2 + 4x = 10$

3. $2 - 3x = 17$

4. $6t - 1 = 35$

5. $4 = 2x + 10$

6. $9 - x = 7$

7. $8b - 12 = 20$

8. $-3w - 7 = -10$

9. $14 - 5x = -11$

10. $26 - 13y = 13$

B

11. $-5x + 15 = 7$

12. $23 - 8x = 13$

13. $3.5x - 8 = 6$

14. $7 = 5t - 12$

15. $-3.2 = 6.8 - 2.5x$

16. $\frac{2}{3}a - 5 = 7$

17. $\frac{3}{4}t + 17 = 2$

18. $24 = 7 - 2.5x$

19. $-15 = -8 - 3.5x$

20. $\frac{5x}{2} - 3 = 9$

21. $-\frac{4x}{5} + \frac{3}{5} = 5$

22. $4.5 = 7 - \frac{3x}{2}$

Section 4.5

Solve the following linear equations and check your answers.

1. $12x + 18 = 14x$

2. $10m - 15 = 7m$

3. $4x + 7 = x + 10$

4. $6 + 5y = 8y - 6$

5. $2y + 3 = y - 15$

6. $25 - 7x = 4 - 10x$

7. $-7 + 21y = 35$

8. $3x + 6 = 2x - 9$

9. $32 - 8x = 3 - 9x$

10. $5 + 3z = 8 - 2z$

Section 4.6

Remove the parentheses then simplify each expression.

A

1. $3(5a + 7)$

2. $7(2x + 6)$

3. $2(9t - 5)$

4. $4(5x - 6)$

5. $6(-x - 5)$

6. $3(-5y - 8)$

7. $-3(2b + 4)$

8. $-5(3w + 2)$

9. $-4(5x - 7)$

10. $-8(3x - 6)$

11. $-5(-2a - 3)$

12. $-12(-4b - 1)$

B

13. $2(2y - 1) - 5y$

14. $1 + 4(3 + x)$

15. $-3(2y + 1) - 7$

16. $-2(x - 3) + 1$

17. $7 - 4(x + 2)$

18. $3x - (2 - x)$

19. $(2x - 1) - (x + 1)$

20. $-3(x + 7) + (x - 1)$

21. $5x - (-6x - 3)$

22. $7x - 5 - (8x - 2)$

23. $-(3t - 5) - 5(-2t + 7)$

24. $8(2x - 1) - 4(-3x + 2)$

Section 4.7

Solve the following linear equations and check your answers.

A

1. $3(2x + 1) = 21$

2. $10 = 5(y - 20)$

3. $9 = 3(4y - 1)$

4. $7(2t - 1) = -35$

5. $2(3 + 4x) - 5 = 49$

6. $5(2 + 3x) - 1 = -6$

7. $5y - (2y + 8) = 16$

8. $6x - (3x + 8) = 10$

9. $3x - 4(x + 2) = 5$

10. $2x - (x + 1) = 0$

B

11. $3(2x - 1) = 5(x + 2)$

12. $3x - 5 = -7(2x - 3) + 8$

13. $5 - 2(3t - 6) = 7t - (8t - 2)$

14. $5(2a + 3) - 21 = -4(3a - 5) - 4$

15. $17 - 3(-2x - 7) = 5(-x + 5) - 2x$

ANSWERS TO PRACTICE PROBLEMS

Section 4.0

1. yes **2.** no **3.** yes **4.** yes
5. yes **6.** no **7.** yes

Section 4.1

1. 4 **2.** 7 **3.** -13 **4.** -20
5. 8 **6.** -4 **7.** 8 **8.** -10
9. 3 **10.** -3 **11.** -10 **12.** -11
13. 2 **14.** -1 **15.** $-\frac{1}{2}$ **16.** 19.05
17. $-9\frac{19}{24}$

Section 4.2

1. 4 **2.** 4 **3.** 9 **4.** 4
5. -4 **6.** -6 **7.** 4 **8.** 3
9. -12 **10.** $\frac{1}{8}$ **11.** $\frac{1}{12}$ **12.** $-\frac{3}{16}$
13. $-\frac{1}{9}$ **14.** 1.1 **15.** -12.4 **16.** $-.8$
17. 77 **18.** -45 **19.** $-\frac{5}{2}$ **20.** 6
21. 20 **22.** $-\frac{4}{5}$ **23.** 16 **24.** -12

Section 4.3

1. 4 **2.** 1.55 **3.** -2 **4.** $-5\frac{1}{3}$
5. 2 **6.** -51 **7.** -3.5 **8.** $3\frac{7}{8}$
9. -25 **10.** -15 **11.** 20 **12.** $4\frac{4}{5}$

Section 4.4

1. 25 **2.** -2 **3.** 6 **4.** -3

Section 4.5

1. 3 **2.** -3 **3.** 4 **4.** $-\frac{3}{2}$
5. $-\frac{3}{5}$

Section 4.6

1. $12x + 21$ **2.** $14x + 21$ **3.** $30x + 40$
4. $-12x - 18$ **5.** $-35x - 7$ **6.** $6x - 14$
7. $-20x + 30$ **8.** $15x + 6$ **9.** $-35x + 40$
10. $18x + 30$ **11.** $13x + 2$ **12.** $10x - 2$
13. $-10x + 15$ **14.** $-29x - 51$ **15.** $-6x - 7$
16. $9x - 1$ **17.** $3x + 34$ **18.** $49x + 39$
19. $-6x - 3$ **20.** $-31x + 36$ **21.** $42x - 2$
22. $-11x + 22$

Section 4.7

1. 1 **2.** -2 **3.** -2 **4.** -5
5. 8 **6.** 2 **7.** -1 **8.** 2

C H A P T E R 5

Integer Exponents

The objective of this chapter is:

1. **To know the basic rules of exponents and be able to apply them.**

In Section 1.4, we introduced exponents and have used them throughout our discussion. In this chapter, we will look at exponents in more detail. This study will give you the opportunity to see how exponents fit as a part of mathematics. The various parts of mathematics, like a fine work of art or piece of music, fit together beautifully.

5.1 NONNEGATIVE INTEGER EXPONENTS

In arithmetic, repeated additions such as

$$5 + 5 + 5 + 5 + 5 + 5 + 5$$

$$7 + 7 + 7 + 7 + 7 + 7 + 7 + 7$$

and

$$3 + 3 + 3 + 3 + 3$$

are used quite often. Mathematicians decided to develop an easier way to write such sums. Instead of writing

$$7 + 7 + 7 + 7 + 7 + 7 + 7 + 7$$

they counted how many sevens there were in the sum, and wrote

$$8(7)$$

This is how **multiplication** started. In the notation 8(7), 8 is often called the **coefficient** of 7. In the algebraic expression

$$3x^2 + 4x + 2$$

3 is the coefficient of x^2

and

4 is the coefficient of x

Example 1 **Write the following addition problem as a multiplication problem.**

$$7 + 7 + 7 + 7$$

Solution: Count how many sevens there are in the sum. There are 4 sevens in the sum; thus,

$$7 + 7 + 7 + 7 = 4(7)$$ ■

Example 2 **Write the following multiplication problem as an addition problem.**

$$6(2)$$

Solution: The coefficient, 6, tells us that there are 6 twos in the sum; thus,

$$6(2) = 2 + 2 + 2 + 2 + 2 + 2$$ ■

Example 3 **Write the following addition problem as a multiplication problem.**

$$x + x + x + x + x + x + x$$

Solution: Count how many x's there are in the sum. There are 7 x's in the sum; thus,

$$x + x + x + x + x + x + x = 7x$$ ■

That's pretty neat. All we did was count the number of x's *in the sum and write it down. Using coefficients could save a lot of writing.*

That's true, Charlie. It would take a long time to write the sum of 100 x's but it is no trouble to write $100x$.

Example 4 **Write the following multiplication problem as an addition problem.**

$$3y$$

Solution: The coefficient, 3, tells us that there are three y's in the sum; thus,

$$3y = y + y + y$$ ■

So, the coefficient of y *just tells us how many* y's *we have in a sum.*

That's correct, Charlie. Instead of writing a repeated addition, you write the number or letter that is added; then, in front of this number or letter you write how many times it appears in the sum.

Practice 1: Write the following addition problem as a multiplication problem.

$$5 + 5 + 5 + 5 + 5 + 5$$

1. ____________

Practice 2: Write the following addition problem as a multiplication problem.

$$a + a + a$$

2. ____________

Practice 3: Write the following multiplication problem as an addition problem.

$$5(8)$$

3. ____________

Practice 4: Write the following multiplication problem as an addition problem.

$$4x$$

4. ____________

Mathematicians found the same problem with repeated multiplication. For example, what about

$$7 \times 7 \times 7 \times 7 \times 7 \times 7 \times 7 \times 7$$

It seems that solving one math problem often creates another.

That's sometimes true, Charlie. However, solving one math problem often teaches you how to solve another. To solve the multiplication problem, we use the same technique that was used for addition. Write the number that is multiplied, 7, then write how many times it appears in the product, 8.

That's great, but where are we going to put the 8?

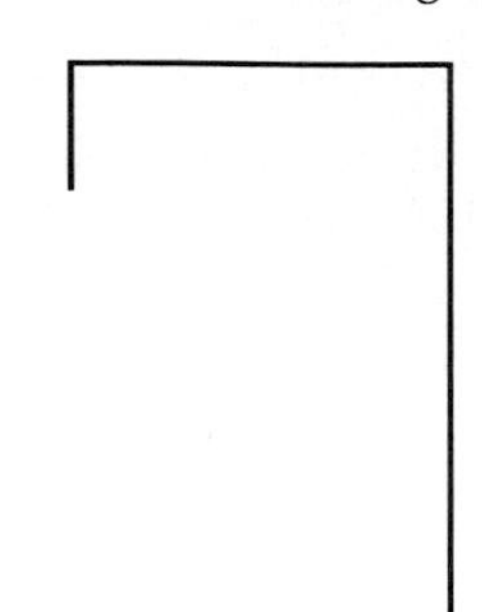

It looks like there is no place to put the 8; however, mathematicians are clever. They put the 8 on the upper right side of the 7. Now,

$$7 \times 7 \times 7 \times 7 \times 7 \times 7 \times 7 \times 7$$

is written as

$$7^8$$

where 7 is called the **base** and 8 the **exponent.**

Example 5 **Write the following multiplication problem using exponents.**

$$3 \times 3 \times 3 \times 3 \times 3$$

Solution: Count how many threes there are in the product. There are 5 threes in the product; thus,

$$3 \times 3 \times 3 \times 3 \times 3 = 3^5$$ ■

Gee! Exponents are the same as coefficients. All we did was count the number of threes in the product and write it down.

That's true, Charlie. We just write how many threes we have in the product on the upper right side of 3.

Example 6 **Write the following multiplication problem without using exponents.**

$$4^6$$

Solution: The exponent, 6, tells us that there are 6 fours in the product; thus,

$$4^6 = 4 \times 4 \times 4 \times 4 \times 4 \times 4$$ ■

A N S W E R

So, the exponent of 4 just tells us how many fours we have in a product. Using exponents could save me a lot of writing.

That's true, Charlie. It would take a long time to write the product of 100 fours, but it is no trouble to write 4^{100}.

Example 7 **Write the following multiplication problem using exponents.**

$$x \times x \times x \times x \times x$$

Solution: Count how many x's there are in the product. There are 5 x's in the product; thus,

$$x \times x \times x \times x \times x = x^5$$ ■

I always thought x^5 *was something really strange; but, it's just a way of writing the product of 5 x's.*

That's true, Charlie. Exponents are just a short way of writing a product. Let's do some practice problems.

Practice 5: Write the following multiplication problem using exponents.

$$5 \times 5 \times 5 \times 5 \times 5 \times 5$$

5. ____________________

Practice 6: Write the following multiplication problem using exponents.

$$a \times a \times a$$

6. ____________________

Practice 7: Write the following multiplication problem without using exponents.

$$8^5$$

7. ____________________

Practice 8: Write the following multiplication problem without using exponents.

8. ____________________

$$x^4$$

It's amazing how this math fits together.

Let's look at another example that shows the similarity between coefficients and exponents.

Example 8

Write the following sum using coefficients.

$$3 + 5 + 3 + 4 + 5 + 3 + 3 + 5 + 4$$

Step 1: Count the number of threes; there are 4 of them. Thus, we write 4(3).

Step 2: Count the number of fours; there are 2 of them. Thus, we write 2(4).

Step 3: Count the number of fives; there are 3 of them. Thus, we write 3(5).

Step 4: Thus,

$$3 + 5 + 3 + 4 + 5 + 3 + 3 + 5 + 4 = 4(3) + 2(4) + 3(5)$$

Write the following product using exponents.

$$3 \times 5 \times 3 \times 4 \times 5 \times 3 \times 3 \times 5 \times 4$$

Step 1: Count the number of threes; there are 4 of them. Thus, we write 3^4.

Step 2: Count the number of fours; there are 2 of them. Thus, we write 4^2.

Step 3: Count the number of fives; there are 3 of them. Thus, we write 5^3.

Step 4: Thus,

$$3 \times 5 \times 3 \times 4 \times 5 \times 3 \times 3 \times 5 \times 4 = 3^4 \times 4^2 \times 5^3$$ ■

Those two problems are basically the same. If you can count, you can use coefficients and exponents.

That's correct, Charlie. Let's look at another example.

Example 9

Write the following sum using coefficients.

$$x + y + x + x + y + z$$

Step 1: Count the number of x's; there are 3 of them. Thus, we write $3x$.

Step 2: Count the number of y's; there are 2 of them. Thus, we write $2y$.

Write the following product using exponents.

$$x \times y \times x \times x \times y \times z$$

Step 1: Count the number of x's; there are 3 of them. Thus, we write x^3.

Step 2: Count the number of y's; there are 2 of them. Thus, we write y^2.

ANSWER

Step 3: Count the number of z's; there is 1 of them. Thus, we write $1z$ or z.

Step 4: Thus,

$$x + y + x + x + y + z = 3x + 2y + 1z = 3x + 2y + z$$

Step 3: Count the number of z's; there is 1 of them. Thus, we write z^1 or z.

Step 4: Thus,

$$x \times y \times x \times x \times y \times z = x^3 \times y^2 \times z^1 = x^3 \times y^2 \times z$$

■

Note: When a letter does not have a written coefficient, the coefficient is understood to be 1. For example, $z = 1z$. When a letter does not have a written exponent, the exponent is understood to be 1. For example, $z = z^1$.

These exponents are as simple as counting 1, 2, 3.

That's right, Charlie. You should be ready to work some practice problems. When working the following practice problems be sure to note the similarity of the multiplication problem to the previous addition problem. Should you have trouble writing one of the products using exponents, refer to the previous addition problem and compare how you wrote the sum using coefficients.

Practice 9: Write the following sum using coefficients.

$$4 + 3 + 4 + 4 + 5 + 3 + 5$$

9. ______________

Practice 10: Write the following product using exponents.

$$4 \times 3 \times 4 \times 4 \times 5 \times 3 \times 5$$

10. ______________

Practice 11: Write the following sum using coefficients.

$$2 + 5 + 5 + 2 + 2 + 2 + 5 + 2 + 5 + 2$$

11. ______________

Practice 12: Write the following product using exponents.

$$2 \times 5 \times 5 \times 2 \times 2 \times 2 \times 5 \times 2 \times 5 \times 2$$

12. ______________

Practice 13: Write the following sum using coefficients.

$$x + y + x + x + y$$

13. ______________

Practice 14: Write the following product using exponents.

$$x \times y \times x \times x \times y$$

14. ______________

How did you do on the practice problems, Charlie?

I did well on the practice problems. All I did was count the number of twos, threes, fours, etc. in the sum or product and write it down.

Good! Let's look at another example.

Example 10 **Write the following sum using coefficients.**

$$2 + 2 + 5 + 4 + 4 + 2 + 5 + 5 + 5$$

Solution: Count the number of twos. There are 3 of them; therefore, we write 3(2). Now, count the number of fours. There are 2 of them; therefore, we write 2(4). Then count the number of fives. There are 4 of them; therefore, we write 4(5). Thus,

$$2 + 2 + 5 + 4 + 4 + 2 + 5 + 5 + 5 = 3(2) + 2(4) + 4(5)$$

I know how many twos, fours, and fives are in the sum. But how many threes are there?

There are no threes in the sum.

Well, how am I supposed to know that if you don't say so?

That's a good point, Charlie. If there are no threes in the sum, then there is zero of them. If you want to show this, then write 0(3). Thus,

$$\begin{aligned} 2 + 2 + 5 + 4 + 4 + 2 + 5 + 5 + 5 &= 3(2) + 2(4) + 4(5) \\ &= 3(2) + \mathbf{0(3)} + 2(4) + 4(5) \end{aligned}$$

Good! Now I can tell that you do not have any threes. But are the last two expressions the same?

Yes, they are Charlie, since $0(3) = 0$ and any number added to zero is that number $(0 + a = a)$. ■

Example 11 **Write the following product using exponents.**

$$2 \times 2 \times 5 \times 4 \times 4 \times 2 \times 5 \times 5 \times 5$$

Solution: Count the number of twos. There are 3 of them; therefore, we write 2^3. Now count the number of fours. There are 2 of them; therefore, we write 4^2. Then count the number of fives. There are 4 of them; therefore, we write 5^4. Thus,

$$2 \times 2 \times 5 \times 4 \times 4 \times 2 \times 5 \times 5 \times 5 = 2^3 \times 4^2 \times 5^4$$ ■

Notice, there are no threes in the product; therefore, there is zero of them. If you want to show this, then write 3^0. Thus,

$$\begin{aligned} 2 \times 2 \times 5 \times 4 \times 4 \times 2 \times 5 \times 5 \times 5 &= 2^3 \times 4^2 \times 5^4 \\ &= 2^3 \times 3^0 \times 4^2 \times 5^4 \end{aligned}$$

Remember, 0 is the only number that does not change the value of a sum $(0 + a = 0)$. What is the only number that does not change the value of a product?

One is the only number that does not change the value of a product.

That's correct, Charlie. One does not change the value of a product since $1 \times a = a$ for all real numbers a. Therefore, 3^0 is equal to 1. In fact, any nonzero number to the zero power is equal to 1.

Definition: $a^0 = 1, a \neq 0$.

Example 12 **Find the value of the following product.**

$$2^3 \times 5^0 \times 6^2$$

Solution:

$$2^3 = 2 \times 2 \times 2 = 8$$

$$5^0 = 1$$

$$6^2 = 6 \times 6 = 36$$

Thus,

$$\begin{aligned} 2^3 \times 5^0 \times 6^2 &= 8 \times 1 \times 36 \\ &= 8 \times 36 \\ &= 288 \end{aligned}$$ ■

When working the following practice problems, be sure to note the similarity of Practice 16 to 15 and Practice 18 to 17.

A N S W E R

15. ____________

Practice 15: Find the value of the following expression.

$$2(5) + 3(2) + 0(4)$$

16. ____________

Practice 16: Find the value of the following expression.

$$5^2 \times 2^3 \times 4^0$$

17. ____________

Practice 17: Find the value of the following expression.

$$3(2) + 0(12) + 2(3)$$

18. ____________

Practice 18: Find the value of the following expression.

$$2^3 \times (12)^0 \times 3^2$$

Once again, the similarity of exponents to coefficients was shown in the preceding practice problems. Let's look at two more examples that highlight this similarity.

Example 13 **Simplify the following sum.**

$$3x + 0(xy) + 2y$$

Solution:

$$\begin{aligned} 3x + 0(xy) + 2y &= 3x + 0 + 2y && \textbf{Multiplication property of zero.} \\ &= 3x + 2y && \textbf{Addition property of zero.} \end{aligned}$$

Thus, $3x + 0(xy) + 2y = 3x + 2y$. ■

Example 14 **Simplify the following product.**

$$x^3 \times (xy)^0 \times y^2$$

Solution:

A N S W E R

$$x^3 \times (xy)^0 \times y^2 = x^3 \times 1 \times y^2$$ **$a^0 = 1$: Any nonzero number to the 0 power is equal to 1.**

$$= x^3 \times y^2$$ **Multiplication property of one.**

Thus, $x^3 \times (xy)^0 \times y^2 = x^3 \times y^2$. ■

So, when you have zero xy's *you can just leave* xy *out of the sum or product.*

That's correct, Charlie:

$$3x + 0(xy) + 2y = 3x + 2y$$

and

$$x^3 \times (xy)^0 \times y^2 = x^3 \times y^2$$

Practice 19: Simplify the following sum.

$$2x + 0(y) + 5z$$

19. ______________

Practice 20: Simplify the following product.

$$x^2 \times y^0 \times z^5$$

20. ______________

Practice 21: Simplify the following sum.

$$5x + 7y + 0(x + y)$$

21. ______________

Practice 22: Simplify the following product.

$$x^5 \times y^7 \times (x + y)^0$$

22. ______________

Practice 23: Simplify the following sum.

$$x^2 + 3y + 0(x^2 + 5y)$$

23. ______________

Practice 24: Simplify the following product.

$$x^2 \times y^3 \times (x^2 + 5y)^0$$

24. ______________

Now, let's look at some examples to help us develop some rules for working with exponents.

Example 15

Simplify the following product.

$$a^3 \times a^2$$

Solution: $a^3 = a \times a \times a$

$a^2 = a \times a$

Thus,

$$\begin{aligned} a^3 \times a^2 &= (a \times a \times a) \times (a \times a) \\ &= a \times a \times a \times a \times a \\ &= a^5 \end{aligned}$$

Simplify the following sum.

$$3a + 2a$$

Solution: $3a = a + a + a$

$2a = a + a$

Thus,

$$\begin{aligned} 3a + 2a &= (a + a + a) + (a + a) \\ &= a + a + a + a + a \\ &= 5a \end{aligned}$$

■

Wow! Exponents work the same way as coefficients.

Yes, Charlie. In fact, when we write a rule for exponents, we will also write the corresponding rule for coefficients to show the similarity. Also, many of the examples will have a corresponding addition problem to reinforce the similarity between exponents and coefficients. Let's write Rule 1 for exponents and the corresponding rule for coefficients. This rule is based on the last example.

Rule 1 (exponents): $x^m \times x^n = x^{m+n}$

Rule 1 (coefficients): $mx + nx = (m + n)x$

Note: Rule 1 (exponents) only works when the bases are the same.

Example 16

Simplify the following product.

$$a^5 \times a^3$$

Solution: Using Rule 1 (exponents),

$x^m \times x^n = x^{m+n}$, we have

$$\begin{aligned} a^5 \times a^3 &= a^{5+3} \\ &= a^8 \end{aligned}$$

Simplify the following sum.

$$5a + 3a$$

Solution: Using Rule 1 (coefficients),

$mx + nx = (m + n)x$, we have

$$\begin{aligned} 5a + 3a &= (5 + 3)a \\ &= 8a \end{aligned}$$

■

The left column of Example 16 makes sense. If you have a product of 5 a's *times a product of 3* a's, *then combine them, you have a product of 8* a's.

That's correct, Charlie. Let's try some practice problems.

Practice 25: Simplify the following product.

$$x^2 \times x$$

Practice 26: Simplify the following product.

$$a^7 \times a^8$$

Practice 27: Simplify the following product.

$$y^5 \times y^3$$

A N S W E R

25. ______

26. ______

27. ______

Let's look at another example.

Example 17

Simplify the following product.

$$(a \times b \times b \times b \times a) \times (a \times a \times a \times b)$$

Step 1: The first product has 2 a's, and 3 b's. Thus,

$$(a \times b \times b \times b \times a) = (a^2b^3)$$

Step 2: The second product has 3 a's and 1 b. Thus,

$$(a \times a \times a \times b) = (a^3b^1)$$

Step 3: Thus,

$$(a \times b \times b \times b \times a) \times (a \times a \times a \times b) = (a^2b^3)(a^3b^1)$$

Step 4: Group like bases.

$$(a^2b^3)(a^3b^1) = (a^2a^3)(b^3b^1)$$

Step 5: Using Rule 1 (exponents), $x^m \times x^n = x^{m+n}$, we have

$$(a^2a^3)(b^3b^1) = a^{(2+3)}b^{(3+1)} = a^5b^4$$

Thus,

$$(a \times b \times b \times b \times a) \times (a \times a \times a \times b) = a^5b^4$$

Simplify the following sum.

$$(a + b + b + b + a) + (a + a + a + b)$$

Step 1: The first sum has 2 a's and 3 b's. Thus,

$$(a + b + b + b + a) = (2a + 3b)$$

Step 2: The second sum has 3 a's and 1 b. Thus,

$$(a + a + a + b) = (3a + 1b)$$

Step 3: Thus,

$$(a + b + b + b + a) + (a + a + a + b) = (2a + 3b) + (3a + 1b)$$

Step 4: Group like terms.

$$(2a + 3b) + (3a + 1b) = (2a + 3a) + (3b + 1b)$$

Step 5: Using Rule 1 (coefficients), $mx + nx = (m + n)x$, we have

$$(2a + 3a) + (3b + 1b) = (2 + 3)a + (3 + 1)b = 5a + 4b$$

Thus,

$$(a + b + b + b + a) + (a + a + a + b) = 5a + 4b$$

■

A N S W E R

All we did in Example 17 was count the number of a's *and* b's *in the products or sums.*

That's correct, Charlie.

28. ____________________

Practice 28: Simplify the following product.

$$(a \times b \times a \times a \times b)(a \times a \times b \times a \times b \times a)$$

29. ____________________

Practice 29: Simplify the following product.

$$(x \times x \times y)(x \times x \times y \times x \times y)$$

We will work the next example in a casual manner, then work it in a more formal way. It is important to reason through problems, as we do in Solution 1 (following), and it is also important to apply your rules, as we do in Solution 2 (following).

Example 18 **Simplify the following product.**

$$(8a^5b^3)(3a^2b^4)$$

Solution 1: There are 5 a's in the first product and 2 a's in the second product, for a total of 7 a's ($a^5 \times a^2 = a^{5+2} = a^7$). There are 3 b's in the first product and 4 b's in the second product, for a total of 7 b's ($b^3 \times b^4 = b^{3+4} = b^7$). Also, $8 \times 3 = 24$. Thus,

$$(8a^5b^3)(3a^2b^4) = 24a^7b^7$$

Solution 2:

Step 1: Using the commutative and associative properties of multiplication, group the coefficients together and the like bases together.

$$(8a^5b^3)(3a^2b^4) = (8 \times 3)(a^5a^2)(b^3b^4)$$

Step 2: Multiply the coefficients and use Rule 1 (exponents), $x^m \times x^n = x^{m+n}$, to combine like bases.

$$(8 \times 3)(a^5a^2)(b^3b^4) = 24a^{(5+2)}b^{(3+4)}$$
$$= 24a^7b^7$$

Thus, $(8a^5b^3)(3a^2b^4) = 24a^7b^7$. ■

Work the following practice problems using either method from the last example.

Practice 30: Simplify the following product.

$$(3a^2b^3)(5a^4b^2)$$

ANSWER

30. ____________

Practice 31: Simplify the following product.

$$(4x^3y^5)(2x^3y^2)$$

31. ____________

Practice 32: Simplify the following product.

$$(2pq^6)(3p^2q)$$

32. ____________

Practice 33: Simplify the following product.

$$(-5a^2b^3c)(2ab^4c^5)$$

33. ____________

Example 19 **Simplify the following product.**

$$5a^2b^3(3a^3b + 7a^2b^2)$$

Step 1: Use the distributive property.

$$5a^2b^3(3a^3b + 7a^2b^2) = 5a^2b^3(3a^3b) + 5a^2b^3(7a^2b^2)$$

Step 2: Use the commutative and associative properties of multiplication to group the coefficients and like bases together.

$$5a^2b^3(3a^3b) + 5a^2b^3(7a^2b^2) = (5 \times 3)(a^2a^3)(b^3b) + (5 \times 7)(a^2a^2)(b^3b^2)$$

Step 3: Multiply the coefficients together and use Rule 1 (exponents), $x^m \times x^n = x^{m+n}$, to combine like bases.

$$\begin{aligned}(5 \times 3)(a^2a^3)(b^3b) + (5 \times 7)(a^2a^2)(b^3b^2) &= 15a^{2+3}b^{3+1} + 35a^{2+2}b^{3+2} \\ &= 15a^5b^4 + 35a^4b^5\end{aligned}$$

Thus, $5a^2b^3(3a^3b + 7a^2b^2) = 15a^5b^4 + 35a^4b^5$. ■

A N S W E R

34. ____________

Practice 34: Simplify the following product.

$$5x^2(3x^3 + 7x)$$

35. ____________

Practice 35: Simplify the following product.

$$4x^3y(2x^2y^4 + 3x^3y^3)$$

36. ____________

Practice 36: Simplify the following product.

$$2a^4b^2(2a^3b^2 - 5a^2b^3)$$

37. ____________

Practice 37: Simplify the following product.

$$8x^4y^3(x^3y^2 + 5x^2y^3 + 2xy^4)$$

Now, let's look at some examples to help us develop some additional rules for exponents.

Example 20

Simplify the following product.

$$a^2 \times a^2 \times a^2 \times a^2$$

Solution: a^2 is written 4 times, so using short notation:

$$a^2 \times a^2 \times a^2 \times a^2 = (a^2)^4$$

There are 4 groups of a^2.
Now each group contains two a's ($a^2 = a \times a$); thus, there is a total of 8 a's ($4 \times 2 = 8$). Hence,

$$a^2 \times a^2 \times a^2 \times a^2 = (a^2)^4 = a^8$$

Simplify the following sum.

$$2a + 2a + 2a + 2a$$

Solution: $2a$ is written 4 times, so using short notation:

$$2a + 2a + 2a + 2a = 4(2a)$$

There are 4 groups of $2a$.
Now each group contains two a's ($2a = a + a$); thus, there is a total of 8 a's ($4 \times 2 = 8$). Hence,

$$2a + 2a + 2a + 2a = 4(2a) = 8a$$ ■

The above example leads us to Rule 2.

A N S W E R

Rule 2 (exponents): $(x^m)^n = x^{nm}$	**Rule 2 (coefficients):** $n(mx) = nmx$

Example 21

Simplify the following expression.

$$(x^5)^3$$

Solution: Using Rule 2 (exponents), $(x^m)^n = x^{nm}$, we have

$$\begin{aligned}(x^5)^3 &= x^{(3\times5)}\\ &= x^{15}\end{aligned}$$

Simplify the following expression.

$$3(5x)$$

Solution: Using Rule 2 (coefficients), $n(mx) = nmx$, we have

$$\begin{aligned}3(5x) &= (3\times5)x\\ &= 15x\end{aligned}$$

■

In the left column of Example 21, we had three products and each product contained 5 x's; thus, combined we had the product of 15 (3 × 5 = 15) x's.

$$\begin{aligned}(x^5)^3 &= (x^5)\times(x^5)\times(x^5)\\ &= (x\times x\times x\times x\times x)\times(x\times x\times x\times x\times x)\times(x\times x\times x\times x\times x)\\ &= (x\times x\times x\times x\times x\times x\times x\times x\times x\times x\times x\times x\times x\times x\times x)\\ &= x^{15}\end{aligned}$$

That's correct, Charlie. You are really beginning to understand exponents.

Practice 38: Simplify the following expression.

$$(a^2)^3$$

38. ____________

Practice 39: Simplify the following expression.

$$(x^5)^2$$

39. ____________

Practice 40: Simplify the following expression.

$$(y^4)^7$$

40. ____________

Practice 41: Simplify the following expression.

$$(p^3)^9$$

41. ____________

Example 22

Simplify the following product.

$$(ab)(ab)(ab)$$

Solution: ab is written 3 times, so using short notation:

$$(ab)(ab)(ab) = (ab)^3$$

There are 3 groups of ab. Each group contains one a and one b; thus, there is a total of 3 a's $(3 \times 1 = 3)$ and 3 b's $(3 \times 1 = 3)$. Hence,

$$(ab)(ab)(ab) = (ab)^3 = a^3b^3$$

Simplify the following sum.

$$(a + b) + (a + b) + (a + b)$$

Solution: $a + b$ is written 3 times, so using short notation:

$$(a + b) + (a + b) + (a + b) = 3(a + b)$$

There are 3 groups of $a + b$. Each group contains one a and one b; thus, there is a total of 3 a's $(3 \times 1 = 3)$ and 3 b's $(3 \times 1 = 3)$. Hence,

$$(a + b) + (a + b) + (a + b) = 3(a + b) = 3a + 3b$$ ■

This leads us to Rule 3.

Rule 3 (exponents): $(xy)^n = x^ny^n$ **Rule 3 (coefficients):** $n(x + y) = nx + ny$

Example 23

Simplify the following product.

$$(a^4b^2)^2$$

Step 1: Using Rule 3 (exponents), $(xy)^n = x^ny^n$, we have

$$(a^4b^2)^2 = (a^4)^2(b^2)^2$$

Step 2: Using Rule 2 (exponents), $(x^m)^n = x^{nm}$, we have

$$(a^4)^2(b^2)^2 = a^{(2\times4)}b^{(2\times2)} = a^8b^4$$

Thus, $(a^4b^2)^2 = a^8b^4$

Simplify the following sum.

$$2(4a + 2b)$$

Step 1: Using Rule 3 (coefficients), $n(x + y) = nx + ny$, we have

$$2(4a + 2b) = 2(4a) + 2(2b)$$

Step 2: Using Rule 2 (coefficients), $n(mx) = nmx$, we have

$$2(4a) + 2(2b) = (2 \times 4)a + (2 \times 2)b = 8a + 4b$$

Thus, $2(4a + 2b) = 8a + 4b$ ■

In the left column of Example 23, we had two groups of a^4b^2, *each group contained 4* a*'s and 2* b*'s; thus, combined we had 8 (2 × 4 = 8)* a*'s and 4 (2 × 2 = 4)* b*'s.*

That's correct. Charlie.

Practice 42: Simplify the following product.

$$(x^2y)^2$$

A N S W E R

42. ________

Practice 43: Simplify the following product.

$$(a^3b^5)^2$$

43. ________

Practice 44: Simplify the following product.

$$(a^5b^3)^4$$

44. ________

Practice 45: Simplify the following product.

$$(p^3q^7)^5$$

45. ________

Example 24 **Simplify the following product.**

$$(-2x^2y)^3$$

Step 1: Using Rule 3 (exponents), $(xy)^n = x^ny^n$, we have

$$(-2x^2y)^3 = (-2)^3(x^2)^3(y)^3$$

Step 2: Using Rule 2 (exponents), $(x^m)^n = x^{nm}$, we have

$$(-2)^3(x^2)^3(y)^3 = (-2)^3x^6y^3$$
$$= -8x^6y^3 \longleftrightarrow (-2)^3 = (-2)(-2)(-2) = -8$$

Thus, $(-2x^2y)^3 = -8x^6y^3$. ■

Example 25 **Simplify the following product.**

$$(-x^3y^2)^5$$

Step 1: Insert 1.

$$(-x^3y^2)^5 = (-1x^3y^2)^5$$

A N S W E R

Step 2: Using Rule 3 (exponents), $(xy)^n = x^n y^n$, we have

$$(-1x^3y^2)^5 = (-1)^5(x^3)^5(y^2)^5$$

Step 3: Using Rule 2 (exponents), $(x^m)^n = x^{nm}$, we have

$$\begin{aligned}(-1)^5(x^3)^5(y^2)^5 &= (-1)^5x^{5\times3}y^{5\times2}\\ &= -1x^{15}y^{10}\\ &= -x^{15}y^{10}\end{aligned}$$

$$\begin{aligned}(-1)^5 &= (-1)(-1)(-1)(-1)(-1)\\ &= -1\end{aligned}$$

Thus, $(-x^3y^2)^5 = -x^{15}y^{10}$. ■

46. ____________________

Practice 46: Simplify the following product.

$$(-3ab^2)^2$$

47. ____________________

Practice 47: Simplify the following product.

$$(-5p^3q^4)^3$$

48. ____________________

Practice 48: Simplify the following product.

$$(2x^6y^3)^5$$

49. ____________________

Practice 49: Simplify the following product.

$$(-a^4b^2c)^3$$

Here is a list of the rules for exponents that we have so far.

Rule 1 (exponents): $x^m \times x^n = x^{m+n}$
Rule 2 (exponents): $(x^m)^n = x^{nm}$
Rule 3 (exponents): $(xy)^n = x^n y^n$

EXERCISE 5.1

PART 1

Replace each ? with the appropriate value.

1. $a \times a \times a \times a \times a \times a \times a = a^?$
2. $a + a + a + a + a + a + a = ?a$
3. $x^6 = x^4 \times x^?$
4. $(ab)^? = 1$
5. $(x^3)^? = x^6$
6. $(xy^?)^2 = x^2y^4$
7. $y \times y \times y \times y \times y = y^?$
8. $(x^?)^4 = x^{12}$
9. $x^{10} = (x^?)^5$
10. $(2y)^4 = ?y^4$
11. $x^a x^b = x^?$
12. $(x^a)^b = x^?$
13. $a^5 a^6 = a^?$
14. $(a^? b^3)^2 = a^{10}b^6$
15. $(5x^3y)^? = 1$

PART 2

Simplify the following expressions.

A

1. $a \times a \times a$
2. $x \times y \times y \times x \times x$
3. $a \times b \times b \times b \times a \times a \times b \times a$
4. $(2a)(2a)(2a)(2a)$
5. $x \times x \times x \times y \times z \times z \times z \times y \times z$
6. $a \times a \times c \times c \times c \times b \times b \times a \times a \times b \times c$
7. $(y^3)y$
8. $(2w)(w^2)$

9. $(x^2)(x^3)$

10. $(a^3)(a^2)$

11. $(b^5)(b^4)$

12. $(t^3)(t^7)$

13. $(2a^2)(3a^4)$

14. $(5x^4)(2x^3)$

15. $(x^2y)(x^2)$

16. $(a^3b)(ab^3)$

17. $a^3b^2a^2b^4$

18. $(2a)(4a^3)$

19. $(x^0y^3)(x^2y^5)$

20. $(5x^2)(7x^3y)$

21. $2x^2y(3xy^2 + 5x^2y)$

22. $7x^3y^2(x^4y^3 + 6x^3y^4)$

23. $4a^3b^4(8ab^5 - 7a^4b^2)$

24. $3xy^5(4x^5y^3 - 9xy^6)$

B

25. $(x^2)^2$

26. $(5a^3)^0$

27. $(a^4)^2$

28. $(b^5)^3$

29. $(5a^2b)(-2ab^3)$

30. $(-4a^3)^2$

31. $(a^2y)^3$

32. $(2x)(4y)(2xy)$

33. $3ab^3c(2a^4b^2 - 5a^2bc^4)$

34. $-5x^3y^4(2xy^5 - 3x^4y^2)$

35. $5xy^3(7x^4y - 8x^4y^5 + 7)$

36. $3a^2b^4(5ab^5 - 7a^2b^4 + 8a^5b)$

37. $(x^2y^3)^4$

38. $(3x \times 2y)^3$

39. $(5a^3b^0)^2$

40. $(8x^2y)(-3x^3y^2)$

41. $(a^2b^3)(ab^4)(a^5b^2)$

42. $(3xy^3)(-2y^5)(2x^2y^3)$

43. $(-5a^2b)(-3a^6b^2)(-a^2)$

44. $(3x^2y^3)^4$

C

45. $[(a^2b^3)^2]^3$

46. $[(x^3y^4)^0]^5$

47. $(2x^3y^2)^2(-3x^2y^3)^2$

48. $(5ab^4)(-2a^23b^5)^3$

49. $[(2t^3)^5]^2$

50. $[(-2xy^2)^3]^2$

5.2 NEGATIVE INTEGER EXPONENTS

Let's look at the following which will help us arrive at our next definition.

$$a^8\left(\frac{1}{a^5}\right) = \frac{a^8}{a^5} = \frac{a^5 \times a^3}{a^5} = \frac{a^5}{a^5} \times a^3 = 1 \times a^3 = a^3$$

Also, according to Rule 1 (exponents), $x^m x^n = x^{m+n}$, we have

$$\begin{aligned} a^8a^{-5} &= a^{8+(-5)} \\ &= a^3 \end{aligned}$$

Now, $a^8(\frac{1}{a^5})$ and $a^8(a^{-5})$ both equal a^3. Since there is only one number which multiplied by a^8 equals a^3, $a^{-5} = \frac{1}{a^5}$. This leads us to the following definition.

Definition: $x^{-n} = \frac{1}{x^n}$, where $x \neq 0$ and n is a positive integer.

Example 1

$$2^{-3} = \frac{1}{2^3} = \frac{1}{8} \qquad (2^3 = 2 \times 2 \times 2 = 8)$$

$$x^{-2} = \frac{1}{x^2}$$

$$b^{-1} = \frac{1}{b^1} = \frac{1}{b}$$

$$(2a^2b)^{-5} = \frac{1}{(2a^2b)^5}$$

■

Do the rules of exponents from Section 5.1 work for negative exponents?

Yes, Charlie. All the rules and discussions concerning exponents from Section 5.1 are true for negative exponents. Now that we have negative exponents, we can develop our next rule. Let's look at simplifying $\frac{a^7}{a^3}$.

$$\begin{aligned} \frac{a^7}{a^3} &= a^7\left(\frac{1}{a^3}\right) && \frac{x}{y} = x\left(\frac{1}{y}\right) \\ &= a^7a^{-3} && x^{-n} = \frac{1}{x^n} \\ &= a^{7+(-3)} && \textbf{Rule 1: } x^mx^n = x^{m+n} \\ &= a^{7-3} && a + (-b) = a - b \\ &= a^4 \end{aligned}$$

Now, let's simplify $\frac{a^7}{a^3}$ another way and see if we get the same answer.

$$\frac{a^7}{a^3} = \frac{a^3 \times a^4}{a^3} = \frac{a^3}{a^3} \times a^4 = 1 \times a^4 = a^4$$

We got the same answer; so, I could work the problem either way.

That's correct, Charlie. However, the first method leads to a very useful rule for division.

Rule 4 (exponents): $\frac{x^m}{x^n} = x^{m-n}, x \neq 0.$

A N S W E R

Example 2 **Simplify the following expression leaving only a positive exponent in the answer.**

$$\frac{a^6}{a^4}$$

Solution:

$$\frac{a^6}{a^4} = a^{6-4} \qquad \text{Rule 4: } \frac{x^m}{x^n} = x^{m-n}$$

$$= a^2$$

■

Work the following practice problems using Rule 4 (exponents).

Practice 1: Simplify the following expression leaving only a positive exponent in the answer.

$$\frac{x^5}{x^2}$$

1. ____________________

Practice 2: Simplify the following expression leaving only a positive exponent in the answer.

$$\frac{a^9}{a^3}$$

2. ____________________

Practice 3: Simplify the following expression leaving only a positive exponent in the answer.

$$\frac{y^8}{y^4}$$

3. ____________________

Example 3 **Simplify the following expression leaving only a positive exponent in the answer.**

$$\frac{a^4}{a^9}$$

Solution:

$$\frac{a^4}{a^9} = a^{4-9} \qquad \text{Rule 4: } \frac{x^m}{x^n} = x^{m-n}$$

$$= a^{-5}$$

$$= \frac{1}{a^5} \qquad x^{-n} = \frac{1}{x^n}$$

■

A N S W E R

4. ____________________

Practice 4: Simplify the following expression leaving only a positive exponent in the answer.

$$\frac{y^3}{y^6}$$

5. ____________________

Practice 5: Simplify the following expression leaving only a positive exponent in the answer.

$$\frac{a^2}{a^7}$$

6. ____________________

Practice 6: Simplify the following expression leaving only a positive exponent in the answer.

$$\frac{a^5}{a^6}$$

Example 4 **Simplify the following expression leaving only a positive exponent in the answer.**

$$\frac{a^2a^{-4}}{a^3}$$

Solution:

$$\frac{a^2a^{-4}}{a^3} = \frac{a^{2+(-4)}}{a^3} \qquad \textbf{Rule 1: } x^m x^n = x^{m+n}$$

$$= \frac{a^{-2}}{a^3}$$

$$= a^{-2-3} \qquad \textbf{Rule 4: } \frac{x^m}{x^n} = x^{m-n}$$

$$= a^{-5}$$

$$= \frac{1}{a^5} \qquad x^{-n} = \frac{1}{x^n}$$

■

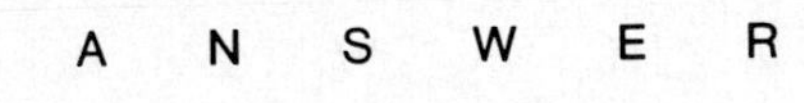

Practice 7: Simplify the following expression leaving only a positive exponent in the answer.

$$\frac{a^{-3}}{a^5}$$

7. ____________

Practice 8: Simplify the following expression leaving only a positive exponent in the answer.

$$\frac{a^{-4}}{a^3}$$

8. ____________

Practice 9: Simplify the following expression leaving only a positive exponent in the answer.

$$\frac{a^{-3}a^2}{a^5}$$

9. ____________

Practice 10: Simplify the following expression leaving only a positive exponent in the answer.

$$\frac{p^{-5}p^3}{p^5p^{-2}}$$

10. ____________

Example 5 **Simplify the following expression leaving only positive exponents in the answer.**

$$\left(\frac{a}{b}\right)^3$$

Solution: $\left(\frac{a}{b}\right)^3 = \left(\frac{a}{b}\right)\left(\frac{a}{b}\right)\left(\frac{a}{b}\right)$ **Definition of exponent.**

$= \frac{a \times a \times a}{b \times b \times b}$ **Definition of multiplication of fractions.**

$= \frac{a^3}{b^3}$ **Definition of exponent.** ■

ANSWER

Example 5 showed us that cubing a fraction was the same as cubing the numerator and denominator. Now, let's write the general case of this example as Rule 5 for exponents.

Rule 5 (exponents): $\left(\frac{x}{y}\right)^n = \frac{x^n}{y^n}$, $y \neq 0$.

Example 6 **Simplify the following expression leaving only positive exponents in the answer.**

$$\left(\frac{ab}{c}\right)^2$$

Solution:

$$\left(\frac{ab}{c}\right)^2 = \frac{(ab)^2}{c^2} \qquad \text{Rule 5: } \left(\frac{x}{y}\right)^n = \frac{x^n}{y^n}$$

$$= \frac{a^2b^2}{c^2} \qquad \text{Rule 3: } (xy)^n = x^n y^n$$

■

Work the following practice problems using Rule 5 (exponents).

11. ____________

Practice 11: Simplify the following expression leaving only positive exponents in the answer.

$$\left(\frac{a}{bc}\right)^2$$

12. ____________

Practice 12: Simplify the following expression leaving only positive exponents in the answer.

$$\left(\frac{a^2}{b}\right)^3$$

13. ____________

Practice 13: Simplify the following expression leaving only positive exponents in the answer.

$$\left(\frac{a^2}{b^2}\right)^2$$

A N S W E R

Example 7 **Simplify the following expression leaving only positive exponents in the answer.**

$$\left(-\frac{2a^3}{b^4}\right)^2$$

Solution:

$$\left(-\frac{2a^3}{b^4}\right)^2 = \frac{(-2a^3)^2}{(b^4)^2} \qquad \text{Rule 5: } \left(\frac{x}{y}\right)^n = \frac{x^n}{y^n}$$

$$= \frac{(-2)^2(a^3)^2}{(b^4)^2} \qquad \text{Rule 3: } (xy)^n = x^n y^n$$

$$= \frac{4(a^3)^2}{(b^4)^2}$$

$$= \frac{4a^{2\times 3}}{b^{2\times 4}} \qquad \text{Rule 2: } (x^m)^n = x^{nm}$$

$$= \frac{4a^6}{b^8}$$

■

Practice 14: Simplify the following expression leaving only positive exponents in the answer.

$$\left(\frac{a^2}{b^3}\right)^2$$

14. ____________________

Practice 15: Simplify the following expression leaving only positive exponents in the answer.

$$\left(\frac{a^5}{b^4}\right)^3$$

15. ____________________

Practice 16: Simplify the following expression leaving only positive exponents in the answer.

$$\left(\frac{2x^2}{y^3}\right)^2$$

16. ____________________

A N S W E R

17. ____________________

Practice 17: Simplify the following expression leaving only positive exponents in the answer.

$$\left(-\frac{3x^3}{y^4}\right)^2$$

18. ____________________

Practice 18: Simplify the following expression leaving only positive exponents in the answer.

$$\left(-\frac{2a^3}{3b}\right)^3$$

19. ____________________

Practice 19: Simplify the following expression leaving only positive exponents in the answer.

$$\left(-\frac{5a^2b}{2c}\right)^2$$

Let's list the rules of exponents along with a definition.

Rule 1 (exponents): $x^m x^n = x^{m+n}$
Rule 2 (exponents): $(x^m)^n = x^{nm}$
Rule 3 (exponents): $(xy)^n = x^n y^n$
Definition: $x^{-n} = \frac{1}{x^n}$, where $x \neq 0$, and n is a positive integer.
Rule 4 (exponents): $\frac{x^m}{x^n} = x^{m-n}$, $x \neq 0$.
Rule 5 (exponents): $\left(\frac{x}{y}\right)^n = \frac{x^n}{y^n}$, $y \neq 0$.

EXERCISE 5.2

Part 1

Replace each ? with the appropriate value.

A

1. $x^m x^n = x^?$

2. $x^? x^2 = x^5$

3. $(x^m)^n = x^?$

4. $(x^?)^3 = x^6$

5. $(xy)^n = x^? y^n$

6. $(xy)^? = x^3 y^3$

7. $x^{-n} = \dfrac{1}{x^?}$

8. $2^? = \dfrac{1}{8}$

9. $\dfrac{x^m}{x^n} = x^?$

10. $\dfrac{x^?}{x^3} = x^2$

11. $\left(\dfrac{x}{y}\right)^n = \dfrac{x^n}{y^?}$

12. $\left(\dfrac{x^2}{y^3}\right)^? = \dfrac{x^6}{y^9}$

B

13. $a^{-5} = \dfrac{1}{a^?}$

14. $x^? = \dfrac{1}{x^3}$

15. $3^? - \dfrac{1}{9}$

16. $5^? + \dfrac{1}{125}$

17. $\left(\dfrac{a}{b}\right)^5 = \dfrac{a^5}{b^?}$

18. $\dfrac{b^3}{a^3} = \left(\dfrac{b}{a}\right)^?$

19. $\dfrac{x^4}{x^4} = x^? = ?$

20. $\left(\dfrac{x^3}{y^?}\right)^2 = \dfrac{x^6}{y^8}$

21. $\left(\dfrac{x^2}{y^6}\right)^2 = \dfrac{x^?}{y^{12}}$

22. $\dfrac{x^?}{x^3} = x^7$

23. $\dfrac{(ab)^4}{(ab)^3} = (ab)^?$

24. $\dfrac{(ab)^2}{(ab)^6} = (ab)^? = \dfrac{1}{(ab)^?}$

25. $\left(\dfrac{2a^2}{3b}\right)^3 = \dfrac{?a^6}{27b^3}$

26. $\left(\dfrac{?a^3}{b^2}\right)^3 = \dfrac{-8a^9}{b^6}$

27. $\left(\dfrac{-3a^4}{2b}\right)^3 = \dfrac{?a^{12}}{8b^3}$

PART 2

Simplify the following expressions leaving only positive exponents in the answer.

A

1. $x^3 \times x^{-2}$
2. $(a^3)^{-2}$
3. $x^3 \times x^{-3}$
4. $(a^2)^0$
5. $(x^2y^3)^{-1}$
6. $(a^{-2})^4$
7. $\dfrac{x^3}{x^5}$
8. $\dfrac{(3y)^7}{(3y)^2}$
9. $\dfrac{(-2a)^5}{-2a}$
10. $\left(\dfrac{x^2}{y}\right)^3$
11. $\left(\dfrac{2x}{y^3}\right)^2$
12. $\left(\dfrac{-3p^2}{q}\right)^2$
13. $\left(\dfrac{x^2}{y^3}\right)^2$
14. $(4x^{-2})^3$
15. $\left(\dfrac{5a}{b^2}\right)^2$
16. $(x^{-3}y^{-1})^4$
17. $(a^{-5})^{-2}$
18. $(2x^{-3})^{-2}$

19. $\dfrac{2x^2}{4x^3}$

20. $\left(\dfrac{2a^3}{3b^2}\right)^3$

21. $\left(\dfrac{4x^2}{3y^3}\right)^2$

22. $\dfrac{3y^3}{21y^4}$

5.3 SIMPLIFYING EXPRESSIONS INVOLVING INTEGER EXPONENTS

Note: The rules of exponents in Sections 5.1 and 5.2 apply to multiplication and division. These rules do not always work when addition and subtraction are involved.

Let's look at some examples that show this point. First, let's look at an example involving only multiplication.

Example 1 **Simplify the following expression leaving only positive exponents in the answer.**

$$(ab)^2$$

Solution: Use Rule 3 (exponents), $(xy)^n = x^n y^n$.

$$(ab)^2 = a^2b^2$$ ■

Now, let's look at an example where multiplication is replaced by addition.

Example 2 **Simplify the following expression leaving only positive exponents in the answer.**

$$(a + b)^2$$

Solution: We cannot apply Rule 3 (exponents), $(xy)^n = x^n y^n$, since we have the sum of a and b; however, we can multiply as in Section 3.3.

$$(a + b)^2 = (a + b)(a + b) = a^2 + 2ab + b^2$$ ■

Note: $(a + b)^2$ does *not* equal $a^2 + b^2$

In Examples 3 and 4, we will once again show the difference between using the rules of exponents in a multiplication problem and an addition problem. First, let's look at an example involving only multiplication.

Example 3 **Simplify the following expression leaving only positive exponents in the answer.**

$$(a^{-1}b^{-1})^2$$

Step 1: Use Rule 3 (exponents), $(xy)^n = x^n y^n$.

$$(a^{-1}b^{-1})^2 = (a^{-1})^2(b^{-1})^2$$

Step 2: Use Rule 2 (exponents), $(x^m)^n = x^{nm}$.

$$\begin{aligned}(a^{-1})^2(b^{-1})^2 &= a^{2\times(-1)}b^{2\times(-1)}\\ &= a^{-2}b^{-2}\end{aligned}$$

Step 3: Use the definition, $x^{-n} = \frac{1}{x^n}$.

$$\begin{aligned}a^{-2}b^{-2} &= \left(\frac{1}{a^2}\right)\left(\frac{1}{b^2}\right)\\ &= \frac{1}{a^2b^2}\end{aligned}$$

Thus, $(a^{-1}b^{-1})^2 = \frac{1}{a^2b^2}$. ■

Now, let's look at an example where multiplication is replaced by addition.

Example 4 **Simplify the following expression leaving only positive exponents in the answer.**

$$(a^{-1} + b^{-1})^2$$

Step 1: Use the definition, $x^{-n} = \frac{1}{x^n}$.

$$(a^{-1} + b^{-1})^2 = \left(\frac{1}{a} + \frac{1}{b}\right)^2$$

Step 2: Find the least common denominator and add the fractions.

$$\begin{aligned}\left(\frac{1}{a} + \frac{1}{b}\right)^2 &= \left(\frac{b}{ab} + \frac{a}{ab}\right)^2\\ &= \left(\frac{b + a}{ab}\right)^2\\ &= \left(\frac{a + b}{ab}\right)^2\end{aligned}$$

Step 3: Use Rule 4 (exponents), $\left(\frac{x}{y}\right)^n = \frac{x^n}{y^n}$

$$\left(\frac{a + b}{ab}\right)^2 = \frac{(a + b)^2}{(ab)^2}$$

Step 4: Square the numerator.

$$\frac{(a + b)^2}{(ab)^2} = \frac{a^2 + 2ab + b^2}{(ab)^2}$$

A N S W E R

Step 5: Use Rule 3 (exponents), $(xy)^n = x^n y^n$.

$$\frac{a^2 + 2ab + b^2}{(ab)^2} = \frac{a^2 + 2ab + b^2}{a^2b^2}$$

Thus, $(a^{-1} + b^{-1})^2 = \frac{a^2 + 2ab + b^2}{a^2b^2}$. ■

Note: You must be careful when using the rules of exponents in problems involving addition or subtraction.

Practice 1: Simplify the following expression leaving only positive exponents in the answer.

$$(a + b)^{-2}$$

1. ____________________

Practice 2: Simplify the following expression leaving only positive exponents in the answer.

$$x^{-2} - y^{-2}$$

2. ____________________

Practice 3: Simplify the following expression leaving only positive exponents in the answer.

$$(x^{-1} + y)^2$$

3. ____________________

ANSWER

Practice 4: Simplify the following expression leaving only positive exponents in the answer.

$$(a - b)^{-2}$$

4. ____________________

EXERCISE 5.3

Simplify the following expressions leaving only positive exponents in the answer.

1. $(a + b^{-1})^2$
2. $(x^{-1} + y^{-1})^2$
3. $(2x^{-1} + 1)^2$
4. $(2a^{-1} + 2b)^2$
5. $(x^{-1} - y^{-1})^{-1}$
6. $(x^{-1} + y)^{-2}$
7. Explain why $(2x + y)^2$ does not equal $4x^2 + y^2$

USEFUL INFORMATION

1. A short cut in writing repeated multiplication problems is through the use of a base and an exponent. For example,

$$5 \times 5 \times 5 \times 5 = 5^4$$

where 5 is the **base** and 4 is the **exponent.**

2. Any nonzero number to the zero power is equal to 1 ($x^0 = 1, x \neq 0$).

3. **Rule 1 (exponents):** $x^m \times x^n = x^{m+n}$
When you multiply two terms with the same base, you write the base and add the exponents. For example,

$$a^2 \times a^6 = a^{2+6} = a^8$$

4. **Rule 2 (exponents):** $(x^m)^n = x^{nm}$
When using Rule 2 (exponents), you write the same base and multiply the exponents. For example,

$$(y^4)^3 = y^{3\times 4} = y^{12}$$

5. **Rule 3 (exponents):** $(xy)^n = x^n y^n$
When the base is written as a product of terms, you distribute the exponent to each term. For example,

$$(a^5b^2)^3 = (a^5)^3(b^2)^3$$

6. **Definition:** $x^{-n} = \frac{1}{x^n}$, where $x \neq 0$ and n is a positive integer.
For example,

$$3^{-4} = \frac{1}{3^4}$$

7. **Rule 4 (exponents):** $\frac{x^m}{x^n} = x^{m-n}, x \neq 0.$
When you divide two terms with the same base, you write the base then subtract the exponent which was in the denominator from the exponent in the numerator. For example,

$$\frac{a^6}{a^4} = a^{6-4} = a^2$$

8. **Rule 5 (exponents):** $(\frac{x}{y})^n = \frac{x^n}{y^n}, y \neq 0.$
When the base is written as a quotient of terms, you distribute the exponent to each term. For example,

$$\left(\frac{a^2}{b^3}\right)^2 = \frac{(a^2)^2}{(b^3)^2}$$

9. Rules 1 through 5 for exponents apply to multiplication and division. These rules do *not* always work when addition and subtraction are involved. For example,

$$(a^{-1} + b^{-1})^3 \neq a^{-3} + b^{-3}$$

whereas

$$(a^{-1}b^{-1})^3 = a^{-3}b^{-3}$$

10. **Summary of rules for exponents.**
Rule 1. $x^m x^n = x^{m+n}$
Rule 2. $(x^m)^n = x^{nm}$
Rule 3. $(xy)^n = x^n y^n$
Rule 4. $\frac{x^m}{x^n} = x^{m-n}, x \neq 0$
Rule 5. $(\frac{x}{y})^n = \frac{x^n}{y^n}, y \neq 0$

REVIEW PROBLEMS

Section 5.1

Simplify the following expressions.

A

1. $x \times x^6$

2. $(2a)(3a)$

3. $(5a)(-2a)$

4. $(-3x)(-4x)$

5. $(3a)(-3a^2)$

6. $(-2x^2)(3x^3)$

7. $(x^3)^4$

8. $(a^2)^5$

9. $(2a^3)^2$

10. $(-2xy^2)^3$

11. $2x(3x^2 + 4x)$

12. $-5x^2(2x^3 - 3x^2)$

13. $-6x^3(2x^2 - 2x)$

14. $3x^3(-2x^3 + 4x^2)$

B

15. $(x^3y)(xy^2)$

16. $(5xy^2)(5x^2y^3)$

17. $(-10ab^2)(-2a^2b)$

18. $(6xy^2)(3xy^3)$

19. $(2x^3y)^3$

20. $(-7x^2)^2$

21. $(a^3b^4)^2$

22. $(2x^3)^3$

23. $(-2x^3)^2$

24. $(-x^2y)^3$

25. $(3a^2b^3)^2$

26. $(-2a^3b^4)^4$

27. $5x^2y(3x^3y^2 - 4x^5y^4)$

28. $-4x^2y^3(5x^3y^4 - x^2y^3)$

29. $-2x^3y^2(3x^4y - 5x^6)$

30. $3x^4y^3(2x^5 + 3xy^3)$

C

31. $(-5x^2y^3)^0$

32. $(6a^2b)^2\,(2a^3b^4)$

33. $(2a^2b)(-3a^4b^5)\,(-2ab^3)$

34. $(-x^2y^5)(-2x^3y)(-xy^4)$

35. $(-ab^3c^2)^3$

36. $(8x^3y)^3(-2xy)$

Section 5.2

Simplify the following expressions leaving only positive exponents in the answer.

A

1. a^{-2}

2. 3^{-4}

3. x^{-5}

4. a^2b^{-3}

5. $\frac{a^7}{a^6}$

6. $\frac{x^5}{x^9}$

7. $\frac{y^2}{y^3}$

8. $\frac{b^8}{b^6}$

9. $2x^{-4}y^3$

10. x^7x^{-5}

11. a^3a^{-5}

12. $3x^{-4}x^3$

13. $5a^{-3}a^{-2}$

14. $7x^{-2}x^{-5}$

15. $(b^2)^{-2}$

16. $(x^{-3})^2$

17. $\left(\frac{2x}{3y^2}\right)^2$

18. $\left(\frac{5a^2}{b^3}\right)^2$

19. $\left(\frac{-2x^2}{3y^2}\right)^3$

20. $\left(\frac{2x^3}{3y^4}\right)^3$

B

21. $\frac{y^4}{y^{-2}}$

22. $\frac{2a^3}{a^{-5}}$

23. $(a^{-1}b^{-1})^2$

24. $\left(\frac{2x^4}{y^2}\right)^2$

25. $\left(\frac{-3a^2b^3}{2c}\right)^2$

26. $(2a^{-2})^2$

27. $(3x^3)^{-2}$

28. $(5y^{-2})^{-4}$

29. $(2x^{-3})^{-1}$

30. $\left(\frac{2y^3y^{-2}}{5x}\right)^3$

Section 5.3

Simplify the following expressions leaving only positive exponents in the answer.

1. $(x^{-1} + 1)^2$

2. $(a^3 + b^3)^2$

3. $\left(\frac{a}{b} - b^{-1}\right)^2$

4. $\left(\frac{a - b}{a}\right)^2$

5. $(a^2 - b)^{-2}$

6. $(a + b)^{-2}$

ANSWERS TO PRACTICE PROBLEMS

Section 5.1

1. $6(5)$
2. $3a$
3. $8 + 8 + 8 + 8 + 8$
4. $x + x + x + x$
5. 5^6
6. a^3
7. $8 \times 8 \times 8 \times 8 \times 8$
8. $x \times x \times x \times x$
9. $2(3) + 3(4) + 2(5)$
10. $3^2 \times 4^3 \times 5^2$
11. $6(2) + 4(5)$
12. $2^6 \times 5^4$
13. $3x + 2y$
14. $x^3 \times y^2$
15. 16
16. 200
17. 12
18. 72
19. $2x + 5z$
20. $x^2 \times z^5$
21. $5x + 7y$
22. $x^5 \times y^7$
23. $x^2 + 3y$
24. $x^2 \times y^3$
25. x^3
26. a^{15}
27. y^8
28. $a^7 \times b^4$
29. $x^5 \times y^3$
30. $15a^6b^5$
31. $8x^6y^7$
32. $6p^3q^7$
33. $-10a^3b^7c^6$
34. $15x^5 + 35x^3$
35. $8x^5y^5 + 12x^6y^4$
36. $4a^7b^4 - 10a^6b^5$
37. $8x^7y^5 + 40x^6y^6 + 16x^5y^7$
38. a^6
39. x^{10}
40. y^{28}
41. p^{27}
42. x^4y^2
43. a^6b^{10}
44. $a^{20}b^{12}$
45. $p^{15}q^{35}$
46. $9a^2b^4$
47. $-125p^9q^{12}$
48. $32x^{30}y^{15}$
49. $-a^{12}b^6c^3$

Section 5.2

1. x^3
2. a^6
3. y^4
4. $\frac{1}{y^3}$
5. $\frac{1}{a^5}$
6. $\frac{1}{a}$
7. $\frac{1}{a^8}$
8. $\frac{1}{a^7}$
9. $\frac{1}{a^6}$
10. $\frac{1}{p^5}$
11. $\frac{a^2}{b^2c^2}$
12. $\frac{a^6}{b^3}$
13. $\frac{a^4}{b^4}$
14. $\frac{a^4}{b^6}$
15. $\frac{a^{15}}{b^{12}}$
16. $\frac{4x^4}{y^6}$
17. $\frac{9x^6}{y^8}$
18. $\frac{-8a^9}{27b^3}$
19. $\frac{25a^4b^2}{4c^2}$

Section 5.3

1. $\frac{1}{a^2 + 2ab + b^2}$
2. $\frac{y^2 - x^2}{x^2y^2}$
3. $\frac{1 + 2xy + x^2y^2}{x^2}$
4. $\frac{1}{a^2 - 2ab + b^2}$

C H A P T E R 6

Special Products and Factoring

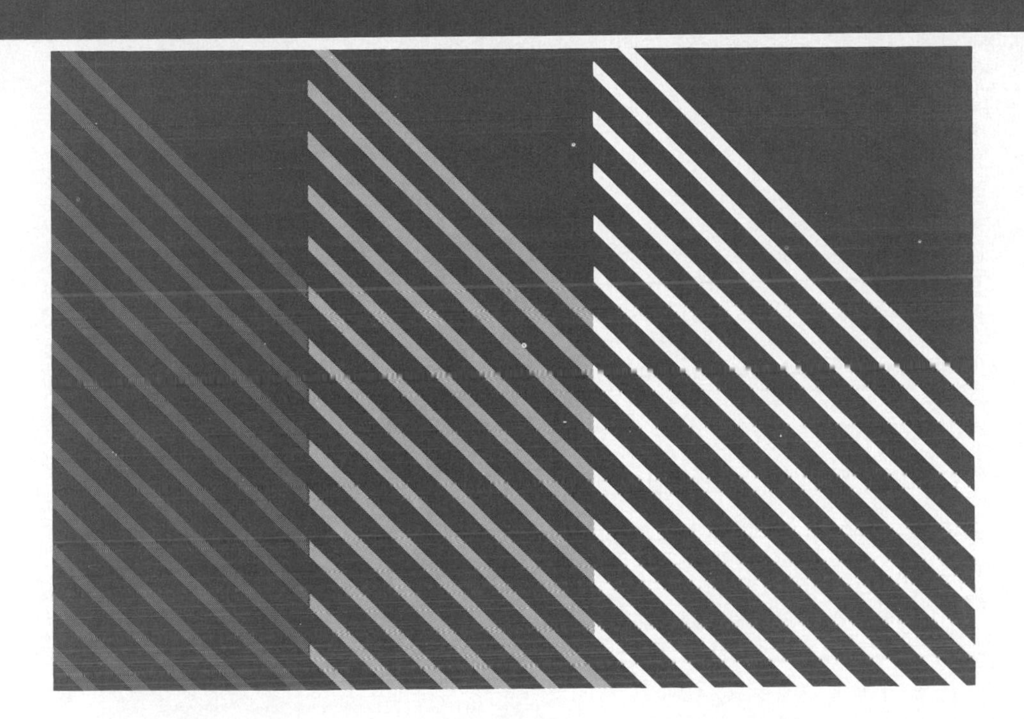

The objectives of this chapter are:

1. **To be able to multiply certain polynomials using the special products rules.**
2. **To be able to factor certain polynomials using the special products rules.**

A N S W E R

In this chapter, we will look at some timesaving short cuts that can be used when multiplying certain polynomials. These special products will serve as a multiplication table for these polynomials and should also help you when factoring polynomials. As you will discover, factoring has many uses in algebra.

6.1 SPECIAL PRODUCTS

Let's start by squaring a binomial of the form $a + b$; that is, we will multiply $(a + b)(a + b)$.

$$\begin{array}{r} a + b \\ a + b \\ \hline ab + b^2 \\ a^2 + ab \\ \hline a^2 + 2ab + b^2 \end{array}$$

This gives us the first special product.

Special Product 1. $(a + b)^2 = (a + b)(a + b) = a^2 + 2ab + b^2$

What do a *and* b *represent?*

That's a good question, Charlie. They represent the two terms of the binomial being squared (multiplied by itself). The first term of the binomial is represented by a and the second term of the binomial is represented by b. Let's look at an example.

Example 1 **Select the terms in the following binomials that correspond to *a* and *b* in the first special product.**

$(2x + 3)^2$ **$a = 2x$ and $b = 3$**

$(5x + 7)^2$ **$a = 5x$ and $b = 7$**

$(3y + 1)^2$ **$a = 3y$ and $b = 1$** ■

1. ________

Practice 1: Select the terms in the following binomial that correspond to a and b in the first special product.

$$(x + 4)^2$$

2. ________

Practice 2: Select the terms in the following binomial that correspond to a and b in the first special product.

$$(7x + 6)^2$$

ANSWER

Practice 3: Select the terms in the following binomial that correspond to a and b in the first special product.

$$(2x + 3y)^2$$

3. ____________

Now that you understand what a and b represent in Special Product 1, let's use the rule to multiply a binomial by itself.

Example 2 **Use Special Product 1 to carry out the following multiplication.**

$$(2x + 3)^2$$

Solution: Now, $2x$ corresponds to a and 3 corresponds to b. Thus,

$$\begin{array}{lcl} (a \quad + b)^2 & = & a^2 \quad + 2 \quad a \quad b \quad + b^2 \\ | \qquad | & & | \qquad\quad | \quad | \quad | \qquad | \\ (2x + 3)^2 & = & (2x)^2 + 2(2x)(3) + 3^2 \\ & = & 4x^2 + \quad 12x \quad + 9 \end{array}$$

■

As you gain more experience you will be able to omit the middle step.

Practice 4: Use Special Product 1 to carry out the following multiplication.

$$(x + 2)^2$$

4. ____________

Practice 5: Use Special Product 1 to carry out the following multiplication.

$$(3x + 4)^2$$

5. ____________

Practice 6: Use Special Product 1 to carry out the following multiplication.

$$(8x + 7)^2$$

6. ____________

Practice 7: Use Special Product 1 to carry out the following multiplication.

$$(2x + 3y)^2$$

7. ____________

Practice 8: Use Special Product 1 to carry out the following multiplication.

$$(3p + 5q)^2$$

8. ____________

Next, let's look at squaring a binomial of the form $a - b$; that is, multiplying $(a - b)(a - b)$.

$$\begin{array}{rrr} & a & - \; b \\ & a & - \; b \\ \hline & - \; ab & + \; b^2 \\ a^2 & - \; ab & \\ \hline a^2 & - \; 2ab & + \; b^2 \end{array}$$

This gives us the second special product.

A N S W E R

Special Product 2. $(a - b)^2 = (a - b)(a - b) = a^2 - 2ab + b^2$

Example 3 **Use Special Product 2 to carry out the following multiplication.**

$$(5x - 2)^2$$

Solution: Now, $5x$ corresponds to a and 2 corresponds to b. Thus,

$$\begin{aligned} (a - b)^2 &= a^2 - 2\,a\,b + b^2 \\ (5x - 2)^2 &= (5x)^2 - 2(5x)(2) + 2^2 \\ &= 25x^2 - 20x + 4 \end{aligned}$$

■

Once again, as you gain more experience you will be able to omit the middle step.

9. ____________

Practice 9: Use Special Product 2 to carry out the following multiplication.

$$(3x - 5)^2$$

10. ____________

Practice 10: Use Special Product 2 to carry out the following multiplication.

$$(4x - 6)^2$$

11. ____________

Practice 11: Use Special Product 2 to carry out the following multiplication.

$$(3x - 2y)^2$$

12. ____________

Practice 12: Use Special Product 2 to carry out the following multiplication.

$$(5p - 2q)^2$$

The next special product is of the form $(a + b)(a - b)$.

$$\begin{array}{rr} & a - b \\ & a + b \\ \hline & ab - b^2 \\ a^2 - ab & \\ \hline a^2 & - b^2 \end{array}$$

This gives us the third special product.

Special Product 3. $(a + b)(a - b) = (a - b)(a + b) = a^2 - b^2$

The product of the sum and difference of two terms is the square of the first term minus the square of the second term.

A N S W E R

Example 4 **Use Special Product 3 to carry out the following multiplication.**

$$(2x + 5)(2x - 5)$$

Solution: Now, $2x$ corresponds to a and 5 corresponds to b.

$$\begin{aligned}(a + b)(a - b) &= a^2 - b^2 \\ (2x + 5)(2x - 5) &= (2x)^2 - 5^2 \\ &= 4x^2 - 25\end{aligned}$$

■

Once again, as you gain more experience you will be able to omit the middle step.

Practice 13: Use Special Product 3 to carry out the following multiplication.

$$(3x + 2)(3x - 2)$$

13. ____________

Practice 14: Use Special Product 3 to carry out the following multiplication.

$$(4x + 7)(4x - 7)$$

14. ____________

Practice 15: Use Special Product 3 to carry out the following multiplication.

$$(5x - 3)(5x + 3)$$

15. ____________

Practice 16: Use Special Product 3 to carry out the following multiplication.

$$(2x + y)(2x - y)$$

16. ____________

Practice 17: Use Special Product 3 to carry out the following multiplication.

$$(7n + 6)(7n - 6)$$

17. ____________

The last special product is of the form $(ax + b)(cx + d)$.

$$\begin{array}{rr} & ax + b \\ & cx + d \\ \hline & adx + bd \\ acx^2 + & bcx \\ \hline acx^2 + (ad + bc)x & + bd \end{array}$$

This gives us the fourth special product.

Special Product 4—The FOIL rule. $(ax + b)(cx + d) = acx^2 + (ad + bc)x + bd$

Let's look at why Special Product 4 is called the **FOIL** rule.

THE FOIL RULE

Step 1: The F in FOIL refers to the product of the *first* terms.

$$F \times F \longrightarrow F$$

$$(ax + b)(cx + d) = acx^2 + ?$$

Step 2: The O in FOIL refers to the *outer* product. The I in FOIL refers to the *inner* product.

$$\begin{matrix} O \times O \\ + \\ I \times I \end{matrix} \longrightarrow O \times O + I \times I$$

$$\begin{aligned}(ax + b)(cx + d) &= acx^2 + adx + bcx + ? \\ &= acx^2 + (ad + bc)x + ?\end{aligned}$$

Step 3: The L in FOIL refers to the product of the *last* terms.

$$L \times L \longrightarrow L$$

$$(ax + b)(cx + d) = acx^2 + (ad + bc)x + bd$$

Why do we have to learn all these special products? Why can't we just multiply the way we did in Chapter 3?

You can. It just takes longer.

I would rather take a little longer and not have to learn all these special products.

Well, Charlie, I might agree with you if the only reason for learning these special products was to save time when multiplying. However, the main use of these special product rules is to factor polynomials. It would be very difficult to factor without these rules. The time you spend learning these rules now will save you time and effort when factoring polynomials.

So, I need to learn these special product rules to help factor as well as multiply polynomials.

That's correct, Charlie.

Example 5 **Use the FOIL rule to carry out the following multiplication.**

$$(2x + 3)(3x + 1)$$

Step 1:

$F \times F \rightarrow F$

$$(\mathbf{2x} + 3)(\mathbf{3x} + 1) = \mathbf{6x^2} + ?$$

Step 2:

$O \times O + I \times I \rightarrow O \times O + I \times I$

$$(2x + \mathbf{3})(\mathbf{3x} + 1) = 6x^2 + 2x + \mathbf{9x} + ?$$
$$= 6x^2 + 11x + ?$$

Step 3:

$L \times L \rightarrow L$

$$(2x + \mathbf{3})(3x + \mathbf{1}) = 6x^2 + 11x + \mathbf{3}$$

Thus, $(2x + 3)(3x + 1) = 6x^2 + 11x + 3$ ■

Example 6 **Use the FOIL rule to carry out the following multiplication.**

$$(5x - 2)(2x + 4)$$

Step 1:

$F \times F \rightarrow F$

$$(\mathbf{5x} - 2)(\mathbf{2x} + 4) = \mathbf{10x^2} + ?$$

A N S W E R

Step 2:

$O \times O + I \times I$

$(5x - \mathbf{2})(\mathbf{2x} + 4) = 10x^2 + 20x + (-\mathbf{4x}) + ?$

$= 10x^2 + 16x + ?$

Step 3:

$L \times L \rightarrow L$

$(5x - \mathbf{2})(2x + \mathbf{4}) = 10x^2 + 16x - \mathbf{8}$

Thus, $(5x - 2)(2x + 4) = 10x^2 + 16x - 8$ ■

18. ______________

Practice 18: Use the FOIL rule to carry out the following multiplication.

$$(2x + 1)(3x + 2)$$

19. ______________

Practice 19: Use the FOIL rule to carry out the following multiplication.

$$(2x + 5)(2x + 3)$$

20. ______________

Practice 20: Use the FOIL rule to carry out the following multiplication.

$$(4x - 1)(2x + 5)$$

21. ______________

Practice 21: Use the FOIL rule to carry out the following multiplication.

$$(3x + 4)(2x - 3)$$

Practice 22: Use the FOIL rule to carry out the following multiplication.

$$(2x - 3)(4x - 2)$$

A N S W E R

22. ____________________

Practice 23: Use the FOIL rule to carry out the following multiplication.

$$(x - 5)(7x - 3)$$

23. ____________________

Practice 24: Use the FOIL rule to carry out the following multiplication.

$$(2x - 5)(3x + 4)$$

24. ____________________

Practice 25: Use the FOIL rule to carry out the following multiplication.

$$(2x - 3y)(x + 2y)$$

25. ____________________

Let's list the four special product rules.

Special Product 1. $(a + b)^2 = (a + b)(a + b) = a^2 + 2ab + b^2$
Special Product 2. $(a - b)^2 = (a - b)(a - b) = a^2 - 2ab + b^2$
Special Product 3. $(a + b)(a - b) = (a - b)(a + b) = a^2 - b^2$
Special Product 4—The FOIL rule. $(ax + b)(cx + d) = acx^2 + (ad + bc)x + bd$

EXERCISE 6.1

Use the special product rules to multiply the following binomials.

A

1. $(x - 2)^2$

2. $(x + 2)^2$

3. $(2x + 1)^2$

4. $(x - 2)(x + 2)$

5. $(2x - 3)^2$

6. $(2x - 5)(2x + 5)$

7. $(3x + 2)(3x - 2)$

8. $(4t - 1)(4t + 1)$

9. $(5x - 1)^2$

10. $(6x + 2)^2$

11. $(4a + 7)^2$

12. $(6y - 3)^2$

13. $(2a - 7)(2a + 7)$

14. $(7y - 8)(7y + 8)$

15. $(x + 3)(x + 2)$

16. $(y + 4)(y + 7)$

17. $(a - 3)(a + 2)$

18. $(x - 4)(x - 6)$

19. $(2c - 5)(2c + 5)$

20. $(3x - 6)(x + 2)$

21. $(y - 12)(y + 2)$

22. $(5x + 3)(x + 1)$

23. $(2x - 9)(5x + 3)$

24. $(2a + 3)(3a + 5)$

25. $(7x + 3)(x - 5)$

26. $(4y - 3)(2y - 5)$

27. $(2a - 4)(2a + 3)$

28. $(10y + 3)(5y - 9)$

B

29. $(a - 2b)(a + 2b)$

30. $(m - 3n)^2$

31. $(a + 3b)^2$

32. $(2x - y)^2$

33. $(2x + 3y)(2x - 3y)$

34. $(a - 5b)^2$

35. $(3x + 4y)^2$

36. $(5x - 4y)(5x + 4y)$

37. $(2a + b)(a + 3b)$

38. $(x - 5y)(2x + y)$

39. $(3x - 2y)(4x + 3y)$

40. $(7a - b)(3a - 2b)$

6.2 COMMON MONOMIAL FACTORS

The remainder of this chapter will deal with **factoring polynomials**—that is, writing polynomials as products. Factoring has many uses in algebra. Two examples are (1) solving quadratic equations and (2) performing arithmetic operations on algebraic fractions.

Factoring polynomials is the opposite of multiplying polynomials; therefore, we will use the rules of multiplication in reverse when factoring. Let's start with the simplest factorization—that is, removing a *common* monomial factor from a polynomial. This process makes use of the distributive property. Now, writing the distributive property in reverse gives us the first method for factoring.

Distributive property. $ab + ac = a(b + c)$

Note: When factoring using the distributive property, remove all the common monomial factors or remove the greatest common monomial factor.

A N S W E R

Example 1 **Factor the following polynomial and check the answer by multiplying.**

$$6x + 30$$

Solution: Two, three, and six are the common monomial factors, and six is the greatest common monomial factor. Thus, using the distributive property we have

$$6x + 30 = 6(x + 5)$$

Check: $6(x + 5) = 6(x) + 6(5) = 6x + 30$ ■

What happens if I don't take out the greatest common factor?

Well, Charlie, you have not taken out all the common factors; thus, you will have to continue the factoring process. Suppose in Example 1 that we had thought that the greatest common factor of $6x + 30$ was 2 and we removed 2 giving us

$$6x + 30 = 2(3x + 15)$$

We should now check $3x + 15$ for a common factor; thus, we discover that 3 is also a common factor. Now we remove 3 getting

$$\begin{aligned} 6x + 30 &= 2(3x + 15) \\ &= 2[3(x + 5)] \end{aligned}$$

Multiplying the common factors 2 and 3 we get

$$\begin{aligned} 6x + 30 &= 2[3(x + 5)] \\ &= (2 \times 3)(x + 5) \\ &= 6(x + 5) \end{aligned}$$

which is the same answer that we got when we removed the greatest common factor 6. After removing what you believe to be the greatest common factor from a polynomial, always check the resulting polynomial for a common factor. It is quite easy to overlook common factors when factoring polynomials.

1. ____________________

Practice 1: Factor the following polynomial and check your answer by multiplying.

$$2x + 6$$

2. ____________________

Practice 2: Factor the following polynomial and check your answer by multiplying.

$$12x + 8$$

3. ____________________

Practice 3: Factor the following polynomial and check your answer by multiplying.

$$15x - 45$$

ANSWER

Example 2 **Factor the following polynomial and check the answer by multiplying.**

$$-8x^2 + 20x - 12$$

Solution: Two and four are the common monomial factors, and four is the greatest common factor. Thus, using the distributive property we have

$$-8x^2 + 20x - 12 = -4(2x^2 - 5x + 3)$$

Check:
$$-4(2x^2 - 5x + 3) = -4(2x^2) + (-4)(-5x) + (-4)3$$
$$= -8x^2 + 20x - 12$$ ■

Why did you remove -4 instead of 4?

Well, Charlie, we want the coefficient of x^2 to be positive; thus, we removed -4 from each term. In general, we want the coefficient of the largest power of x to be positive.

Practice 4: Factor the following polynomial and check your answer by multiplying.

$$6x^2 + 15x + 9$$

4. ____________

Practice 5: Factor the following polynomial and check your answer by multiplying.

$$-8x^2 - 16x + 32$$

5. ____________

Example 3 **Factor the following polynomial and check the answer by multiplying.**

$$7x^3y - 21x^2y^3$$

Solution: Seven, x^2, and y are common monomial factors, and $7x^2y$ is the greatest common monomial factor. Thus, using the distributive property we have

$$7x^3y - 21x^2y^3 = 7x^2y(x - 3y^2)$$

Check:
$$7x^2y(x - 3y^2) = 7x^2y(x) + 7x^2y(-3y^2)$$
$$= 7x^3y - 21x^2y^3$$ ■

Practice 6: Factor the following polynomial and check your answer by multiplying.

$$32a^3 - 48a^2b$$

6. ____________

ANSWER

7. ____________

Practice 7: Factor the following polynomial and check your answer by multiplying.

$$6x^3 + 9x^2 - 12x$$

8. ____________

Practice 8: Factor the following polynomial and check your answer by multiplying.

$$-16a^3 - 50ab^2$$

EXERCISE 6.2

Factor the following polynomials and check your answer by multiplying.

A

1. $4x + 8$
2. $7a - 14$
3. $8y + 24$
4. $12b + 18$
5. $10x - 15$
6. $21a - 14$
7. $19y - 38$
8. $15x + 9$
9. $27a - 36$
10. $72x + 18$

B

11. $8y^2 - 16y$
12. $8x^3 + 6x$
13. $-x^3 - x^2 + x$
14. $12a^2 + 11a$
15. $16x^2 - 7$
16. $25x^3 + 15x^2$
17. $5a^4 + 25a^2$
18. $26x^3 - 13x^2 + 39x$
19. $16x^3 + 48x^2 + 24x$
20. $-8a^6 - 18a^3 + 12a^2$

C

21. $x^2y^2 + xy$

22. $-5xy^4 - 10x^3y^2$

23. $27a^3y^2 + 45ay^2$

24. $a^2 + 2a + 2$

25. $90x^4 - 45x^3 + 180x^2$

26. $-15x^3y^3 + 5x^2y^2 - 10xy$

27. $-15a^2b^3c - 35a^4bc^3$

28. $18x^4y^3 - 96x^2y^6$

29. $28x^4y^4 + 14x^3y^3 + 35x^2y^2$

30. $2a^3(a - 5) + 5(a - 5)$

6.3 FACTORING THE DIFFERENCE OF TWO SQUARES

When a number or expression is multiplied by itself, the product is a **square**. When a whole number is multiplied by itself, the product is called a **perfect square.** When a monomial with a whole number coefficient is multiplied by itself, the product is called a **perfect square monomial.**

Example 1

$$1 \times 1 = 1$$
$$3 \times 3 = 9$$
$$5 \times 5 = 25$$
$$9 \times 9 = 81$$
$$x \times x = x^2$$
$$2y \times 2y = 4y^2$$
$$7a^2 \times 7a^2 = 49a^4$$

Therefore, 1, 9, 25, 81, x^2, $4y^2$, and $49a^4$ are perfect squares. ■

In this section, when factoring the difference of two squares, we will actually be factoring the difference of two perfect squares. Thus, you will need to be able to tell when a monomial is a perfect square. Let's look at some examples.

Example 2 **Determine whether or not the following monomial is a perfect square.**

$$65x^2$$

Step 1: Determine whether or not there is a whole number which multiplied by itself equals 65. Now,

$$8 \times 8 = 64$$

and

$$9 \times 9 = 81$$

Since $8 \times 8 = 64$ and 64 is less than 65, and $9 \times 9 = 81$ and 81 is greater than 65, there is no whole number which multiplied by itself equals 65. Thus, 65 is not a perfect square. Therefore, $65x^2$ is not a perfect square. ■

Example 3 **Determine whether or not the following monomial is a perfect square.**

$$36x^4$$

Step 1: Determine whether or not there is a whole number which multiplied by itself equals 36. Now,

$$6 \times 6 = 36$$

and 36 is a perfect square.

Step 2: Determine whether or not x^4 can be written as the product of a quantity times itself. Now,

$$x^4 = x \times x \times x \times x = (x \times x) \times (x \times x) = x^2 \times x^2$$

and x^4 is a perfect square.

Therefore, $36x^4$ is a perfect square.

$$36x^4 = 6x^2 \times 6x^2$$ ■

Example 4 **Determine whether or not the following monomial is a perfect square.**

$$121a^3$$

Step 1: Determine whether or not there is a whole number which multiplied by itself equals 121. Now,

$$11 \times 11 = 121$$

and 121 is a perfect square.

Step 2: Determine whether or not a^3 can be written as a product of a quantity times itself. Now,

$$a^3 = a \times a \times a$$

$$= a \times (a \times a)$$

$$= (a \times a) \times a.$$

So, a^3 cannot be written as a product of a quantity times itself; thus, a^3 is not a perfect square. Therefore, $121a^3$ is not a perfect square. ■

A N S W E R

Note: An easy way to tell if x^n is a perfect square is to look at the exponent (n). If n is even (2, 4, 6, 8, 10, etc.), then x^n is a perfect square and $x^n = x^{n/2}x^{n/2}$. If n is odd (1, 3, 5, 7, etc.), then x^n is not a perfect square.

Example 5 **Which of the following are perfect squares?**

$$x^8$$
$$a^5$$
$$y^7$$
$$(ab)^6$$

Solution: Since 8 and 6 are even, x^8 and $(ab)^6$ are perfect squares.

$$x^8 = x^4x^4$$

$$(ab)^6 = (ab)^3(ab)^3$$

■

Practice 1: Determine whether or not the following monomial is a perfect square.

$$a^9$$

1. ____________

Practice 2: Determine whether or not the following monomial is a perfect square.

$$x^{10}$$

2. ____________

Practice 3: Determine whether or not the following monomial is a perfect square.

$$4x^2$$

3. ____________

Practice 4: Determine whether or not the following monomial is a perfect square.

$$24x^2$$

4. ____________

Practice 5: Determine whether or not the following monomial is a perfect square.

$$49a^3$$

5. ____________

Practice 6: Determine whether or not the following monomial is a perfect square.

$$49a^4$$

6. ____________

Now, writing Special Product 3 (Section 6.1) in reverse gives us the second method for factoring.

Special Product 3. $a^2 - b^2 = (a + b)(a - b)$

If we have a binomial that is the *difference* of two perfect squares, then we can use Special Product 3 to factor the binomial.

Example 6 **Factor the following binomial.**

$$9x^2 - 16$$

Step 1: Check to find if the two terms of the binomial are perfect squares. Now, $9x^2 = 3x \times 3x = (3x)^2$ and $9x^2$ is a perfect square, and $16 = 4 \times 4 = 4^2$ and 16 is a perfect square. Hence, $9x^2 - 16 = (3x)^2 - 4^2$.

Step 2: Factor using Special Product 3.

$$\begin{array}{ccccc} a^2 & - b^2 = & (a + b) & (a - b) \\ | & | & | \quad | & | \quad | \\ (3x)^2 & - 4^2 = & (3x + 4) & (3x - 4) \end{array}$$

■

So, we find what a *and* b *equal then substitute these quantities into Special Product 3.*

That's correct, Charlie. Now, a is the quantity which multiplied by itself equals the first term of the binomial and b is the quantity which multiplied by itself equals the second term of the binomial.

What happens if one of the terms of the binomial is not a perfect square?

Well, Charlie, unless both terms are perfect squares, we cannot factor using this method.

Example 7 **Factor the following binomial.**

$$36x^2 - 49y^2$$

Step 1: Check to find if the two terms of the binomial are perfect squares. Now, $36x^2 = 6x \times 6x = (6x)^2$ and $36x^2$ is a perfect square, and $49y^2 = 7y \times 7y = (7y)^2$ and $49y^2$ is a perfect square. Hence, $36x^2 - 49y^2 = (6x)^2 - (7y)^2$.

A N S W E R

Step 2: Factor using Special Product 3.

$$\begin{array}{ccccccc} a^2 & - & b^2 & = & (a + b)(a - b) \\ | & & | & & | \quad | \quad | \quad | \\ (6x)^2 & - & (7y)^2 & = & (6x + 7y)(6x - 7y) \end{array}$$

■

Practice 7: Factor the following binomial.

$$4x^2 - 9$$

7. ______________________

Practice 8: Factor the following binomial.

$$49x^2 - 100$$

8. ______________________

Practice 9: Factor the following binomial.

$$16x^2 - 9y^2$$

9. ______________________

Practice 10: Factor the following binomial.

$$144n^2 - 121$$

10. ______________________

Practice 11: Factor the following binomial.

$$81x^4 - 225y^2$$

11. ______________________

EXERCISE 6.3

Factor the following binomials.

A

1. $x^2 - 4$

2. $a^2 - 9$

3. $x^2 - 25$

4. $4x^2 - 1$

5. $16y^2 - 1$

6. $9y^2 - 4$

7. $64b^2 - 25$

8. $25a^2 - 16$

9. $81a^2 - 121$

10. $49y^2 - 81$

B

11. $x^4 - 225$

12. $x^6 - 1$

13. $t^6 - 9$

14. $a^4 - b^6$

15. $4x^{10} - 9$

16. $4a^4 - 25$

17. $x^2y^2 - 36$

18. $25a^8 - 16b^4$

19. $81x^4 - 169y^6$

20. $121a^2b^4 - 49$

6.4 FACTORING TRINOMIALS

To factor certain trinomials, we can use the FOIL rule (Special Product 4).

> **FOIL rule:** $(ax + b)(cx + d) = acx^2 + (ad + bc)x + bd$

Example 1 **Factor the following trinomial.**

$$x^2 + 5x + 6$$

Solution: Assume $x^2 + 5x + 6$ equals the product of two binomials, then

$$x^2 + 5x + 6 = (\_x + \_)(\_x + \_)$$

Step 1: Find the first term of each binomial. According to the FOIL rule, we have

$$\underbrace{(\_x + \_)(\_x + \_)}_{F \times F} = \underset{\to F}{x^2} + 5x + 6$$

That is, the product of the first terms of the binomials equals x^2. Now, $1x \times 1x = 1x^2 = x^2$; thus, the first term of each binomial is $1x$.

$$\overbrace{(1x + \_)(1x}^{F \times F} + \_) = \underbrace{x^2}_{F} + 5x + 6$$

<u>*Step 2:*</u> Find the last term of each binomial. According to the FOIL rule, we have

$$(1x + \overbrace{\_)(1x + \_}^{L \times L}) = x^2 + 5x + \underbrace{6}_{L}$$

That is, the product of the last terms of the binomials equals 6. Now, there are four pairs of factors whose product is 6.

$$2 \times 3 = 6 \qquad (-2) \times (-3) = 6$$

$$1 \times 6 = 6 \qquad (-1) \times (-6) = 6$$

However, only one of these pairs will give us the correct answer. Since all the terms of the trinomial have positive signs, the factors of 6 must be positive. Let's try 1 and 6 as the last terms of the binomials. Now, we must find whether the sum of the outer and inner products of the binomials gives us the middle term of the trinomial.

<u>*Step 3:*</u> Add the outer and inner products; this sum must equal $5x$. According to the FOIL rule, we have

$O \times O$ + $I \times I$ → $O \times O + I \times I$

$$\begin{aligned}(1x + 1)(1x + 6) &= x^2 + 6x + 1x + 6 \\ &= x^2 + 7x + 6\end{aligned}$$

We did *not* get $5x$ as the middle term; thus, we have to choose another pair of factors and repeat the process.

<u>*Step 4:*</u> Select another pair of factors as last terms for the binomials. There is only one other pair of positive factors whose product is 6, 2 and 3. Thus, we will try 2 and 3 as the last terms of the binomials.

<u>*Step 5:*</u> Add the outer and inner products; this sum must equal $5x$. According to the FOIL rule, we have

$O \times O$ + $I \times I$ → $O \times O + I \times I$

$$\begin{aligned}(1x + 2)(1x + 3) &= x^2 + 3x + 2x + 6 \\ &= x^2 + 5x + 6\end{aligned}$$

Therefore, $x^2 + 5x + 6 = (x + 2)(x + 3)$. ■

Let's look at $x^2 + 5x + 6 = (x + 2)(x + 3)$ and develop some guidelines to help you factor similar trinomials.

A N S W E R

> **Case 1:** Factoring $x^2 + bx + c$ into the product of two binomials, where b and c are positive.
> The last term of each binomial factor will have a positive sign since each term of $x^2 + bx + c$ has a positive sign.

Note: When selecting the last term of each binomial, if the sum of the outer and inner products does not equal the middle term of the trinomial, then you must choose another pair of factors (last term of each binomial) and repeat the process. For this reason this process is sometimes called factoring by trial and error.

What happens if we can't get the sum of the outer and inner products of the binomials to equal the middle term of the trinomial for any pair of factors.

Then we cannot factor the trinomial using the FOIL method; thus, the trinomial cannot be factored. However, be sure to check each pair of factors twice before saying that the trinomial cannot be factored.

1. ____________________

Practice 1: Factor the following trinomial.

$$x^2 + 7x + 10$$

2. ____________________

Practice 2: Factor the following trinomial.

$$x^2 + 7x + 12$$

3. ____________________

Practice 3: Factor the following trinomial.

$$x^2 + 14x + 48$$

Practice 4: Factor the following trinomial.

$$x^2 + 13x + 42$$

A N S W E R

4. ______________

Example 2 **Factor the following trinomial.**

$$x^2 - 8x + 15$$

Solution: Assume $x^2 - 8x + 15$ equals the product of two binomials, then

$$x^2 - 8x + 15 = (\_x + \_)(\_x + \_)$$

<u>*Step 1:*</u> Find the first term of each binomial. According to the FOIL rule, we have

$$\underbrace{(\_x + \_)(\_x + \_)}_{F \times F} = \underbrace{x^2}_{\to F} + 2x - 8$$

That is, the product of the first terms of the binomials equals x^2. Now, $1x \times 1x = x^2$; thus, the first term of each binomial is $1x$.

$$\underbrace{(1x + \_)(1x + \_)}_{F \times F} = \underbrace{x^2}_{\to F} + 2x - 8$$

<u>*Step 2:*</u> Find the last term of each binomial. According to the FOIL rule, we have

$$\underbrace{(1x + \_)(1x + \_)}_{L \times L} = x^2 + 2x - \underbrace{8}_{\to L}$$

That is, the product of the last terms of the binomials equals 15. Now, there are four pairs of factors whose product is 15.

$$3 \times 5 = 15 \qquad -3 \times (-5) = 15$$

$$1 \times 15 = 15 \qquad -1 \times (-15) = 15$$

However, only one of these pairs will give us the correct answer. Since the middle term of the trinomial, $-8x$, has a negative sign, the factors of 15 must be negative. Let's try -3 and -5 as the last terms of the binomials. Now, we must find if the sum of the outer and inner products of the binomials gives us the middle term of the trinomial.

<u>*Step 3:*</u> Add the outer and inner products; this sum must equal $-8x$.

$O \times O$ + $I \times I$ $\to$ $O \times O + I \times I$

$$(1x - 3)(1x - 5) = x^2 - 5x + (-3x) + 24$$
$$= x^2 - 8x + 24$$

Therefore, $x^2 - 8x + 15 = (x - 3)(x - 5)$. ■

A N S W E R

Let's look at $x^2 - 8x + 15 = (x - 3)(x - 5)$ and develop some guidelines to help you factor similar trinomials.

Case 2: Factoring $x^2 - bx + c$ into the product of two binomials, where b and c are positive.
The last term of each binomial factor will have a negative sign since the last term of $x^2 - bx + c$ is positive and the middle term has a negative sign.

Remember when selecting the last term of each binomial, if the sum of the outer and inner products does not equal the middle term of the trinomial, then you must choose another pair of factors (last term of each binomial) and repeat the process.

This factoring is a lot of work.

Well, Charlie, that depends on how you look at it. Do you like to solve crossword puzzles?

Sure, I like crossword puzzles; they are a lot of fun.

Well, Charlie, factoring a polynomial is like solving a crossword puzzle. You are looking for some missing numbers and you are given hints on how to find these numbers. If you enjoy solving crossword puzzles, you should enjoy factoring polynomials.

I never thought of factoring as being similar to solving a crossword puzzle. I will remember that when I work the practice problems.

5. ____________________

Practice 5: Factor the following trinomial.

$$x^2 - 7x + 12$$

Practice 6: Factor the following trinomial.

$$x^2 - 11x + 30$$

ANSWER

6. ______

Practice 7: Factor the following trinomial.

$$x^2 - 15x + 56$$

7. ______

Practice 8: Factor the following trinomial.

$$x^2 - 15x + 54$$

8. ______

Example 3 **Factor the following trinomial.**

$$x^2 + 2x - 8$$

Solution: Assume $x^2 + 2x - 8$ equals the product of two binomials, then

$$x^2 + 2x - 8 = (\_x + \_)(\_x + \_).$$

Step 1: Find the first term of each binomial. According to the FOIL rule, we have

$$\underbrace{(\_x + \_)(\_x + \_)}_{F \times F \to F} = x^2 - 8x + 15$$

That is, the product of the first terms of the binomials equals x^2. Now, $1x \times 1x = x^2$; thus, the first term of each binomial is $1x$.

$$\underbrace{(1x + \_)(1x + \_)}_{F \times F \to F} = x^2 - 8x + 15$$

A N S W E R

Step 2: Find the last term of each binomial. According to the FOIL rule, we have

$$\underbrace{(1x + \_)(1x + \_)}_{L \times L \;\to\; L} = x^2 - 8x + 15$$

That is, the product of the last terms of the binomials equals -8. Now, there are four pairs of factors whose product is -8.

$$-2 \times 4 = -8 \qquad 2 \times (-4) = -8$$

$$-1 \times 8 = -8 \qquad 1 \times (-8) = -8$$

However, only one of these pairs will give us the correct answer. Since the middle term of the trinomial, $2x$, has a positive sign, the larger factor (without regard to sign) should have a positive sign. Let's try -2 and 4 as the last terms of the binomials. Now, we must find if the sum of the outer and inner products of the binomials gives us the middle term of the trinomial.

Step 3: Add the outer and inner products; this sum must equal $2x$.

$$O \times O + I \times I \;\to\; O \times O + I \times I$$

$$\begin{aligned}(1x - 2)(1x + 4) &= x^2 + 4x + (-2x) - 8 \\ &= x^2 + 2x - 8\end{aligned}$$

Therefore, $x^2 + 2x - 8 = (x - 2)(x + 4)$. ■

Let's look at $x^2 + 2x - 8 = (x - 2)(x + 4)$ and develop some guidelines to help you factor similar trinomials.

> **Case 3:** Factoring $x^2 + bx - c$ into the product of two binomials, where b and c are positive.
> The last terms of the binomial factors will have opposite signs since the last term of $x^2 + bx - c$ is negative. Also the larger (without regard to sign) of the last terms of the binomials will have a positive sign since the middle term of the trinomial has a positive sign.

Remember when selecting the last term of each binomial, if the sum of the outer and inner products does not equal the middle term of the trinomial, then you must choose another pair of factors (last term of each binomial) and repeat the process.

9. ______________

Practice 9: Factor the following trinomial.

$$x^2 + x - 6$$

Practice 10: Factor the following trinomial.

$$x^2 + 2x - 15$$

A N S W E R

10. ______

Practice 11: Factor the following trinomial.

$$x^2 + x - 12$$

11. ______

Practice 12: Factor the following trinomial.

$$x^2 + 5x - 24$$

12. ______

Example 4 **Factor the following trinomial.**

$$x^2 - 7x - 18$$

Solution: Assume $x^2 - 7x - 18$ equals the product of two binomials, then

$$x^2 - 7x - 18 = (\_x + \_)(\_x + \_)$$

Step 1: Find the first term of each binomial. According to the FOIL rule, we have

$$\underbrace{(\_x + \_)(\_x + \_)}_{F \times F \;\rightarrow\; F} = x^2 - 7x - 18$$

That is, the product of the first terms of the binomials equals x^2. Now $1x \times 1x = x^2$; thus, the first term of each binomial is $1x$.

$$\underbrace{(1x + \_)(1x + \_)}_{F \times F \;\rightarrow\; F} = x^2 - 7x - 18$$

Step 2: Find the last term of each binomial. According to the FOIL rule, we have

$$\overbrace{(1x + \_)(1x + \_)}^{L \times L} = x^2 - 7x - \underbrace{18}_{L}$$

That is, the product of the last terms of the binomials equals -18. Now, there are six pairs of factors whose product is -18;

$$\begin{array}{ll} -3 \times 6 = -18 & 3 \times (-6) = -18 \\ -2 \times 9 = -18 & 2 \times (-9) = -18 \\ -1 \times 18 = -18 & 1 \times (-18) = -18 \end{array}$$

Since the middle term of the trinomial, $-7x$, has a negative sign, the larger factor (without regard to sign) should have a negative sign. Let's try 3 and -6 as the last terms of the binomials. Now, we must find if the sum of the outer and inner products of the binomials gives us the middle term of the trinomial.

Step 3: Add the outer and inner products; this sum must equal $-7x$. According to the FOIL rule, we have

$O \times O + I \times I \rightarrow O \times O + I \times I$

$$\begin{aligned}(1x + 3)(1x - 6) &= x^2 - 6x + 3x - 18 \\ &= x^2 - 3x - 18\end{aligned}$$

We did *not* get $-7x$ as the middle term; thus, we have to choose another pair of factors and repeat the process.

Step 4: Select another pair of factors as the last terms for the binomials. Since the middle term of the trinomials, $-7x$, has a negative sign, the larger factor (without regard to sign) should still have a negative sign. Let's try 2 and -9 as last terms for the binomials.

Step 5: Add the outer and inner products; this sum must equal $-7x$. According to the FOIL rule, we have

$O \times O + I \times I \rightarrow O \times O + I \times I$

$$\begin{aligned}(1x + 2)(1x - 9) &= x^2 - 9x + 2x - 18 \\ &= x^2 - 7x - 18\end{aligned}$$

Therefore, $x^2 - 7x - 18 = (x + 2)(x - 9)$. ■

Let's look at $x^2 - 7x - 18 = (x + 2)(x - 9)$ and develop some guidelines to help you factor similar trinomials.

ANSWER

Case 4: Factoring $x^2 - bx - c$ into the product of two binomials, where b and c are positive.
The last terms of the binomial factors will have oppositive signs since the last term of $x^2 - bx - c$ is negative. Also the larger (without regard to sign) of the last terms of the binomials will have a negative sign since the middle term of the trinomial has a negative sign.

Practice 13: Factor the following trinomial.

$$x^2 - 4x - 32$$

13. ____________________

Practice 14: Factor the following trinomial.

$$x^2 - 7x - 30$$

14. ____________________

Practice 15: Factor the following trinomial.

$$x^2 - x - 30$$

15. ____________________

Practice 16: Factor the following trinomial.

$$x^2 - 9x - 36$$

16. ____________________

Example 5 **Factor the following trinomial.**

$$6x^2 + 7x - 20$$

Solution: Assume $6x^2 + 7x - 20$ equals the product of two binomials, then

$$6x^2 + 7x - 20 = (\_x + \_)(\_x + \_)$$

Step 1: Find the first term of each binomial. According to the FOIL rule, we have

$$\underbrace{(\_x + \_)(\_x + \_)}_{F \times F} = \underbrace{6x^2}_{\to F} + 7x - 20$$

That is, the product of the first terms of the binomials equals $6x^2$. Now, there are two pairs of positive numbers whose product is 6.

$$2 \times 3 = 6$$

$$1 \times 6 = 6$$

However, only one of these pairs will give us the correct answer. Let's try 2 and 3 as the coefficients of the first terms of the binomials; that is, $2x$ and $3x$ will be the first terms of the binomials. We may discover later that 2 and 3 were not the correct choice and we will have to try 1 and 6. But, we will start with $2x$ and $3x$ as the first terms of the binomials.

$$\underbrace{(2x + \_)(3x + \_)}_{F \times F} = \underbrace{6x^2}_{\to F} + 7x - 20$$

Step 2: Find the last term of each binomial. According to the FOIL rule, we have

$$(2x + \underbrace{\_)(3x + \_}_{L \times L}) = 6x^2 + 7x - \underbrace{20}_{\to L}$$

That is, the product of the last terms of the binomials equals -20. Now, there are six pairs of factors whose product is -20.

$$-4 \times 5 = -20 \qquad 4 \times (-5) = -20$$

$$-2 \times 10 = -20 \qquad 2 \times (-10) = -20$$

$$-1 \times 20 = -20 \qquad 1 \times (-20) = -20$$

Let's try -4 and 5 as the last terms of the binomials. Since the first term of each binomial is different, we must decide where to place -4 and 5. Let's place -4 in the first binomial and 5 in the second binomial; if this does not work, then we will switch -4 and 5. Now, we must find if the sum of the outer and inner products of the binomials gives us the middle term of the trinomial.

ANSWER

Step 3: Add the outer and inner products; this sum must equal $7x$. According to the FOIL rule, we have

$$O \times O + I \times I$$

$$(3x - 4)(2x + 5) = 6x^2 + 15x + (-8x) - 20$$
$$= 6x^2 + 7x - 20$$

Therefore, $6x^2 + 7x - 20 = (3x - 4)(2x + 5)$. ■

Let's look at $6x^2 + 7x - 20$ and develop some guidelines to help you factor similar trinomials.

Case 5: Factoring $ax^2 + bx + c$ into the product of two binomials, where a is positive.
There may be more than one choice for the first term of each binomial since the first term of the trinomial is ax^2. Always select a positive first term for each binomial. After selecting the first term for each binomial, work through the process used in cases one through four.

Note: It may be that your choice of a first term for each binomial is not the correct choice. If this happens you will eventually discover it and make another selection of first terms.

You are never sure about your choices when factoring trinomials using the FOIL rule. It could take a long time to find all the right parts for the binomials.

That's true, Charlie. However, uncertainty is what makes life interesting. Who wants to watch a replay of a ballgame?

Practice 17: Factor the following trinomial.

$$2x^2 + x - 3$$

17. ______________

A N S W E R

18. ____________________

Practice 18: Factor the following trinomial.

$$3x^2 - x - 14$$

19. ____________________

Practice 19: Factor the following trinomial.

$$10x^2 - 19x - 15$$

20. ____________________

Practice 20: Factor the following trinomial.

$$8x^2 + 6x - 35$$

21. ____________________

Practice 21: Factor the following trinomial.

$$9x^2 + 30x + 25$$

Practice 22: Factor the following trinomial.

$$16x^2 - 56x + 49$$

A N S W E R

22. ______________

If a trinomial has a first and last term which are perfect squares, such as those in Practice 21 and Practice 22, then you may be able to factor the trinomial using Special Product 1 or 2.

Special Product 1. $(a + b)^2 = a^2 + 2ab + b^2$

Special Product 2. $(a - b)^2 = a^2 - 2ab + b^2$

Example 6 **Factor the following trinomial.**

$$4x^2 + 12x + 9$$

Step 1: Since $2x \times 2x = 4x^2$ and $3 \times 3 = 9$, $4x^2$ and 9 are perfect squares. Thus, $4x^2 + 12x + 9 = (2x)^2 + 12x + 3^2$.

Step 2: Factor using Special Product 1.

$$\begin{array}{ccccccccc} a^2 & + & 2ab & + & b^2 & = & (a & + & b)^2 \\ | & & & & | & ? & | & & | \\ (2x)^2 & + & 12x & + & 3^2 & = & (2x & + & 3)^2 \end{array}$$

Step 3: Even though the first and last terms match, we are not sure that the middle terms are equal. Thus, we must check to find if the sum of the outer and inner products ($2ab$) equals $12x$. Now,

$$\begin{array}{ccc} 2 & a & b \\ | & | & | \\ 2 & (2x) & (3) \end{array} = 12x$$

and the middle terms are equal. Therefore, $4x^2 + 12x + 9 = (2x + 3)^2$. ■

Example 7 **Factor the following trinomial.**

$$9x^2 - 30x + 16$$

Step 1: Since $3x \times 3x = 9x^2$ and $4 \times 4 = 16$, $9x^2$ and 16 are perfect squares. Thus, $9x^2 - 30x + 16 = (3x)^2 - 30x + 4^2$.

ANSWER

Step 2: Factor using Special Product 2.

$$\begin{array}{c} a^2 - 2ab + b^2 = (a - b)^2 \\ (3x)^2 - 30x + 4^2 \stackrel{?}{=} (3x - 4)^2 \end{array}$$

Step 3: Even though the first and last terms match, we are not sure that the middle terms are equal. Thus, we must check to find if the sum of the outer and inner products ($-2ab$) equals $-30x$. Now,

$$\begin{array}{c} -2\ a\ b \\ -2(3x)(4) = -24x \end{array}$$

and the middle terms are *not* equal. Therefore,

$$9x^2 - 30x + 16 \neq (3x - 4)^2.$$

Thus, $9x^2 - 30x + 16$ cannot be factored using Special Product 2; however, it can be factored using the FOIL rule. For this reason, some students prefer to always use the FOIL rule when factoring trinomials. However, let's work the following practice problems using Special Product 1 or 2. ■

23. ____________

Practice 23: Factor the following trinomial using Special Product 1 or 2.

$$25x^2 + 20x + 4$$

24. ____________

Practice 24: Factor the following trinomial using Special Product 1 or 2.

$$36x^2 - 12x + 1$$

25. ____________

Practice 25: Factor the following trinomial using Special Product 1 or 2.

$$64y^2 - 80y + 25$$

EXERCISE 6.4

Factor the following trinomials.

A

1. $x^2 + 6x + 8$

2. $x^2 + 5x + 6$

3. $x^2 + 4x - 5$

4. $x^2 - 2x - 3$

5. $x^2 - 4x + 3$

6. $x^2 - 6x + 5$

7. $t^2 - 12t + 35$

8. $y^2 - 2y - 35$

9. $x^2 + 2x - 15$

10. $y^2 - 10y + 21$

11. $x^2 + 6x + 9$

12. $y^2 - 10y + 25$

13. $a^2 - 12a + 36$

14. $b^2 + 8b + 16$

B

15. $2x^2 + x - 3$

16. $2x^2 + 7x + 3$

17. $6x^2 - 13x + 5$

18. $2y^2 + 13y + 15$

19. $2t^2 - 9t - 5$

20. $2a^2 + 13a - 24$

21. $2y^2 - 15y + 25$

22. $3x^2 + 10x - 8$

23. $9x^2 + 6x + 1$

24. $25y^2 + 20y + 4$

25. $4x^2 - 20x + 25$

6.5 COMPLETE FACTORIZATION OF A POLYNOMIAL

A polynomial is **completely factored** when it is written as a product of factors that cannot be factored further. When factoring a polynomial, we will follow the steps outlined in the diagram below.

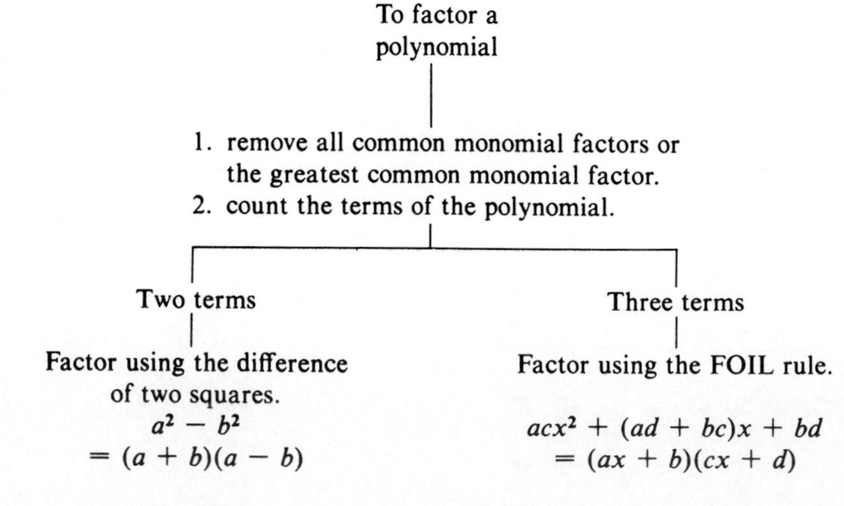

A N S W E R

Are there other ways to factor polynomials?

Yes, Charlie, there are other ways to factor polynomials. The diagram shows the methods that you have studied. When you learn a new method, you can add it to the diagram.

Note: When we say factor a polynomial, we mean completely factor the polynomial.

Example 1 **Factor the following polynomial.**

$$18x^2 - 50$$

Step 1: Remove all common monomial factors.

$$18x^2 - 50 = 2(9x^2 - 25)$$

Step 2: Now, $9x^2 - 25$ contains two terms and is the difference of two squares. Therefore, we continue to factor.

$$\begin{aligned} 18x^2 - 50 &= 2(9x^2 - 25) \\ &= 2[(3x)^2 - 5^2] \\ &= 2(3x + 5)(3x - 5) \end{aligned}$$

Thus, $18x^2 - 50 = 2(3x + 5)(3x - 5)$. ■

What would we have done if $9x^2 - 25$ had not been the difference of two squares?

We would not have been able to factor further.

Practice 1: Factor the following polynomial.

$$15x^2 - 60$$

1. ______________________

ANSWER

2. ______________

Practice 2: Factor the following polynomial.

$$32x^2 - 72$$

3. ______________

Practice 3: Factor the following polynomial.

$$50x^3 - 98x$$

4. ______________

Practice 4: Factor the following polynomial.

$$48a^4 - 75a^2$$

Example 2 **Factor the following polynomial.**

$$12y^3 + 26y^2 - 10y$$

Step 1: Remove all common monomial factors.

$$12y^3 + 26y^2 - 10y = 2y(6y^2 + 13y - 5)$$

Step 2: Now, $6y^2 + 13y - 5$ contains three terms; therefore, we continue to factor using the FOIL rule.

$$\begin{aligned} 12y^3 + 26y^2 - 10y &= 2y(6y^2 + 13y - 5) \\ &= 2y(3y - 1)(2y + 5) \end{aligned}$$

Thus, $12y^3 + 26y^2 - 10y = 2y(3y - 1)(2y + 5)$. ■

ANSWER

Practice 5: Factor the following polynomial.

$$6x^2 - 6x - 72$$

5. ____________________

Practice 6: Factor the following polynomial.

$$3x^2 + 27x + 42$$

6. ____________________

Practice 7: Factor the following polynomial.

$$36x^3 - 42x^2 - 18x$$

7. ____________________

Practice 8: Factor the following polynomial.

$$24y^3 + 6y^2 - 45y$$

8. ____________________

A N S W E R

Example 3 **Factor the following polynomial.**

$$6x^2y^2 + 24x^2$$

Step 1: Remove all common monomial factors.

$$6x^2y^2 + 24x^2 = 6x^2(y^2 + 4)$$

Step 2: Now, $y^2 + 4$ contains two terms; however, it is not the difference of two squares. Therefore, we cannot factor further. Thus,

$$6x^2y^2 + 24x^2 = 6x^2(y^2 + 4)$$

■

Example 4 **Factor the following polynomial.**

$$5x^4 + 25x^3 + 10x^2$$

Step 1: Remove all common monomial factors.

$$5x^4 + 25x^3 + 10x^2 = 5x^2(x^2 + 5x + 2)$$

Step 2: Now, $x^2 + 5x + 2$ contains three terms; therefore, we attempt to factor using the FOIL rule.

$$5x^4 + 25x^3 + 10x^2 = 5x^2(x^2 + 5x + 2)$$

Now, applying the FOIL rule to $x^2 + 5x + 2$, we find that the only possible binomial factors are $(x + 1)$ and $(x + 2)$. But,

$$(x + 1)(x + 2) = x^2 + 3x + 2$$

and the middle terms ($5x$ and $3x$) are not equal. Therefore, we cannot factor further. Thus,

$$5x^4 + 25x^3 + 10x^2 = 5x^2(x^2 + 5x + 2)$$

■

9. ______________

Practice 9: Factor the following polynomial.

$$6y^3 + 36y^2 + 18y$$

Practice 10: Factor the following polynomial.

$$9x^3 + 45x$$

ANSWER

10. ______________________

Practice 11: Factor the following polynomial.

$$6x^3y - 6xy^2$$

11. ______________________

Practice 12: Factor the following polynomial.

$$10x^3 - 40x^2 + 5x$$

12. ______________________

EXERCISE 6.5

Factor the following polynomials.

A

1. $2x^2 + 6x + 4$

2. $3x^2 - 6x - 9$

3. $3a^2 - 3a - 60$

4. $8x^2 - 32$

5. $2x^2 - 18x + 28$

6. $6x^2 - 54$

7. $12x^2 - 60$

8. $7x^2 - 35x - 21$

9. $72x^2 - 8$

10. $72x^2 - 30x - 12$

B

11. $4t^3 + 16t^2 - 48t$

12. $3t^3 - 3t^2 - 18t$

13. $2a^3 + 12a^2 + 18a$

14. $x^3 - 5x^2 - 6x$

15. $6x^3 + 9x^2 - 9x$

16. $5y^3 + 45y^2 + 100y$

17. $2x^3 + 4x^2 - 30x$

18. $2y^4 - 6y^3 - 8y^2$

19. $3a^3 - 3a^2 - 6a$

20. $2x^2 + 6$

21. $x^5 + 4x^4 + 4x^3$

22. $3x^4 - 12x^3 - 15x^2$

23. $5a^4 - 10a^3 + 5a^2$

24. $4y^4 - 48y^3 + 144y^2$

6.6 SOLVING A QUADRATIC EQUATION BY FACTORING

In this section, we will look at one application of factoring a polynomial (solving a quadratic equation). An equation which can be written in the form $ax^2 + bx + c = 0$, $a \neq 0$, is a **quadratic equation** in one variable.

Example 1 **The following equations are quadratic equations in one variable.**

$$2x^2 + 3x + 5 = 0$$

$$5x^2 - 8x - 6 = 0$$

$$3x^2 - 6 = 0$$

$$8x^2 - 2x = 9$$

$$3x^2 - 5 = 7x + 6$$

■

When solving quadratic equations by factoring, we will use the following rule.

Zero Product Rule. If $ab = 0$, then $a = 0$ or $b = 0$.

In terms of polynomials, the rule states that if the product of two polynomials equals zero, then one or both of the polynomials equals zero.

Will solving quadratic equations be difficult?

No, Charlie. You have already learned the skills that are needed to solve a quadratic equation by factoring: factoring a polynomial and solving a linear equation.

Great! I can relax for a while.

Let's look at an example before you become too relaxed.

Example 2 **Solve the following quadratic equation.**

$$x^2 - x = 6$$

Step 1: Write the equation in the form $ax^2 + bx + c = 0$.

$$x^2 - x = 6$$
$$x^2 - x - 6 = 0$$

Step 2: Factor the left side of the equation.

$$x^2 - x - 6 = 0$$
$$(x - 3)(x + 2) = 0$$

Step 3: Set each factor equal to zero (zero product rule).

$$x - 3 = 0 \quad \text{or} \quad x + 2 = 0$$

Step 4: Solve the resulting linear equations.

$$x - 3 = 0 \quad \text{or} \quad x + 2 = 0$$
$$x = 0 + 3 \qquad x = 0 - 2$$
$$x = 3 \qquad x = -2$$

Check: Substitute 3 and -2 for x in the original equation.

$x^2 - x = 6$ **Replace x with 3.**

$$3^2 - 3 \stackrel{?}{=} 6$$
$$9 - 3 \stackrel{?}{=} 6$$
$$6 = 6$$

$x^2 - x = 6$ **Replace x with -2.**

$$(-2)^2 - (-2) \stackrel{?}{=} 6$$
$$4 + 2 \stackrel{?}{=} 6$$
$$6 = 6$$

Thus, 3 and -2 are the solutions. ■

To solve a quadratic equation by factoring:
1. Write the quadratic equation in the form $ax^2 + bx + c = 0$.
2. Factor $ax^2 + bx + c$.
3. Set each factor equal to zero.
4. Solve the resulting linear equations.

Example 3 **Solve the following quadratic equation.**

$$2x^2 = -3x + 20$$

Step 1: Write the equation in the form $ax^2 + bx + c = 0$.

$$2x^2 = -3x + 20$$
$$2x^2 + 3x - 20 = 0$$

ANSWER

Step 2: Factor the left side of the equation.

$$2x^2 + 3x - 20 = 0$$
$$(2x - 5)(x + 4) = 0$$

Step 3: Set each factor equal to zero (zero product rule).

$$2x - 5 = 0 \quad \text{or} \quad x + 4 = 0$$

Step 4: Solve the resulting linear equations.

$$2x - 5 = 0 \quad \text{or} \quad x + 4 = 0$$
$$2x = 0 + 5 \qquad x = 0 - 4$$
$$2x = 5 \qquad x = -4$$
$$x = \frac{5}{2}$$

Check: Substitute $\frac{5}{2}$ and -4 for x in the original equation.

Replace x with $\frac{5}{2}$.

$$2x^2 = -3x + 20$$
$$2\left(\frac{5}{2}\right)^2 \stackrel{?}{=} -3\left(\frac{5}{2}\right) + 20$$
$$2\left(\frac{25}{4}\right) \stackrel{?}{=} -\frac{15}{2} + 20$$
$$\frac{25}{2} \stackrel{?}{=} -\frac{15}{2} + \frac{40}{2}$$
$$\frac{25}{2} \stackrel{?}{=} \frac{-15 + 40}{2}$$
$$\frac{25}{2} = \frac{25}{2}$$

Replace x with -4.

$$3x^2 = -3x + 20$$
$$2(-4)^2 \stackrel{?}{=} -3(-4) + 20$$
$$2(16) \stackrel{?}{=} 12 + 20$$
$$32 = 32$$

Thus, $\frac{5}{2}$ and -4 are the solutions. ■

Practice 1: Solve the following quadratic equation.

$$3x^2 - 6x = 0$$

1. ____________________

A N S W E R

Practice 2: Solve the following quadratic equation.

$$2x^2 - x = 15$$

2. ______________

3. ______________

Practice 3: Solve the following quadratic equation.

$$x^2 + 2 = 6$$

4. ______________

Practice 4: Solve the following quadratic equation.

$$12x^2 = 31x - 20$$

EXERCISE 6.6

Solve the following quadratic equations.

A

1. $x^2 + 2x = 0$

2. $2x^2 - 4x = 0$

3. $x^2 - 10x + 25 = 0$

4. $y^2 - y - 6 = 0$

5. $t^2 - 2t - 15 = 0$

6. $a^2 - 8a + 7 = 0$

7. $x^2 - 11x + 18 = 0$

8. $x^2 - 9 = 0$

9. $x^2 - 25 = 0$

10. $x^2 - 36 = 0$

B

11. $x^2 = 16$

12. $y^2 - 2y = 35$

13. $y^2 - 16y = -63$

14. $x^2 = -2x + 35$

15. $x^2 = -14x - 40$

16. $2x^2 + x - 3 = 0$

17. $5x^2 - 2x - 7 = 0$

18. $49y^2 - 1 = 0$

19. $64x^2 - 25 = 0$

20. $8x^2 + 2x - 3 = 0$

C

21. $2x^2 = 13x - 6$

22. $3x^2 = 48$

23. $x - 3 = -14x^2$

24. $12x^2 - 35 = -x$

25. $16x^2 = 10x + 21$

6.7 STRATEGIES FOR SOLVING APPLICATIONS

The word problems in this section will involve setting up and solving a quadratic equation. Once the equation is set up, you should be able to solve it by using the information in Section 6.6.

The suggestions for setting up a quadratic equation to solve a word problem are basically the same as those for setting up a linear equation to solve a word problem. You may want to review the linear equation suggestions in Chapter 4.

Example 1 **The width of a rectangle is 4 inches less than the length. The area is 96 in.². Find the length and width of the rectangle.**

Step 1: Let L = length. Then, $L - 4$ = width since it is 4 inches less than the length.

Step 2: Draw the rectangle.

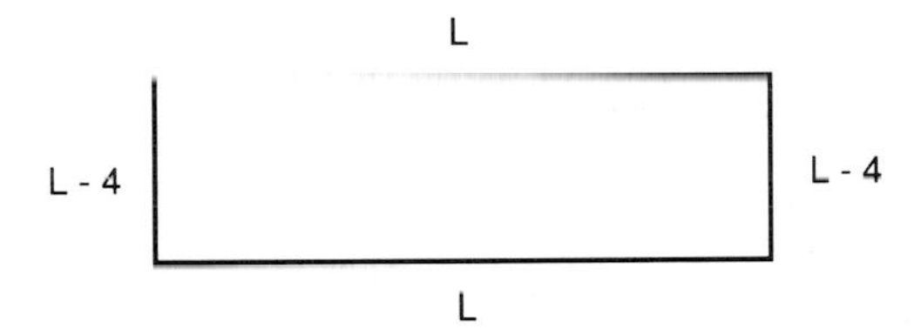

Step 3: Set up the equation. We know that the area of a rectangle is length times width. Therefore,

$$\text{Area} = (L - 4)L$$

The area is also equal to 96 in.². Thus,

$$(L - 4)L = 96$$

Step 4: Solve the equation.

$$(L - 4)L = 96$$

a. Use the distributive property.

$$L^2 - 4L = 96$$

b. Subtract 96 from both sides.

$$L^2 - 4L - 96 = 0$$

c. Factor.

$$(L - 12)(L + 8) = 0$$

d. Solve the resulting linear equations.

$$L - 12 = 0 \qquad L + 8 = 0$$

$$L = 12 \qquad L = -8$$

Since the length has to be a positive number, -8 is not a solution to the original problem. Therefore,

$$L = 12 \text{ inches is the length}$$

and $L - 4 = 12 - 4 = 8$ inches is the width ■

Example 2 **The product of two consecutive odd integers is 143. Find the integers.**

Step 1: Let x = first odd integer. Then $x + 2$ = next consecutive odd integer. For example, 5 is an odd integer; the next consecutive odd integer is $5 + 2 = 7$.

Step 2: Set up the equation. The product of two consecutive odd integers is $x(x + 2)$. Also the product is 143. Therefore,

$$x(x + 2) = 143$$

Step 3: Solve the equation.

$$x(x + 2) = 143$$

a. Use the distributive property.

$$x^2 + 2x = 143$$

b. Subtract 143 from both sides.

$$x^2 + 2x - 143 = 0$$

c. Factor.

$$(x + 13)(x - 11) = 0$$

d. Solve the resulting linear equations.

$$x + 13 = 0 \qquad x - 11 = 0$$

$$x = -13 \qquad x = 11$$

If $x = -13$, then $x + 2 = -13 + 2 = -11$ and one pair of consecutive odd integers whose product is 143 is $(-13,-11)$.
If $x = 11$, then $x + 2 = 13$ and a second pair of consecutive odd integers whose product is 143 is $(11,13)$. Therefore, $(-13,-11)$ and $(11,13)$ are the two pairs of consecutive odd integers whose product is 143. ■

Example 3 **If two times the square of a number minus three times the number is 27, find the number(s).**

Step 1: Let x = the number.

Step 2: Set up the equation.

$$\text{Square of the number} = x \times x = x^2$$

$$\text{Two times the square of the number} = 2x^2$$

$$\text{Three times the number} = 3x$$

Now, two times the square of a number minus three times the number is $2x^2 - 3x$ which equals 27. Therefore, $2x^2 - 3x = 27$.

Step 3: Solve the equation.

$$2x^2 - 3x = 27$$

a. Subtract 27 from both sides.

$$2x^2 - 3x - 27 = 0$$

b. Factor.

$$(2x - 9)(x + 3) = 0$$

c. Solve the resulting linear equations.

$$2x - 9 = 0 \qquad x + 3 = 0$$
$$2x = 9 \qquad x = -3$$
$$x = \frac{9}{2}$$

Therefore, there are two numbers that work: $\frac{9}{2}$ and -3. ■

Example 4 **The area of a triangle is 30 in.^{2}. If the base is 4 inches more than the height, find the base and height of the triangle.**

Step 1: Let h = height. Then $h + 4$ = base since the base is 4 inches more than the height.

Step 2: Draw the triangle.

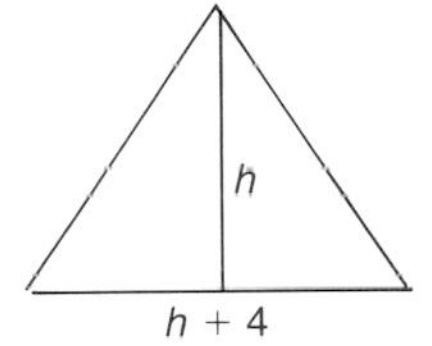

Step 3: Set up the equation.

$$\text{Area of a triangle} = \frac{1}{2}\,\text{base} \times \text{height}$$
$$30 = \frac{1}{2}(h + 4)(h)$$

Step 4: Solve the equation.

$$\frac{1}{2}(h + 4)(h) = 30$$

a. Multiply both sides by 2.

$$2\left[\frac{1}{2}(h + 4)(h)\right] = 2(30)$$
$$(h + 4)(h) = 60$$

b. Use the distributive property.

$$h^2 + 4h = 60$$

c. Subtract 60 from both sides.

$$h^2 + 4h - 60 = 0$$

d. Factor.

$$(h + 10)(h - 6) = 0$$

e. Solve the resulting linear equations.

$$h + 10 = 0 \qquad h - 6 = 0$$
$$h = -10 \qquad h = 6$$

The value -10 is not a solution to the original problem since the height of a triangle must be a positive number. Therefore, the height of the triangle is 6 inches and the base is 6 inches + 4 inches = 10 inches. ■

APPLICATION EXERCISES

1. The square of a positive number is increased by the number and the result is 20. Find the number.
2. If the area of a rectangle is 64 m² and the length is four times the width, find the width and length of the rectangle.
3. The product of two consecutive positive integers is 72. Find the integers.
4. The square of a positive number less three times the number is 10. Find the number.
5. The area of a triangle is 20 in.². If the base is 6 inches less than the height, what are the base and height?
6. The sum of a negative integer and its square is 42. Find the negative integer.
7. Find two consecutive positive even integers whose product is 224.
8. Find two numbers whose square is 196.
9. The square of a positive number is five more than four times the number. Find the number.
10. Adding 9 to the square of a number is the same as subtracting 3 from seven times the number. Find the number.
11. The length of each side of a square is extended 2 inches. The area of the resulting square is 144 in.². Find the length of a side of the original square.
12. An object is thrown downward from the top of a building 320 feet high with an initial speed of 16 feet per second. How many seconds later will the object hit the ground? Use the equation $d = vt + 16t^2$, where d is the distance in feet, v is the initial speed, and t is the time in seconds.
13. The length of the base of a triangle is three times the height. The area of the triangle is 24 in.². Find the length of the base.
14. John wants to construct a dog pen in the most economical fashion possible. He learns that a square pen is the most economical. He also learns that a pen must contain at least 144 square feet to maintain a healthy dog. How much wire will John need to buy in order to build a square pen with an area of 144 square feet?

*15. A 25,434 square foot lake is approximately circular in shape. How far is it across the lake—that is, what is the diameter of the lake? Use $A = \pi r^2$; $d = 2r$, and $\pi = 3.14$.

USEFUL INFORMATION

1. **Special Product 1.** $(a + b)^2 = (a + b)(a + b) = a^2 + 2ab + b^2$. For example,

$$(2x + 3)^2 = (2x)^2 + 2(2x)(3) + 3^2 = 4x^2 + 12x + 9$$

2. **Special Product 2.** $(a - b)^2 = (a - b)(a - b) = a^2 - 2ab + b^2$. For example,

$$(2x - 3)^2 = (2x)^2 - 2(2x)(3) + 3^2 = 4x^2 - 12x + 9$$

3. If a trinomial has a first term and a last term that are perfect squares, then you may be able to factor the trinomial using Special Product 1 or 2. For example, $x^2 + 6x + 9$ has x^2 and 9 that are perfect squares. Using special product 1,

$$x^2 + 6x + 9 = (x + 3)(x + 3)$$

4. **Special Product 3.** $(a + b)(a - b) = a^2 - b^2$. For example,

$$(2x + 3)(2x - 3) = (2x)^2 - 3^2 = 4x^2 - 9$$

5. Special Product 3 allows you to factor a binomial that is the difference of two squares. For example,

$$9y^2 - 4 = (3y + 2)(3y - 2)$$

6. **Special Product 4—The FOIL rule.** $(ax + b)(cx + d) = acx^2 + (ad + bc)x + bd$. For example,

$$(2x + 3)(3x + 1) = 6x^2 + (2 + 9)x + 3 = 6x^2 + 11x + 3$$

7. Special Product 4 (FOIL rule) allows you to factor certain trinomials. For example,

$$x^2 - 7x - 18 = (x + 2)(x - 9)$$

8. When factoring a polynomial, make sure you *first* factor out all common monomial factors, or the greatest common factor; then count the terms of the polynomial. If there are two terms, you may be able to factor using the difference of two squares. If there are three terms, you may be able to factor using the FOIL rule.

9. To solve a quadratic equation by factoring:
   a. Write the equation in the form

$$ax^2 + bx + c = 0$$

   b. Factor $ax^2 + bx + c$.
   c. Set each factor equal to zero.
   d. Solve the resulting linear equations.
   e. The solutions to the two linear equations are the solutions to the quadratic equation.

REVIEW PROBLEMS

Section 6.1

Use the special product rules to multiply the following binomials.

A

1. $(x - 1)(x - 1)$

2. $(t - 4)(t - 4)$

3. $(2y + 3)(2y + 3)$

4. $(x - 3)(x - 2)$

5. $(y - 5)(y + 5)$

6. $(2a - 1)(2a + 1)$

7. $(3t + 1)(t - 4)$

8. $(2x - 5)(x - 3)$

9. $(5a + 11)(5a + 11)$

10. $(9t - 15)(9t + 15)$

11. $(8t - 1)(2t + 3)$

12. $(5y - 3)(2y - 7)$

13. $(7p + 15)(2p + 3)$

14. $(3a + 12)(2a - 12)$

B

15. $(a - 2b)(a - 2b)$

16. $(5x + 3y)(5x - 3y)$

17. $(6a + b)(7a - 3b)$

18. $(3x + 4y)(3x + 4y)$

19. $(7a - 6b)(7a + 6b)$

20. $(6x + 4y)(3x - 2y)$

Section 6.2

Factor the following polynomials.

A

1. $12x + 24$

2. $18m - 9$

3. $15a - 12$

4. $26t - 52$

5. $32y - 40$

6. $18x + 42$

7. $35x + 49$

8. $77a - 22$

9. $63b - 42$

10. $72x + 81$

B

11. $9a^2 - 18a$

12. $5y^4 + 25y^3 - 20y^2$

13. $3x^4 - 15$

14. $2x^2 - 4x + 16$

15. $6t^2 - 12t + 4$

16. $5a^3 - 25a^2$

17. $7x^3 - 14x^2 + 7x$

18. $-12a^3 - 6a^2 - 18a$

19. $-3x^4 - 6x^2$

20. $-6x^3 + 15x^2 - 12x$

C

21. $15x^3y^3 + 25x^2y^2 + 40xy$

22. $-8x^3y^2 - 12x^2y$

23. $30a^2b - 10ab^2 - 15ab$

24. $9x^6 - 6x^4 + 3x^2$

25. $-26a^3b^4 + 39a^2b^2 - 52ab^2$

Section 6.3

Factor the following binomials.

A

1. $x^2 - 64$

2. $4a^2 - 25$

3. $9b^2 - 16$

4. $25y^2 - 81$

5. $16x^2 - 49$

6. $49x^2 - 121$

7. $9a^2 - 169$

8. $36x^2 - 49$

9. $64x^2 - 121$

10. $81a^2 - 169$

B

11. $t^2 - m^2$

12. $x^2 - 4y^2$

13. $9y^2 - 1$

14. $144a^2 - 25$

15. $x^{10} - 1$

16. $225x^2 - 4$

17. $t^4 - 49$

18. $1 - x^2$

19. $a^6 - b^4$

20. $81 - 4x^2$

Section 6.4

Factor the following trinomials.

A

1. $x^2 + 6x + 5$

2. $x^2 - 9x + 14$

3. $x^2 + 11x + 30$

4. $a^2 - 14a + 48$

5. $y^2 + 4y - 21$

6. $x^2 + 7x - 18$

7. $x^2 - 13x - 30$

8. $y^2 + 8y - 33$

9. $a^2 - 13a - 48$

10. $x^2 + 16x + 55$

B

11. $4x^2 + 4x + 1$

12. $9x^2 + 12x + 4$

13. $4x^2 + 12x + 9$

14. $9t^2 - 6t + 1$

15. $6a^2 - 7a - 3$

16. $5x^2 + x - 6$

17. $4x^2 + 15x + 9$

18. $3a^2 + 10a + 7$

19. $12t^2 - 7t - 12$

20. $12x^2 - x - 6$

21. $12x^2 - 13x - 35$

22. $2a^2 + 3a - 2$

23. $2b^2 + b - 3$

24. $3x^2 - 17x + 10$

25. $6y^2 + y - 70$

26. $2x^2 + 5x - 3$

Section 6.5

Factor the following polynomials.

A

1. $18x^2 + 12x + 2$

2. $2t^2 - 8t + 6$

3. $12x^2 + 6x - 60$

4. $5x^2 - 25x - 120$

5. $90x^2 - 105x + 30$

6. $90y^2 - 63y - 108$

7. $6x^2 + 150$

8. $44y^2 - 99$

9. $48x^2 - 147$

10. $45x^2 - 245$

B

11. $6x^4 - 16x^3 + 10x^2$

12. $12a^3 + 51a^2 - 45a$

13. $10x^3 - 35x^2 - 25x$

14. $-12y^2 - 18y - 6$

15. $4x^3 - 16x$

16. $8y^3 - 16y^2 + 6y$

17. $6y^3 + 14y^2 + 4y$

18. $6x^4 - 24x^2$

19. $16x^5 - x^3$

20. $28x^3 - 63x$

Section 6.6

Solve the following quadratic equations.

A

1. $x^2 - 2x - 15 = 0$

2. $x^2 + 9x + 14 = 0$

3. $x^2 - 10x + 24 = 0$

4. $x^2 - 11x + 24 = 0$

5. $6x^2 - x - 1 = 0$

6. $6x^2 + 19x - 7 = 0$

7. $6x^2 - 19x + 15 = 0$

8. $10x^2 + 13x - 3 = 0$

9. $8x^2 - 18x - 5 = 0$

10. $15x^2 - 31x + 10 = 0$

B

11. $x^2 + 6x = -8$

12. $t^2 - 7t = 8$

13. $y^2 + 5y = 6$

14. $3a^2 + 2a = 8$

15. $30 + x^2 = 11x$

16. $4y^2 = 49$

17. $y^2 = 9$

18. $25a^2 - 16 = 0$

19. $6x^2 - 15x - 9 = 0$

20. $2x^2 + 10x - 12 = 0$

ANSWERS TO PRACTICE PROBLEMS

Section 6.1

1. $a = x$ and $b = 4$
2. $a = 7x$ and $b = 6$
3. $a = 2x$ and $b = 3y$
4. $x^2 + 4x + 4$
5. $9x^2 + 24x + 16$
6. $64x^2 + 112x + 49$
7. $4x^2 + 12xy + 9y^2$
8. $9p^2 + 30pq + 25q^2$
9. $9x^2 - 30x + 25$
10. $16x^2 - 48x + 36$
11. $9x^2 - 12xy + 4y^2$
12. $25p^2 - 20pq + 4q^2$
13. $9x^2 - 4$
14. $16x^2 - 49$
15. $25x^2 - 9$
16. $4x^2 - y^2$
17. $49n^2 - 36$
18. $6x^2 + 7x + 2$
19. $4x^2 + 16x + 15$
20. $8x^2 + 18x - 5$
21. $6x^2 - x - 12$
22. $8x^2 - 16x + 6$
23. $7x^2 - 38x + 15$
24. $6x^2 - 7x - 20$
25. $2x^2 + xy - 6y^2$

Section 6.2

1. $2(x + 3)$
2. $4(3x + 2)$
3. $15(x - 3)$
4. $3(2x^2 + 5x + 3)$
5. $-8(x^2 + 2x - 4)$
6. $16a^2(2a - 3b)$
7. $3x(2x^2 + 3x - 4)$
8. $-2a(8a^2 + 25b^2)$

Section 6.3

1. no
2. yes, x^5x^5
3. yes, $2x \times 2x$
4. no
5. no
6. yes, $7a^2 \times 7a^2$
7. $(2x + 3)(2x - 3)$
8. $(7x + 10)(7x - 10)$
9. $(4x + 3y)(4x - 3y)$
10. $(12n + 11)(12n - 11)$
11. $(9x^2 + 15y)(9x^2 - 15y)$

Section 6.4

1. $(x + 5)(x + 2)$
2. $(x + 4)(x + 3)$
3. $(x + 6)(x + 8)$
4. $(x + 7)(x + 6)$
5. $(x - 4)(x - 3)$
6. $(x - 5)(x - 6)$
7. $(x - 8)(x - 7)$
8. $(x - 9)(x - 6)$
9. $(x + 3)(x - 2)$
10. $(x + 5)(x - 3)$
11. $(x + 4)(x - 3)$
12. $(x + 8)(x - 3)$
13. $(x - 8)(x + 4)$
14. $(x - 10)(x + 3)$
15. $(x - 6)(x + 5)$
16. $(x - 12)(x + 3)$
17. $(2x + 3)(x - 1)$
18. $(3x - 7)(x + 2)$
19. $(5x + 3)(2x - 5)$
20. $(4x - 7)(2x + 5)$
21. $(3x + 5)(3x + 5)$
22. $(4x - 7)(4x - 7)$
23. $(5x + 2)(5x + 2)$
24. $(6x - 1)(6x - 1)$
25. $(8y - 5)(8y - 5)$

Section 6.5

1. $15(x + 2)(x - 2)$
2. $8(2x + 3)(2x - 3)$
3. $2x(5x + 7)(5x - 7)$
4. $3a^2(4a + 5)(4a - 5)$
5. $6(x - 4)(x + 3)$
6. $3(x + 7)(x + 2)$
7. $6x(3x + 1)(2x - 3)$
8. $3y(4y - 5)(2y + 3)$
9. $6y(y^2 + 6y + 3)$
10. $9x(x^2 + 5)$
11. $6xy(x^2 - y)$
12. $5x(2x^2 - 8x + 1)$

Section 6.6

1. $0, 2$
2. $-\frac{5}{2}, 3$
3. $-2, 2$
4. $\frac{5}{4}, \frac{4}{3}$

CHAPTER 7

Algebraic Fractions

The objective of this chapter is:

1. **To be able to add, subtract, multiply, and divide algebraic fractions.**

A fraction which contains at least one algebraic expression is an **algebraic fraction.**

Example 1 **The following are algebraic fractions.**

$$\frac{2x+3}{5}, \frac{2x}{3x^2+2x-5}, \frac{2x^2-5x+6}{4x^2+6x+8}, \frac{x^2y}{5x^4y^3}$$ ■

In this chapter, we will examine algebraic fractions in which the algebraic expressions are polynomials.

The rules for adding, subtracting, multiplying, and dividing algebraic fractions are the same as the rules for arithmetic fractions.

7.1 REDUCING ALGEBRAIC FRACTIONS

The procedure for reducing algebraic fractions is the same as that for arithmetic fractions. That is, we completely factor the numerator and denominator of the fraction and remove the common factors. An algebraic fraction in which all the common factors have been removed is in reduced (simplified) form.

Let's look at how we remove common factors from fractions.

$$\frac{ac}{bc} = \frac{a}{b} \times \frac{c}{c}$$ **Definition of multiplication of fractions.**

$$= \frac{a}{b} \times 1$$ **Any nonzero number divided by itself is 1.**

$$= \frac{a}{b}$$ **Multiplication property of 1.**

Remember that a can be 1. For example,

$$\frac{3}{6} = \frac{1 \times 3}{2 \times 3} = \frac{1}{2} \times \frac{3}{3} = \frac{1}{2} \times 1 = \frac{1}{2}$$

Example 1 **Write the following fraction in reduced form.**

$$\frac{5x^2y^3}{35x^4y^2}$$

Step 1: Completely factor the numerator and denominator.

$$\frac{5x^2y^3}{5 \times 7x^4y^2}$$

Step 2: Remove common factors.

a. Remove the 5 from the numerator and the 5 from the denominator.

$$\frac{5x^2y^3}{5 \times 7x^4y^2} = \frac{x^2y^3}{7x^4y^2}$$

b. Remove the two x's in the numerator and two of the four x's in the denominator, leaving two x's in the denominator.

$$\frac{x^2y^3}{7x^4y^2} = \frac{y^3}{7x^2y^2}$$

c. Remove two of the three y's in the numerator and the two y's in the denominator, leaving one y in the numerator.

A N S W E R

$$\frac{y^3}{7x^2y^2} = \frac{y}{7x^2}$$

Thus,
$$\frac{5x^2y^3}{35x^4y^2} = \frac{y}{7x^2}$$

Let's look at how we removed the two y's in Part c.

$$\frac{y^3}{7x^2y^2} = \frac{yy^2}{7x^2y^2} = \frac{y}{7x^2} \times \frac{y^2}{y^2} = \frac{y}{7x^2} \times 1 = \frac{y}{7x^2}$$ ■

Practice 1: Write the following fraction in reduced form.

$$\frac{3x^2y}{6x^3y}$$

1. ______________

Practice 2: Write the following fraction in reduced form.

$$\frac{-2a^3b^4}{8a^3b^2}$$

2. ______________

Practice 3: Write the following fraction in reduced form.

$$\frac{32a^4b^2c^5}{16a^3b^2c^7}$$

3. ______________

Practice 4: Write the following fraction in reduced form.

$$\frac{13x^2y^3z^2}{26x^3y^4z^2}$$

4. ______________

Example 2 **Write the following fractions in reduced form.**

Algebra	Arithmetic
$\dfrac{x^2 - x - 2}{2x^2 + 7x + 5}$	$\dfrac{30}{45}$
Step 1: Completely factor the numerator and denominator. $\dfrac{(x + 1)(x - 2)}{(x + 1)(2x + 5)}$	*Step 1:* Completely factor the numerator and denominator. $\dfrac{2 \times 3 \times 5}{3 \times 3 \times 5}$
Step 2: Remove common factors. $\dfrac{\cancel{(x + 1)}(x - 2)}{\cancel{(x + 1)}(2x + 5)} = \dfrac{x - 2}{2x + 5}$	*Step 2:* Remove common factors. $\dfrac{2 \times \cancel{3} \times \cancel{5}}{3 \times \cancel{3} \times \cancel{5}} = \dfrac{2}{3}$ ■

Once again, the algebra problem is worked using the same procedure as the corresponding arithmetic problem.

So, reducing algebraic fractions is just like reducing arithmetic fractions. You factor the numerator and denominator, then remove the common factors.

That's correct, Charlie. Since you already know how to factor, you should be able to reduce algebraic fractions. Let's look at another example and then work some practice problems.

Example 3 **Write the following fraction in reduced form.**

$$\frac{4x^2 - 9}{2x^2 + x - 3}$$

Step 1: Completely factor the numerator and denominator.

$$\frac{4x^2 - 9}{2x^2 + x - 3} = \frac{(2x - 3)(2x + 3)}{(x - 1)(2x + 3)}$$

Step 2: Remove common factors.

$$\frac{(2x - 3)\cancel{(2x + 3)}}{(x - 1)\cancel{(2x + 3)}} = \frac{2x - 3}{x - 1}$$

Thus,

$$\frac{4x^2 - 9}{2x^2 + x - 3} = \frac{2x - 3}{x - 1}$$

■

ANSWER

Practice 5: Write the following fraction in reduced form.

$$\frac{x^2 - 3x + 2}{x^2 + 2x - 3}$$

5. ______________

Practice 6: Write the following fraction in reduced form.

$$\frac{x^2 - 4}{3x^2 - 5x - 2}$$

6. ______________

Practice 7: Write the following fraction in reduced form.

$$\frac{6x^2 + 5x - 6}{8x^2 + 2x - 15}$$

7. ______________

Practice 8: Write the following fraction in reduced form.

$$\frac{12x^2 + 5x - 3}{16x^2 - 9}$$

8. ______________

EXERCISE 7.1

Write the following fractions in reduced form.

A

1. $\dfrac{x}{3x}$

2. $\dfrac{2y^2}{10y^3}$

3. $\dfrac{12a^3b^2}{16a^2b}$

4. $\dfrac{72x^4y^3}{45x^7y^2}$

5. $\dfrac{-26xy^3z}{-39x^4yz^3}$

6. $\dfrac{-18a^5b^2c^3}{27a^2b^3c^5}$

7. $\dfrac{a(a+1)}{6(a+1)}$

8. $\dfrac{x+2}{x^2-4}$

9. $\dfrac{a+3}{a^2-9}$

10. $\dfrac{w^2+7w}{w^2-49}$

11. $\dfrac{2y+3}{2y-3}$

12. $\dfrac{x^2+10x+25}{x+5}$

13. $\dfrac{a^2+2ab+b^2}{a+b}$

14. $\dfrac{x+8}{x^2+2x-48}$

B

15. $\dfrac{a^2 - 7ab + 10b^2}{a - 2b}$

16. $\dfrac{x^2 - 3x - 10}{x^2 - 8x + 15}$

17. $\dfrac{y^2 - 7y + 6}{y^2 - 2y + 1}$

18. $\dfrac{y^2 + 7y + 6}{y^2 + 12y + 36}$

19. $\dfrac{m^2 + 2m - 8}{2m^2 - m - 6}$

20. $\dfrac{3x^2 - 4x - 4}{3x^2 - 7x - 6}$

21. $\dfrac{2x^2 + 7x - 15}{2x^2 + 5x - 12}$

22. $\dfrac{x^2 - 4x + 3}{x^2 - 2x - 3}$

23. $\dfrac{x - y}{y - x}$

24. $\dfrac{x^2 - 1}{1 - x}$

7.2 MULTIPLYING ALGEBRAIC FRACTIONS

The procedure for multiplying algebraic fractions is the same as that for arithmetic fractions. Let's write the definition of multiplication for fractions.

Definition of Multiplication of Fractions. For fractions $\frac{a}{b}$ and $\frac{c}{d}$, $\frac{a}{b} \times \frac{c}{d} = \frac{a \times c}{b \times d}$.

Example 1 **Multiply the following fractions and reduce, if possible.**

$$\frac{5x^2}{14y^3} \times \frac{35y^5}{15x^3}$$

Step 1: Using the definition of multiplication write the fractions as one fraction indicating the multiplications.

$$\frac{5x^2}{14y^3} \times \frac{35y^5}{15x^3} = \frac{5x^2 \times 35y^5}{14y^3 \times 15x^3}$$

Step 2: Reduce the fraction, if possible.

a. Completely factor the numerator and denominator.

$$\frac{5x^2 \times 35y^5}{14y^3 \times 15x^3} = \frac{5x^2 \times 5 \times 7y^5}{2 \times 7y^3 \times 3 \times 5x^3}$$

b. Remove common factors.

i. Remove the 7 from the numerator and the 7 from the denominator.

$$\frac{5x^2 \times 5 \times 7y^5}{2 \times 7y^3 \times 3 \times 5x^3} = \frac{5x^2 \times 5y^5}{2y^3 \times 3 \times 5x^3}$$

ii. Remove one of the two 5's from the numerator and the 5 from the denominator.

$$\frac{5x^2 \times 5y^5}{2y^3 \times 3 \times 5x^3} = \frac{5x^2y^5}{2y^3 \times 3x^3}$$

iii. Remove the two x's in the numerator and two of the three x's in the denominator, leaving one x in the denominator.

$$\frac{5x^2y^5}{2y^3 \times 3x^3} = \frac{5y^5}{2y^3 \times 3x}$$

iv. Remove three of the five y's in the numerator and the three y's in the denominator, leaving two y's in the numerator.

$$\frac{5y^5}{2y^3 \times 3x} = \frac{5y^2}{2 \times 3x}$$

Step 3: Perform the multiplications.

$$\frac{5y^2}{2 \times 3x} = \frac{5y^2}{6x}$$

Thus,

$$\frac{5x^2}{14y^3} \times \frac{35y^5}{15x^3} = \frac{5y^2}{6x}.$$

■

A N S W E R

Practice 1: Multiply the following fractions and reduce, if possible.

$$\frac{6x^2}{10y} \times \frac{5y^3}{3x^5}$$

1. ____________

Practice 2: Multiply the following fractions and reduce, if possible.

2. ____________

$$\frac{14x^5}{18y^3} \times \frac{9y^2}{49x^3}$$

Practice 3: Multiply the following fractions and reduce, if possible.

3. ____________

$$\frac{-4a^3}{15b^3} \times \frac{5b^4}{44a^7}$$

Example 2 **Multiply the following fractions and reduce, if possible.**

Algebra	**Arithmetic**
$\frac{x+2}{2x^2+5x-3} \times \frac{4x^2-1}{x^2+4x+4}$	$\frac{2}{15} \times \frac{9}{14}$
Step 1: Using the definition of multiplication, write the fractions as one fraction indicating the multiplications.	*Step 1:* Using the definition of multiplication, write the fractions as one fraction indicating the multiplications.
$\frac{(x+2) \times (4x^2-1)}{(2x^2+5x-3) \times (x^2+4x+4)}$	$\frac{2 \times 9}{15 \times 14}$

Step 2: Reduce the fraction, if possible.

a. Completely factor the numerator and denominator.

$$\frac{(x+2) \times (2x+1)(2x-1)}{(2x-1)(x+3) \times (x+2)(x+2)}$$

b. Remove common factors.

$$\frac{\cancel{(x+2)} \times (2x+1)\cancel{(2x-1)}}{\cancel{(2x-1)}(x+3) \times \cancel{(x+2)}(x+2)} = \frac{(2x+1)}{(x+3) \times (x+2)}$$

Step 3: Perform the multiplications.

$$\frac{(2x+1)}{(x+3) \times (x+2)} = \frac{2x+1}{x^2+5x+6}$$

Step 2: Reduce the fraction, if possible.

a. Completely factor the numerator and denominator.

$$\frac{2 \times (3 \times 3)}{(3 \times 5) \times (2 \times 7)}$$

b. Remove common factors.

$$\frac{\cancel{2} \times (3 \times \cancel{3})}{(\cancel{3} \times 5) \times (\cancel{2} \times 7)} = \frac{3}{5 \times 7}$$

Step 3: Perform the multiplications.

$$\frac{3}{5 \times 7} = \frac{3}{35}$$

■

To multiply algebraic fractions:

1. Use the definition of multiplication to write the fractions as one fraction indicating the multiplications. (Do not actually perform the multiplications.)
2. Completely factor the numerator and denominator.
3. Remove common factors.
4. Perform the multiplications.

Example 3 **Multiply the following fractions and reduce, if possible.**

$$\frac{x^2-25}{3x^2+5x-2} \times \frac{x^2-4}{x^2+2x-15}$$

Step 1: Write the fractions as one fraction indicating the multiplications.

$$\frac{(x^2-25) \times (x^2-4)}{(3x^2+5x-2) \times (x^2+2x-15)}$$

Step 2: Reduce the fraction, if possible.

a. Completely factor the numerator and denominator.

$$\frac{(x+5)(x-5) \times (x+2)(x-2)}{(3x-1)(x+2) \times (x+5)(x-3)}$$

b. Remove common factors.

$$\frac{\cancel{(x+5)}(x-5) \times \cancel{(x+2)}(x-2)}{(3x-1)\cancel{(x+2)} \times \cancel{(x+5)}(x-3)} = \frac{(x-5)(x-2)}{(3x-1)(x-3)}$$

ANSWER

Step 3: Perform the multiplications.

$$\frac{(x-5)(x-2)}{(3x-1)(x-3)} = \frac{x^2-7x+10}{3x^2-10x+3}$$

Thus, $$\frac{x^2-25}{3x^2+5x-2} \times \frac{x^2-4}{x^2+2x-15} = \frac{x^2-7x+10}{3x^2-10x+3}$$ ■

Practice 4: Multiply the following fractions and reduce, if possible.

$$\frac{5x}{2x-4} \times \frac{x-2}{5x+15}$$

4. ____________

Practice 5: Multiply the following fractions and reduce, if possible.

$$\frac{x^2-x-6}{x^2-4} \times \frac{x^2-x-2}{x^2+2x-15}$$

5. ____________

Practice 6: Multiply the following fractions and reduce, if possible.

$$\frac{6a^2+a-2}{4a^2-1} \times \frac{a-1}{9a^2+12a+4}$$

6. ____________

A N S W E R

7. ____________

Practice 7: Multiply the following fractions and reduce, if possible.

$$\frac{2y^2 - 3y - 2}{6y^2 + y - 15} \times \frac{2y^2 + y - 6}{y^2 - 4}$$

EXERCISE 7.2

Multiply the following fractions in each problem and reduce, if possible.

A

1. $\frac{9m^2}{16} \times \frac{4}{3m}$

2. $\frac{21z^4}{8} \times \frac{12}{7z^3}$

3. $\frac{9x^5}{5y^6} \times \frac{30y^3}{27x^7}$

4. $\frac{52a^9}{35b^3} \times \frac{49b^6}{39a^5}$

5. $\frac{-7x^7}{22y^3} \times \frac{33y^2}{28x^5}$

6. $\frac{-9a^2}{39b^3} \times \frac{-65b}{15a^3}$

7. $\frac{a+b}{2} \times \frac{12}{(a+b)^2}$

8. $\frac{3(x-1)}{y} \times \frac{2y}{5(x-1)}$

9. $\frac{5m + 25}{10} \times \frac{12}{6m + 30}$

10. $\frac{9y - 18}{6y + 12} \times \frac{3y + 6}{15y - 30}$

11. $\frac{3a + 12}{8} \times \frac{16a}{9a + 36}$

12. $\frac{a^2 - 9}{a^2 - 4a + 4} \times \frac{a - 2}{a + 3}$

13. $\frac{z^3}{z + 3} \times \frac{z^2 - 9}{z^2 - 3z}$

14. $\frac{5x + 5}{x - 2} \times \frac{x^2 - 4x + 4}{x^2 - 1}$

B

15. $\frac{4x + 4}{4x - 4} \times \frac{x^2 - 2x + 1}{x^2 - 1}$

16. $\frac{y^2 - 5y + 6}{y^2 - 9} \times \frac{y + 2}{y - 2}$

17. $\frac{x^2 + 2x + 1}{9x^2} \times \frac{3x^3}{x^2 - 1}$

18. $\frac{12x^4}{x^2 + 8x + 16} \times \frac{x^2 - 16}{3x^2}$

19. $\dfrac{y^2 + 6y + 5}{7y^2 - 63} \times \dfrac{7y + 21}{y + 5}$

20. $\dfrac{x^2 - 16}{5x - 1} \times \dfrac{5x^2 - x}{x^2 - 8x + 16}$

21. $\dfrac{2x - 8}{x^2 - 4} \times \dfrac{x^2 + 6x + 8}{x - 4}$

22. $\dfrac{x - 1}{x^2 - x - 6} \times \dfrac{x^2 + 5x + 6}{x^2 - 1}$

23. $\dfrac{x^2 - 13x + 42}{x^2 + 10x + 21} \times \dfrac{x^2 + x - 6}{x^2 - 4}$

24. $\dfrac{x^2 - 1}{x^2 + 7x + 10} \times \dfrac{x^2 + 5x + 6}{x + 1} \times \dfrac{x + 5}{x^2 + 2x - 3}$

7.3 DIVIDING ALGEBRAIC FRACTIONS

The procedure for dividing algebraic fractions is the same as that for arithmetic fractions. Let's write the definition of division for fractions.

Definition of Division of Fractions. If $\frac{c}{d}$ is not equal to 0, then $\frac{a}{b} \div \frac{c}{d} = \frac{a}{b} \times \frac{d}{c}$.

Example 1 **Divide the following fractions and reduce, if possible.**

Algebra

$$\frac{x + 2}{9x^2 + 15x + 4} \div \frac{x^2 - 6x - 16}{9x^2 - 16}$$

Step 1: Using the definition of division, change the division to multiplication.

$$\frac{x + 2}{9x^2 + 15x + 4} \times \frac{9x^2 - 16}{x^2 - 6x - 16}$$

Step 2: Multiply the resulting fractions.

$$\frac{(x + 2) \times (9x^2 - 16)}{(9x^2 + 15x + 4) \times (x^2 - 6x - 16)}$$

$$= \frac{\cancel{(x + 2)} \times \cancel{(3x + 4)}(3x - 4)}{(3x + 1)\cancel{(3x + 4)} \times \cancel{(x + 2)}(x - 8)}$$

$$= \frac{(x + 2) \times (3x + 4)(3x - 4)}{(3x + 1)(3x + 4) \times (x + 2)(x - 8)}$$

$$= \frac{3x - 4}{(3x + 1) \times (x - 8)}$$

$$= \frac{3x - 4}{3x^2 - 23x - 8}$$

Arithmetic

$$\frac{2}{21} \div \frac{22}{35}$$

Step 1: Using the definition of division, change the division to multiplication.

$$\frac{2}{21} \times \frac{35}{22}$$

Step 2: Multiply the resulting fractions.

$$\frac{2 \times 35}{21 \times 22}$$

$$= \frac{\cancel{2} \times (5 \times \cancel{7})}{(3 \times \cancel{7}) \times (\cancel{2} \times 11)}$$

$$= \frac{2 \times (5 \times 7)}{(3 \times 7) \times (2 \times 11)}$$

$$= \frac{5}{3 \times 11} = \frac{5}{33}$$

■

To divide two algebraic fractions:

1. Use the definition of division to change the division to multiplication.
2. Multiply the resulting fractions (see Section 7.2).

So, if you can multiply fractions, then you should be able to divide fractions.

Example 2 **Divide the following fractions and reduce, if possible.**

$$\frac{2x + 3}{10x^2 + 13x - 3} \div \frac{3x - 7}{5x^2 + 9x - 2}$$

ANSWER

Step 1: Using the definition of division, change the division to multiplication.

$$\frac{2x + 3}{10x^2 + 13x - 3} \times \frac{5x^2 + 9x - 2}{3x - 7}$$ **Definition of division of fractions.**

Step 2: Multiply the resulting fractions.

$$\frac{(2x + 3) \times (5x^2 + 9x - 2)}{(10x^2 + 13x - 3) \times (3x - 7)}$$ **Definition of multiplication of fractions.**

$$= \frac{(2x + 3) \times (5x - 1)(x + 2)}{(5x - 1)(2x + 3) \times (3x - 7)}$$ **Factor the numerator and denominator.**

$$= \frac{\cancel{(2x + 3)} \times \cancel{(5x - 1)}(x + 2)}{\cancel{(5x - 1)}\cancel{(2x + 3)} \times (3x - 7)}$$ **Remove common factors.**

$$= \frac{x + 2}{3x - 7}$$

■

1. ____________________

Practice 1: Divide the following fractions and reduce, if possible.

$$\frac{2x - 3}{x^2 - x - 6} \div \frac{x^2 - 16}{x^2 + x - 12}$$

2. ____________________

Practice 2: Divide the following fractions and reduce, if possible.

$$\frac{4x^2 - 25}{4x^2 + 4x - 15} \div \frac{2x - 5}{2x^2 + 3x - 9}$$

3. ____________________

Practice 3: Divide the following fractions and reduce, if possible.

$$\frac{p^2 - 3p}{4p^2 - 1} \div \frac{p^2 - 9}{4p^2 + 4p + 1}$$

Practice 4: Divide the following fractions and reduce, if possible.

$$\frac{8x^2 + 2x - 15}{4x^2 + 16x + 15} \div \frac{4x^2 + 23x - 35}{6x^2 + 13x - 5}$$

ANSWER

4. ______________

EXERCISE 7.3

Divide the following fractions in each problem and reduce, if possible.

A

1. $\frac{2x^2}{5y^2} \div \frac{8x^3}{10y}$

2. $\frac{24a^5}{33b^4} \div \frac{32a^3}{22b^6}$

3. $\frac{x}{x + 2} \div \frac{x}{3}$

4. $\frac{2a}{a + 1} \div \frac{5a}{a + 1}$

5. $\frac{x + 5}{3x} \div \frac{x + 5}{9x}$

6. $\frac{6(a + 4)}{5} \div \frac{4(a + 4)}{35}$

7. $\frac{5y^4}{y^2 - 1} \div \frac{5y^3}{y^2 + 2y + 1}$

8. $\frac{6a^2}{a^2 - 25} \div \frac{3a^3}{5a - 25}$

9. $\dfrac{z^2 - 4z + 4}{z^2 - 1} \div \dfrac{z - 2}{5z + 5}$

10. $\dfrac{x^2 + 4x + 4}{x^2 - 4} \div \dfrac{x + 2}{3x - 6}$

11. $\dfrac{x^2 - 36}{x^2 - 6x} \div (x + 6)$

12. $\dfrac{y^2 - y - 20}{y^2 + 7y + 12} \div \dfrac{y^2 - 25}{y + 3}$

B

13. $\dfrac{2r + 2p}{8r} \div \dfrac{r^2 + rp}{72}$

14. $\dfrac{4y + 12}{2y - 10} \div \dfrac{y^2 - 9}{y^2 - y - 20}$

15. $\dfrac{8 - x}{8 + x} \div \dfrac{x - 8}{x + 8}$

16. $\dfrac{x^2 - x - 6}{x^2 + x - 12} \div \dfrac{x^2 + 2x - 3}{x^2 + 3x - 4}$

17. $\dfrac{y^2 + y - 2}{y^2 + 3y - 4} \div \dfrac{y + 2}{y + 3}$

18. $\dfrac{2m^2 - 5m - 12}{m^2 - 10m + 24} \div \dfrac{4m^2 - 9}{m^2 - 9m + 18}$

19. $\dfrac{a^2 + a - 2}{a^2 + 5a + 24} \div \dfrac{a - 1}{a}$

20. $\dfrac{y - 5}{y^2 + 3y + 2} \div \dfrac{y^2 - 5y + 6}{y^2 - 2y - 3}$

21. $\dfrac{2a^2 + 7a + 3}{a^2 - 16} \div \dfrac{4a^2 + 8a + 3}{2a^2 - 5a - 12}$

22. $\dfrac{5y^2 - 6y + 1}{y^2 - 1} \div \dfrac{16y^2 - 9}{4y^2 + 7y + 3}$

7.4 ADDING AND SUBTRACTING ALGEBRAIC FRACTIONS

The procedure for adding and subtracting algebraic fractions is the same as that for arithmetic fractions. Let's look at the addition of two fractions with like denominators.

$$\frac{a}{b} + \frac{c}{b} = a\left(\frac{1}{b}\right) + c\left(\frac{1}{b}\right)$$ **Definition of multiplication of fractions.**

$$= (a + c)\left(\frac{1}{b}\right)$$ **Distributive property.**

$$= \frac{a + c}{b}$$ **Definition of multiplication of fractions.**

So, to add two fractions with a common denominator, we simply add the numerators. Let's look at an example.

Example 1 **Add the following fractions and reduce, if possible.**

Algebra	Arithmetic
$\dfrac{5a + 6}{2a + 1} + \dfrac{3a - 2}{2a + 1}$	$\dfrac{7}{15} + \dfrac{3}{15}$

ANSWER

Step 1: Since the denominators are the same, simply add the numerators.

$$\frac{(5a + 6) + (3a - 2)}{2a + 1}$$

$$= \frac{(5a + 3a) + (6 - 2)}{2a + 1}$$

$$= \frac{8a + 4}{2a + 1}$$

Step 2: Reduce.

$$\frac{8a + 4}{2a + 1} = \frac{4(2a + 1)}{2a + 1}$$

$$= \frac{4\cancel{(2a + 1)}}{\cancel{2a + 1}}$$

$$= \frac{4}{1} = 4$$

Step 1: Since the denominators are the same, simply add the numerators.

$$\frac{7 + 3}{15} = \frac{10}{15}$$

Step 2: Reduce.

$$\frac{10}{15} = \frac{2 \times 5}{3 \times 5}$$

$$= \frac{2 \times \cancel{5}}{3 \times \cancel{5}}$$

$$= \frac{2}{3}$$

■

1. ______________

Practice 1: Add the following fractions and reduce, if possible.

$$\frac{3y - 2}{y + 1} + \frac{5y + 7}{y + 1}$$

2. ______________

Practice 2: Add the following fractions and reduce, if possible.

$$\frac{5x - 3}{x^2} + \frac{2x + 3}{x^2}$$

3. ______________

Practice 3: Add the following fractions and reduce, if possible.

$$\frac{2p + 5}{3p^2 + 5p + 2} + \frac{p - 3}{3p^2 + 5p + 2}$$

To add two fractions in which the denominators are not the same we first get a common denominator. Let's review the rule that allows us to get a common denominator by inserting common factors into a fraction.

$$\frac{a}{b} = \frac{a}{b} \times 1$$ **Multiplication property of 1.**

$$= \frac{a}{b} \times \frac{c}{c}$$ **Any nonzero number divided by itself is 1.**

$$= \frac{ac}{bc}$$ **Definition of multiplication of fractions.**

For example,

$$\frac{7}{x^3y} = \frac{7}{x^3y} \times 1 = \frac{7}{x^3y} \times \frac{y^3}{y^3} = \frac{7 \times y^3}{x^3y \times y^3} = \frac{7y^3}{x^3y^4}$$

Now, let's look at an example in which we add two fractions where the denominators are not the same.

Example 2 **Add the following fractions and reduce, if possible.**

Algebra

$$\frac{7}{x^3y} + \frac{5}{x^2y^4}$$

Step 1: Find the least common denominator.

a. Completely factor the denominators.

$$\frac{7}{x^3y} + \frac{5}{x^2y^4}$$

b. Get the same denominators.

i. The first denominator has three x's and the second denominator has two x's; thus, we put one x in the second denominator and numerator. This gives us three x's in the second denominator.

$$\frac{7}{x^3y} + \frac{5x}{x^3y^4}$$

Arithmetic

$$\frac{7}{24} + \frac{5}{324}$$

Step 1: Find the least common denominator.

a. Completely factor the denominators.

$$\frac{7}{2 \times 2 \times 2 \times 3} + \frac{5}{2 \times 2 \times 3 \times 3 \times 3 \times 3}$$

b. Get the same denominators.

i. The first denominator has three 2's and the second denominator has two 2's; thus, we put one 2 in the second denominator and numerator. This gives us three 2's in the second denominator.

$$\frac{7}{2 \times 2 \times 2 \times 3} + \frac{2 \times 5}{2 \times 2 \times 2 \times 3 \times 3 \times 3 \times 3}$$

ANSWER

ii. The second denominator has four y's and the first denominator has one y; thus, we put three y's in the first denominator and numerator. This gives us four y's in the first denominator.

$$\frac{7y^3}{x^3y^4} + \frac{5x}{x^3y^4}$$

ii. The second denominator has four 3's and the first denominator has one 3; thus, we put three 3's in the first denominator and numerator. This gives us four 3's in the first denominator.

$$\frac{7 \times 3 \times 3 \times 3}{2 \times 2 \times 2 \times 3 \times 3 \times 3 \times 3} + \frac{2 \times 5}{2 \times 2 \times 2 \times 3 \times 3 \times 3 \times 3}$$

Step 2: Add the fractions and reduce, if possible.

$$\frac{7y^3 + 5x}{x^3y^4}$$

Step 2: Add the fractions and reduce, if possible.

$$\frac{(7 \times 3 \times 3 \times 3) + (2 \times 5)}{2 \times 2 \times 2 \times 3 \times 3 \times 3 \times 3}$$

$$\frac{189 + 10}{648}$$

$$\frac{199}{648}$$

■

4. ______________

Practice 4: Add the following fractions and reduce, if possible.

$$\frac{2}{x^2} + \frac{1}{x^4}$$

5. ______________

Practice 5: Add the following fractions and reduce, if possible.

$$\frac{5x}{x^3y} + \frac{7y}{x^2y^2}$$

Practice 6: Add the following fractions and reduce, if possible.

$$\frac{y^3}{12x^2y^5} + \frac{x^2}{15x^4y^2}$$

ANSWER

6. ______________

Example 3 **Add the following fractions and reduce, if possible.**

Algebra

$$\frac{2}{3x^2 - 8x - 3} + \frac{1}{3x^2 + 7x + 2}$$

Step 1: Find the least common denominator.

a. Completely factor the denominators.

$$\frac{2}{(3x + 1)(x - 3)} + \frac{1}{(3x + 1)(x + 2)}$$

b. Get the same denominators.

$$\frac{2(\mathbf{x} + \mathbf{2})}{(3x + 1)(x - 3)(\mathbf{x} + \mathbf{2})} + \frac{1(\mathbf{x} - \mathbf{3})}{(3x + 1)(x + 2)(\mathbf{x} - \mathbf{3})}$$

$$= \frac{2x + 4}{(3x + 1)(x - 3)(x + 2)} + \frac{x - 3}{(3x + 1)(x + 2)(x - 3)}$$

Step 2: Add the fractions and reduce, if possible.

$$\frac{(2x + 4) + (x - 3)}{(3x + 1)(x - 3)(x + 2)}$$

$$= \frac{3x + 1}{(3x + 1)(x - 3)(x + 2)}$$

$$= \frac{\cancel{3x + 1}}{\cancel{(3x + 1)}(x - 3)(x + 2)}$$

$$= \frac{1}{(x - 3)(x + 2)}$$

Arithmetic

$$\frac{8}{15} + \frac{3}{10}$$

Step 1: Find the least common denominator.

a. Completely factor the denominators.

$$\frac{8}{3 \times 5} + \frac{3}{2 \times 5}$$

b. Get the same denominators.

$$\frac{8 \times \mathbf{2}}{3 \times 5 \times \mathbf{2}} + \frac{3 \times \mathbf{3}}{2 \times 5 \times \mathbf{3}}$$

$$= \frac{16}{3 \times 5 \times 2} + \frac{9}{2 \times 5 \times 3}$$

Step 2: Add the fractions and reduce, if possible.

$$\frac{16 + 9}{3 \times 5 \times 2} = \frac{25}{3 \times 5 \times 2}$$

$$= \frac{\cancel{5} \times 5}{3 \times \cancel{5} \times 2}$$

$$= \frac{5}{3 \times 2}$$

$$= \frac{5}{6}$$

■

A N S W E R

Example 4 **Add the following fractions and reduce, if possible.**

$$\frac{3}{x^2 - 49} + \frac{5}{x^2 + 6x - 7}$$

Step 1: Find the least common denominator.

a. Completely factor the denominators.

$$\frac{3}{(x + 7)(x - 7)} + \frac{5}{(x + 7)(x - 1)}$$

b. Get the same denominators.

$$\frac{3(\mathbf{x} - \mathbf{1})}{(x + 7)(x - 7)(\mathbf{x} - \mathbf{1})} + \frac{5(\mathbf{x} - \mathbf{7})}{(x + 7)(x - 1)(\mathbf{x} - \mathbf{7})}$$

$$= \frac{3x - 3}{(x + 7)(x - 7)(x - 1)} + \frac{5x - 35}{(x + 7)(x - 1)(x - 7)}$$

Step 2: Add the fractions and reduce, if possible.

$$\frac{(3x - 3) + (5x - 35)}{(x + 7)(x - 7)(x - 1)} = \frac{8x - 38}{(x + 7)(x - 7)(x - 1)}$$ ■

The procedure for adding algebraic fractions is the same as that for arithmetic fractions. Any time you are uncertain as to what to do, just think about adding arithmetic fractions. With this in mind, let's work some practice problems.

7. ______________________

Practice 7: Add the following fractions and reduce, if possible.

$$\frac{-a + 6}{a^2 - 4} + \frac{2}{a + 2}$$

8. ______________________

Practice 8: Add the following fractions and reduce, if possible.

$$\frac{2}{x^2} + \frac{1}{x^2 - 2x}$$

Practice 9: Add the following fractions and reduce, if possible.

$$\frac{3}{2p^2 + 7p - 4} + \frac{-1}{2p^2 + p - 1}$$

A N S W E R

9. ________________

Practice 10: Add the following fractions and reduce, if possible.

$$\frac{-2}{x^2 - 9} + \frac{2x}{x^3 + 6x^2 + 9x}$$

10. ________________

When subtracting fractions, we follow the same procedure that we use when adding fractions. The only difference is that once we find the least common denominator, we subtract the numerators instead of adding them. Let's look at an example.

Example 5 **Subtract the following fractions and reduce, if possible.**

Algebra	**Arithmetic**
$\frac{x}{x^2 + 6x + 9} - \frac{x + 1}{x^2 - 9}$	$\frac{17}{20} - \frac{5}{12}$
Step 1: Find the least common denominator. a. Completely factor the denominators.	*Step 1:* Find the least common denominator. a. Completely factor the denominators.
$\frac{x}{(x + 3)(x + 3)} - \frac{x + 1}{(x + 3)(x - 3)}$	$\frac{17}{2 \times 2 \times 5} - \frac{5}{2 \times 2 \times 3}$

b. Get the same denominators.

$$\frac{x(\mathbf{x-3})}{(x+3)(x+3)(\mathbf{x-3})} - \frac{(x+1)(\mathbf{x+3})}{(x+3)(x-3)(\mathbf{x+3})}$$

$$= \frac{x^2-3x}{(x+3)(x+3)(x-3)} - \frac{x^2+4x+3}{(x+3)(x-3)(x+3)}$$

Step 2: Subtract the fractions and reduce, if possible.

$$\frac{(x^2-3x)-(x^2+4x+3)}{(x+3)(x+3)(x-3)}$$

$$= \frac{x^2-3x-x^2-4x-3}{(x+3)(x+3)(x-3)}$$

$$= \frac{-7x-3}{(x+3)(x+3)(x-3)}$$

b. Get the same denominators.

$$\frac{17 \times \mathbf{3}}{2 \times 2 \times 5 \times \mathbf{3}} - \frac{5 \times \mathbf{5}}{2 \times 2 \times 3 \times \mathbf{5}}$$

$$= \frac{51}{2 \times 2 \times 5 \times 3} - \frac{25}{2 \times 2 \times 3 \times 5}$$

Step 2: Subtract the fractions and reduce, if possible.

$$\frac{51-25}{2 \times 2 \times 3 \times 5}$$

$$= \frac{26}{2 \times 2 \times 3 \times 5}$$

$$= \frac{2 \times 13}{2 \times 2 \times 3 \times 5} = \frac{13}{30}$$ ■

Example 6 **Subtract the following fractions and reduce, if possible.**

$$\frac{7}{3x^2+7x+2} - \frac{2}{9x^2-1}$$

Step 1: Find the least common denominator.

a. Completely factor the denominators.

$$\frac{7}{(3x+1)(x+2)} - \frac{2}{(3x+1)(3x-1)}$$

b. Get the same denominators.

$$\frac{7(\mathbf{3x-1})}{(3x+1)(x+2)(\mathbf{3x-1})} - \frac{2(\mathbf{x+2})}{(3x+1)(3x-1)(\mathbf{x+2})}$$

$$= \frac{21x-7}{(3x+1)(x+2)(3x-1)} - \frac{2x+4}{(3x+1)(3x-1)(x+2)}$$

Step 2: Subtract the fractions and reduce, if possible.

$$\frac{(21x-7)-(2x+4)}{(3x+1)(3x-1)(x+2)} = \frac{21x-7-2x-4}{(3x+1)(3x-1)(x+2)}$$

$$= \frac{19x-11}{(3x+1)(3x-1)(x+2)}$$ ■

Now, let's try some practice problems.

Practice 11: Subtract the following fractions and reduce, if possible.

$$\frac{5}{x} - \frac{2x + 10}{x^2 + 2x}$$

ANSWER

11. ____________________

Practice 12: Subtract the following fractions and reduce, if possible.

$$\frac{2x}{2x + 1} - \frac{3x}{2x^2 - 5x - 3}$$

12. ____________________

Practice 13: Subtract the following fractions and reduce, if possible.

$$\frac{3}{10x^2 + 13x - 3} - \frac{4}{5x^2 + 9x - 2}$$

13. ____________________

Practice 14: Subtract the following fractions and reduce, if possible.

$$\frac{-2x + 6}{4x^2 - 25} - \frac{-x + 5}{2x^2 - 5x - 25}$$

14. ____________________

EXERCISE 7.4

PART 1

Add the fractions in each problem and reduce, if possible.

1. $\dfrac{3x}{x+1} + \dfrac{2x}{x+1}$

2. $\dfrac{2}{3y} + \dfrac{1}{4y}$

3. $\dfrac{y}{x} + \dfrac{1}{x^3}$

4. $\dfrac{4+2x}{5} + \dfrac{2+x}{10}$

5. $\dfrac{m^2+2}{m^3n^2} + \dfrac{n+3}{4mn^3}$

6. $\dfrac{9}{x^2} + \dfrac{4}{xy}$

7. $\dfrac{1}{m^2-9} + \dfrac{1}{m+3}$

8. $\dfrac{x-1}{x^2+2x-8} + \dfrac{3x-2}{x^2+3x-4}$

9. $\dfrac{x}{x^2 - 1} + \dfrac{x}{x + 1}$

10. $\dfrac{2}{y^2 - 1} + \dfrac{1}{y^2 + 2y - 3}$

11. $\dfrac{7}{3(y + 1)(y - 6)} + \dfrac{-4}{3(y + 1)(y - 3)}$

12. $\dfrac{5x}{x + y} + \dfrac{6y - 2x}{x - y}$

PART 2

Subtract the fractions in each problem and reduce, if possible.

1. $\dfrac{x}{6y} - \dfrac{2}{3x}$

2. $\dfrac{4}{3x^4} - \dfrac{1}{x}$

3. $\dfrac{4b^2}{6ab^3} - \dfrac{a^3}{3a^4b}$

4. $\dfrac{5 - 4y}{8} - \dfrac{2 - 3y}{6}$

5. $\frac{6}{x^2} - \frac{2}{x}$

6. $\frac{8}{x - 2} - \frac{4}{x + 2}$

7. $\frac{1}{a - b} - \frac{a}{4a - 4b}$

8. $\frac{6}{y^2 - 4} - \frac{3}{2y + 4}$

9. $\frac{1}{m^2 - 1} - \frac{1}{m^2 + 3m + 2}$

10. $\frac{y}{y^2 - 1} - \frac{y + 2}{y^2 + y - 2}$

11. $\frac{x + 1}{2x^2 + 9x - 5} - \frac{x - 1}{2x^2 + 11x + 5}$

12. $\dfrac{1}{x(x - y)} - \dfrac{2}{y(x + y)}$

7.5 SOLVING EQUATIONS CONTAINING FRACTIONS

First, let's look at equations that contain fractions in which the variable is in the numerator.

Example 1 **Solve the following equation.**

$$\frac{x}{2} = \frac{x}{3} + 5$$

Step 1: Find the least common denominator of the fractions. The least common denominator of 2 and 3 is 6.

Step 2: Multiply both sides of the equation by the least common denominator, 6, to clear the equation of fractions.

$\frac{x}{2} = \frac{x}{3} + 5$	**Original equation.**
$6\left(\frac{x}{2}\right) = 6\left(\frac{x}{3} + 5\right)$	**Multiply both sides by 6.**
$6\left(\frac{x}{2}\right) = 6\left(\frac{x}{3}\right) + 6(5)$	**Distributive property.**
$3x = 2x + 30$	**Perform the multiplications.**

Step 3: Solve the resulting equation.

$3x = 2x + 30$	**Equation from Step 2.**
$3x - 2x = 2x + 30 - 2x$	**Subtract $2x$ from both sides.**
$x = 30$	**Simplify.**

Check: Substitute 30 for x in the original equation.

$\frac{x}{2} = \frac{x}{3} + 5$	**Original equation.**
$\frac{\mathbf{30}}{2} \stackrel{?}{=} \frac{\mathbf{30}}{3} + 5$	**Substitute 30 for x.**
$15 \stackrel{?}{=} 10 + 5$	**Perform the divisions.**
$15 = 15$	**Perform the addition.**

Thus, 30 is the solution. ■

> **To solve an equation containing fractions in which the variable is in the numerator:**
> 1. Find the least common denominator of the fractions.
> 2. Multiply both sides of the equation by the least common denominator to clear the equation of fractions.
> 3. Solve the resulting equation.

Example 2 **Solve the following equation.**

$$\frac{x}{15} + \frac{x}{10} = 2$$

Step 1: Find the least common denominator of the fractions. Now, $15 = 3 \times 5$ and $10 = 2 \times 5$; thus, the least common denominator is $2 \times 3 \times 5 = 30$.

Step 2: Multiply both sides of the equation by the least common denominator, 30, to clear the equation of fractions.

$\frac{x}{15} + \frac{x}{10} = 2$	**Original equation.**
$30\left(\frac{x}{15} + \frac{x}{10}\right) = 30(2)$	**Multiply both sides by 30.**
$30\left(\frac{x}{15}\right) + 30\left(\frac{x}{10}\right) = 30(2)$	**Distributive property.**
$2x + 3x = 60$	**Perform the multiplications.**

Step 3: Solve the resulting equation.

$2x + 3x = 60$	**Equation from Step 2.**
$5x = 60$	**Combine like terms.**
$x = 12$	**Divide both sides by 5.**

Check: Substitute 12 for x in the original equation.

$\frac{x}{15} + \frac{x}{10} = 2$	**Original equation.**
$\frac{\mathbf{12}}{15} + \frac{\mathbf{12}}{10} \stackrel{?}{=} 2$	**Substitute 12 for x.**

$$\frac{4}{5} + \frac{6}{5} \stackrel{?}{=} 2$$ **Reduce the fractions.**

$$\frac{10}{5} \stackrel{?}{=} 2$$ **Add the fractions.**

$$2 = 2$$ **Reduce.**

Thus, 12 is the solution. ■

ANSWER

Practice 1: Solve the following equation.

$$2x = \frac{x}{4} + 7$$

1. ______

Practice 2: Solve the following equation.

$$\frac{a}{3} + \frac{a}{2} = 10$$

2. ______

Practice 3: Solve the following equation.

$$\frac{x}{10} = \frac{-x}{6} - 4$$

3. ______

Practice 4: Solve the following equation.

$$\frac{y}{12} + \frac{y}{3} = \frac{y}{4} + 3$$

4. ______

Now, let's look at equations that contain fractions in which the variable is in the denominator.

Note: Since the denominator of a fraction cannot be zero, the variable in the denominator cannot take on a value which makes the denominator zero.

Example 3 **In the equation $\frac{3}{x-1} = 5$ you cannot have x equal 1 since 1 makes the denominator of $\frac{3}{x-1}$ zero.** ■

What is wrong with the denominator of $\frac{3}{x-1}$ being zero?

Well, Charlie, $\frac{3}{x-1}$ is really 3 divided by $x - 1$ and division by zero is not defined.

Example 4 **In the equation $\frac{5}{x^2-4} = 6x$ you cannot have x equal 2 or −2, since 2 and −2 make the denominator of $\frac{5}{x^2-4}$ zero.** ■

Example 5 **Solve the following equation.**

$$\frac{3}{10x} - \frac{1}{6x} = \frac{1}{15}$$

Step 1: Find the least common denominator of the fractions. Now, $10x = 2 \times 5 \times x$, $6x = 2 \times 3 \times x$, and $15 = 3 \times 5$; thus, the least common denominator is $2 \times 3 \times 5 \times x = 30x$.

Step 2: Multiply both sides of the equation by the least common denominator, $30x$, to clear the equation of fractions.

$$\frac{3}{10x} - \frac{1}{6x} = \frac{1}{15} \qquad \text{Original equation.}$$

$$30x\left(\frac{3}{10x} - \frac{1}{6x}\right) = 30x\left(\frac{1}{15}\right) \qquad \text{Multiply both sides by } 30x.$$

$$30x\left(\frac{3}{10x}\right) - 30x\left(\frac{1}{6x}\right) = 30x\left(\frac{1}{15}\right) \qquad \text{Distributive property.}$$

$$\frac{90x}{10x} - \frac{30x}{6x} = \frac{30x}{15} \qquad \text{Perform the multiplications.}$$

$$9 - 5 = 2x \qquad \text{Reduce the fractions.}$$

Step 3: Solve the resulting equation.

$9 - 5 = 2x$	**Equation from Step 2.**
$4 = 2x$	**Perform the subtraction.**
$2x = 4$	**Symmetric property of equality.**
$x = 2$	**Divide both sides by 2.**

Step 4: Check to find if 2 makes any of the denominators in the *original* equation zero. If it does, then it is not a solution of the original equation since you cannot have zero in the denominator of a fraction.

$\frac{3}{10x} - \frac{1}{6x} = \frac{1}{15}$	**Original equation.**
$\frac{3}{10 \times 2} - \frac{1}{6 \times 2} \stackrel{?}{=} \frac{1}{15}$	**Substitute 2 for x.**
$\frac{3}{20} - \frac{1}{12} \stackrel{?}{=} \frac{1}{15}$	**Perform the multiplications.**

Now, none of the denominators are zero; thus, 2 is the solution of $\frac{3}{10x} - \frac{1}{6x} = \frac{1}{15}$.
Note: In Step 4 it is only necessary to check that none of the denominators equal zero when 2 is substituted for x. However, you may want to complete the check so that you will be sure that 2 is the correct answer. ■

To solve an equation containing fractions in which the variable is in the denominator:

1. Find the least common denominator of the fractions.
2. Multiply both sides of the equation by the least common denominator to clear the equation of fractions.
3. Solve the resulting equation.
4. Check each answer to find if it makes any of the denominators in the *original* equation zero. If it does, then it is not a solution of the original equation.

Do I always have to check my answers?

Well, Charlie, you have to check each answer to find if it makes any of the denominators in the original equation zero. It would also be a good idea to check each answer to be sure it satisfies the original equation.

Example 6 **Solve the following equation.**

$$\frac{x}{x-2} - \frac{1}{x+2} = \frac{8}{x^2-4}$$

Step 1: Find the least common denominator of the fractions. Now, $x^2 - 4 = (x-2)(x+2)$; thus, the least common denominator is $(x+2)(x-2) = x^2 - 4$.

Step 2: Multiply both sides of the equation by the least common denominator, $x^2 - 4 = (x-2)(x+2)$, to clear the equation of fractions.

$$\frac{x}{x-2} - \frac{1}{x+2} = \frac{8}{x^2-4}$$ **Original equation.**

$$(x^2-4)\left(\frac{x}{x-2} - \frac{1}{x+2}\right) = (x^2-4)\left(\frac{8}{x^2-4}\right)$$ **Multiply both sides by $x^2 - 4$.**

$$(x^2-4)\left(\frac{x}{x-2}\right) - (x^2-4)\left(\frac{1}{x+2}\right) = (x^2-4)\left(\frac{8}{x^2-4}\right)$$ **Distributive property.**

$$(x+2)x - (x-2)1 = 8$$ **Perform the multiplications.**

$$(x^2+2x) - (x-2) = 8$$ **Distributive property.**

$$x^2 + 2x - x + 2 = 8$$ **Remove parentheses.**

$$x^2 + x + 2 = 8$$ **Combine like terms.**

Step 3: Solve the resulting equation.

$$x^2 + x + 2 = 8$$ **Equation from Step 2.**

$$x^2 + x + 2 - 8 = 8 - 8$$ **Subtract 8 from both sides.**

$$x^2 + x - 6 = 0$$ **Simplify.**

$$(x+3)(x-2) = 0$$ **Factor the left side.**

$$x + 3 = 0 \quad \text{or} \quad x - 2 = 0$$
$$x = -3 \qquad\qquad x = 2$$
Set the factors equal to zero and solve the resulting linear equations.

Step 4: Check to find if -3 or 2 makes any of the denominators in the original equation zero. If one does, then it is not a solution to the original equation.

Substitute -3 for x in the original equation.

$$\frac{x}{x-2} - \frac{1}{x+2} = \frac{8}{x^2-4}$$ **Original equation.**

$$\frac{-3}{-3-2} - \frac{1}{-3+2} \stackrel{?}{=} \frac{8}{[(-3^2)-4]}$$ **Substitute -3 for x.**

$$\frac{-3}{-5} - \frac{1}{-1} \stackrel{?}{=} \frac{8}{9-4}$$ **Simplify.**

$$\frac{-3}{-5} - \frac{1}{-1} \stackrel{?}{=} \frac{8}{5}$$

A N S W E R

Now, none of the denominators are zero; thus, -3 is a solution of the original equation.

Substitute 2 for x in the original equation.

$$\frac{x}{x-2} - \frac{1}{x+2} = \frac{8}{x^2-4} \quad \textbf{Original equation.}$$

$$\frac{2}{2-2} - \frac{2}{2+2} \stackrel{?}{=} \frac{8}{2^2-4} \quad \textbf{Substitute 2 for } x.$$

$$\frac{2}{0} - \frac{2}{4} \stackrel{?}{=} \frac{8}{0} \quad \textbf{Simplify.}$$

Since at least one denominator equals zero, 2 is *not* a solution of the original equation. Therefore, -3 is the only solution. ■

Practice 5: Solve the following equation.

$$\frac{3}{2x} + \frac{1}{x} = 1$$

5. ______________

Practice 6: Solve the following equation.

$$\frac{5}{x+2} - \frac{1}{x} = \frac{2}{x+2}$$

6. ______________

A N S W E R

Practice 7: Solve the following equation.

$$1 - \frac{2}{2x - 1} = \frac{-3}{2x + 1}$$

7. ____________________

8. ____________________

Practice 8: Solve the following equation.

$$\frac{5}{y - 5} - \frac{y}{y - 5} = -4$$

How did you do on the practice problems Charlie?

I missed Practice 8. The answer at the end of the chapter is no solution and I have 5 for a solution.

You solved the equation properly but you forgot to check 5 to find if it made any of the denominators in the original equation zero. Now, 5 makes at least one of the denominators zero and thus, is not a solution. Therefore, the equation has no solution.

EXERCISE 7.5

Solve the following equations. Be sure to check your answers for possible zeros in the denominators.

A

1. $\frac{1}{4} = \frac{x}{2}$

2. $\frac{2}{x} = \frac{5}{12}$

3. $\frac{x}{2} - \frac{x}{4} = 6$

4. $\frac{4}{y} + \frac{2}{3} = 1$

5. $\frac{x+1}{2} = \frac{x+2}{3}$

6. $\frac{m-2}{4} + \frac{m+1}{3} = \frac{10}{3}$

7. $\frac{a+7}{8} - \frac{a-2}{3} = \frac{4}{3}$

8. $\frac{a}{2} - \frac{17+a}{5} = 2a$

9. $\frac{5-y}{y} + \frac{3}{4} = \frac{7}{y}$

10. $\frac{x}{x-4} = \frac{2}{x-4} + 5$

B

11. $\frac{5y}{3} - \frac{2y - 1}{4} = \frac{1}{4}$

12. $\frac{8x + 3}{x} = 3x$

13. $\frac{2}{x} = \frac{x}{5x - 12}$

14. $\frac{1}{y + 5} - \frac{3}{y - 5} = \frac{-10}{y^2 - 25}$

15. $\frac{3x}{x - 3} - 3 = \frac{1}{x}$

16. $\frac{4}{x - 5} - \frac{5}{x + 3} = 0$

17. $\frac{x - 3}{x - 2} = \frac{x - 2}{x + 3}$

18. $\frac{10}{x^2 - 25} + 1 = \frac{x}{x + 5}$

19. $\frac{7}{a^2 - 4} = \frac{4}{a^2 + 2a}$

20. $\frac{2}{a^2 - 9} = \frac{3}{a^2 + a - 12}$

7.6 STRATEGIES FOR SOLVING APPLICATIONS

Setting up the equation in a word problem is one of the most difficult skills in algebra. Using the suggestions in the application sections of Chapters 4 and 6 and reading through the following examples should help you set up the equations in this section.

Example 1 **If the same number is added to both the numerator and denominator of the fraction $\frac{3}{4}$, the result is $\frac{5}{6}$. Find the number.**

Step 1: Let x = the unknown number.

Step 2: Set up the equation. The same number added to the numerator and denominator of $\frac{3}{4}$ means $\frac{x+3}{x+4}$. The result, which is $\frac{x+3}{x+4}$, is also equal to $\frac{5}{6}$. Thus,

$$\frac{x+3}{x+4} = \frac{5}{6}$$

Step 3: Solve

$$\frac{x+3}{x+4} = \frac{5}{6}$$

a. Multiply both sides of the equation by $6(x + 4)$, the least common denominator of the fractions.

$$\frac{x+3}{x+4} = \frac{5}{6}$$

$$6(x+4)\left(\frac{x+3}{x+4}\right) = 6(x+4)\left(\frac{5}{6}\right)$$

$$6(x+3) = (x+4)5$$

b. Solve the resulting equation.

$$6(x+3) = (x+4)5$$

$$6x + 18 = 5x + 20$$

$$6x - 5x = 20 - 18$$

$$x = 2$$

Step 4: Check the solution.

$$\frac{x+3}{x+4} = \frac{5}{6}$$

$$\frac{2+3}{2+4} \stackrel{?}{=} \frac{5}{6}$$

$$\frac{5}{6} = \frac{5}{6}$$

None of the denominators are zero; thus, 2 is the number. ■

Example 2 **If three times a number is added to twice its reciprocal, the answer is 5. Find the number(s).**

Step 1: Let x = the unknown number.

Step 2: Set up the equation. Three times the number is $3x$. The reciprocal of the number is $\frac{1}{x}$. Twice its reciprocal is $2(\frac{1}{x}) = \frac{2}{x}$. Now, three times the number added to twice its reciprocal is $3x + \frac{2}{x}$. Thus,

$$3x + \frac{2}{x} = 5$$

Step 3: Solve

$$3x + \frac{2}{x} = 5$$

a. Multiply both sides of the equation by x, the least common denominator of the fractions.

$$x\left(3x + \frac{2}{x}\right) = x(5)$$

$$x(3x) + x\left(\frac{2}{x}\right) = x(5)$$

$$3x^2 + 2 = 5x$$

b. Solve the resulting equation.

$$3x^2 + 2 = 5x$$

$$3x^2 - 5x + 2 = 0$$

$$(3x - 2)(x - 1) = 0$$

$$3x - 2 = 0 \quad \text{or} \quad x - 1 = 0$$

$$3x = 2 \qquad x = 1$$

$$x = \frac{2}{3}$$

Step 4: Check the solutions.

$$3x + \frac{2}{x} = 5 \qquad 3x + \frac{2}{x} = 5$$

$$3\left(\frac{2}{3}\right) + \frac{2}{\frac{2}{3}} \stackrel{?}{=} 5 \qquad 3(1) + \frac{2}{1} \stackrel{?}{=} 5$$

None of the denominators are zero; thus, 1 and $\frac{2}{3}$ are the numbers. ■

Example 3 **Machine A can do a certain job in seven hours and machine B takes 12 hours. How long will it take the two machines working together to do the job?**

Step 1: Let $x =$ the time in hours it will take the two machines to do the job together.

Step 2: Set up the equation. If machine A can do the job in seven hours, then $\frac{1}{7}$ is the amount of the job it can do in one hour. If machine B can do the job in 12 hours, then $\frac{1}{12}$ is the amount of the job it can do in one hour. Therefore, $\frac{1}{7} + \frac{1}{12}$ is the amount of the job both machines can do together in one hour. Since x is the time in hours both machines can do the job together, $\frac{1}{x}$ is also the amount of the job both machines can do together in 1 hour. Thus,

$$\frac{1}{7} + \frac{1}{12} = \frac{1}{x}$$

Step 3: Solve

$$\frac{1}{7} + \frac{1}{12} = \frac{1}{x}$$

a. Multiply both sides of the equation by, $7(12)(x)$, the least common denominator of the fractions.

$$7(12)(x)\left(\frac{1}{7} + \frac{1}{12}\right) = 7(12)(x)\left(\frac{1}{x}\right)$$

$$7(12)(x)\left(\frac{1}{7}\right) + 7(12)(x)\left(\frac{1}{12}\right) = 7(12)(x)\left(\frac{1}{x}\right)$$

$$12x + 7x = 84$$

b. Solve the resulting equation.

$$12x + 7x = 84$$

$$19x = 84$$

$$x = \frac{84}{19} = 4\frac{8}{19}$$

Step 4: Check the solution.

$$\frac{1}{7} + \frac{1}{12} = \frac{1}{x}$$

$$\frac{1}{7} + \frac{1}{12} \stackrel{?}{=} \frac{1}{\frac{84}{19}}$$

None of the denominators are zero; thus, it takes $\frac{84}{19}$ hours or $4\frac{8}{19}$ hours for both machines to do the job together. ■

Example 4 **If a family drinks 5 gallons of milk in two weeks, how many gallons of milk will be consumed in a year (52 weeks)?**

Step 1: Let $x =$ the number of gallons of milk consumed in 52 weeks.

Step 2: Set up the equation. Now, $\frac{x}{52}$ is the number of gallons of milk consumed in 1 week. Since the family drinks 5 gallons of milk in 2 weeks, they would drink $\frac{5}{2}$ gallons in 1 week. Thus,

$$\frac{x}{52} = \frac{5}{2}$$

Step 3: Solve

$$\frac{x}{52} = \frac{5}{2}$$

a. Multiply both sides of the equation by 52, the least common denominator of the fractions.

$$52\left(\frac{x}{52}\right) = 52\left(\frac{5}{2}\right)$$

$$x = 26(5)$$

$$x = 130$$

Step 4: Check the solution.

$$\frac{x}{52} = \frac{5}{2}$$

$$\frac{130}{52} \stackrel{?}{=} \frac{5}{2}$$

None of the denominators are zero; thus, 130 gallons is the amount of milk consumed by the family in 52 weeks. ■

Example 5 **Don can row 4 miles per hour in still water. It takes as long to row 8 miles upstream as 24 miles downstream. What is the rate of the current?**

Step 1: Let $x =$ the rate of the current.

Step 2: Set up the equation. This type of problem involves the distance formula, $d = rt$, where $d =$ distance, $r =$ rate, and $t =$ time. Since Don can row 4 miles per hour in still water, $4 - x$ would be the rate of the boat upstream and $4 + x$ would be the rate of the boat downstream. Let's set up a chart.

	d	r	t
Upstream	8	$4 - x$	
Downstream	24	$4 + x$	

Using

$$d = rt$$

divide both sides of the equation by r to find t.

$$\frac{d}{r} = \frac{rt}{r}$$

$$\frac{d}{r} = t$$

Now, let's complete the chart. The time it takes to go upstream is $\frac{8}{4-x}$. The time it takes to go downstream is $\frac{24}{4+x}$.

	d	r	t
Upstream	8	$4 - x$	$\frac{8}{4 - x}$
Downstream	24	$4 + x$	$\frac{24}{4 + x}$

From the information in the example, the times are equal; thus,

$$\frac{8}{4 - x} = \frac{24}{4 + x}$$

Step 3: Solve

$$\frac{8}{4 - x} = \frac{24}{4 + x}$$

a. Multiply both sides of the equation by $(4 - x)(4 + x)$, the least common denominator of the fractions.

$$\frac{(4 - x)(4 + x)8}{4 - x} = \frac{(4 - x)(4 + x)24}{4 + x}$$

$$(4 + x)8 = (4 - x)24$$

$$32 + 8x = 96 - 24x$$

b. Solve the resulting equation.

$$32 + 8x = 96 - 24x$$

$$8x + 24x = 96 - 32$$

$$32x = 64$$

$$x = 2$$

Step 4: Check the solution.

$$\frac{8}{4 - x} = \frac{24}{4 + x}$$

$$\frac{8}{4 - 2} \stackrel{?}{=} \frac{24}{4 + 2}$$

$$\frac{8}{2} \stackrel{?}{=} \frac{24}{6}$$

None of the denominators are zero; thus, the rate of the current is 2 miles per hour. ■

APPLICATION EXERCISES

1. The numerator of a fraction is 3 more than the denominator, and the value of the fraction is $\frac{4}{3}$. Find the fraction.

2. If nine hot dogs cost \$4.23, find the price of 15 hot dogs.

3. Find the number when added to both the numerator and denominator of $\frac{3}{11}$ will equal $\frac{1}{2}$.

4. The reciprocal of 4 less than a number is three times the reciprocal of the number. Find the number.

5. The sum of a whole number and its reciprocal is $\frac{5}{2}$. Find the number.

6. One half of a number is 3 more than $\frac{1}{6}$ of the same number. What is the number?

7. George can mow a yard in two hours. His friend Daniel can mow the same yard in three hours. How long would it take them to mow the yard if they worked together?

8. If Ann can type eight pages in 1.5 hours, how many hours will it take her to type 32 pages?

9. On a road map, 2 inches represents 50 miles. How many inches would represent a distance of 40 miles?

10. If 12 balls cost \$200, what is the cost of 18 balls?

11. A car wash can wash ten cars in 50 minutes. How many cars can be washed in $1\frac{1}{2}$ hours?

12. If Kenneth is three years older than Tim and the ratio of their ages is $\frac{5}{6}$, how old is each?

13. A car uses 10 gallons of gasoline in 6 hours. How many gallons of gasoline are used in nine hours?

14. Marie won the mayor's race by a 3 to 2 margin. If she received 15,000 votes, how many votes did her opponent receive?

15. If 50 feet of wire weighs 15 pounds, what would 80 feet of the same wire weigh?

16. If it takes a small pipe three times longer to fill a gasoline tank than a larger pipe and together they can fill it in six hours, how many hours would it take the larger pipe to fill the tank?

17. A river flows at a rate of 3 miles per hour. A boat takes as long to go 12 miles downstream as to go 8 miles upstream. What is the speed of the boat in still water?

18. The sum of the ages of Mary and Susan is 45 years. Mary's age divided by Susan's age is $\frac{5}{4}$. How old is each?

USEFUL INFORMATION

1. A fraction which contains at least one algebraic expression is an **algebraic fraction.**

2. The following procedure allows us to **remove common factors** from a fraction.

$$\frac{ac}{ab} = \frac{a}{b} \times \frac{c}{c} = \frac{a}{b} \times 1 = \frac{a}{b}$$

3. When **reducing an algebraic fraction**, you should first completely factor the numerator and denominator; then remove the common factors.

4. **Definition of multiplication of fractions.** For fractions $\frac{a}{b}$ and $\frac{c}{d}$, $\frac{a}{b} \times \frac{c}{d} = \frac{a \times c}{b \times d}$.

5. When **multiplying algebraic fractions** use the following steps.
   a. Use the definition of multiplication to write the fractions as one fraction indicating the multiplications (do not actually perform the multiplications).
   b. Completely factor the numerator and denominator.
   c. Remove common factors.
   d. Perform the multiplications.

6. **Definition of division of fractions.** If $\frac{c}{d}$ is not equal to zero, then $\frac{a}{b} \div \frac{c}{d} = \frac{a}{b} \times \frac{d}{c}$.

7. When **dividing algebraic fractions** use the definition of division to change the division to multiplication and then multiply using your rules for multiplication.

8. When **adding fractions** with like denominators we simply add the numerators.

$$\frac{a}{b} + \frac{c}{b} = \frac{a + c}{b}$$

9. When **subtracting fractions** with like denominators we simply subtract the numerators.

$$\frac{a}{b} - \frac{c}{b} = \frac{a - c}{b}$$

10. When adding or subtracting algebraic fractions with unlike denominators, you first find the least common denominator.

11. **To find the least common denominator of two fractions:**
    a. Completely factor the denominators.
    b. Insert any factors necessary so that the two denominators contain the same factors.

$$\frac{a}{b} = \frac{a}{b} \times 1 = \frac{a}{b} \times \frac{c}{c} = \frac{ac}{bc}$$

12. **To solve an equation containing algebraic fractions:**
    a. Find the least common denominator of the fractions.
    b. Multiply both sides of the equation by the least common denominator to clear the equation of fractions.
    c. Solve the resulting equation.
    d. Check each answer to find if it makes any of the denominators in the *original* equation zero. If one does, then it is not a solution of the original equation.

REVIEW PROBLEMS

Section 7.1

Write the following fractions in reduced form.

A

1. $\frac{5y}{y^2}$

2. $\frac{8a^2}{12a^3}$

3. $\frac{x + 3}{5x + 15}$

4. $\frac{2x + 10}{5x + 25}$

5. $\frac{x - 2}{x^2 - 4}$

6. $\frac{x - 3}{x^2 - 6x + 9}$

7. $\frac{n^2 + n}{n}$

8. $\frac{y - 2}{2y^2 + y - 10}$

9. $\frac{z^2 - z}{8z - 8}$

10. $\frac{4x - 2}{4x^2 - 1}$

B

11. $\dfrac{m^2 - 1}{m^2 + 2m - 3}$

12. $\dfrac{x^2 + 5x + 6}{x^2 - 9}$

13. $\dfrac{2n^2 + 3n - 2}{4n^2 - 1}$

14. $\dfrac{n^2 - 16}{n + 4}$

15. $\dfrac{x^2 - y^2}{x^2 - 2xy + y^2}$

16. $\dfrac{4m^2 + 11m - 3}{4m^2 + 7m - 2}$

17. $\dfrac{6x^2 - 8x - 8}{3x^2 - 13x - 10}$

18. $\dfrac{4x^2 - 4x - 35}{6x^2 - 19x - 7}$

19. $\dfrac{15x^2 - 7x - 2}{25x^2 - 1}$

20. $\dfrac{7x^2 - 21x - 70}{4x^2 - 100}$

Section 7.2

Multiply the fractions in each of the following problems and reduce, if possible.

A

1. $\frac{10x}{2} \times \frac{3}{5x^2}$

2. $\frac{3a^2}{5b} \times \frac{10ab}{3}$

3. $\frac{w}{5w - 2} \times \frac{3w + 1}{w}$

4. $\frac{x + 2}{x + 3} \times \frac{1}{6x + 12}$

5. $\frac{5x - 10}{x + 2} \times \frac{6x + 12}{7x - 14}$

6. $\frac{4x + 6}{15x - 3} \times \frac{10x - 2}{6x + 2}$

7. $\frac{3x - 12}{2x + 6} \times \frac{3x + 4}{x - 4}$

8. $\frac{x + 4}{x^2 - 4} \times \frac{x + 2}{5x + 20}$

9. $\frac{2x + 6}{8x - 4} \times \frac{4x^2 - 1}{x + 3}$

10. $\frac{18x - 45}{15x + 10} \times \frac{15}{2x^2 - x - 10}$

B

11. $\dfrac{x^2 - 7x + 12}{x^2 - 16} \times \dfrac{x^2 + 5x + 4}{x^2 - 2x - 3}$

12. $\dfrac{y}{y^2 - 9} \times \dfrac{y^2 - 6y + 9}{y + 3}$

13. $\dfrac{3x^2 - 7x + 2}{4x^2 - 4} \times \dfrac{3x^2 + x - 4}{9x^2 - 1}$

14. $\dfrac{21x - 18}{4x^2 - 23x - 35} \times \dfrac{8x^2 + 6x - 5}{7x^2 + x - 6}$

15. $\dfrac{36x^2 - 9}{21x^2 + 4x - 1} \times \dfrac{49x^2 - 1}{6x^2 + 9x - 6}$

16. $\dfrac{x^3 + 3x^2}{30x^2 + 7x - 2} \times \dfrac{30x^2 - 5x}{4x^2 + 17x + 15}$

17. $\dfrac{2x - 10}{3x - 9} \times \dfrac{x - 3}{x - 5} \times \dfrac{5}{x^2}$

Section 7.3

Divide the fractions in each of the following problems and reduce, if possible.

1. $\frac{2a}{3b} \div \frac{7a}{9b}$

2. $\frac{8x^2}{7y} \div x^2$

3. $\frac{x}{x^2 - 1} \div \frac{1}{x + 1}$

4. $\frac{x^2 - 9}{x^2 - 4} \div \frac{x^2 + 6x + 9}{x^2 + 4x + 4}$

5. $\frac{x^3 - x^2}{y^2 - 1} \div \frac{x^3}{y - 1}$

6. $\frac{h + 2}{h} \div \frac{h^2 + 5h + 4}{2}$

7. $\frac{2x^2 - 3x - 14}{9x + 3} \div \frac{4x^2 - 8x - 21}{3x + 1}$

8. $\frac{y^3 + 6y^2}{2y^2 - 4y} \div \frac{3y^2 + 18y}{6y - 12}$

9. $\frac{z^2 - 10z + 16}{z^2 - 8z - 20} \div \frac{z^2 - 7z - 8}{z^2 - 9z - 10}$

10. $5x^2 - 20 \div \frac{x - 2}{3}$

Section 7.4

PART 1

Add the fractions in each of the following problems and reduce, if possible.

1. $\frac{5}{x} + \frac{2}{x}$

2. $\frac{2}{x-3} + \frac{7}{x-3}$

3. $\frac{1}{x} + \frac{1}{y}$

4. $\frac{1}{x^2} + 8xy^2$

5. $\frac{9}{4x+8} + \frac{7}{x+2}$

6. $\frac{x}{x+4} + \frac{1}{x+3}$

7. $\frac{1}{x^2-x-2} + \frac{3}{x^2-4x+4}$

8. $\frac{1}{a^2-4} + \frac{1}{a-2}$

9. $\frac{a}{a^2-9} + \frac{3}{a^2+6a+9}$

10. $\frac{3}{3x^2-4x+1} + \frac{x}{x^2-1}$

PART 2

Subtract the fractions in each of the following problems and reduce, if possible.

1. $\frac{6}{8x} - \frac{3}{8x}$

2. $\frac{15y}{4} - \frac{3}{4}$

3. $\frac{3w}{w-2} - \frac{6}{w-2}$

4. $\frac{z+1}{z-3} - \frac{4}{z-3}$

5. $\frac{x+2}{x^2-1} - \frac{2}{x+1}$

6. $\frac{5}{x} + \frac{2}{x} - \frac{3}{x}$

7. $\frac{x}{x^2-4} - \frac{3}{x^2+5x-14}$

8. $\frac{3x}{x^2+5x+6} - \frac{2x-1}{x^2+6x+9}$

9. $\frac{2x}{x^2-16} - \frac{2}{x+4}$

10. $\frac{2}{x-2} + \frac{3}{x^2-4} - \frac{1}{x+2}$

Section 7.5

Solve the following equations. Be sure to check your answers for possible zeros in the denominators.

1. $\frac{1}{x} = \frac{2}{3}$

2. $7 = \frac{14}{x - 3}$

3. $\frac{2}{x} + \frac{1}{3} = 1$

4. $\frac{2}{x} + \frac{8}{x^2} = 1$

5. $\frac{3}{4} + \frac{x}{x + 1} = \frac{3}{2}$

6. $1 - \frac{2}{x + 2} = \frac{5}{x + 2}$

7. $\frac{x}{2} = \frac{3x + 1}{x + 7}$

8. $\frac{5x}{3 - 2x} = -3$

9. $\frac{x + 1}{x} + \frac{2}{x - 1} = \frac{1}{x}$

10. $\frac{3}{x + 2} - \frac{1}{2x + 1} = \frac{2}{2x^2 + 5x + 2}$

ANSWERS TO PRACTICE PROBLEMS

Section 7.1

1. $\frac{1}{2x}$

2. $\frac{-b^2}{4}$

3. $\frac{2a}{c^2}$

4. $\frac{1}{2xy}$

5. $\frac{x - 2}{x + 3}$

6. $\frac{x + 2}{3x + 1}$

7. $\frac{3x - 2}{4x - 5}$

8. $\frac{3x - 1}{4x - 3}$

Section 7.2

1. $\frac{y^2}{x^3}$

2. $\frac{x^2}{7y}$

3. $\frac{-b}{33a^4}$

4. $\frac{x}{2x + 6}$

5. $\frac{x + 1}{x + 5}$

6. $\frac{a - 1}{6a^2 + 7a + 2}$

7. $\frac{2y + 1}{3y + 5}$

Section 7.3

1. $\frac{2x - 3}{x^2 - 2x - 8}$

2. $x + 3$

3. $\frac{2p^2 + p}{2p^2 + 5p - 3}$

4. $\frac{3x - 1}{x + 7}$

Section 7.4

1. $\frac{8y + 5}{y + 1}$

2. $\frac{7}{x}$

3. $\frac{1}{p + 1}$

4. $\frac{2x^2 + 1}{x^4}$

5. $\frac{12}{x^2y}$

6. $\frac{3}{20x^2y^2}$

7. $\frac{1}{a - 2}$

8. $\frac{3x - 4}{x^3 - 2x^2}$

9. $\frac{1}{p^2 + 5p + 4}$

10. $\frac{-12}{(x - 3)(x + 3)(x + 3)}$

11. $\frac{3}{x + 2}$

12. $\frac{2x^2 - 9x}{2x^2 - 5x - 3}$

13. $\frac{-5x - 6}{(5x - 1)(2x + 3)(x + 2)}$

14. $\frac{1}{4x^2 - 25}$

Section 7.5

1. 4

2. 12

3. -15

4. 18

5. $\frac{5}{2}$

6. 1

7. $\frac{-3}{2}$, 1

8. No solution

C H A P T E R 8

Fractional Exponents and Radicals

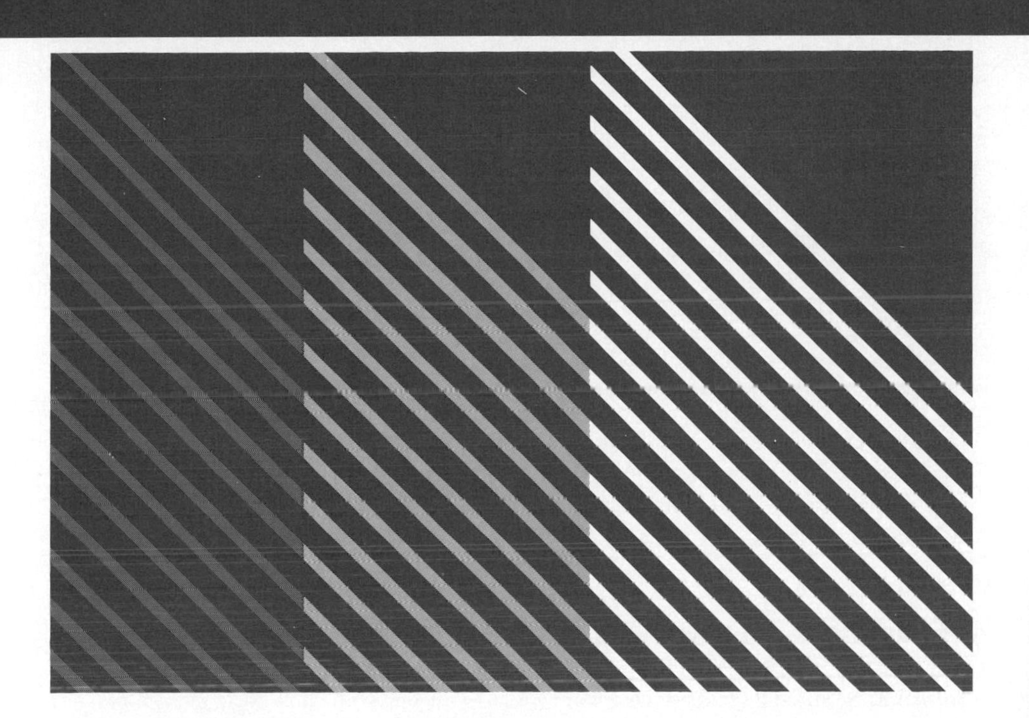

The objectives of this chapter are:

1. To know the rules of fractional exponents and radicals.
2. To be able to use these rules when working problems containing fractional exponents or radicals.

ANSWER

So far we have discussed exponents that are integers. Now, let's consider fractional exponents. The rules for integer exponents will also hold for fractional exponents. The similarity between exponents and coefficients will continue for fractional exponents and fractional coefficients. Let's list the rules for exponents from Chapter 5.

Rule 1 (exponents): $x^m x^n = x^{m+n}$
Rule 2 (exponents): $(x^m)^n = x^{nm}$
Rule 3 (exponents): $(xy)^n = x^n y^n$
Rule 4 (exponents): $\frac{x^m}{x^n} = x^{m-n}, x \neq 0.$
Rule 5 (exponents): $(\frac{x}{y})^n = \frac{x^n}{y^n}, y \neq 0.$

8.1 POSITIVE FRACTIONAL EXPONENTS AND RADICALS

In mathematics, we often want to answer questions such as the following.

1. What number added to itself equals 8?
2. What number multiplied by itself equals 9?

The answer to Question 1 is 4, since $4 + 4 = 8$, and the answer to Question 2 is 3 or -3, since $3 \times 3 = 9$ and $(-3)(-3) = 9$. Questions of this type are not always easy to answer. In fact, a major part of this section will deal with questions similar to Question 2. Let's look at some examples.

Example 1 **Find a number which added to itself equals 16.**

Solution: 8, since $8 + 8 = 16$ ■

Example 2 **Find a number which multiplied by itself equals 16.**

Solution: 4 or -4, since $4 \times 4 = 16$ and $(-4)(-4) = 16$ ■

Example 3 **Find a number which multiplied by itself equals 49.**

Solution: 7 or -7, since $7 \times 7 = 49$ and $(-7)(-7) = 49$ ■

1. ____________

2. ____________

3. ____________

Practice 1: Find a number which multiplied by itself equals 25.

Practice 2: Find a number which multiplied by itself equals 36.

Practice 3: Find a number which multiplied by itself equals 121.

A N S W E R

Example 4 **Find a number which added to itself equals 7.**

Solution: There is no integer which added to itself equals 7. However, we can find a solution by using fractional coefficients.

Now, $\left(\frac{1}{2}\right)7 + \left(\frac{1}{2}\right)7 = 2\left[\left(\frac{1}{2}\right)7\right]$ **Definition of multiplication.**

$= \left[2\left(\frac{1}{2}\right)\right]7$ **Associative property.**

$= (1)7$ $x \times \frac{1}{x} = 1$

$= 7$ $1 \times x = x$

Thus, $(\frac{1}{2})7$ is the number which added to itself equals 7. ■

Practice 4: Find a number which added to itself equals 5.

Practice 5: Find a number which added to itself equals 11.

4. ________

5. ________

Example 5 **Find a term which added to itself equals a.**

Solution: $\left(\frac{1}{2}\right)a + \left(\frac{1}{2}\right)a = 2\left[\left(\frac{1}{2}\right)a\right]$ **Definition of multiplication.**

$= \left[2\left(\frac{1}{2}\right)\right]a$ **Associative property.**

$= 1a$ $x \times \frac{1}{x} = 1$

$= a$ $1 \times x = x$

Thus, $(\frac{1}{2})a$ is the term which added to itself equals a. ■

Practice 6: Find a term which added to itself equals x.

Practice 7: Find a term which added to itself equals t.

Practice 8: Find a term which added to itself equals b.

6. ________

7. ________

8. ________

Now, let's look at some more examples similar to Question 2 at the beginning of this section.

Example 6 **Find a number which multiplied by itself equals 7.**

Solution: There is no rational number which multiplied by itself equals 7. Let's solve this problem similar to Example 4—that is, let's use fractional exponents. Now,

$7^{1/2} \times 7^{1/2} = (7^{1/2})^2$ **Definition of exponent.**

$= 7^{2(1/2)}$ $(x^m)^n = x^{nm}$

$= 7^1$ $x \times \frac{1}{x} = 1$

$= 7$ $x^1 = x$

Thus, $7^{1/2}$ is a number which multiplied by itself equals 7. ■

A N S W E R

What is $7^{1/2}$?

Well, Charlie, $7^{1/2}$ is simply a number which multiplied by itself equals 7.

It looks like $7^{1/2}$ is a number that someone invented.

That's true, Charlie. All the numbers that we use are numbers that someone invented. At one time, we did not have negative numbers. However, we needed an answer for $4 - 7$; so, -3 was used and negative numbers were introduced. Now, we need an answer to "what number multiplied by itself equals 7?" Thus, we introduced the number $7^{1/2}$. You will find numbers like $7^{1/2}$ to be very useful.

Using Example 6 as a guide, work the following practice problems.

9. ____________

Practice 9: Find a number which multiplied by itself equals 5.

10. ____________

Practice 10: Find a number which multiplied by itself equals 11.

11. ____________

Practice 11: Find a number which multiplied by itself equals 21.

Example 7 **Find a term which multiplied by itself equals *a*, where *a* is a positive number.**

Solution:

$a^{1/2} \times a^{1/2} = (a^{1/2})^2$	**Definition of exponent.**
$= a^{2(1/2)}$	$(x^m)^n = x^{nm}$
$= a^1$	$x \times \frac{1}{x} = 1$
$= a$	$x^1 = x$

Thus, $a^{1/2}$ is a term which multiplied by itself equals a. ■

Using Example 7 as a guide, work the following practice problems.

12. ____________

Practice 12: Find a term which multiplied by itself equals x, where x is a positive number.

13. ____________

Practice 13: Find a term which multiplied by itself equals t, where t is a positive number.

14. ____________

Practice 14: Find a term which multiplied by itself equals b, where b is a positive number.

When we find a term which multiplied by itself equals a, we are finding a square root of a; that is, a **square root** of a is a term which multiplied by itself equals a. Thus, c is a square root of a, if $c \times c = a$ or $c^2 = a$. Now, $a^{1/2}$, where a is a positive real number, is a square root of a and we often write $a^{1/2}$ as $\sqrt{a}$. We will call an expression of the form $\sqrt{a}$ *a* **radical,** the symbol $\sqrt{\ }$ is called a **radical sign** and a is called the **radicand.** Now, $\sqrt{a} = a^{1/2}$ represents the **principal square root** or positive square root of a.

Note: Throughout this chapter, with the exception of Section 8.7, all letters appearing under a radical sign or in a base with an exponent of $\frac{1}{2}$ represent positive real numbers.

What is $\sqrt{a}$*?*

Well, Charlie, $\sqrt{a}$ is just a way of writing the positive number which multiplied by itself equals a.

So $\sqrt{a} = a^{1/2}$ *and is the positive number which multiplied by itself equals a. That is,* $\sqrt{a}\sqrt{a} = a$.

That's correct, Charlie. Let's look at this using exponents.

$$\begin{aligned} \sqrt{a}\sqrt{a} &= (\sqrt{a})^2 && \textbf{Definition of exponent.} \\ &= (a^{1/2})^2 && \sqrt{a} = a^{1/2} \\ &= a^{2(1/2)} && (x^m)^n = x^{nm} \\ &= a^1 && x \times (\tfrac{1}{x}) = 1 \\ &= a && x^1 = x \end{aligned}$$

Let's write this result as the first rule of radicals.

Rule 1 (radicals): $(\sqrt{a})^2 = a$, where a is a positive real number.

Now, let's look at $\sqrt{a^2}$. Using exponents we have

$$\begin{aligned} \sqrt{a^2} &= (a^2)^{1/2} && \sqrt{x} = x^{1/2} \\ &= a^{(1/2)2} && (x^m)^n = x^{nm} \\ &= a^1 && x \times \tfrac{1}{x} = 1 \\ &= a && x^1 = x \end{aligned}$$

Let's write this as the second rule of radicals.

Rule 2 (radicals): $\sqrt{a^2} = a$, where a is a positive real number.

A N S W E R

Rule 2 (radicals) makes sense because $\sqrt{a^2}$ represents the positive number which multiplied by itself equals a^2 and we know that $a \times a = a^2$.

That's true, Charlie, $a \times a = a^2$; thus, $\sqrt{a^2} = a$.

How many square roots can a number have?

Well, Charlie, every positive number has two square roots, a positive square root (principal square root) and a negative square root. For example, $5 \times 5 = 25$ and $(-5) \times (-5) = 25$; thus, 5 and -5 are both square roots of 25 (5 is the principal square root of 25).

Now, a negative number does not have a real square root since a positive number times a positive number is a positive number and a negative number times a negative number is also a positive number. For example, there is no real number which multiplied by itself equals -25.

15. ____________________ Practice 15: What is the only number that has exactly one square root?

16. ____________________ Practice 16: Find the two square roots of 64.

17. ____________________ Practice 17: Find the two square roots of $\frac{1}{4}$.

18. ____________________ Practice 18: Find the two square roots of 225.

19. ____________________ Practice 19: Find the principal square root of 121 ($\sqrt{121}$).

20. ____________________ Practice 20: Find the principal square root of 144 ($\sqrt{144}$).

21. ____________________ Practice 21: Find the principal square root of 81 ($\sqrt{81}$).

Since zero has only one square root (Practice 15), the principal square root of zero is zero; that is, $\sqrt{0} = 0$.

Example 8 **Write the following expression using the radical sign.**

$$6^{1/2}$$

Solution:

$$6^{1/2} = \sqrt{6}$$

■

ANSWER

Example 9 **Write the following expresion using the radical sign.**

$$5x^{1/2}$$

Solution:

$$5x^{1/2} = 5\sqrt{x}$$ ■

Example 10 **Write the following expression using the radical sign.**

$$(5x)^{1/2}$$

Solution:

$$(5x)^{1/2} = \sqrt{5x}$$ ■

Practice 22: Write the following expression using the radical sign.

$$3^{1/2}$$

22. ______________

Practice 23: Write the following expression using the radical sign.

$$y^{1/2}$$

23. ______________

Practice 24: Write the following expression using the radical sign.

$$(xy)^{1/2}$$

24. ______________

Practice 25: Write the following expression using the radical sign.

$$(6x)^{1/2}$$

25. ______________

Practice 26: Write the following expression using the radical sign.

$$11x^{1/2}$$

26. ______________

Example 11 **Write the following radical expression using a fractional exponent.**

$$\sqrt{5}$$

Solution:

$$\sqrt{5} = 5^{1/2}$$ ■

Example 12 **Write the following radical expression using a fractional exponent.**

$$\sqrt{2x}$$

Solution:

$$\sqrt{2x} = (2x)^{1/2}$$ ■

A N S W E R

Example 13 **Write the following radical expression using a fractional exponent.**

$$8\sqrt{x}$$

Solution:

$$8\sqrt{x} = 8x^{1/2}$$ ■

27. ______________

Practice 27: Write the following radical expression using a fractional exponent.

$$\sqrt{15}$$

28. ______________

Practice 28: Write the following radical expression using a fractional exponent.

$$\sqrt{11x}$$

29. ______________

Practice 29: Write the following radical expression using a fractional exponent.

$$7\sqrt{x}$$

30. ______________

Practice 30: Write the following radical expression using a fractional exponent.

$$\sqrt{2y^3}$$

31. ______________

Practice 31: Write the following radical exression using a fractional exponent.

$$5\sqrt{x^3y}$$

A **square root radical expression in simplest form** does not have a perfect square factor under a radical sign. Let's look at how we will use Rule 2 in simplifying radical expressions.

Example 14 **Simplify the following radical expression.**

$$\sqrt{x^4}$$

Solution:

$$\begin{aligned} \sqrt{x^4} &= \sqrt{(x^2)^2} && a^{nm} = (a^m)^n \\ &= x^2 && \sqrt{a^2} = a : \text{Rule 2 (radicals).} \end{aligned}$$ ■

Example 15 **Simplify the following radical expression.**

$$\sqrt{4x^2}$$

Solution:

$$\begin{aligned} \sqrt{4x^2} &= \sqrt{2^2x^2} && 4 = 2^2 \\ &= \sqrt{(2x)^2} && a^nb^n = (ab)^n \\ &= 2x && \sqrt{a^2} = a : \text{Rule 2 (radicals).} \end{aligned}$$ ■

32. ______________

Practice 32: Simplify the following radical expression.

$$\sqrt{y^4}$$

A N S W E R

Practice 33: Simplify the following radical expression.

$$\sqrt{9x^2}$$

33. ____________

Practice 34: Simplify the following radical expression.

$$\sqrt{16y^4}$$

34. ____________

Practice 35: Simplify the following radical expression.

$$\sqrt{x^2y^2}$$

35. ____________

Practice 36: Simplify the following radical expression.

$$\sqrt{121x^6}$$

36. ____________

Example 16 **Simplify the following expression.**

$$(3^{1/2})^4$$

Solution:

$$\begin{aligned} (3^{1/2})^4 &= 3^{4(1/2)} && (x^m)^n = x^{nm} \\ &= 3^2 && 4(\tfrac{1}{2}) = 2 \\ &= 9 \end{aligned}$$

■

Example 17 **Simplify the following expression.**

$$(4x^6)^{1/2}$$

Solution:

$$\begin{aligned} (4x^6)^{1/2} &= 4^{1/2}(x^6)^{1/2} && (ab)^n = a^nb^n \\ &= 2(x^6)^{1/2} && 4^{1/2} = \sqrt{4} = 2 \\ &= 2x^{(1/2)6} && (a^m)^n = a^{nm} \\ &= 2x^3 && (\tfrac{1}{2})6 = 3 \end{aligned}$$

■

Practice 37: Simplify the following expression.

$$(5^{1/2})^6$$

37. ____________

Practice 38: Simplify the following expression.

$$[(5x)^{1/2}]^4$$

38. ____________

A N S W E R

39. ____________

Practice 39: Simplify the following expression.

$$(5x^{1/2})^2$$

40. ____________

Practice 40: Simplify the following expression.

$$(16x^2)^{1/2}$$

41. ____________

Practice 41: Simplify the following expression.

$$(9y^4)^{1/2}$$

42. ____________

Practice 42: Simplify the following expression.

$$(25x^2y^2)^{1/2}$$

Let's list the rules of exponents and radicals.

Rule 1 (exponents): $x^m x^n = x^{m+n}$
Rule 2 (exponents): $(x^m)^n = x^{nm}$
Rule 3 (exponents): $(xy)^n = x^n y^n$
Rule 4 (exponents): $\frac{x^m}{x^n} = x^{m-n}$, $x \neq 0$.
Rule 5 (exponents): $(\frac{x}{y})^n = \frac{x^n}{y^n}$, $y \neq 0$.

Rule 1 (radicals): $(\sqrt{a})^2 = a$, where a is a positive real number.
Rule 2 (radicals): $\sqrt{a^2} = a$, where a is a positive real number.

EXERCISE 8.1

PART 1

Find the two square roots of each number.

1. 25

2. 144

3. 225

4. 400

5. 625

6. 169

7. 196

8. 256

PART 2

Find the value for each of the following.

1. $\sqrt{4}$
2. $\sqrt{49}$
3. $-\sqrt{64}$
4. $\sqrt{169}$
5. $-\sqrt{81}$
6. $\sqrt{289}$
7. $-\sqrt{900}$
8. $\sqrt{10,000}$
9. $\sqrt{.01}$
10. $-\sqrt{14,400}$

PART 3

Write the following expressions using a radical sign.

1. $8^{1/2}$
2. $20^{1/2}$
3. $5^{1/2}$
4. $x^{1/2}$
5. $y^{1/2}$
6. $(2a)^{1/2}$
7. $(y^3)^{1/2}$
8. $(x + 1)^{1/2}$
9. $(a - b)^{1/2}$
10. $\left(\frac{2}{x}\right)^{1/2}$

PART 4

Write the following radical expressions using fractional exponents.

1. $\sqrt{8}$
2. $\sqrt{x}$
3. $\sqrt{x^2}$
4. $\sqrt{2a}$
5. $\sqrt{x - 1}$
6. $\sqrt{\frac{x}{2}}$
7. $\sqrt{5y^2}$
8. $\sqrt{a^2 + 1}$
9. $\sqrt{x^3}$
10. $\sqrt{(x - 1)^5}$

PART 5

Simplify the following radical expressions.

1. $\sqrt{x^2}$
2. $\sqrt{9x^2}$
3. $\sqrt{25x^2}$
4. $\sqrt{16x^2y^2}$
5. $\sqrt{169x^4}$
6. $\sqrt{4x^6y^2}$
7. $\sqrt{9x^2y^4}$
8. $\sqrt{x^4y^6}$
9. $\sqrt{49x^8}$
10. $\sqrt{81x^2y^8}$

PART 6

Simplify the following expressions.

1. $(2^{1/2})^8$
2. $(3^{1/2})^6$
3. $(2x^{1/2})^4$
4. $(x^{10})^{1/2}$
5. $(x^4)^{1/2}$
6. $(x^6)^{1/2}$
7. $(4x^2)^{1/2}$
8. $(16x^4)^{1/2}$
9. $(9^4)^{1/2}$
10. $(x^6y^4)^{1/2}$

8.2 SIMPLIFYING RADICAL EXPRESSIONS USING THE PRODUCT RULE

In order to simplify radical expressions, we will need some additional rules for radicals. The rules of exponents will help us develop the rules for radicals.

$$\begin{aligned} \sqrt{ab} &= (ab)^{1/2} && \sqrt{x} = x^{1/2} \\ &= a^{1/2}b^{1/2} && (xy)^n = x^n y^n \\ &= \sqrt{a}\sqrt{b} && \sqrt{x} = x^{1/2} \end{aligned}$$

This gives us the following product rule for radicals.

A N S W E R

Rule 3 (radicals): The Product Rule. $\sqrt{ab} = \sqrt{a}\sqrt{b}$, where a and b are positive real numbers.

Therefore, the square root of ab is the square root of a times the square root of b. For example,

$$\begin{aligned}\sqrt{4 \times 25} &= \sqrt{4} \times \sqrt{25} \\ &= 2 \times 5 \\ &= 10\end{aligned}$$

Example 1 **Using the product rule for radicals simplify the following radical expression.**

$$\sqrt{9 \times 2}$$

Solution:

$$\begin{aligned}\sqrt{9 \times 2} &= \sqrt{9} \times \sqrt{2} && \sqrt{ab} = \sqrt{a}\sqrt{b}\text{: Product rule for radicals.} \\ &= 3\sqrt{2} && \sqrt{9} = 3\end{aligned}$$

■

Practice 1: Using the product rule for radicals, simplify the following radical expression.

$$\sqrt{4 \times 9}$$

1. ______________________

Practice 2: Using the product rule for radicals, simplify the following radical expression.

$$\sqrt{4 \times 36}$$

2. ______________________

Practice 3: Using the product rule for radicals, simplify the following radical expression.

$$\sqrt{9 \times 3}$$

3. ______________________

Practice 4: Using the product rule for radicals, simplify the following radical expression.

$$\sqrt{36 \times 2}$$

4. ______________________

Practice 5: Using the product rule for radicals, simplify the following radical expression.

$$\sqrt{81 \times 5}$$

5. ______________________

Example 2 **Simplify the following radical expression.**

$$\sqrt{360}$$

Step 1: Completely factor 360.

$$\begin{aligned}360 &= 2 \times 180\\ &= 2 \times 2 \times 90\\ &= 2 \times 2 \times 2 \times 45\\ &= 2 \times 2 \times 2 \times 3 \times 15\\ &= 2 \times 2 \times 2 \times 3 \times 3 \times 5\end{aligned}$$

Step 2: Find the largest perfect square that is a factor of 360.

$$\begin{aligned}360 &= 2 \times 2 \times 2 \times 3 \times 3 \times 5\\ &= 2 \times 2^2 \times 3^2 \times 5\\ &= (2^2 \times 3^2)(2 \times 5)\\ &= (2 \times 3)^2(2 \times 5)\\ &= 6^2(10)\\ &= 36 \times 10\end{aligned}$$

Step 3: Use the product rule for radicals.

$\sqrt{360} = \sqrt{36 \times 10}$	**Factor out the largest perfect square from the radicand.**
$= \sqrt{36}\sqrt{10}$	$\sqrt{ab} = \sqrt{a}\sqrt{b}$**: Product rule for radicals.**
$= 6\sqrt{10}$	$\sqrt{a^2} = a$**: Rule 2 (radicals).**

Thus, $\sqrt{360} = 6\sqrt{10}$. ■

Instead of completely factoring, can we just find the largest perfect square factor?

Yes, Charlie. In fact, that is often easier. Let's redo Example 2 your way.

Example 2A **Alternative Simplify the following radical expression.**

$$\sqrt{360}$$

Step 1: Find the largest perfect square that is a factor of 360.

$$360 = 36 \times 10$$

Step 2: Use the product rule for radicals.

$\sqrt{360} = \sqrt{36 \times 10}$	**Factor out the largest perfect square from the radicand.**
$= \sqrt{36}\sqrt{10}$	$\sqrt{ab} = \sqrt{a}\sqrt{b}$**: Product rule for radicals.**
$= 6\sqrt{10}$	$\sqrt{a^2} = a$**: Rule 2 (radicals).**

Thus, $\sqrt{360} = 6\sqrt{10}$. ■

Will you list the first few perfect squares?

Sure, the first sixteen perfect squares are 0, 1, 4, 9, 16, 25, 36, 49, 64, 81, 100, 121, 144, 169, 196, and 225.

Example 3 **Simplify the following radical expression.**

$$\sqrt{96}$$

Step 1: Find the largest perfect square that is a factor of 96.

$$96 = 16 \times 6$$

Step 2: Use the product rule for radicals.

$$\begin{aligned} \sqrt{96} &= \sqrt{16 \times 6} && \textbf{Factor out the largest perfect square from the radicand.} \\ &= \sqrt{16}\sqrt{6} && \sqrt{ab} = \sqrt{a}\sqrt{b}\textbf{: Product rule for radicals.} \\ &= 4\sqrt{6} && \sqrt{a^2} = a\textbf{: Rule 2 (radicals).} \end{aligned}$$

Thus, $\sqrt{96} = 4\sqrt{6}$. ■

I thought 4 was the largest perfect square that was a factor of 96 and wrote $96 = 4 \times 24$ in Step 1.

Well, Charlie, 4 is a perfect square factor of 96 but it is not the largest one. Thus, when you write 96 as 4×24, the other factor, 24, still contains a perfect square. So, when you remove what you believe to be the largest perfect square, check the other factor and make sure that it does not contain a perfect square. If it does, continue the process until there are no perfect squares in the remaining factor. Now,

$$\begin{aligned} \sqrt{96} &= \sqrt{4 \times 24} \\ &= \sqrt{4}\sqrt{24} \\ &= 2\sqrt{24} \\ &= 2\sqrt{4 \times 6} \\ &= 2\sqrt{4}\sqrt{6} \\ &= 2 \times 2\sqrt{6} \\ &= 4\sqrt{6}. \end{aligned}$$

If you have difficulty finding the largest perfect square factor, then you can use the method of complete factorization from Example 2. You can also remove what you believe to be the largest perfect square and check the remaining factor for a perfect square. If there is one, you continue the process until there are no perfect squares in the remaining factor. Always check the remaining factor for a perfect square, it is quite easy to overlook a perfect square factor. Let's try some practice problems.

A N S W E R

6. ______________

Practice 6: Simplify the following radical expression.

$$\sqrt{18}$$

7. ______________

Practice 7: Simplify the following radical expression.

$$\sqrt{75}$$

8. ______________

Practice 8: Simplify the following radical expression.

$$\sqrt{300}$$

9. ______________

Practice 9: Simplify the following radical expression.

$$\sqrt{450}$$

Example 4 **Simplify the following radical expression.**

$$\sqrt{12x^3}$$

Step 1: Find the largest perfect square that is a factor of 12.

$$12 = 4 \times 3$$

Step 2: Find the largest perfect square that is a factor of x^3.

$$x^3 = x^2x$$

Step 3: Group the factors that are perfect squares.

$$\begin{aligned} 12x^3 &= 4 \times 3 \times x^2x \\ &= (4x^2)(3x) \end{aligned}$$

Step 4: Use the product rule for radicals.

$\sqrt{12x^3} = \sqrt{(4x^2)(3x)}$	**Factor out the largest perfect square from the radicand.**
$= \sqrt{4x^2}\sqrt{3x}$	**$\sqrt{ab} = \sqrt{a}\sqrt{b}$: Product rule for radicals.**
$= 2x\sqrt{3x}$	**$\sqrt{a^2} = a$: Rule 2 (radicals).**

Thus, $\sqrt{12x^3} = 2x\sqrt{3x}$. ■

How did you find the largest perfect square factor of x^3 *in Step 2?*

Well, Charlie, it is easy to find the largest perfect square factor when working with exponents. Let's look at x^n. If n is even, then x^n is a perfect square. For example, $x^8 = (x^4)^2$. If n is greater than 1 and not even, then x^n is not a perfect square but x^{n-1} is a perfect square and we factor x^n into $x^{n-1}x$. For example, $x^7 = x^6x = (x^3)^2x$.

Example 5 **Simplify the following radical expression.**

$$\sqrt{8x^4y^3}$$

Step 1: Find the largest perfect square that is a factor of 8.

$$8 = 4 \times 2$$

Step 2: Find the largest perfect square that is a factor of x^4. Since the exponent is even, x^4 is a perfect square.

$$x^4 = (x^2)^2$$

Step 3: Find the largest perfect square that is a factor of y^3. Since the exponent is not even $y^{3-1} = y^2$ is a perfect square.

$$y^3 = y^2y$$

Step 4: Group the factors that are perfect squares.

$$\begin{aligned} 8x^4y^3 &= 4 \times 2 \times x^4y^2y \\ &= (4x^4y^2)(2y) \end{aligned}$$

Step 5: Use the product rule for radicals.

$\sqrt{8x^4y^3} = \sqrt{(4x^4y^2)(2y)}$	**Factor out the largest perfect square from the radicand.**
$= \sqrt{4x^4y^2}\sqrt{2y}$	**$\sqrt{ab} = \sqrt{a}\sqrt{b}$: Product rule for radicals.**
$= 2x^2y\sqrt{2y}$	**$\sqrt{a^2} = a$: Rule 2 (radicals).**

Thus, $\sqrt{8x^4y^3} = 2x^2y\sqrt{2y}$. ■

ANSWER

Practice 10: Simplify the following radical expression.

$$\sqrt{27a^3}$$

10. ______________

11. ______________

Practice 11: Simplify the following radical expression.

$$\sqrt{24x^2y}$$

12. ______________

Practice 12: Simplify the following radical expression.

$$\sqrt{16p^6q^3}$$

13. ______________

Practice 13: Simplify the following radical expression.

$$\sqrt{125x^{15}y^7}$$

Let's list the rules for radicals.

> **Rule 1 (radicals):** $(\sqrt{a})^2 = a$, where a is a positive real number.
> **Rule 2 (radicals):** $\sqrt{a^2} = a$, where a is a positive real number.
> **Rule 3 (radicals): The Product Rule.** $\sqrt{ab} = \sqrt{a}\sqrt{a}$, where a and b are positive real numbers.

EXERCISE 8.2

Simplify the following radical expressions.

A

1. $\sqrt{25}$

2. $\sqrt{20}$

3. $\sqrt{90}$

4. $\sqrt{27}$

5. $\sqrt{64}$

6. $\sqrt{200}$

7. $\sqrt{160}$

8. $\sqrt{84}$

B

9. $\sqrt{4x^3}$

10. $\sqrt{x^2y^3}$

11. $\sqrt{18x^5}$

12. $\sqrt{16a^6b^5}$

13. $\sqrt{125xy^3}$

14. $\sqrt{40x^3y^2}$

15. $\sqrt{27xy^4}$

C

16. $\sqrt{27xy^3}$

17. $\sqrt{64a^6b^4}$

18. $\sqrt{125x^4y^2}$

19. $\sqrt{16x^4y^8}$

20. $\sqrt{32x^5y^2}$

21. $\sqrt{32m^5n^9}$

22. $\sqrt{64x^5y^{12}}$

ANSWER

8.3 MULTIPLYING RADICAL EXPRESSIONS

To multiply radical expressions, we will use the product rule for radicals from Section 8.2 and the special product rules from Chapter 6. You may need to review the special product rules in Section 6.1 before continuing.

> **Rule 3 (radicals): The Product Rule.** $\sqrt{a}\sqrt{b} = \sqrt{ab}$, where a and b are positive real numbers.

Example 1 **Multiply the following radical expressions and simplify, if possible.**

$$\sqrt{10}\sqrt{15}$$

Step 1: Use the product rule for radicals to multiply the two expressions.

$$\begin{aligned} \sqrt{10}\sqrt{15} &= \sqrt{10 \times 15} && \sqrt{a}\sqrt{b} = \sqrt{ab}\text{: Product rule for radicals.} \\ &= \sqrt{150} && \text{Perform the multiplication.} \end{aligned}$$

Step 2: Use the product rule for radicals to simplify the answer.

$$\begin{aligned} \sqrt{150} &= \sqrt{25 \times 6} && \text{Factor out the largest perfect square from the radicand.} \\ &= \sqrt{25}\sqrt{6} && \sqrt{a}\sqrt{b} = \sqrt{ab}\text{: Product rule for radicals.} \\ &= 5\sqrt{6} && \sqrt{a^2} = a\text{: Rule 2 (radicals).} \end{aligned}$$

Thus, $\sqrt{10}\sqrt{15} = 5\sqrt{6}$. ■

So, we used the product rule to multiply and then we used the product rule to simplify the answer.

That's correct, Charlie. We will use the product rule to multiply and simplify radical expressions. Let's try some practice problems.

1. ______________

Practice 1: Multiply the following radical expressions and simplify, if possible.

$$\sqrt{2}\sqrt{6}$$

2. ______________

Practice 2: Multiply the following radical expressions and simplify, if possible.

$$\sqrt{3}\sqrt{6}$$

Practice 3: Multiply the following radical expressions and simplify, if possible.

$$\sqrt{6}\sqrt{15}$$

ANSWER

3. ______

Practice 4: Multiply the following radical expressions and simplify, if possible.

$$\sqrt{14}\sqrt{21}$$

4. ______

Example 2 **Multiply the following radical expressions and simplify, if possible.**

$$\sqrt{2a}\sqrt{2ab}$$

Step 1: Use the product rule for radicals to multiply the two expressions.

$\sqrt{2a}\sqrt{2ab} = \sqrt{(2a)(2ab)}$ — **$\sqrt{x}\sqrt{y} = \sqrt{xy}$: Product rule for radicals.**

$= \sqrt{4a^2b}$ — **Perform the multiplication.**

Step 2: Use the product rule for radicals to simplify the answer.

$\sqrt{4a^2b} = \sqrt{(4a^2)b}$ — **Factor out the largest perfect square from the radicand.**

$= \sqrt{4a^2}\sqrt{b}$ — **$\sqrt{xy} = \sqrt{x}\sqrt{y}$: Product rule for radicals.**

$= 2a\sqrt{b}$ — **$\sqrt{x^2} = x$: Rule 2 (radicals).**

Thus, $\sqrt{2a}\sqrt{2ab} = 2a\sqrt{b}$. ■

Practice 5: Multiply the following radical expressions and simplify, if possible.

$$\sqrt{a}\sqrt{ab}$$

5. ______

Practice 6: Multiply the following radical expressions and simplify, if possible.

$$\sqrt{2x}\sqrt{6x}$$

6. ______

A N S W E R

7. ____________________

Practice 7: Multiply the following radical expressions and simplify, if possible.

$$\sqrt{a}\sqrt{a^3}$$

8. ____________________

Practice 8: Multiply the following radical expressions and simplify, if possible.

$$\sqrt{3pq}\sqrt{6p^3q}$$

Example 3 **Multiply the following radical expressions and simplify, if possible.**

$$\sqrt{a}(2\sqrt{a} + 3)$$

Solution:

$$\sqrt{a}(2\sqrt{a} + 3) = \sqrt{a}(2\sqrt{a}) + (\sqrt{a})3$$ **Distributive property.**

$$= 2\sqrt{a^2} + 3\sqrt{a}$$ **Perform the multiplications.**

$$= 2a + 3\sqrt{a}$$ **$\sqrt{a^2} = a$: Rule 2 (radicals).** ■

9. ____________________

Practice 9: Multiply the following radical expressions and simplify, if possible.

$$\sqrt{2}(x + \sqrt{2})$$

10. ____________________

Practice 10: Multiply the following radical expressions and simplify, if possible.

$$\sqrt{5x}\sqrt{x} + 2$$

Practice 11: Multiply the following radical expressions and simplify, if possible.

$$\sqrt{6x}(\sqrt{3x} + \sqrt{6})$$

A N S W E R

11. ______________

Practice 12: Multiply the following radical expressions and simplify, if possible.

$$2\sqrt{2x}(\sqrt{2x} + \sqrt{6})$$

12. ______________

Example 4 **Multiply the following radical expressions and simplify, if possible.**

$$(\sqrt{2x} - 5)(\sqrt{2x} + 5)$$

Solution: Use the special product rule $(a - b)(a + b) = a^2 - b^2$.

$$(\sqrt{2x} - 5)(\sqrt{2x} + 5) = (\sqrt{2x})^2 - 5^2 \qquad (a - b)(a + b) = a^2 - b^2$$
$$= 2x - 25 \qquad (\sqrt{a})^2 = a\text{: Rule 1 (radicals).}$$ ■

So, we can use the special product rules when multiplying radical expressions.

Yes, Charlie. In fact, the special product rules are very important when working with radical expressions.

Example 5 **Square the following radical expression and simplify, if possible.**

$$(3\sqrt{x} + 4)^2$$

Solution: Use the special product rule $(a + b)^2 = a^2 + 2ab + b^2$.

$$(3\sqrt{x} + 4)^2 = (3\sqrt{x})^2 + 2(3\sqrt{x})(4) + 4^2 \qquad (a + b)^2 = a^2 + 2ab + b^2.$$
$$= 9(\sqrt{x})^2 + 24\sqrt{x} + 16 \qquad \text{Perform the multiplications.}$$
$$= 9x + 24\sqrt{x} + 16 \qquad (\sqrt{a})^2 = a\text{: Rule 1 (radicals).}$$ ■

A N S W E R

13. ____________________

Practice 13: Multiply the following radical expressions and simplify, if possible.

$$(x - \sqrt{2})(x + \sqrt{2})$$

14. ____________________

Practice 14: Square the following radical expression and simplify, if possible.

$$(\sqrt{x} - 3)^2$$

15. ____________________

Practice 15: Multiply the following radical expressions and simplify, if possible.

$$(\sqrt{3x} - \sqrt{5})(\sqrt{3x} + \sqrt{5})$$

16. ____________________

Practice 16: Square the following radical expression and simplify, if possible.

$$(\sqrt{2x} + 7)^2$$

17. ____________________

Practice 17: Multiply the following radical expressions and simplify, if possible.

$$(2\sqrt{y} + 3)(\sqrt{y} - 2)$$

EXERCISE 8.3

Multiply the following radical expressions and simplify, if possible.

A

1. $\sqrt{2}\sqrt{3}$
2. $\sqrt{8}\sqrt{4}$
3. $\sqrt{9}\sqrt{4}$
4. $\sqrt{10}\sqrt{20}$
5. $\sqrt{3}\sqrt{6}$
6. $\sqrt{28}\sqrt{7}$
7. $\sqrt{15}\sqrt{3}$
8. $\sqrt{x}\sqrt{xy^2}$
9. $\sqrt{ab}\sqrt{b}$
10. $\sqrt{2x}\sqrt{2y}$
11. $\sqrt{3a^3}\sqrt{6a}$
12. $\sqrt{2x}\sqrt{6x^5}$

B

13. $\sqrt{3}(\sqrt{15} - \sqrt{3})$
14. $\sqrt{6}(\sqrt{3} + \sqrt{2})$
15. $\sqrt{x}(\sqrt{2x} - 3)$
16. $\sqrt{5x}(\sqrt{10x} - \sqrt{5})$
17. $\sqrt{2ab}(\sqrt{2a} + \sqrt{ab})$
18. $(\sqrt{a} + \sqrt{b})^2$
19. $(\sqrt{x} - 5)(\sqrt{x} + 5)$
20. $(y - \sqrt{2})(y + \sqrt{2})$

21. $(\sqrt{a} - 1)^2$

22. $(\sqrt{2x} + 1)^2$

23. $(\sqrt{y} - 2)(\sqrt{y} + 3)$

24. $(2\sqrt{x} + 3)(2\sqrt{x} - 3)$

25. $(3\sqrt{x} - \sqrt{y})(3\sqrt{x} + \sqrt{y})$

8.4 SIMPLIFYING RADICAL EXPRESSIONS USING THE QUOTIENT RULE AND RATIONALIZING THE DENOMINATOR

In order to simplify radical expressions involving division, we will need a division rule for radicals. Let's look at the development of such a rule.

$$\sqrt{\frac{a}{b}} = \left(\frac{a}{b}\right)^{1/2} \qquad \sqrt{x} = x^{1/2}$$

$$= \frac{a^{1/2}}{b^{1/2}} \qquad \left(\frac{x}{y}\right)^n = \frac{x^n}{y^n}$$

$$= \frac{\sqrt{a}}{\sqrt{b}} \qquad \sqrt{x} = x^{1/2}$$

> **Rule 4 (radicals): The Quotient Rule.** $\sqrt{\frac{a}{b}} = \frac{\sqrt{a}}{\sqrt{b}}$, where a and b are positive real numbers.

The square root of $\frac{a}{b}$ is the square root of a divided by the square root of b. For example,

$$\sqrt{\frac{36}{25}} = \frac{\sqrt{36}}{\sqrt{25}}$$

$$= \frac{6}{5}$$

In Section 8.1 we stated that a square root radical expression in simplest form does not have a perfect square factor under a radical sign. Also, a **radical expression in simplest form** does not have a fraction under a radical sign or a radical in the denominator of a fraction. For example, $\sqrt{\frac{1}{2}}$ is not in simplest form since there is a fraction under the radical sign and $\frac{1}{\sqrt{2}}$ is not in simplest form since there is a radical in the denominator. Removing the radicals from a denominator gives us a denominator that is a rational number and is called **rationalizing the denominator.**

Rule 4, the quotient rule for radicals, will be used to simplify radical expressions as well as to divide radical expressions. Let's look at some examples.

Example 1 **Simplify the following radical expression.**

$$\sqrt{\frac{15}{54}}$$

Step 1: Reduce the fraction.

$$\sqrt{\frac{15}{54}} = \sqrt{\frac{3 \times 5}{3 \times 18}}$$

$$= \sqrt{\frac{5}{18}}$$

Step 2: Use the quotient rule for radicals.

$$\sqrt{\frac{5}{18}} = \frac{\sqrt{5}}{\sqrt{18}}$$

$\sqrt{\frac{a}{b}} = \frac{\sqrt{a}}{\sqrt{b}}$**: Quotient rule for radicals.**

Step 3: Use the product rule for radicals.

$$\frac{\sqrt{5}}{\sqrt{18}} = \frac{\sqrt{5}}{\sqrt{9 \times 2}}$$

Factor out the largest perfect squares from the radicands.

$$= \frac{\sqrt{5}}{\sqrt{9}\sqrt{2}}$$

$\sqrt{ab} = \sqrt{a}\sqrt{b}$**: Product rule for radicals.**

$$= \frac{\sqrt{5}}{3\sqrt{2}}$$

$\sqrt{9} = 3$

Step 4: Rationalize the denominator (remove the radical, $\sqrt{2}$, from the denominator).

$$\frac{\sqrt{5}}{3\sqrt{2}} = \frac{\sqrt{5}\sqrt{2}}{3\sqrt{2}\sqrt{2}}$$

Multiply the numerator and denominator by $\sqrt{2}$.

$$= \frac{\sqrt{10}}{3(\sqrt{2})^2}$$

Perform the multiplications.

$$= \frac{\sqrt{10}}{3(2)}$$

$(\sqrt{a})^2 = a$**: Rule 1 (radicals).**

$$= \frac{\sqrt{10}}{6}$$

Perform the multiplication.

Thus, $\sqrt{\frac{5}{18}} = \frac{\sqrt{10}}{6}$. ■

Why did we multiply the numerator and denominator of $\frac{\sqrt{5}}{3\sqrt{2}}$ by $\sqrt{2}$ in Step 4?

Well, Charlie, a radical expression in simplest form does not have a radical in the denominator. Now, $\frac{\sqrt{5}}{3\sqrt{2}}$ contains a radical, $\sqrt{2}$, in the denominator. Since $\sqrt{2}\sqrt{2} = 2$, we multiplied the numerator and denominator by $\sqrt{2}$ in order to remove the radical from the denominator. Now, $\frac{\sqrt{2}}{\sqrt{2}}$ is 1; thus, we do not change the value of the expression and as you can see the final expression $\frac{\sqrt{10}}{6}$ does not contain a radical in the denominator.

Note: After simplifying a radical expression by using the rules of radicals, should the denominator consist of a single term and contain a factor such as $\sqrt{a}$, you rationalize the denominator by multiplying *both* the numerator and denominator by $\sqrt{a}$. Since $\sqrt{a}\sqrt{a} = (\sqrt{a})^2 = a$, the radical is removed from the denominator.

Example 2 **Simplify the following radical expression.**

$$\sqrt{\frac{8}{3x}}$$

Step 1: Reduce the fraction.

$\frac{8}{3x}$ is in reduced form.

Step 2: Use the quotient rule for radicals.

$$\sqrt{\frac{8}{3x}} = \frac{\sqrt{8}}{\sqrt{3x}}$$

$\sqrt{\frac{a}{b}} = \frac{\sqrt{a}}{\sqrt{b}}$: Quotient rule for radicals.

Step 3: Use the product rule for radicals.

$$\frac{\sqrt{8}}{\sqrt{3x}} = \frac{\sqrt{4 \times 2}}{\sqrt{3x}}$$

Factor out the largest perfect squares from the radicands.

$$= \frac{\sqrt{4}\sqrt{2}}{\sqrt{3x}}$$

$\sqrt{ab} = \sqrt{a}\sqrt{b}$: Product rule for radicals.

$$= \frac{2\sqrt{2}}{\sqrt{3x}}$$

$\sqrt{4} = 2$

Step 4: Rationalize the denominator (remove the radical from the denominator).

$$\frac{2\sqrt{2}}{\sqrt{3x}} = \frac{2\sqrt{2}\sqrt{3x}}{\sqrt{3x}\sqrt{3x}}$$

Multiply the numerator and denominator by $\sqrt{3x}$.

$$= \frac{2\sqrt{6x}}{3x}$$

Perform the multiplications.

Thus, $\sqrt{\frac{8}{3x}} = \frac{2\sqrt{6x}}{3x}$. ■

Practice 1: Simplify the following radical expression.

$$\sqrt{\frac{4}{6}}$$

A N S W E R

1. ____________________

Practice 2: Simplify the following radical expression.

$$\sqrt{\frac{12}{3a}}$$

2. ____________________

Practice 3: Simplify the following radical expression.

$$\sqrt{\frac{27a}{b}}$$

3. ____________________

Practice 4: Simplify the following radical expression.

$$\sqrt{\frac{16a^2}{ab^3}}$$

4. ____________________

EXERCISE 8.4

Simplify the following radical expressions.

A

1. $\sqrt{\frac{3}{2}}$

2. $\sqrt{\frac{2}{6}}$

3. $\sqrt{\frac{5}{24}}$

4. $\sqrt{\frac{8}{30}}$

5. $\sqrt{\frac{12}{3x}}$

6. $\sqrt{\frac{5a}{60}}$

7. $\sqrt{\frac{32y}{xy}}$

8. $\sqrt{\frac{a}{2b}}$

9. $\sqrt{\frac{8x}{34x}}$

10. $\sqrt{\frac{3xy}{4xy}}$

11. $\sqrt{\frac{1}{2x}}$

12. $\sqrt{\frac{6}{2y}}$

B

13. $\sqrt{\frac{5}{5x^2}}$

14. $\sqrt{\frac{8a}{b^3}}$

15. $\sqrt{\dfrac{12x^2}{4y}}$

16. $\sqrt{\dfrac{1}{x-1}}$

17. $\sqrt{\dfrac{a}{b+1}}$

18. $\sqrt{\dfrac{32a^2b}{50b^2}}$

19. $\sqrt{\dfrac{8x^5}{y^2}}$

20. $\sqrt{\dfrac{30y^2}{6x^2y}}$

8.5 DIVIDING RADICAL EXPRESSIONS AND RATIONALIZING THE DENOMINATOR

The division of radical expressions will often involve rationalizing the denominator since a radical expression in simplest form does not have a radical in the denominator or a fraction under a radical sign. In the last section, you rationalized denominators that consisted of a single term. Later in this section, we will also rationalize denominators that consist of two terms.

Example 1 **Divide the following radical expressions and simplify, if possible.**

$$\frac{\sqrt{8x^3y}}{\sqrt{2x}}$$

Step 1: Since the numerator and denominator contain common factors, rewrite as one radical using the quotient rule.

$$\frac{\sqrt{8x^3y}}{\sqrt{2x}} = \sqrt{\frac{8x^3y}{2x}} \qquad \frac{\sqrt{a}}{\sqrt{b}} = \sqrt{\frac{a}{b}}\text{: Quotient rule for radicals.}$$

Step 2: Reduce the fraction.

$$\sqrt{\frac{8x^3y}{2x}} = \sqrt{4x^2y}$$

A N S W E R

Step 3: Now that the fraction has been reduced and we no longer have a quotient, we use the product rule to simplify the answer.

$$\begin{aligned}\sqrt{4x^2y} &= \sqrt{(4x^2)y} && \textbf{Factor out the largest perfect square from the radicand.}\\ &= \sqrt{4x^2}\sqrt{y} && \sqrt{ab} = \sqrt{a}\sqrt{b}\textbf{: Product rule for radicals.}\\ &= 2x\sqrt{y} && \sqrt{a^2} = a\textbf{: Rule 2 (radicals).}\end{aligned}$$

Thus, $\frac{\sqrt{8x^3y}}{\sqrt{2x}} = 2x\sqrt{y}$. ■

1. ____________

Practice 1: Divide the following radical expressions and simplify, if possible.

$$\frac{\sqrt{15}}{\sqrt{5}}$$

2. ____________

Practice 2: Divide the following radical expressions and simplify, if possible.

$$\frac{\sqrt{35a^2}}{\sqrt{7a}}$$

3. ____________

Practice 3: Divide the following radical expressions and simplify, if possible.

$$\frac{\sqrt{32x^3y^2}}{\sqrt{2x^2y}}$$

Example 2 **Simplify the following radical expression.**

$$\frac{\sqrt{45x^2y^3}}{\sqrt{10x^3y}}$$

Step 1: Since the numerator and denominator contain common factors, rewrite as one radical using the quotient rule.

$$\frac{\sqrt{45x^2y^3}}{\sqrt{10x^3y}} = \sqrt{\frac{45x^2y^3}{10x^3y}} \qquad \frac{\sqrt{a}}{\sqrt{b}} = \sqrt{\frac{a}{b}}\textbf{: Quotient rule for radicals.}$$

Step 2: Reduce the fraction.

$$\sqrt{\frac{45x^2y^3}{10x^3y}} = \sqrt{\frac{9y^2}{2x}}$$

A N S W E R

Step 3: Now that the fraction has been reduced and we still have a quotient, we rewrite as two radicals using the quotient rule.

$$\sqrt{\frac{9y^2}{2x}} = \frac{\sqrt{9y^2}}{\sqrt{2x}} \qquad \sqrt{\frac{a}{b}} = \frac{\sqrt{a}}{\sqrt{b}}\text{: Quotient rule for radicals.}$$

$$= \frac{3y}{\sqrt{2x}} \qquad \sqrt{a^2} = a\text{: Rule 2 (radicals).}$$

Step 4: Rationalize the denominator.

$$\frac{3y}{\sqrt{2x}} = \frac{3y\sqrt{2x}}{\sqrt{2x}\sqrt{2x}} \qquad \text{Multiply the numerator and denominator by } \sqrt{2x}.$$

$$= \frac{3y\sqrt{2x}}{2x} \qquad \sqrt{a}\sqrt{a} = (\sqrt{a})^2 = a\text{: Rule 1 (radicals).}$$

Thus, $\frac{\sqrt{45x^2y^3}}{\sqrt{10x^3y}} = \frac{3y\sqrt{2x}}{2x}$. ■

Practice 4: Simplify the following radical expression.

$$\frac{\sqrt{18x^3y^2}}{\sqrt{2x^4y}}$$

4. ____________________

Practice 5: Simplify the following radical expression.

$$\frac{\sqrt{6x^2y}}{\sqrt{12x^4y^2}}$$

5. ____________________

Practice 6: Simplify the following radical expression.

$$\frac{\sqrt{50a^2b^5}}{\sqrt{2a^5b^2}}$$

6. ____________________

A N S W E R

7. ____________________

Practice 7: Simplify the following radical expression.

$$\frac{\sqrt{26a^3b^2}}{\sqrt{13a^4b^3}}$$

Example 3 **Rationalize the denominator and simplify, if possible.**

$$\frac{\sqrt{6}+3}{\sqrt{3}}$$

Solution: Multiply *both* the numerator and denominator by $\sqrt{3}$ and simplify.

$$\frac{\sqrt{6}+3}{\sqrt{3}} = \frac{(\sqrt{6}+3)\sqrt{3}}{\sqrt{3}\sqrt{3}}$$ **Multiply the numerator and denominator by $\sqrt{3}$.**

$$= \frac{\sqrt{6}\sqrt{3}+3\sqrt{3}}{\sqrt{3}\sqrt{3}}$$ **Distributive property.**

$$= \frac{\sqrt{6}\sqrt{3}+3\sqrt{3}}{3}$$ **$\sqrt{a}\sqrt{a} = (\sqrt{a})^2 = a$: Rule 1 (radicals).**

$$= \frac{\sqrt{6 \times 3}+3\sqrt{3}}{3}$$ **$\sqrt{a}\sqrt{b} = \sqrt{ab}$: Product rule for radicals.**

$$= \frac{\sqrt{18}+3\sqrt{3}}{3}$$ **Perform the multiplication.**

$$= \frac{\sqrt{9 \times 2}+3\sqrt{3}}{3}$$ **Factor out the largest perfect square.**

$$= \frac{\sqrt{9}\sqrt{2}+3\sqrt{3}}{3}$$ **$\sqrt{ab} = \sqrt{a}\sqrt{b}$: Product rule for radicals.**

$$= \frac{3\sqrt{2}+3\sqrt{3}}{3}$$ **$\sqrt{a^2} = a$: Rule 2 (radicals).**

$$= \frac{3(\sqrt{2}+\sqrt{3})}{3}$$ **Factor the numerator.**

$$= \frac{\sqrt{2}+\sqrt{3}}{1}$$ **Reduce.**

$$= \sqrt{2}+\sqrt{3}$$ **$\frac{a}{1} = a$**

Thus, $\frac{\sqrt{6}+3}{\sqrt{3}} = \sqrt{2}+\sqrt{3}$. ■

Practice 8: Rationalize the denominator and simplify, if possible.

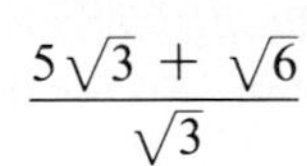

$$\frac{5\sqrt{3} + \sqrt{6}}{\sqrt{3}}$$

ANSWER

8. ____________________

Practice 9: Rationalize the denominator and simplify, if possible.

$$\frac{5\sqrt{10} + \sqrt{30}}{\sqrt{10}}$$

9. ____________________

Practice 10: Rationalize the denominator and simplify, if possible.

$$\frac{5\sqrt{x} + \sqrt{2x}}{\sqrt{x}}$$

10. ____________________

Practice 11: Rationalize the denominator and simplify, if possible.

$$\frac{3x + \sqrt{x}}{\sqrt{x}}$$

11. ____________________

When the denominator consists of two terms, with at least one containing a radical, we will use the special product rule

$$(a - b)(a + b) = a^2 - b^2$$

to rationalize the denominator.

To rationalize a denominator that is the *sum* of two terms, multiply *both* the numerator and denominator by the *difference* of the two terms. To rationalize a denominator that is the *difference* of two terms, multiply *both* the numerator and denominator by the *sum* of the two terms.

Why do we rationalize this way?

Well, Charlie, when you square a and b in the special product $(a - b)(a + b) = a^2 - b^2$, you remove any square root radicals that they contain. Let's look at an example.

Example 4 **Rationalize the denominator and simplify, if possible.**

$$\frac{2 + \sqrt{3}}{2 - \sqrt{3}}$$

Step 1: Multiply *both* the numerator and denominator by $2 + \sqrt{3}$.

$$\frac{2 + \sqrt{3}}{2 - \sqrt{3}} = \frac{(2 + \sqrt{3})(\mathbf{2} + \sqrt{\mathbf{3}})}{(2 - \sqrt{3})(\mathbf{2} + \sqrt{\mathbf{3}})}$$

Step 2: Multiply the denominator using the special product rule $(a - b)(a + b) = a^2 - b^2$.

$$\begin{aligned}\frac{2 + \sqrt{3}}{2 - \sqrt{3}} &= \frac{(2 + \sqrt{3})(2 + \sqrt{3})}{(2 - \sqrt{3})(2 + \sqrt{3})} \\ &= \frac{(2 + \sqrt{3})(2 + \sqrt{3})}{2^2 - (\sqrt{3})^2} \\ &= \frac{(2 + \sqrt{3})(2 + \sqrt{3})}{4 - 3} \\ &= \frac{(2 + \sqrt{3})(2 + \sqrt{3})}{1} \\ &= (2 + \sqrt{3})(2 + \sqrt{3})\end{aligned}$$

Step 3: Use the special product rule $(a + b)^2 = (a + b)(a + b) = a^2 + 2ab + b^2$ to multiply the numerator.

$$\begin{aligned}(2 + \sqrt{3})(2 + \sqrt{3}) &= 2^2 + 2(2)(\sqrt{3}) + (\sqrt{3})^2 \\ &= 4 + 4\sqrt{3} + 3 \\ &= 7 + 4\sqrt{3}\end{aligned}$$

Thus, $\frac{2 + \sqrt{3}}{2 - \sqrt{3}} = 7 + 4\sqrt{3}$. ■

A N S W E R

Example 5 **Rationalize the denominator and simplify, if possible.**

$$\frac{5}{2\sqrt{x} + 3}$$

Step 1: Multiply *both* the numerator and denominator by $2\sqrt{x} - 3$.

$$\frac{5}{2\sqrt{x} + 3} = \frac{5(\mathbf{2\sqrt{x} - 3})}{(2\sqrt{x} + 3)(\mathbf{2\sqrt{x} - 3})}$$

Step 2: Multiply the denominator using the special product rule $(a - b)(a + b) = a^2 - b^2$.

$$\frac{5}{2\sqrt{x} + 3} = \frac{5(2\sqrt{x} - 3)}{(2\sqrt{x} + 3)(2\sqrt{x} - 3)}$$

$$= \frac{5(2\sqrt{x} - 3)}{(2\sqrt{x})^2 - 3^2}$$

$$= \frac{5(2\sqrt{x} - 3)}{4x - 9}$$

Step 3: Multiply the numerator using the distributive property.

$$\frac{5(2\sqrt{x} - 3)}{4x - 9} = \frac{10\sqrt{x} - 15}{4x - 9}$$

Thus, $\frac{5}{2\sqrt{x} + 3} = \frac{10\sqrt{x} - 15}{4x - 9}$. ■

Practice 12: Rationalize the denominator and simplify, if possible.

$$\frac{\sqrt{2} - 1}{2\sqrt{2} - 3}$$

12. ____________________

Practice 13: Rationalize the denominator and simplify, if possible.

$$\frac{3 + \sqrt{5}}{3 - \sqrt{5}}$$

13. ____________________

A N S W E R

Practice 14: Rationalize the denominator and simplify, if possible.

$$\frac{2}{\sqrt{x} - 5}$$

14. ____________

15. ____________

Practice 15: Rationalize the denominator and simplify, if possible.

$$\frac{\sqrt{a} - \sqrt{b}}{\sqrt{a} + \sqrt{b}}$$

EXERCISE 8.5

Rationalize the denominator and simplify, if possible.

A

1. $\frac{\sqrt{75}}{\sqrt{3}}$

2. $\frac{\sqrt{72}}{\sqrt{6}}$

3. $\frac{\sqrt{55x}}{\sqrt{11x}}$

4. $\frac{\sqrt{32a}}{\sqrt{8a}}$

A N S W E R

8.6 ADDING AND SUBTRACTING RADICAL EXPRESSIONS

We call an expression of the form $\sqrt{a}$ *a* **radical**, the symbol $\sqrt{\ }$ is called the **radical sign,** and *a* is called the **radicand.** Square root radicals are **like radicals** if they have the same radicand.

Example 1 **The following radical expressions contain like radicals.**

a. $2\sqrt{5}$ and $7\sqrt{5}$

b. $3\sqrt{x}$ and $5\sqrt{x}$

c. $6\sqrt{xy}$ and $3\sqrt{xy}$ ■

Example 2 **The following radicals are *not* like radicals.**

a. $\sqrt{6}$ and $\sqrt{3}$

b. $\sqrt{x}$ and $\sqrt{y}$

c. $\sqrt{x}$ and $\sqrt{2x}$ ■

The basic rule of addition which states that only like things can be added is also true for radicals. Thus, *you can only add or subtract square root radicals that have the same radicand.*

Example 3 **Simplify the following radical expression by combining like radicals.**

$$5\sqrt{3} + 7\sqrt{2} - 3\sqrt{3} + 4\sqrt{2}$$

Solution:

$$5\sqrt{3} + 7\sqrt{2} - 3\sqrt{3} + 4\sqrt{2} = (5\sqrt{3} - 3\sqrt{3}) + (7\sqrt{2} + 4\sqrt{2})$$

Group like radicals.

$$= 2\sqrt{3} + 11\sqrt{2}$$

Combine like radicals. ■

Practice 1: Simplify the following radical expression by combining like radicals.

$$2\sqrt{5} - 3\sqrt{7} + 8\sqrt{5} - 4\sqrt{7}$$

1. ____________________

Practice 2: Simplify the following radical expression by combining like radicals.

$$3\sqrt{2} - 5\sqrt{3} + 8\sqrt{3} - 4\sqrt{2}$$

2. ____________________

Practice 3: Simplify the following radical expression by combining like radicals.

$$2\sqrt{6} + 7\sqrt{5} - 9\sqrt{6} - 12\sqrt{5}$$

3. ____________________

A N S W E R

In many problems, the radicands are not the same; however, after simplifying the radicals, you will find that some of the radicands become the same and you are then able to combine like radicals. Let's look at an example.

Example 4 **Simplify the following radical expression.**

$$\sqrt{20} - \sqrt{45} + \sqrt{80}$$

Solution:

$\sqrt{20} - \sqrt{45} + \sqrt{80} = \sqrt{4 \times 5} - \sqrt{9 \times 5} + \sqrt{16 \times 5}$ — **Factor out the largest perfect squares from the radicands.**

$= \sqrt{4}\sqrt{5} - \sqrt{9}\sqrt{5} + \sqrt{16}\sqrt{5}$ — **$\sqrt{ab} = \sqrt{a}\sqrt{b}$: Product rule for radicals.**

$= 2\sqrt{5} - 3\sqrt{5} + 4\sqrt{5}$ — **$\sqrt{a^2} = a$: Rule 2 (radicals).**

$= 3\sqrt{5}$ — **Combine like radicals.** ■

Example 5 **Simplify the following radical expression.**

$$\sqrt{4x} + \sqrt{9x} - \sqrt{25x^4} + x\sqrt{36x^2}$$

Solution:

$\sqrt{4x} + \sqrt{9x} - \sqrt{25x^4} + x\sqrt{36x^2}$

$= \sqrt{4 \times x} + \sqrt{9 \times x} - \sqrt{25x^4} + x\sqrt{36x^2}$ — **Factor out the largest perfect squares from the radicands.**

$= \sqrt{4}\sqrt{x} + \sqrt{9}\sqrt{x} - \sqrt{25x^4} + x\sqrt{36x^2}$ — **$\sqrt{ab} = \sqrt{a}\sqrt{b}$: Product rule for radicals.**

$= 2\sqrt{x} + 3\sqrt{x} - 5x^2 + x(6x)$ — **$\sqrt{a^2} = a$: Rule 2 (radicals).**

$= 2\sqrt{x} + 3\sqrt{x} - 5x^2 + 6x^2$ — **Perform the multiplication.**

$= 5\sqrt{x} + x^2$ — **Combine like terms.** ■

4. 

Practice 4: Simplify the following radical expression.

$$\sqrt{8} + \sqrt{50} - \sqrt{32}$$

5. ______________________

Practice 5: Simplify the following radical expression.

$$\sqrt{54} - \sqrt{24} - \sqrt{96}$$

Practice 6: Simplify the following radical expression.

$$\sqrt{25a} - \sqrt{49a} + \sqrt{16a}$$

A N S W E R

6. ______

Practice 7: Simplify the following radical expression.

$$\sqrt{25x^3y} - \sqrt{36x^3y} + \sqrt{16x^3y}$$

7. ______

EXERCISE 8.6

Simplify the following radical expressions.

A

1. $\sqrt{18} + \sqrt{2}$

2. $\sqrt{90} + 2\sqrt{10}$

3. $\sqrt{20} + \sqrt{45}$

4. $\sqrt{24} - \sqrt{6}$

5. $\sqrt{32} - \sqrt{8}$

6. $\sqrt{12} + \sqrt{27} - \sqrt{75}$

7. $\sqrt{5} + \sqrt{125} + \sqrt{20}$

8. $\sqrt{x} + \sqrt{x^3}$

9. $\sqrt{a^3} - \sqrt{a^5}$

10. $\sqrt{2x} + \sqrt{8x}$

B

11. $3\sqrt{a} - 2\sqrt{a^3}$

12. $\sqrt{y^3} + \sqrt{y^5} + \sqrt{y^7}$

13. $\sqrt{49x} + \sqrt{16x} + \sqrt{64x}$

14. $\sqrt{100xy} - \sqrt{16xy} - \sqrt{25xy}$

15. $\sqrt{x^2y} - 3\sqrt{y}$

16. $\sqrt{a^5b^3} - \sqrt{ab}$

17. $\sqrt{x} + \sqrt{x^3} - \sqrt{xy}$

18. $\sqrt{x - 1} + \sqrt{x - 1}$

19. $\sqrt{1-a} - \sqrt{1-a}$

20. $2\sqrt{x} + \sqrt{9x} + \sqrt{25x^2} + \sqrt{4x^2}$

8.7 SOLVING EQUATIONS CONTAINING RADICALS

To solve an equation containing radicals it is necessary to remove the radicals. Let's look at an example.

Example 1 **Solve the following equation.**

$$\sqrt{x-2} - 3 = 0$$

Step 1: Isolate $\sqrt{x-2}$.

$\sqrt{x-2} - 3 = 0$	**Original equation.**
$\sqrt{x-2} - 3 + 3 = 0 + 3$	**Add 3 to both sides.**
$\sqrt{x-2} = 3$	**Simplify.**

Step 2: Square *both* sides of the resulting equation.

$\sqrt{x-2} = 3$	**Equation from Step 1.**
$(\sqrt{x-2})^2 = 3^2$	**Square both sides.**
$x - 2 = 9$	$(\sqrt{a})^2 = a$

Step 3: Solve the equation from Step 2.

$x - 2 = 9$	**Equation from Step 2.**
$x - 2 + 2 = 9 + 2$	**Add 2 to both sides.**
$x = 11$	**Simplify.**

Step 4: Check 11 in the original equation.

$\sqrt{x-2} - 3 = 0$	**Original equation.**
$\sqrt{11-2} - 3 \stackrel{?}{=} 0$	**Substitute 11 for *x*.**
$\sqrt{9} - 3 \stackrel{?}{=} 0$	
$3 - 3 \stackrel{?}{=} 0$	
$0 = 0$	

Thus, 11 is the solution of the original equation. ■

Why did you isolate $\sqrt{x-2}$?

Well, Charlie, by isolating the radical term $\sqrt{x-2}$ in Step 1, when we squared both sides of the equation in Step 2, the radical was eliminated. Let's look at what would have happened if we had not isolated $\sqrt{x-2}$.

$$\sqrt{x-2} - 3 = 0 \qquad \text{Original equation.}$$
$$(\sqrt{x-2} - 3)^2 = 0^2 \qquad \text{Square both sides.}$$
$$(\sqrt{x-2})^2 - 2(3)\sqrt{x-2} + 9 = 0 \qquad (a-b)^2 = a^2 - 2ab + b^2$$
$$x - 2 - 6\sqrt{x-2} + 9 = 0 \qquad (\sqrt{a})^2 = a$$
$$x - 6\sqrt{x-2} + 7 = 0 \qquad \text{Combine like terms.}$$

You didn't eliminate the radical term $\sqrt{x-2}$; in fact, the equation that you ended with is more difficult than the one you started with.

That's true, Charlie. If we are to eliminate the radical when we square both sides of the equation, then we must first isolate the radical term.

To solve an equation containing radicals:

1. Isolate a radical term on one side of the equation.
2. Square *both* sides of the equation.
3. If a radical term still remains, repeat Steps 1 and 2.
4. Solve the resulting equation.
5. Check each answer in the *original* equation and omit any answer that does not satisfy the original equation.

Step 5 is very important since we often get an extra solution when squaring both sides of an equation. For example,

$$x = 2$$

has one solution, 2. However, when we square both sides of the equation we get

$$x^2 = 4$$

which has two solutions, 2 and −2. Thus, it is necessary to check all answers in the original equation.

Example 2 **Solve the following equation.**

$$3 + \sqrt{x+3} = x$$

ANSWER

Step 1: Isolate $\sqrt{x+3}$.

$3 + \sqrt{x+3} = x$	Original equation.
$3 + \sqrt{x+3} - 3 = x - 3$	Subtract 3 from both sides.
$\sqrt{x+3} = x - 3$	Simplify.

Step 2: Square *both* sides of the resulting equation.

$\sqrt{x+3} = x - 3.$	Equation from Step 1.
$(\sqrt{x+3})^2 = (x-3)^2$	Square both sides.
$x + 3 = (x-3)^2$	$(\sqrt{a})^2 = a$
$x + 3 = x^2 - 6x + 9$	$(a-b)^2 = a^2 - 2ab + b^2$

Step 3: Solve the equation from Step 2.

$x + 3 = x^2 - 6x + 9$	Equation from Step 2.
$x + 3 - x - 3 = x^2 - 6x + 9 - x - 3$	Subtract x and 3 from both sides.
$0 = x^2 - 7x + 6$	Simplify.
$0 = (x-6)(x-1)$	Factor.
$x - 6 = 0$ or $x - 1 = 0$	If $ab = 0$, then $a = 0$ or $b = 0$
$x = 6$ $\quad x = 1$	Isolate x.

Step 4: Check 6 and 1 in the original equation.

Replace x with 6.	Replace x with 1.
$3 + \sqrt{x+3} = x$	$3 + \sqrt{x+3} = x$
$3 + \sqrt{6+3} \stackrel{?}{=} 6$	$3 + \sqrt{1+3} \stackrel{?}{=} 1$
$3 + \sqrt{9} \stackrel{?}{=} 6$	$3 + \sqrt{4} \stackrel{?}{=} 1$
$3 + 3 \stackrel{?}{=} 6$	$3 + 2 \stackrel{?}{=} 1$
$6 = 6$	$5 \neq 1$

Since 1 does not satisfy the original equation, 1 is not a solution to the original equation. Thus, the original equation has only one solution, 6. ■

Practice 1: Solve the following equation.

$$\sqrt{x+1} - 2 = 0$$

1. ____________________

A N S W E R

2. ____________________

Practice 2: Solve the following equation.

$$\sqrt{2x + 3} - 5 = 0$$

3. ____________________

Practice 3: Solve the following equation.

$$\sqrt{3x - 2} = 4$$

4. ____________________

Practice 4: Solve the following equation.

$$\sqrt{2x - 5} = x - 2$$

Example 3 **Solve the following equation.**

$$\sqrt{x + 5} - \sqrt{x} = 1$$

Step 1: Isolate $\sqrt{x + 5}$.

$$\sqrt{x + 5} - \sqrt{x} = 1 \quad \textbf{Original equation.}$$

$$\sqrt{x + 5} - \sqrt{x} + \sqrt{x} = 1 + \sqrt{x} \quad \textbf{Add } \sqrt{x} \textbf{ to both sides.}$$

$$\sqrt{x + 5} = \sqrt{x} + 1 \quad \textbf{Simplify.}$$

Step 2: Square *both* sides of the resulting equation.

$\sqrt{x+5} = \sqrt{x} + 1$	**Equation from Step 1.**
$(\sqrt{x+5})^2 = (\sqrt{x} + 1)^2$	**Square both sides.**
$x + 5 = (\sqrt{x} + 1)^2$	$(\sqrt{a})^2 = a$
$x + 5 = (\sqrt{x})^2 + 2(1)(\sqrt{x}) + 1^2$	$(a + b)^2 = a^2 + 2ab + b^2$
$x + 5 = x + 2\sqrt{x} + 1$	$(\sqrt{a})^2 = a$

What about the radical term $2\sqrt{x}$?

Well, Charlie, in the following steps we will isolate $2\sqrt{x}$ and then eliminate the radical by squaring both sides of the equation again.

Step 3: Isolate $2\sqrt{x}$.

$x + 5 = x + 2\sqrt{x} + 1$	**Equation from Step 2.**
$x + 5 - x - 1 = x + 2\sqrt{x} + 1 - x - 1$	**Subtract x and 1 from both sides.**
$4 = 2\sqrt{x}$	**Simplify.**

Step 4: Square *both* sides of the resulting equation.

$4 = 2\sqrt{x}$	**Equation from Step 3.**
$4^2 = (2\sqrt{x})^2$	**Square both sides.**
$16 = 4x$	$(\sqrt{a})^2 = a$

Step 5: Solve the equation from Step 4.

$16 = 4x$	**Equation from Step 4.**
$4 = x$	**Divide both sides by 4.**

Step 6: Check 4 in the original equation.

$\sqrt{x+5} - \sqrt{x} = 1$	**Original equation.**
$\sqrt{4+5} - \sqrt{4} \stackrel{?}{=} 1$	**Substitute 4 for x.**
$\sqrt{9} - \sqrt{4} \stackrel{?}{=} 1$	
$3 - 2 \stackrel{?}{=} 1$	
$1 = 1$	

Thus, 4 is the solution of the original equation. ■

ANSWER

5. ______________________

Practice 5: Solve the following equation.

$$\sqrt{x+5} = 5 - \sqrt{x}$$

6. ______________________

Practice 6: Solve the following equation.

$$\sqrt{2x+7} = \sqrt{x} + 2$$

7. ______________________

Practice 7: Solve the following equation.

$$\sqrt{5x-3} - \sqrt{3x-1} = 0$$

Practice 8: Solve the following equation.

$$\sqrt{2x + 17} - \sqrt{x + 5} = 2$$

A N S W E R

8. ____________________

EXERCISE 8.7

Solve the following equations (be sure to check your answers in the original equations).

A

1. $\sqrt{x - 1} = 0$

2. $\sqrt{2a + 1} = 0$

3. $\sqrt{x} - 1 = 0$

4. $\sqrt{3x} - 4 = 0$

5. $\sqrt{x - 2} - 1 = 0$

6. $\sqrt{2x + 1} - 3 = 0$

7. $4\sqrt{2a - 1} - 2 = 0$

8. $\sqrt{4x - 3} = 5$

9. $1 + \sqrt{x + 5} = x$

10. $\sqrt{x + 2} = x + 2$

B

11. $3\sqrt{x} = \sqrt{x + 16}$

12. $\sqrt{y + 3} = \sqrt{y + 5}$

13. $\sqrt{3a + 2} - \sqrt{a + 8} = 0$

14. $3\sqrt{x + 7} = 4\sqrt{x}$

15. $\sqrt{x^2 + 5} - x + 5 = 0$

16. $\sqrt{y - 9} + 3 = y$

17. $\sqrt{2x + 1} - \sqrt{x} = 1$

18. $\sqrt{2x - 1} + \sqrt{x} = 2$

19. $\sqrt{x + 4} - \sqrt{x - 1} = 1$

20. $\sqrt{3x + 1} - \sqrt{x - 4} = 3$

21. $\sqrt{a + 2} + \sqrt{3a + 4} = 2$

22. $\sqrt{2x^2 - 1} = \sqrt{x^2 + 3}$

8.8 STRATEGIES FOR SOLVING APPLICATIONS

The suggestions in setting up an equation for each word problem in this section are basically the same as those in Chapters 4, 6, and 7.

There is a new relationship that we will discuss which is used in some of the word problems in this section. A right triangle contains a 90-degree angle (right angle). The side opposite the 90-degree angle is the hypotenuse of the right triangle and the remaining two sides are called the legs.

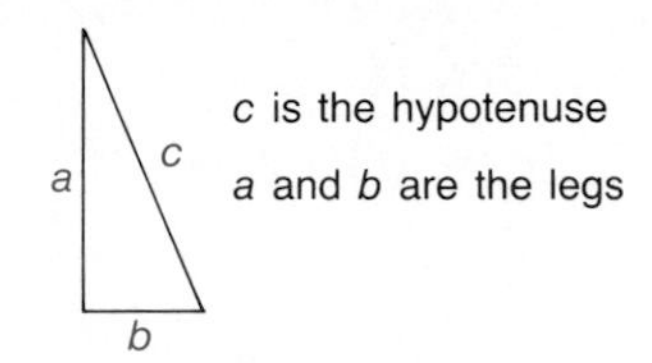

There is a relationship between the hypotenuse and the two legs—that is, $c^2 = a^2 + b^2$. This relationship was proven by the Greek mathematician Pythagoras and is known as the Pythagorean Theorem.

The following three examples show how the Pythagorean Theorem is used.

Example 1 **If the legs of a right triangle are 5 inches and 12 inches, find the hypotenuse.**

Step 1: Draw the figure.

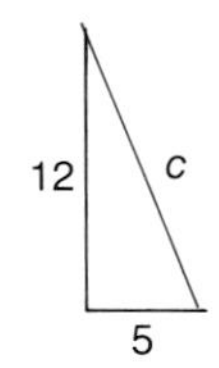

Step 2: Use the Pythagorean Theorem.

$$c^2 = 12^2 + 5^2$$
$$c^2 = 144 + 25$$
$$c^2 = 169$$
$$c^2 = 13 \times 13$$
$$c = 13$$

Thus, the hypotenuse is 13 inches. ■

Before working the next two examples, we need to discuss taking the square root of both sides of an equation. (This is explained in more detail in Chapter 9.) In general, the two solutions of the quadratic equation $x^2 = a$ are $x = -\sqrt{a}$ and $x = \sqrt{a}$. For example, if $x^2 = 15$, then $x = -\sqrt{15}$ or $x = \sqrt{15}$.

Example 2 **Find the length of a diagonal of a square whose sides are 3 centimeters long.**

Step 1: Draw the figure.

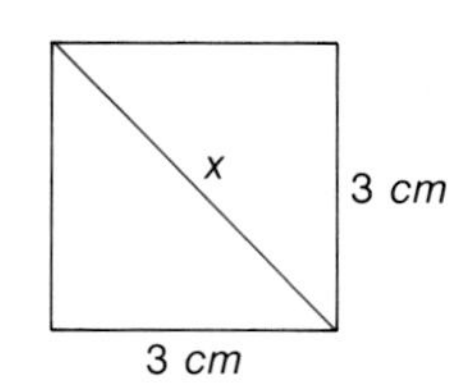

Step 2: Use the Pythagorean Theorem.

$$x^2 = 3^2 + 3^2$$
$$x^2 = 9 + 9$$
$$x^2 = 18$$
$$x = \sqrt{18} \text{ or } -\sqrt{18}$$

Now, $-\sqrt{18}$ is not a solution to the original problem since length is positive. Thus,

$$x = \sqrt{18}$$
$$x = \sqrt{9 \times 2}$$
$$x = \sqrt{9}\,\sqrt{2}$$
$$x = 3\sqrt{2}$$

Thus, the length of the diagonal is $3\sqrt{2}$ centimeters. Looking in the square root table or using a calculator, we find that $\sqrt{2}$ is approximately 1.414; thus, the length of the diagonal is approximately 3(1.414)centimeters = 4.242 centimeters. ■

Example 3 **A 10-foot ladder is leaning against a building. The bottom of the ladder is 6 feet from the building. How high is the top of the ladder?**

Step 1: Draw the figure.

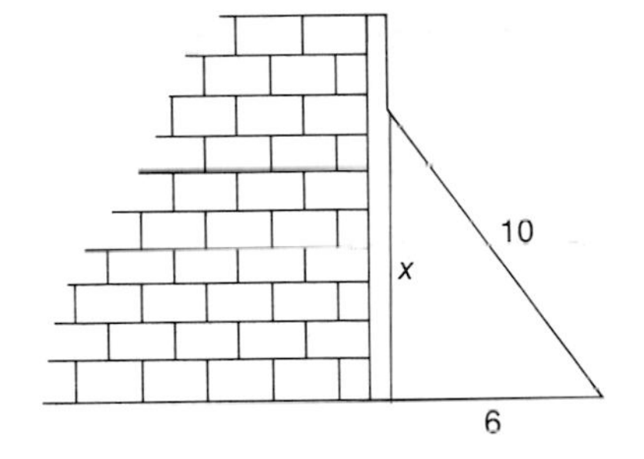

Step 2: Use the Pythagorean Theorem.

$$10^2 = x^2 + 6^2$$
$$100 = x^2 + 36$$
$$100 - 36 = x^2$$
$$64 = x^2$$
$$x^2 = 64$$
$$x = 8 \text{ or } -8$$

Now, $x = -8$ is not a solution to the original problem since length is positive. Thus, the top of the ladder is 8 feet high. ■

Example 4 **The negative square root of a number equals that number decreased by 2. Find the number.**

Step 1: Let x = the number.

Step 2: Set up the equation.

$-\sqrt{x}$ is the negative square root of the number.

$x - 2$ is the number decreased by 2.

Thus, $-\sqrt{x} = x - 2$ is the equation.

Step 3: Solve the equation.

$$-\sqrt{x} = x - 2$$
$$(-\sqrt{x})^2 = (x - 2)^2$$
$$x = x^2 - 4x + 4$$
$$0 = x^2 - 4x + 4 - x$$
$$0 = x^2 - 5x + 4$$
$$x^2 - 5x + 4 = 0$$
$$(x - 1)(x - 4) = 0$$
$$x - 1 = 0 \quad \text{or} \quad x - 4 = 0$$
$$x = 1 \qquad\qquad x = 4$$

Step 4: Check 1 and 4 in the original equation.

Replace x with 4.	Replace x with 1.
$-\sqrt{x} = x - 2$	$-\sqrt{x} = x - 2$
$-\sqrt{4} \stackrel{?}{=} 4 - 2$	$-\sqrt{1} \stackrel{?}{=} 1 - 2$
$-2 \neq 2$	$-1 = -1$

Now, 4 is not a solution of the original equation since -2 does not equal 2; therefore, 1 is the number. ■

APPLICATION EXERCISES

1. If the legs of a right triangle are 3 inches and 4 inches, find the hypotenuse.
2. If the base of a 10-foot ladder is 8 feet from the base of a wall, how far is the top of the ladder from the ground?
3. Find the length of the diagonal of a rectangle if the sides are 40 inches and 30 inches.
4. Twice the square root of a number is 6 less than 12. Find the number.
5. A wire 50 feet long is attached to the top of a tower and to the ground 30 feet from the base of the tower. How tall is the tower?
6. A softball diamond is actually a square 65 feet on each side. How far is it from home plate to second base?
7. One positive number is 3 times another number and the difference of their squares is 200. Find the numbers.
8. A certain number is the same as the square root of the product of 8 and the number. Find the number.
9. The square root of the sum of a number and 4 is 5. Find the number.
10. Three times the square root of 2 equals the square root of the sum of a number and 10. Find the number.

USEFUL INFORMATION

1. A **square root** of a is a term which multiplied by itself equals a. Thus, c is a square root of a, if $c \times c = a$ or $c^2 = a$. Now, $a^{1/2}$, where a is a positive real number, is a square root of a and we often write $a^{1/2}$ as $\sqrt{a}$. We call an expression of the form $\sqrt{a}$ *a* **radical,** the symbol $\sqrt{\ }$ is called a **radical sign** and a is called the **radicand.** Now, $\sqrt{a} = a^{1/2}$ represents the **principal square root** or positive square root of a.
2. Every positive number has two square roots, a positive square root (**principal square root**) and a negative square root. For example, $5 \times 5 = 25$ and $(-5) \times (-5) = 25$; thus, 5 and -5 are both square roots of 25 (5 is the principal square root of 25).
3. A negative number does *not* have a real square root since a positive number times a positive number is a positive number and a negative number times a negative number is also a positive number. For example, there is no real number which multiplied by itself equals -25.
4. The rules for integer exponents also work for fractional exponents.

   Rule 1 (exponents): $x^m \times x^n = x^{m+n}$

   Rule 2 (exponents): $(x^m)^n = x^{nm}$

   Rule 3 (exponents): $(xy)^n = x^n y^n$

   Rule 4 (exponents): $\frac{x^m}{x^n} = x^{m-n}, x \neq 0$

   Rule 5 (exponents): $(\frac{x}{y})^n = \frac{x^n}{y^n}, y \neq 0$
5. **Rule 1 (radicals):** If a is a positive real number, then
$$(\sqrt{a})^2 = a$$
6. **Rule 2 (radicals):** If a is a positive real number, then
$$\sqrt{a^2} = a$$
7. **Rule 3 (radicals): The product rule.** $\sqrt{ab} = \sqrt{a}\sqrt{b}$, where a and b are positive real numbers. That is, the square root of ab is the square root of a times the square root of b.
8. **Rule 4 (radicals): The quotient rule.** $\sqrt{\frac{a}{b}} = \frac{\sqrt{a}}{\sqrt{b}}$, where a and b are positive real numbers. That is, the square root of $\frac{a}{b}$ is the square root of a divided by the square root of b.
9. **A radical expression in simplest form** (1) does not have a perfect square factor under a radical sign (2) does not have a fraction under a radical sign or a radical in the denominator of a fraction. For example, $\sqrt{4x}$ is not in simplest form since 4 is a perfect square, $\sqrt{\frac{1}{2}}$ is not in simplest form since the fraction $\frac{1}{2}$ is under the radical sign and $\frac{1}{\sqrt{2}}$ is not in simplest form since the radical $\sqrt{2}$ is in the denominator. Removing the radicals from a denominator is called **rationalizing the denominator.**
10. If the denominator of a fraction consists of a single term and contains a factor such as $\sqrt{a}$ you rationalize the denominator by multiplying *both* the numerator and denominator by $\sqrt{a}$. Since $\sqrt{a}\sqrt{a} = a$, the radical is removed from the denominator.
11. When the denominator consists of two terms, with at least one term containing a radical, use the special product rule
$$(a - b)(a + b) = a^2 - b^2$$
to rationalize the denominator. That is, to rationalize a denominator that is the *sum* of two terms, multiply *both* the numerator and denominator by the *difference* of the two terms. To rationalize a denominator that is the *difference* of two terms, multiply *both* the numerator and denominator by the *sum* of the two terms.
12. Square root radicals are **like radicals** if they have the same radicand.
13. The basic rule of addition which states that only like things can be added is also true for radicals. Thus, *you can only add or subtract square root radicals that have the same radicand.*
14. In many problems, the radicands are not the same; however, after simplifying the radicals you will find that some of the radicands become the same and you are then able to combine like radicals.
15. **To solve an equation containing radicals,** it is necessary to remove the radicals. Use the following steps when solving an equation containing radicals.
    a. Isolate a radical term on one side of the equation.
    b. Square *both* sides of the equation.
    c. If a radical term still remains, repeat Steps a and b.
    d. Solve the resulting equation.
    e. Check each answer in the *original* equation and omit any answer that does not satisfy the original equation.

REVIEW PROBLEMS

Section 8.1

PART 1

Write the following expressions using the radical sign.

1. $6^{1/2}$
2. $(12)^{1/2}$
3. $a^{1/2}$
4. $y^{1/2}$
5. $(2x)^{1/2}$
6. $(4y)^{1/2}$

7. $5a^{1/2}$

8. $(6x)^{1/2}$

9. $(x - 1)^{1/2}$

10. $(2a - 3b)^{1/2}$

PART 2

Write the following radical expressions using a fractional exponent.

1. $\sqrt{5}$

2. $\sqrt{x}$

3. $\sqrt{2a}$

4. $\sqrt{x^3}$

5. $\sqrt{\frac{y}{5}}$

6. $\sqrt{a + 1}$

7. $\sqrt{a^2 + b^2}$

8. $\sqrt{y^3}$

9. $\sqrt{\frac{y - 2}{3}}$

10. $\sqrt{x^3 - 8}$

PART 3

Simplify the following expressions.

1. $(2^{1/2})^4$

2. $(4^2)^2$

3. $(x^4)^{1/2}$

4. $[(2y)^6]^{1/2}$

5. $(4y^2)^{1/2}$

6. $(25a^6)^{1/2}$

7. $5^{1/2} \times 5^{1/2}$

8. $a^{1/2}a^{1/2}$

9. $(a^{1/2}b^{1/2})^{12}$

10. $[(2x + 1)^6]^{1/2}$

Section 8.2

Simplify the following radical expressions.

1. $\sqrt{36}$

2. $\sqrt{18}$

3. $\sqrt{8}$

4. $\sqrt{150}$

5. $\sqrt{72}$

6. $\sqrt{a^2b}$

7. $\sqrt{x^4y^3}$

8. $\sqrt{12x^2y}$

9. $\sqrt{32ab^2}$

10. $\sqrt{28x^6y^3}$

11. $\sqrt{81x^3y^2}$

12. $\sqrt{16x^5y^6}$

Section 8.3

Multiply the following radical expressions and simplify, if possible.

1. $\sqrt{3}\sqrt{21}$

2. $\sqrt{18}\sqrt{3}$

3. $\sqrt{45}\sqrt{2}$

4. $\sqrt{x}\sqrt{xy^4}$

5. $\sqrt{a^2b}\sqrt{b}$

6. $\sqrt{3x}\sqrt{3y}$

7. $\sqrt{y}\sqrt{y^3}$

8. $\sqrt{8a}\sqrt{a}$

9. $(\sqrt{x} - 1)\sqrt{x}$

10. $(\sqrt{y} + 1)^2$

11. $(\sqrt{x} - 1)(\sqrt{x} + 1)$

12. $(\sqrt{a} - 2)(\sqrt{a} + 1)$

Section 8.4

Simplify the following radical expressions.

1. $\sqrt{\frac{3}{4}}$

2. $\sqrt{\frac{2}{3}}$

3. $\sqrt{\frac{6}{18}}$

4. $\sqrt{\frac{x}{3}}$

5. $\sqrt{\frac{5}{y}}$

6. $\sqrt{\frac{16}{x}}$

7. $\sqrt{\frac{1}{3}}$

8. $\sqrt{\frac{2}{x}}$

9. $\sqrt{\frac{1}{x-1}}$

10. $\sqrt{\frac{a-b}{a+b}}$

11. $\sqrt{\frac{27x}{4y^2}}$

12. $\sqrt{\frac{8y^4}{x^2}}$

Section 8.5

Rationalize the denominator and simplify, if possible.

1. $\frac{\sqrt{3}+1}{\sqrt{2}}$

2. $\frac{\sqrt{5}+\sqrt{3}}{\sqrt{3}}$

3. $\dfrac{\sqrt{y}+1}{\sqrt{y}}$

4. $\dfrac{3\sqrt{x}-\sqrt{2x}}{\sqrt{x}}$

5. $\dfrac{4}{\sqrt{2+1}}$

6. $\dfrac{6}{1-\sqrt{5}}$

7. $\dfrac{1}{\sqrt{2}+\sqrt{3}}$

8. $\dfrac{3}{\sqrt{y}-1}$

9. $\dfrac{x}{1-\sqrt{x}}$

10. $\dfrac{\sqrt{y}+1}{\sqrt{y}-1}$

11. $\dfrac{a-b}{\sqrt{a}-\sqrt{b}}$

12. $\dfrac{\sqrt{2a}+\sqrt{3b}}{\sqrt{2a}-\sqrt{3b}}$

Section 8.6

Simplify the following radical expressions.

1. $\sqrt{5} + \sqrt{20}$

2. $\sqrt{12} - \sqrt{3}$

3. $\sqrt{8} + \sqrt{18} + \sqrt{32}$

4. $\sqrt{x} + \sqrt{x^5}$

5. $2\sqrt{a} - \sqrt{a^3}$

6. $\sqrt{2y} + \sqrt{5y}$

7. $\sqrt{x} + \sqrt{x^3} + \sqrt{x^5}$

8. $\sqrt{16a} + \sqrt{64a} + \sqrt{81a}$

9. $\sqrt{a^2b} - 2\sqrt{b}$

10. $\sqrt{x+1} + \sqrt{x+1}$

11. $4\sqrt{y-1} - \sqrt{y-1}$

12. $\sqrt{m^3n^2} - \sqrt{mn^4}$

Section 8.7

Solve the following equations.

1. $\sqrt{y + 1} = 0$

2. $\sqrt{2x - 1} = 0$

3. $\sqrt{x} - 3 = 0$

4. $\sqrt{2a} - 2 = 0$

5. $\sqrt{y - 3} - 2 = 0$

6. $\sqrt{2a + 1} - 4 = 0$

7. $\sqrt{3x} - 5 = 7$

8. $\sqrt{x + 3} + 4 = 11$

9. $3\sqrt{2x - 5} - \sqrt{x + 23} = 0$

10. $\sqrt{x^2 - 5} - x + 5 = 0$

11. $\sqrt{2x+2} - x + 3 = 0$

12. $\sqrt{2a+1} + \sqrt{3a} = 1$

ANSWERS TO PRACTICE PROBLEMS

Section 8.1

1. 5 or -5
2. 6 or -6
3. 11 or -11
4. $\left(\frac{1}{2}\right)5$
5. $\left(\frac{1}{2}\right)11$
6. $\left(\frac{1}{2}\right)x$
7. $\left(\frac{1}{2}\right)t$
8. $\left(\frac{1}{2}\right)b$
9. $5^{1/2}$
10. $(11)^{1/2}$
11. $(21)^{1/2}$
12. $x^{1/2}$
13. $t^{1/2}$
14. $b^{1/2}$
15. 0
16. 8, -8
17. $\frac{1}{2}, -\frac{1}{2}$
18. 15, -15
19. 11
20. 12
21. 9
22. $\sqrt{3}$
23. $\sqrt{y}$
24. $\sqrt{xy}$
25. $\sqrt{6x}$
26. $11\sqrt{x}$
27. $(15)^{1/2}$
28. $(11x)^{1/2}$
29. $7x^{1/2}$
30. $(2y^3)^{1/2}$
31. $5(x^3y)^{1/2}$
32. y^2
33. $3x$
34. $4y^2$
35. xy
36. $11x^3$
37. 125
38. $25x^2$
39. $25x$
40. $4x$
41. $3y^2$
42. $5xy$

Section 8.2

1. 6
2. 12
3. $3\sqrt{3}$
4. $6\sqrt{2}$
5. $9\sqrt{5}$
6. $3\sqrt{2}$
7. $5\sqrt{3}$
8. $10\sqrt{3}$
9. $15\sqrt{2}$
10. $3a\sqrt{3a}$
11. $2x\sqrt{6y}$
12. $4p^3q\sqrt{q}$
13. $5x^7y^3\sqrt{5xy}$

Section 8.3

1. $2\sqrt{3}$
2. $3\sqrt{2}$
3. $3\sqrt{10}$
4. $7\sqrt{6}$
5. $a\sqrt{b}$
6. $2x\sqrt{3}$
7. a^2
8. $3p^2q\sqrt{2}$
9. $\sqrt{2}x + 2$
10. $\sqrt{5}x + 2\sqrt{5x}$
11. $3x\sqrt{2} + 6\sqrt{x}$
12. $4x + 4\sqrt{3x}$
13. $x^2 - 2$
14. $x - 6\sqrt{x} + 9$
15. $3x - 5$
16. $2x + 14\sqrt{2x} + 49$
17. $2y - \sqrt{y} - 6$

Section 8.4

1. $\frac{\sqrt{6}}{3}$
2. $\frac{2\sqrt{a}}{a}$
3. $\frac{3\sqrt{3ab}}{b}$
4. $\frac{4\sqrt{ab}}{b^2}$

Section 8.5

1. $\sqrt{3}$
2. $\sqrt{5a}$
3. $4\sqrt{xy}$
4. $\frac{3\sqrt{xy}}{x}$
5. $\frac{\sqrt{2y}}{2xy}$
6. $\frac{5b\sqrt{ab}}{a^2}$
7. $\frac{\sqrt{2ab}}{ab}$
8. $5 + \sqrt{2}$
9. $5 + \sqrt{3}$
10. $5 + \sqrt{2}$
11. $3\sqrt{x} + 1$
12. $-(1 + \sqrt{2})$
13. $\frac{7 + 3\sqrt{5}}{2}$
14. $\frac{2\sqrt{x} + 10}{x - 25}$
15. $\frac{a - 2\sqrt{ab} + b}{a - b}$

Section 8.6

1. $10\sqrt{5} - 7\sqrt{7}$
2. $-\sqrt{2} + 3\sqrt{3}$
3. $-7\sqrt{6} - 5\sqrt{5}$
4. $3\sqrt{2}$
5. $-3\sqrt{6}$
6. $2\sqrt{a}$
7. $3x\sqrt{xy}$

Section 8.7

1. 3
2. 11
3. 6
4. 3
5. 4
6. 1, 9
7. 1
8. -4, 4

C H A P T E R 9

Quadratic Equations

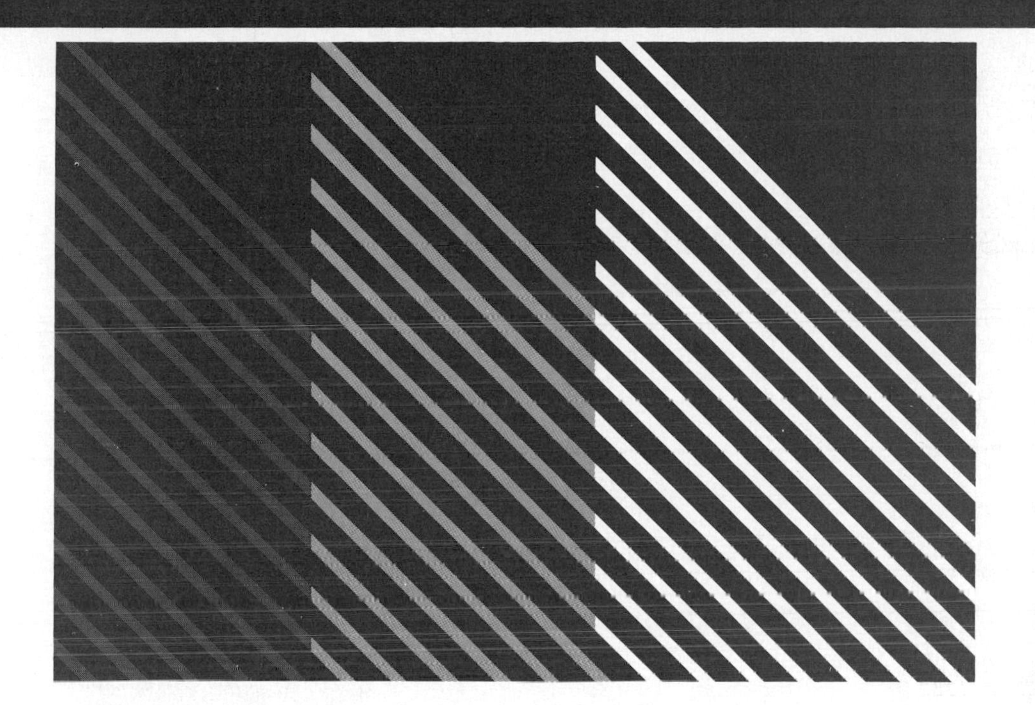

The objective of this chapter is:

1. **To be able to solve quadratic equations in one variable.**

A **quadratic equation** in one variable is an equation which can be written in the form

$$ax^2 + bx + c = 0$$

where a, b, and c represent real numbers and $a \neq 0$. For example,

$$2x^2 + 5x + 3 = 0$$
$$5y^2 - 2y + 7 = 0$$
$$-3x^2 + 7x - 6 = 0$$
$$4x^2 = 5x - 2$$
$$3a^2 = 12$$
$$2x^2 = 5x$$

are quadratic equations in one variable. A quadratic equation in one variable usually has two different real solutions; however, it may have only one or none.

9.1 SOLVING QUADRATIC EQUATIONS BY SQUARE ROOTS

The simplest type of quadratic equation to solve is $ax^2 + c = 0$. Let's look at an example.

Example 1 **Solve the following quadratic equation.**

$$2x^2 - 18 = 0$$

Step 1: Solve the equation for x^2.

$2x^2 - 18 = 0$	**Original equation.**
$2x^2 = 18$	**Add 18 to both sides.**
$x^2 = 9$	**Divide both sides by 2.**

Step 2: Take the square root of *both* sides of the resulting equation.

$x^2 = 9$	**Equation from Step 1.**
$x = \pm\sqrt{9}$	**Take the square root of both sides.**
$x = \pm 3$	

Why did you write $\pm$ in front of $\sqrt{9}$ when you took the square root of both sides of the equation?

Well, Charlie, recall that a positive number has two square roots—a positive square root and a negative square root. When solving a quadratic equation, we want to find both the positive and negative square roots. Now, $\sqrt{9}$ represents only the positive square root of 9 $(+3)$; thus, we must use $\pm\sqrt{9}$ to indicate both the positive $(+3)$ and negative (-3) square roots of 9.

So, ± 3 is really two numbers: $+3$ and -3.

That's correct, Charlie.

Check: Check 3 and -3 in the original equation.

Replace x with 3.	Replace x with -3.
$2x^2 - 18 = 0$	$2x^2 - 18 = 0$
$2(3)^2 - 18 \stackrel{?}{=} 0$	$2(-3)^2 - 18 \stackrel{?}{=} 0$
$2(9) - 18 \stackrel{?}{=} 0$	$2(9) - 18 \stackrel{?}{=} 0$
$18 - 18 \stackrel{?}{=} 0$	$18 - 18 \stackrel{?}{=} 0$
$0 = 0$	$0 = 0$

Thus, the solutions of $2x^2 - 18 = 0$ are 3 and -3. ■

In general, if d is positive, the two solutions of the quadratic equation

$$x^2 = d$$

are

$$x = \sqrt{d} \text{ and } x = -\sqrt{d}$$

which we often write as

$$x = \pm\sqrt{d}$$

Example 2 **Solve the following quadratic equation.**

$$5x^2 - 25 = 0$$

Step 1: Solve the equation for x^2.

$5x^2 - 25 = 0$	**Original equation.**
$5x^2 = 25$	**Add 25 to both sides.**
$x^2 = 5$	**Divide both sides by 5.**

Step 2: Take the square root of *both* sides of the resulting equation.

$x^2 = 5$	**Equation from Step 1.**
$x = \pm\sqrt{5}$	**Take the square root of both sides.**

Thus, the two solutions of the quadratic equation $5x^2 - 25 = 0$ are $\sqrt{5}$ and $-\sqrt{5}$. ■

A N S W E R

Why did you leave the $\sqrt{\ }$ sign in the answer?

Well, Charlie, there is no rational number equal to $\sqrt{5}$; thus, we left the radical sign in the answer. In fact, when solving quadratic equations, most of the solutions will contain radicals.

Example 3 **Solve the following quadratic equation.**

$$2x^2 + 50 = 0$$

Step 1: Solve the equation for x^2.

$2x^2 + 50 = 0$	**Original equation.**
$2x^2 = -50$	**Subtract 50 from both sides.**
$x^2 = -25$	**Divide both sides by 2.**

Now, there is no real number which multiplied by itself equals a negative number. A positive number times a positive number equals a positive number ($5 \times 5 = 25$) and a negative number times a negative number equals a positive number ($(-5) \times (-5) = 25$). Thus, the quadratic equation

$x^2 = -25$ has no solution;

hence, $2x^2 + 50 = 0$ has no solution. ■

1. ____________________

Practice 1: Solve the following quadratic equation.

$$3x^2 = 27$$

2. ____________________

Practice 2: Solve the following quadratic equation.

$$2x^2 - 128 = 0$$

ANSWER

Practice 3: Solve the following quadratic equation.

$$5x^2 - 45 = 0$$

3. ______________

Practice 4: Solve the following quadratic equation.

$$7x^2 - 28 = 0$$

4. ______________

Practice 5: Solve the following quadratic equation.

$$4x^2 - 12 = 0$$

5. ______________

Practice 6: Solve the following quadratic equation.

$$2x^2 - 24 = 0$$

6. ______________

Practice 7: Solve the following quadratic equation.

$$5x^2 + 20 = 0$$

7. ______________

EXERCISE 9.1

Solve the following quadratic equations.

A

1. $x^2 = 16$

2. $2x^2 = 72$

3. $3x^2 - 27 = 0$

4. $6x^2 - 54 = 0$

5. $9x^2 - 36 = 0$

6. $x^2 = 8$

7. $x^2 + 1 = 26$

8. $x^2 - 12 = 0$

9. $(x + 1)^2 = 4$

10. $x^2 + 5 = 30$

B

11. $2x^2 = 100$

12. $4x^2 = 72$

13. $2x^2 - 1 = 49$

14. $3x^2 + 2 = 29$

15. $2x^2 - 7 = 13$

16. $-x^2 = -4$

17. $2 - x^2 = -14$

18. $5 - 2x^2 = 3$

19. $x^2 = \frac{1}{4}$

20. $x^2 - \frac{9}{4} = 0$

9.2 SOLVING QUADRATIC EQUATIONS BY FACTORING

When solving quadratic equations by factoring, we will use the following rule.

> **Zero Product Rule.** If $ab = 0$, then $a = 0$ or $b = 0$.

Therefore, if the product of two polynomials is zero, then one or both of the polynomials is zero.

> **To solve a quadratic equation by factoring:**
> 1. Write the quadratic equation in the form $ax^2 + bx + c = 0$.
> 2. Factor $ax^2 + bx + c$.
> 3. Set each factor equal to zero.
> 4. Solve the two resulting linear equations.

Didn't we solve quadratic equations by factoring earlier?

Yes, Charlie. You solved quadratic equations by factoring when you learned to factor polynomials in Chapter 6. This section is just a review of Section 6.6.

Example 1 **Solve the following quadratic equation.**

$$2x^2 - 7x = 15$$

Step 1: Write the equation in the form $ax^2 + bx + c = 0$.

$$2x^2 - 7x = 15$$
$$2x^2 - 7x - 15 = 0$$

Step 2: Factor the left side of the equation.

$$2x^2 - 7x - 15 = 0$$
$$(2x + 3)(x - 5) = 0$$

Step 3: Set each factor equal to zero.

$$2x + 3 = 0 \quad \text{or} \quad x - 5 = 0$$

Step 4: Solve the resulting linear equations.

$$2x + 3 = 0 \quad \text{or} \quad x - 5 = 0$$
$$2x = -3 \qquad\qquad x = 5$$
$$x = -\frac{3}{2}$$

A N S W E R

Check: Check $-\frac{3}{2}$ and 5 in the original equation.

Replace x with $-\frac{3}{2}$.

$$2x^2 - 7x = 15$$
$$2\left(-\frac{3}{2}\right)^2 - 7\left(-\frac{3}{2}\right) \stackrel{?}{=} 15$$
$$2\left(\frac{9}{4}\right) - 7\left(-\frac{3}{2}\right) \stackrel{?}{=} 15$$
$$\frac{9}{2} + \frac{21}{2} \stackrel{?}{=} 15$$
$$\frac{30}{2} \stackrel{?}{=} 15$$
$$15 = 15$$

Replace x with 5.

$$2x^2 - 7x = 15$$
$$2(5)^2 - 7(5) \stackrel{?}{=} 15$$
$$2(25) - 7(5) \stackrel{?}{=} 15$$
$$50 - 35 \stackrel{?}{=} 15$$
$$15 = 15$$

Thus, the solutions of $2x^2 - 7x = 15$ are $-\frac{3}{2}$ and 5. ■

Practice 1: Solve the following quadratic equation.

$$x^2 - 5x + 6 = 0$$

1. ____________________

Practice 2: Solve the following quadratic equation.

$$6x^2 + x = 2$$

2. ____________________

ANSWER

Practice 3: Solve the following quadratic equation.

$$10x^2 + 23x - 5 = 0$$

3. ______________

4. ______________

Practice 4: Solve the following quadratic equation.

$$2x^2 + 6x = 0$$

EXERCISE 9.2

Solve the following quadratic equations by factoring.

A

1. $2x^2 - x - 1 = 0$

2. $2x^2 - 2x = 12$

3. $5x^2 = 3x$

4. $x^2 - 3x - 28 = 0$

5. $x^2 + x - 6 = 0$

6. $y^2 + 5y + 6 = 0$

7. $y^2 + 3y + 2 = 0$

8. $2x^2 - 8x - 24 = 0$

9. $x^2 = 2x + 35$

10. $y^2 + 36 = -12y$

B

11. $2y^2 - 5y = 3$

12. $6x^2 - 2x - 8 = 0$

13. $3x^2 - 18x = 0$

14. $x^2 - 10x = -25$

15. $y^2 - 8y + 16 = 0$

16. $x^2 - 20 = x$

17. $x^2 - 6 = x$

18. $3 = x(2x + 1)$

19. $3y^2 + 8y = 12y + 15$

20. $x(9 + x) = 4(2x + 5)$

9.3 SOLVING QUADRATIC EQUATIONS BY THE QUADRATIC FORMULA

Many quadratic equations cannot be solved by the methods of Section 9.1 or 9.2. In this section, we will develop a formula that can be used to solve any quadratic equation. We will develop this formula by using a method known as completing the square. Let's look at an example in which we solve a quadratic equation by completing the square.

Example 1 **Solve the following quadratic equation.**

$$x^2 + 4x + 1 = 0$$

Step 1: Rewrite the equation so that the terms containing the variable (terms containing x) are on the left side and the other terms are on the right side.

$$x^2 + 4x = -1$$

Step 2: Find the number which, added to the left side of the resulting equation, makes it a perfect square. We want $x^2 + 4x$ to become a perfect square. That is, we want

$$x^2 + 4x + ? = (x + \_\_)^2$$

The number we are looking for is 4 since

$$x^2 + 4x + 4 = (x + 2)^2$$

Thus, we add 4 to *both* sides of the equation from Step 1.

$$x^2 + 4x = -1$$

$$x^2 + 4x + 4 = -1 + 4$$

Step 3: Write the left side of the equation from Step 2 as a square.

$$x^2 + 4x + 4 = -1 + 4$$
$$(x + 2)^2 = 3$$

Step 4: Take the square root of both sides of the resulting equation and solve for x.

$$(x + 2)^2 = 3$$
$$x + 2 = \pm\sqrt{3}$$
$$x = -2 \pm \sqrt{3}$$

Thus, the solutions of $x^2 + 4x + 1 = 0$ are $-2 + \sqrt{3}$ and $-2 - \sqrt{3}$. ■

The equation $x^2 + 4x + 1 = 0$ could *not* have been solved by the methods of Section 9.1 or 9.2. Thus, it was necessary to complete the square.

Let's look at how we can find the term which added to a polynomial of the form $x^2 + 2ax$ makes it a perfect square. According to a special product rule

$$x^2 + 2ax + a^2 = (x + a)^2$$

Thus, we need to find a^2 to complete the square. Since $2a$ is the coefficient of x, we simply take $\frac{1}{2}$ the coefficient of x, which is a, and square it. This gives us a^2. The following examples will show this procedure.

Example 2 **Solve the following quadratic equation by completing the square.**

$$x^2 + 6x + 2 = 0$$

Step 1: Rewrite the equation so that the terms containing the variable (terms containing x) are on the left side and the other terms are on the right side.

$$x^2 + 6x = -2$$

Step 2: Find the number which added to the left side of the resulting equation makes it a perfect square. We want

$$x^2 + 6x + \underline{\qquad}$$

to be a perfect square. We take $\frac{1}{2}$ of 6, which is 3, and square it, which is 9. Thus, the number we are looking for is 9. That is,

$$x^2 + 6x + 9 = (x + 3)^2$$

Therefore, we add 9 to *both* sides of the equation from Step 1.

$$x^2 + 6x = -2$$
$$x^2 + 6x + 9 = -2 + 9$$

Step 3: Write the left side of the equation from Step 2 as a square.

$$x^2 + 6x + 9 = -2 + 9$$
$$(x + 3)^2 = 7$$

Step 4: Take the square root of both sides of the resulting equation and solve for x.

$$(x + 3)^2 = 7$$
$$x + 3 = \pm\sqrt{7}$$
$$x = -3 \pm \sqrt{7}$$

Thus, the solutions of $x^2 + 6x + 2 = 0$ are $-3 + \sqrt{7}$ and $-3 - \sqrt{7}$. ■

Example 3 **Solve the following quadratic equation by completing the square.**

$$4x^2 + 16x + 13 = 0$$

Step 1: Divide both sides of the equation by 4, making the coefficient of x^2 one.

$$4x^2 + 16x + 13 = 0$$
$$x^2 + 4x + \frac{13}{4} = 0$$

Why did we divide both sides of the equation by 4?

Well, Charlie, we need the coefficient of x^2 to be 1 so that we can complete the square according to $x^2 + 2ax + a^2 = (x + a)^2$.

Step 2: Rewrite the equation from Step 1 so that the terms containing the variable (terms containing x) are on the left side and the other terms are on the right side.

$$x^2 + 4x = -\frac{13}{4}$$

Step 3: Find the number which added to the left side of the resulting equation makes it a perfect square. We want

$$x^2 + 4x + \_\_$$

to be a perfect square. We take $\frac{1}{2}$ of 4 which is 2 and square it which is 4. Thus, the number we are looking for is 4. That is,

$$x^2 + 4x + 4 = (x + 2)^2$$

Therefore, we add 4 to *both* sides of the equation from Step 2.

$$x^2 + 4x = -\frac{13}{4}$$
$$x^2 + 4x + 4 = -\frac{13}{4} + 4$$

ANSWER

Step 4: Write the left side of the equation from Step 3 as a square.

$$x^2 + 4x + 4 = -\frac{13}{4} + 4$$

$$(x + 2)^2 = \frac{3}{4}$$

Step 5: Take the square root of both sides of the resulting equation and solve for x.

$$(x + 2)^2 = \frac{3}{4}$$

$$x + 2 = \pm\sqrt{\frac{3}{4}}$$

$$x + 2 = \pm\frac{\sqrt{3}}{\sqrt{4}}$$

$$x + 2 = \pm\frac{\sqrt{3}}{2}$$

$$x = -2 \pm \frac{\sqrt{3}}{2}$$

Thus, the solutions of $2x^2 + 8x + 5 = 0$ are $-2 + \frac{\sqrt{3}}{2}$ and $-2 - \frac{\sqrt{3}}{2}$. ■

Practice 1: Solve the following quadratic equation by completing the square.

$$x^2 + 6x + 4 = 0$$

1. ____________________

Practice 2: Solve the following quadratic equation by completing the square.

$$x^2 + 8x + 3 = 0$$

2. ____________________

Practice 3: Solve the following quadratic equation by completing the square.

$$3x^2 + 6x - 3 = 0$$

3. ____________________

It is a lot of work to solve quadratic equations by completing the square.

That's true, Charlie. That's why we are going to develop the quadratic formula. However, we have to know how to complete the square in order to understand the development of the quadratic formula which is presented in Example 4.

Example 4 **Solve the following quadratic equations.**

$$ax^2 + bx + c = 0$$

Step 1: Divide both sides of the equation by a.

$$\frac{ax^2 + bx + c}{a} = \frac{0}{a}$$

$$\frac{ax^2}{a} + \frac{bx}{a} + \frac{c}{a} = \frac{0}{a}$$

$$x^2 + \frac{b}{a}x + \frac{c}{a} = 0$$

We are now going to complete the square.

Step 2: Rewrite the resulting equation so that the terms containing the variable (terms containing x) are on the left side and other terms are on the right side.

$$x^2 + \frac{b}{a}x = -\frac{c}{a}$$

Step 3: Find the term which added to the left side of the equation from Step 2 makes it a perfect square. We find $\frac{1}{2}$ of $\frac{b}{a}$ and square it. Now, $\frac{1}{2}$ of $\frac{b}{a}$ is $\frac{b}{2a}$ and $(\frac{b}{2a})^2$ is $\frac{b^2}{4a^2}$. Therefore, the term that we are looking for is $\frac{b^2}{4a^2}$.

$$5x^2 + 9x + 3 = 0$$

Step 1: Divide both sides of the equation by 5.

$$\frac{5x^2 + 9x + 3}{5} = \frac{0}{5}$$

$$\frac{5x^2}{5} + \frac{9x}{5} + \frac{3}{5} = \frac{0}{5}$$

$$x^2 + \frac{9}{5}x + \frac{3}{5} = 0$$

We are now going to complete the square.

Step 2: Rewrite the resulting equation so that the terms containing the variable (terms containing x) are on the left side and other terms are on the right side.

$$x^2 + \frac{9}{5}x = -\frac{3}{5}$$

Step 3: Find the term which added to the left side of the equation from Step 2 makes it a perfect square. We find $\frac{1}{2}$ of $\frac{9}{5}$ and square it. Now, $\frac{1}{2}$ of $\frac{9}{5}$ is $\frac{9}{10}$ and $(\frac{9}{10})^2$ is $\frac{81}{100}$. Therefore, the number that we are looking for is $\frac{81}{100}$.

Thus, we add $\frac{b^2}{4a^2}$ to *both* sides of the equation from Step 2.

$$x^2 + \frac{b}{a}x = -\frac{c}{a}$$

$$x^2 + \frac{b}{a}x + \frac{b^2}{4a^2} = -\frac{c}{a} + \frac{b^2}{4a^2}$$

Step 4: Write the left side of the resulting equation as a square and add the fractions on the right side.

$$x^2 + \frac{b}{a}x + \frac{b^2}{4a^2} = -\frac{c}{a} + \frac{b^2}{4a^2}$$

$$\left(x + \frac{b}{2a}\right)^2 = -\frac{c}{a} + \frac{b^2}{4a^2}$$

$$\left(x + \frac{b}{2a}\right)^2 = \frac{(-c)4a}{a(4a)} + \frac{b^2}{4a^2}$$

$$\left(x + \frac{b}{2a}\right)^2 = \frac{-4ac}{4a^2} + \frac{b^2}{4a^2}$$

$$\left(x + \frac{b}{2a}\right)^2 = \frac{-4ac + b^2}{4a^2}$$

Step 5: Take the square root of both sides of the equation from Step 4 and solve for x.

$$\left(x + \frac{b}{2a}\right)^2 = \frac{-4ac + b^2}{4a^2}$$

$$x + \frac{b}{2a} = \pm\sqrt{\frac{-4ac + b^2}{4a^2}}$$

$$x = \frac{-b}{2a} \pm \sqrt{\frac{-4ac + b^2}{4a^2}}$$

$$x = \frac{-b}{2a} \pm \frac{\sqrt{-4ac + b^2}}{\sqrt{4a^2}}$$

$$x = \frac{-b}{2a} \pm \frac{\sqrt{-4ac + b^2}}{2a}$$

$$x = \frac{-b \pm \sqrt{b^2 - 4ac}}{2a}$$

Thus, we add $\frac{81}{100}$ to *both* sides of the equation from Step 2.

$$x^2 + \frac{9}{5}x = -\frac{3}{5}$$

$$x^2 + \frac{9}{5}x + \frac{81}{100} = -\frac{3}{5} + \frac{81}{100}$$

Step 4: Write the left side of the resulting equation as a square and add the fractions on the right side.

$$x^2 + \frac{9}{5}x + \frac{81}{100} = -\frac{3}{5} + \frac{81}{100}$$

$$\left(x + \frac{9}{10}\right)^2 = -\frac{3}{5} + \frac{81}{100}$$

$$\left(x + \frac{9}{10}\right)^2 = \frac{-3 \times 20}{5 \times 20} + \frac{81}{100}$$

$$\left(x + \frac{9}{10}\right)^2 = \frac{-60}{100} + \frac{81}{100}$$

$$\left(x + \frac{9}{10}\right)^2 = \frac{-60 + 81}{100}$$

$$\left(x + \frac{9}{10}\right)^2 = \frac{21}{100}$$

Step 5: Take the square root of both sides of the equation from Step 4 and solve for x.

$$\left(x + \frac{9}{10}\right)^2 = \frac{21}{100}$$

$$x + \frac{9}{10} = \pm\sqrt{\frac{21}{100}}$$

$$x = -\frac{9}{10} \pm \sqrt{\frac{21}{100}}$$

$$x = -\frac{9}{10} \pm \frac{\sqrt{21}}{\sqrt{100}}$$

$$x = -\frac{9}{10} \pm \frac{\sqrt{21}}{10}$$

$$x = \frac{-9 \pm \sqrt{21}}{10}$$

Thus, the solutions of $ax^2 + bx + c = 0$ are

$$\frac{-b + \sqrt{b^2 - 4ac}}{2a} \text{ and } \frac{-b - \sqrt{b^2 - 4ac}}{2a}$$

Thus, the solutions of $5x^2 + 9x + 3 = 0$ are

$$\frac{-9 + \sqrt{21}}{10} \text{ and } \frac{-9 - \sqrt{21}}{10}$$

■

We now have the quadratic formula for solving quadratic equations.

> **The quadratic formula.** If $ax^2 + bx + c = 0$ $(a \neq 0)$, then
> $$x = \frac{-b \pm \sqrt{b^2 - 4ac}}{2a}$$

The quadratic formula is one of the most important formulas in algebra. Let's look at an example in which the quadratic formula is used to solve a quadratic equation.

Example 5 **Solve the following quadratic equation using the quadratic formula.**

$$x^2 + 3x - 4 = 0$$

Step 1: Identify a, b, and c.

$$ax^2 + bx + c = 0$$

$$1x^2 + 3x - 4 = 0$$

Thus,

$$a = 1$$

$$b = 3$$

$$c = -4$$

Step 2: Substitute the values of a, b, and c into the quadratic formula.

$$x = \frac{-b \pm \sqrt{b^2 - 4ac}}{2a}$$

$$x = \frac{-3 \pm \sqrt{3^2 - 4(1)(-4)}}{2(1)}$$

$$x = \frac{-3 \pm \sqrt{9 + 16}}{2}$$

$$x = \frac{-3 \pm \sqrt{25}}{2}$$

$$x = \frac{-3 \pm 5}{2}$$

Thus, the solutions of $x^2 + 3x - 4 = 0$ are

$$\frac{-3 + 5}{2} = \frac{2}{2} = 1 \quad \text{and} \quad \frac{-3 - 5}{2} = \frac{-8}{2} = -4$$

The quadratic equation $x^2 + 3x - 4 = 0$ could also have been solved by factoring. Let's check the solution by solving $x^2 + 3x - 4 = 0$ by factoring.

$$x^2 + 3x - 4 = 0$$

$$(x - 1)(x + 4) = 0$$

$$x - 1 = 0 \quad \text{or} \quad x + 4 = 0$$

$$x = 1 \qquad\qquad x = -4$$

The solutions are the same when we solve the equation by factoring. ■

Solving the equation by factoring is a lot easier. Why do we need the quadratic formula?

Well, Charlie, factoring is easier and you will want to use the factoring method when possible. However, most quadratic equations cannot be solved by factoring; thus, you will need to use the quadratic formula. Let's look at an example in which the quadratic equation cannot be solved by the factoring method.

Example 6 **Solve the following quadratic equation using the quadratic formula.**

$$2x^2 + 1 = 5x$$

Step 1: Write the equation in the form $ax^2 + bx + c = 0$.

$$2x^2 + 1 = 5x$$

$$2x^2 - 5x + 1 = 0$$

Step 2: Identify a, b, and c.

$$ax^2 + bx + c = 0$$

$$2x^2 - 5x + 1 = 0$$

Thus,

$$a = 2$$

$$b = -5$$

$$c = 1$$

ANSWER

Step 3: Substitute the values of a, b, and c into the quadratic formula.

$$x = \frac{-b \pm \sqrt{b^2 - 4ac}}{2a}$$

$$x = \frac{-(-5) \pm \sqrt{(-5)^2 - 4(2)(1)}}{2(2)}$$

$$x = \frac{5 \pm \sqrt{25 - 8}}{4}$$

$$x = \frac{5 \pm \sqrt{17}}{4}$$

$$x = \frac{5 \pm \sqrt{17}}{4}$$

Thus, the solutions of $2x^2 + 1 = 5x$ are

$$\frac{5 + \sqrt{17}}{4} \quad \text{and} \quad \frac{5 - \sqrt{17}}{4}.$$

■

4. ____________

Practice 4: Solve the following quadratic equation using the quadratic formula.

$$2x^2 - x - 3 = 0$$

5. ____________

Practice 5: Solve the following quadratic equation using the quadratic formula.

$$3x^2 + x = 2$$

6. ____________

Practice 6: Solve the following quadratic equation using the quadratic formula.

$$x^2 + 1 = -5x$$

Example 7 **Solve the following quadratic equation using the quadratic formula.**

$$3x^2 + 2x + 2 = 0$$

Step 1: Identify a, b, and c.

$$ax^2 + bx + c = 0$$

$$3x^2 + 2x + 2 = 0$$

Thus,

$$a = 3$$

$$b = 2$$

$$c = 2$$

Step 2: Substitute the values of a, b, and c into the quadratic formula.

$$x = \frac{-b \pm \sqrt{b^2 - 4ac}}{2a}$$

$$x = \frac{-2 \pm \sqrt{2^2 - 4(3)(2)}}{2(3)}$$

$$x = \frac{-2 \pm \sqrt{4 - 24}}{6}$$

$$x = \frac{-2 \pm \sqrt{-20}}{6}$$

Since there is no real number equal to $\sqrt{-20}$, we say the quadratic equation $3x^2 + 2x + 2 = 0$ has no real solution. ■

Note: When using the quadratic formula, if $b^2 - 4ac$ equals a negative number, the quadratic equation has no real solution.

Example 8 **Solve the following quadratic equation using the quadratic formula.**

$$4x^2 + 12x = -9$$

Step 1: Write the equation in the form $ax^2 + bx + c = 0$

$$4x^2 + 12x = -9$$

$$4x^2 + 12x + 9 = 0$$

Step 2: Identify a, b, and c.

$$ax^2 + bx + c = 0$$

$$4x^2 + 12x + 9 = 0$$

Thus,

$$a = 4$$

$$b = 12$$

$$c = 9$$

A N S W E R

Step 3: Substitute the values of a, b, and c into the quadratic formula.

$$x = \frac{-b \pm \sqrt{b^2 - 4ac}}{2a}$$

$$x = \frac{-12 \pm \sqrt{(12)^2 - 4(4)(9)}}{2(4)}$$

$$x = \frac{-12 \pm \sqrt{144 - 144}}{8}$$

$$x = \frac{-12 \pm \sqrt{0}}{8}$$

$$x = \frac{-12 \pm 0}{8}$$

$$x = \frac{-12}{8}$$

$$x = \frac{-3}{2}$$

Thus, the solution of $4x^2 + 12x + 9 = 0$ is $-\frac{3}{2}$ and the quadratic equation has only one real solution. ■

Note: When using the quadratic formula, if $b^2 - 4ac$ equals zero, the quadratic equation has only one real solution.

7. ______________________

Practice 7: Solve the following quadratic equation using the quadratic formula.

$$x^2 + 10x + 25 = 0$$

8. ______________________

Practice 8: Solve the following quadratic equation using the quadratic formula.

$$3x^2 + 2x = -5$$

Let's summarize what you have learned concerning the number of real solutions to a quadratic equation when using the quadratic formula.

When using the quadratic formula,

$$x = \frac{-b \pm \sqrt{b^2 - 4ac}}{2a}$$

1. If $b^2 - 4ac$ equals a positive number, the quadratic equation has two real solutions.
2. If $b^2 - 4ac$ equals zero, the quadratic equation has one real solution.
3. If $b^2 - 4ac$ equals a negative number, the quadratic equation has no real solution.

EXERCISE 9.3

Solve the following quadratic equations using the quadratic formula.

A

1. $x^2 - 5x + 6 = 0$

2. $x^2 - 9 = 0$

3. $x^2 - 4x - 21 = 0$

4. $x^2 + 5 = 6x$

5. $2x^2 - x = 2$

6. $y^2 + 5y + 5 = 0$

7. $x^2 - 3x + 9 = 0$

8. $x^2 = x - 9$

9. $x^2 + 8x + 3 = 0$

10. $x^2 + 4x = -1$

B

11. $y^2 = 6y + 8$

12. $3y^2 + y - 2 = 0$

13. $7x^2 = -x + 3$

14. $x^2 + 6x = 1$

15. $2y^2 + 6y + 5 = 0$

16. $y^2 + 7y + 12 = 0$

17. $x^2 + (x + 2)^2 = 7$

18. $2x^2 + 3x - 20 = 0$

19. $2x^2 - 5 = 2x$

20. $5x^2 + 1 = -10x$

9.4 STRATEGIES FOR SOLVING APPLICATIONS

Read the following four examples carefully.

Example 1 **The sum of the squares of two consecutive even positive integers is 164. Find the integers.**

Step 1: Let $x =$ the first even positive integer.

Then $x + 2 =$ the next consecutive even positive integer.

Step 2: Set up the equation.

x^2 is the square of the first even integer.

$(x + 2)^2$ is the square of the next consecutive even integer.

$x^2 + (x + 2)^2$ is the sum of their squares.

Thus, $x^2 + (x + 2)^2 = 164$ is the equation.

Step 3: Solve the equation.

$$x^2 + (x + 2)^2 = 164$$
$$x^2 + x^2 + 4x + 4 = 164$$
$$2x^2 + 4x + 4 = 164$$
$$2x^2 + 4x + 4 - 164 = 0$$
$$2x^2 + 4x - 160 = 0$$
$$x^2 + 2x - 80 = 0$$
$$(x + 10)(x - 8) = 0$$
$$x + 10 = 0 \quad \text{or} \quad x - 8 = 0$$
$$x = -10 \qquad\qquad x = 8$$

Since x is positive, $x = 8$ and $x + 2 = 10$. Thus, 8 and 10 are the two consecutive even positive integers. ■

Example 2 **The length of a rectangle is 10 centimeters more than its width. The area of the rectangle is 119 cm². Find the width and length of the rectangle.**

Step 1: Let x = the width of the rectangle. Then $x + 10$ = the length of the rectangle.

Step 2: Set up the equation. The area of a rectangle is length times width. Thus, the equation is

$$(x + 10)x = 119$$

Step 3: Solve the equation.

$$(x + 10)x = 119$$

$$x^2 + 10x = 119$$

$$x^2 + 10x - 119 = 0$$

$$a = 1,\ b = 10, \text{ and } c = -119$$

$$x = \frac{-b \pm \sqrt{b^2 - 4ac}}{2a}$$

$$x = \frac{-10 \pm \sqrt{10^2 - 4(1)(-119)}}{2(1)}$$

$$x = \frac{-10 \pm \sqrt{100 + 476}}{2}$$

$$x = \frac{-10 \pm \sqrt{576}}{2}$$

$$x = \frac{-10 \pm 24}{2}$$

$$x = \frac{-10 + 24}{2} \quad \text{or} \quad x = \frac{-10 - 24}{2}$$

$$x = \frac{14}{2} \qquad x = \frac{-34}{2}$$

$$x = 7 \qquad x = -17$$

Since the width is positive, -17 is not a solution of the original problem. Thus, the width of the rectangle is 7 centimeters and the length is $7 + 10 = 17$ centimeters. ■

Example 3 **The product of 1 more than a number and 1 less than the number is 224. Find the number.**

Step 1: Let x = the number.

Step 2: Set up the equation.

$x + 1$ is 1 more than the number.

$x - 1$ is 1 less than the number.

$(x + 1)(x - 1)$ is the product.

Thus, the equation is $(x + 1)(x - 1) = 224$.

Step 3: Solve the equation.

$$(x + 1)(x - 1) = 224$$
$$x^2 - 1 = 224$$
$$x^2 = 224 + 1$$
$$x^2 = 225$$
$$x = 15 \quad \text{or} \quad x = -15$$

Thus, there are two numbers, 15 and -15, that satisfy the conditions of the problem. ■

Example 4 **The current in a stream moves at a speed of 3 kilometers per hour. A boat travels 45 kilometers upstream and 45 kilometers downstream in a total time of eight hours. What is the speed of the boat in still water?**

Step 1: Let x = the speed of the boat in still water. Then $x + 3$ = the speed of the boat downstream and $x - 3$ = the speed of the boat upstream.

Step 2: Set up a chart.

	d	r	t
Upstream	45	$x - 3$	
Downstream	45	$x + 3$	

Using the formula

$$d = rt$$

divide both sides by r to find t.

$$\frac{d}{r} = \frac{rt}{r}$$

$$\frac{d}{r} = t$$

That is, time equals distance divided by rate. Now, let's complete the chart. The time it would take to go upstream would be $\frac{45}{x-3}$. The time it would take to go downstream would be $\frac{45}{x+3}$.

	d	r	t
Upstream	45	$x - 3$	$\frac{45}{x - 3}$
Downstream	45	$x + 3$	$\frac{45}{x + 3}$

Step 3: Set up the equation. It takes the boat $\frac{45}{x-3}$ hours to go upstream and $\frac{45}{x+3}$ hours to go downstream, which is a total of eight hours. Thus,

$$\frac{45}{x-3} + \frac{45}{x+3} = 8$$

is the equation.

Step 4: Solve the equation.

$$\frac{45}{x-3} + \frac{45}{x+3} = 8$$

$$(x-3)(x+3)\left(\frac{45}{x-3} + \frac{45}{x+3}\right) = (x-3)(x+3)8$$

$$(x-3)(x+3)\left(\frac{45}{x-3}\right) + (x-3)(x+3)\left(\frac{45}{x+3}\right) = (x-3)(x+3)8$$

$$(x+3)45 + (x-3)45 = 8(x-3)(x+3)$$

$$45x + 135 + 45x - 135 = 8(x^2 - 9)$$

$$90x = 8x^2 - 72$$

$$0 = 8x^2 - 90x - 72$$

$$8x^2 - 90x - 72 = 0$$

$$(4x + 3)(x - 12) = 0$$

$$4x + 3 = 0 \quad \text{or} \quad x - 12 = 0$$

$$4x = -3 \qquad x = 12$$

$$x = -\frac{3}{4}$$

Since speed cannot be negative, $-\frac{3}{4}$ is not a solution of the original problem. Thus, the speed of the boat in still water is 12 kilometers per hour. ■

APPLICATION EXERCISES

1. If 5 is added to the square of a number, the result is 41. Find the number.

2. If the square of a number is decreased by 32, the result is four times the number. Find the number.

3. The product of two positive consecutive even integers is 120. Find the integers.

4. Adding 4 to the square of Joyce's age is the same as subtracting 3 from eight times her age. How old is Joyce?

5. Find the length and width of a rectangle if the length is 4 centimeters longer than the width and the area is 140 cm^2.

6. The length of a rectangle is twice the width. The area is 50 m^2. Find the length and width.

7. The hypotenuse of a right triangle is 5 centimeters long. One leg is 2 centimeters longer than the other. Find the lengths of the legs. Hint: Use the Pythagorean Theorem (Applications Chapter 8).

8. The current in a stream moves at a speed of 3 kilometers per hour. A boat travels 40 kilometers upstream and 40 kilometers downstream in a total time of 14 hours. What is the speed of the boat in still water?

9. The side of one square is 5 centimeters longer than the side of a smaller square. The total area of the two squares is 325 cm^2. What is the area of each square?

10. Daniel drove a total of 384 miles round trip in 14 hours. His rate going was 8 miles per hour faster than his rate returning. How many hours did it take him to return? Hint: Use Example 4 as a guide.

USEFUL INFORMATION

1. A quadratic equation in one variable is an equation which can be written in the form

$$ax^2 + bx + c = 0$$

where a, b, and c represent real numbers and $a \neq 0$.

2. If the x term (middle term) is missing from a quadratic equation—that is, if $b = 0$—you can simply solve for x^2 and then take the square root of both sides to find x.

3. To solve a quadratic equation by factoring:

   a. Write the quadratic equation in the form $ax^2 + bx + c = 0$.

   b. Factor $ax^2 + bx + c$.

   c. Set each factor equal to zero.

   d. Solve the two resulting linear equations.

4. The completing the square procedure is important as it will help you understand the development of the quadratic formula. The procedure itself is not used very often to solve quadratic equations since the use of the quadratic formula and factoring are easier procedures.

5. **The Quadratic Formula.** If $ax^2 + bx + c = 0$ $(a \neq 0)$, then

$$x = \frac{-b \pm \sqrt{b^2 - 4ac}}{2a}$$

The quadratic formula is one of the most important formulas in algebra since it will enable you to solve any quadratic equation.

6. When using the quadratic formula to solve a quadratic equation:

   a. If $b^2 - 4ac$ equals a positive number, the equation has two real solutions.

   b. If $b^2 - 4ac$ equals zero, the equation has only one real solution.

   c. If $b^2 - 4ac$ equals a negative number, the equation has no real solution.

REVIEW PROBLEMS

Section 9.1

Solve the following quadratic equations.

1. $x^2 = 25$

2. $2x^2 = 200$

3. $3y^2 = 54$

4. $(x + 2)^2 = 4$

5. $x^2 - 1 = 17$

6. $x^2 + 1 = 29$

7. $2x^2 - 3 = 15$

8. $4x^2 - 5 = 19$

9. $x^2 = \frac{1}{9}$

10. $x^2 = \frac{9}{16}$

Section 9.2

Solve the following quadratic equations by factoring.

1. $x^2 - 5x - 6 = 0$

2. $x^2 + 7x + 6 = 0$

3. $x^2 + 6x - 7 = 0$

4. $2x^2 + 5x + 2 = 0$

5. $2x^2 - 3x - 2 = 0$

6. $4x^2 + 3 = 13x$

7. $6x^2 + 3 = 19x$

8. $6x^2 + 17x = 3$

9. $2x^2 + 7x = 0$

10. $6x^2 + 2 = 8x$

Section 9.3

Solve the following quadratic equations using the quadratic formula.

1. $x^2 + x - 20 = 0$

2. $y^2 - 3y = 10$

3. $y^2 - 10y + 9 = 0$

4. $3x^2 = 7 - 2x$

5. $x^2 - 2x = 2$

6. $x^2 + 6x = 1$

7. $3x^2 = 3 - 2x$

8. $y^2 = 6y + 8$

9. $y^2 = -8 - 3y$

10. $4y^2 + 12y = -9$

ANSWERS TO PRACTICE PROBLEMS

Section 9.1

1. $-3, 3$ **2.** $-8, 8$ **3.** $-3, 3$ **4.** $-2, 2$
5. $-\sqrt{3}, \sqrt{3}$ **6.** $-2\sqrt{3}, 2\sqrt{3}$ **7.** No solution

Section 9.2

1. $2, 3$ **2.** $-\frac{2}{3}, \frac{1}{2}$ **3.** $-\frac{5}{2}, \frac{1}{5}$ **4.** $-3, 0$

Section 9.3

1. $-3 - \sqrt{5}, -3 + \sqrt{5}$ **2.** $-4 - \sqrt{13}, -4 + \sqrt{13}$
3. $-1 - \sqrt{2}, -1 + \sqrt{2}$ **4.** $-1, \frac{3}{2}$
5. $-1, \frac{2}{3}$ **6.** $\frac{-5 - \sqrt{21}}{2}, \frac{-5 + \sqrt{21}}{2}$
7. -5 **8.** No solution

CHAPTER 10

Graphing Linear and Quadratic Equations

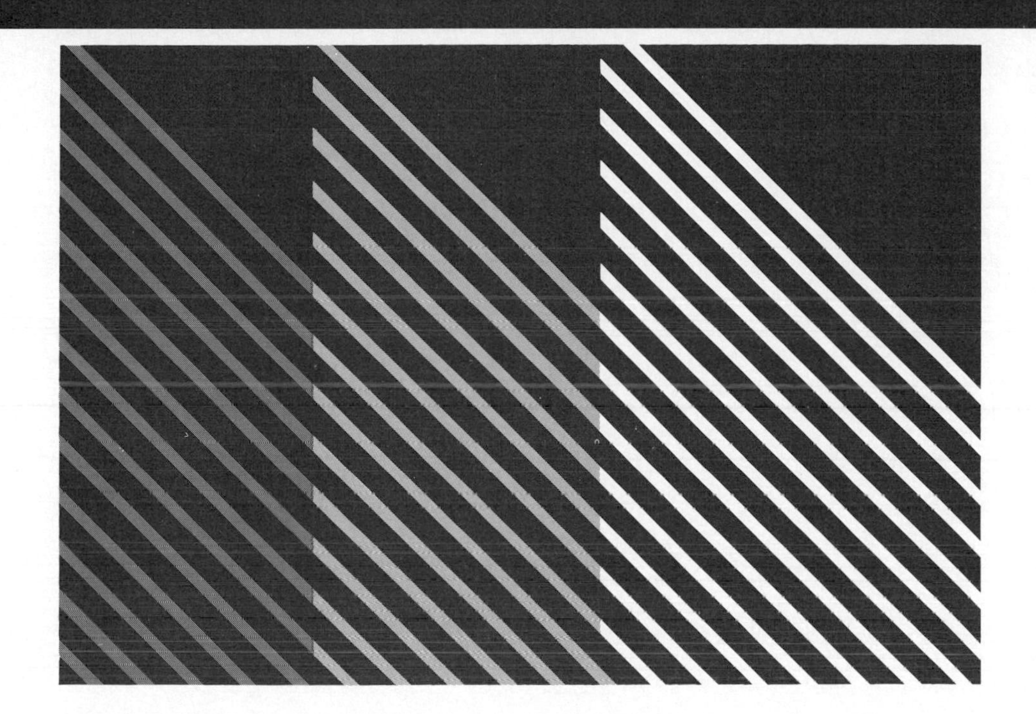

The objective of this chapter is:

1. To be able to graph certain linear and quadratic equations.

The graph of an equation is a diagram or geometric picture of the equation. The graph of a linear equation is a line and the graph of a quadratic equation of the form $y = ax^2 + bx + c$ is a U-shaped curve called a parabola. Graphing linear and quadratic equations will help you find solutions to many algebraic problems. The old saying "a picture is worth a thousand words" is true in algebra.

10.1 THE RECTANGULAR COORDINATE SYSTEM AND GRAPHING LINEAR EQUATIONS

Suppose for a moment that you are visiting Charlie, who lives in the city, and you wish to go from Charlie's house to Albert's house. Charlie, can you give us directions from your house to Albert's?

Sure, you go four blocks east and three blocks north.

Could you show us on a map?

Sure, take a look at the following map. I have placed a C *by my house and an* A *by Albert's house. The arrows indicate how to go from my house to Albert's house.*

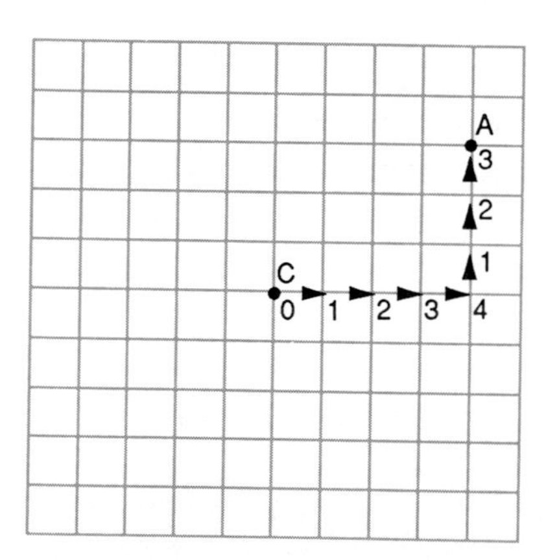

4 blocks east and 3 blocks north

If you understand Charlie's directions then you should have no trouble understanding the rectangular coordinate system. The rectangular coordinate system tells us how to locate a point in the same way that Charlie told us how to locate Albert's house.

The rectangular coordinate system consists of rectangular blocks just like Charlie's map. Just as Charlie's map used his house as a point of origin the rectangular coordinate system has a point of origin. Let's look at a diagram of the rectangular coordinate system.

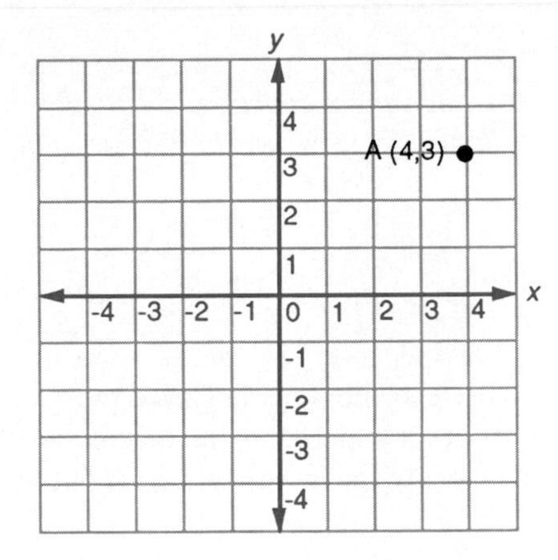

The horizontal line labeled x is called the **x-axis** and the vertical line labeled y is called the **y-axis.** The **origin** of the rectangular coordinate system is the intersection of the x and y axes. With the origin representing Charlie's house, we located Albert's house by going four blocks to the right and three blocks up.

Why did you write (4,3) at Albert's house?

Well, Charlie, (4,3) tells us the location of Albert's house in relation to the origin. You go four blocks to the right and three blocks up to reach point A. In algebra, we call 4 the **x-coordinate** of A and 3 the **y-coordinate** of A.

Example 1 **Locate (−3,2) in the following rectangular coordinate system.**

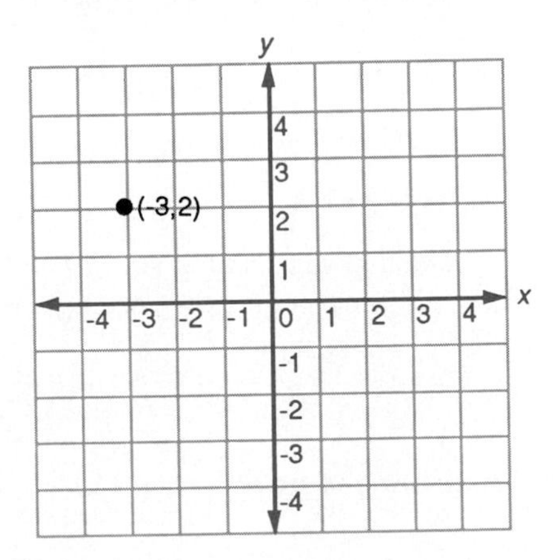

Now, starting at the origin, −3 tells us to go three blocks to the left and 2 tells us to go two blocks up. ■

So, when the x-coordinate is negative, we go to the left. What happens when the y-coordinate is negative?

Well, Charlie, when the y-coordinate is negative, we go down. For example, $(-1,-3)$ tells us to go left 1 unit starting at the origin and then down 3 units.

In Practices 1 through 7 locate the indicated points in the following rectangular coordinate system.

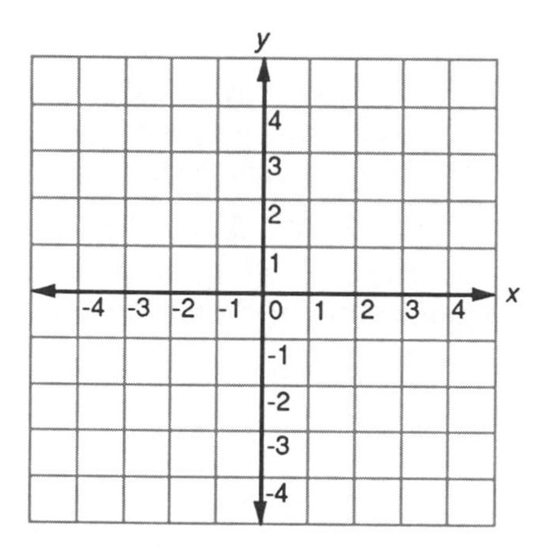

Practice 1: Locate (2,4).

Practice 2: Locate (4,2).

Practice 3: Locate $(-2,4)$.

Practice 4: Locate $(2,-4)$.

Practice 5: Locate $(-2,-4)$.

Practice 6: Locate $(0,-3)$.

Practice 7: Locate (2,0).

In the seven practice problems, you located seven points in a rectangular coordinate system. The location of any point in a rectangular coordinate system is found by an ordered pair.

What is an ordered pair?

Well, Charlie, an ordered pair is two numbers that are written in a certain order. We often use ordered pairs when giving the date. For example, 5–12 represents May 12 and 12–5 represents December 5. Review Practices 1 and 2 noticing that (2,4) and (4,2) represent different points. In algebra, the first number of an ordered pair represents the ***x*-coordinate** (the number of units the point is to the right or left of the origin) and the second number of the ordered pair represents the ***y*-coordinate** (the number of units the point is above or below the origin).

Why are we interested in the location of points in a rectangular coordinate system?

Well, Charlie, in order to graph an equation, we must be able to locate points in a rectangular coordinate system.

How many points will we have to locate in order to graph an equation?

It depends on the equation, Charlie. For example, to graph a linear equation you only have to locate two points. To graph other types of equations it will be necessary to locate more points.

To graph a linear equation we only have to locate two points.

That's correct, Charlie. The graph of a linear equation is a line and two points determine a line. In fact, the name "linear" comes from line. Linear equations in Chapter 4 had only one variable (unknown). In this chapter, we will work with linear equations with two variables.

A **linear equation** in two variables is an equation that can be written in the form $ax + by = c$, where a, b, and c are real numbers with a and b not both equal to zero, and x and y are variables.

The following are linear equations in two variables.

$$2x + 3y = 5$$
$$4x - 7y = 8$$
$$y = 7x - 6$$
$$2x = 3$$
$$4y = 5$$

You made a mistake on the last two equations. The equations $2x = 3$ *and* $4y = 5$ *do not have two variables.*

Well, Charlie, that depends on how you look at the two equations. For example, $2x = 3$ certainly doesn't appear to have two variables. But, $2x = 3$ can be written as $2x + 0y = 3$. Also, $4y = 5$ can be written as $0x + 4y = 5$. In this chapter, we will look at equations like $2x = 3$ and $4y = 5$ as $2x + 0y = 3$ and $0x + 4y = 5$ and say that they have two variables.

How do we graph a linear equation?

Well, Charlie, the graph of a linear equation is a line and it takes two points to determine a line. So, to graph a linear equation we have to find two points on the line.

How do we find a point on the line?

Well, Charlie, a point lies on a line if and only if the x and y coordinates of the point satisfy the linear equation. Thus, to locate two points on a line you need to find two ordered pairs that make the equation true. Let's look at an example.

Example 2 **Graph the following linear equation.**

$$5x + 2y = 9$$

Step 1: Find two ordered pairs of numbers that make the equation true.

a. Pick any number for x and substitute it in the equation. Let $x = 1$ and substitute 1 for x in the equation.

$$5x + 2y = 9$$
$$5(1) + 2y = 9$$
$$5 + 2y = 9$$

b. Solve the resulting equation for y.

$$5 + 2y = 9$$
$$2y = 9 - 5$$
$$2y = 4$$
$$y = 2$$

c. Check to see if (1,2) satisfies the original equation (optional).

$$5x + 2y = 9$$
$$5(1) + 2(2) \stackrel{?}{=} 9$$
$$5 + 4 \stackrel{?}{=} 9$$
$$9 = 9$$

Thus, (1,2) is an ordered pair of numbers that satisfies the equation. To find a second ordered pair, we repeat Steps *a* and *b*.

d. Pick a number other than 1 for x and substitute it in the original equation. Let $x = 3$ and substitute 3 for x in the original equation.

$$5x + 2y = 9$$
$$5(3) + 2y = 9$$
$$15 + 2y = 9$$

e. Solve the resulting equation for y.

$$15 + 2y = 9$$
$$2y = 9 - 15$$
$$2y = -6$$
$$y = -3$$

f. Check to see if $(3,-3)$ satisfies the original equation (optional).

$$5x + 2y = 9$$
$$5(3) + 2(-3) \stackrel{?}{=} 9$$
$$15 + (-6) \stackrel{?}{=} 9$$
$$9 = 9$$

Thus, $(3,-3)$ is a second ordered pair of numbers that satisfies the equation.

When you let x *equal a number and solve for* y *to get an ordered pair, will* y *always be an integer?*

No, Charlie. We always try to choose a number for x so that y will be an integer; however, we do not always succeed.

Step 2: Locate (1,2) and (3,−3) in the following rectangular coordinate system.

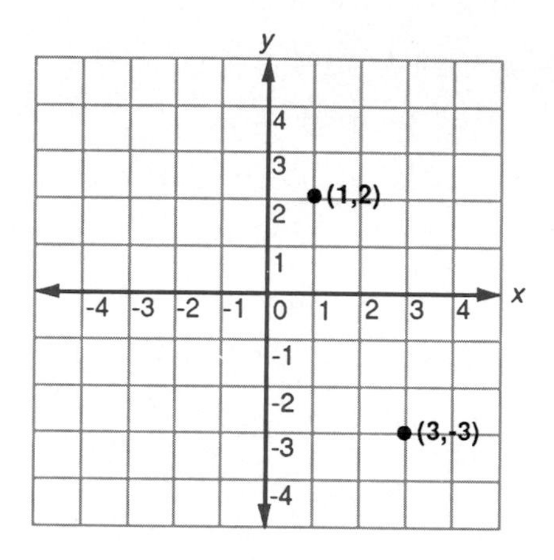

Step 3: Draw the line through the two points located by (1,2) and (3,−3). This line is the graph of $5x + 2y = 9$.

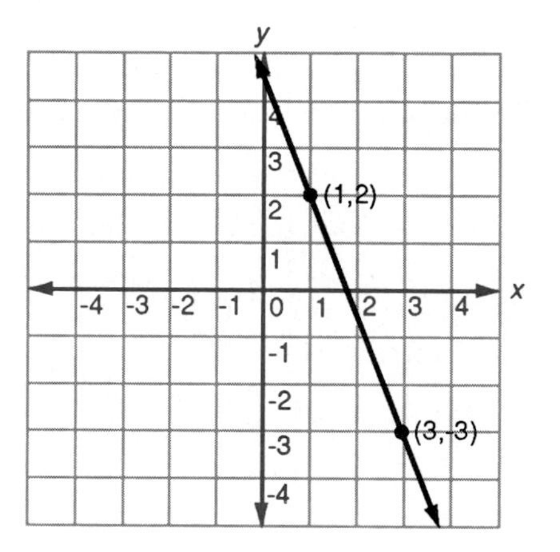

What would have happened if we had selected different values for x *and found two ordered pairs other than (1,2) and (3,−3)?*

Nothing, Charlie. You get the same line regardless of the ordered pairs that you find. In fact, when you work the practice problems, you will likely get ordered pairs different from the ordered pairs given in the answer at the end of the chapter. However, your line should be the same as the one in the answer.

Check: Find a third ordered pair of numbers that makes the equation true and see if the point located by this ordered pair is on the line. If so, you should have the correct line.

Step 1: Find a third ordered pair of numbers that makes the equation true. Let $x = 2$ and substitute 2 for x in the original equation. Then solve for y.

$$5x + 2y = 9$$
$$5(2) + 2y = 9$$
$$10 + 2y = 9$$
$$2y = 9 - 10$$
$$2y = -1$$
$$y = \frac{-1}{2}$$

Now,

$$5x + 2y = 9$$
$$5(2) + 2\left(\frac{-1}{2}\right) \stackrel{?}{=} 9$$
$$10 + (-1) \stackrel{?}{=} 9$$
$$9 = 9$$

Thus, $(2,\frac{-1}{2})$ is a third ordered pair of numbers that satisfies the equation.

Step 2: See if the point located by $(2,\frac{-1}{2})$ is on the line.

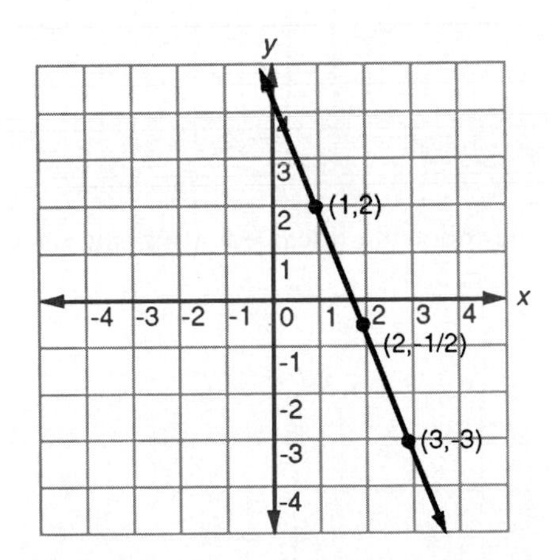

Since the point located by $(2,\frac{-1}{2})$ is on the line, the graph of $5x + 2y = 9$ is correct. ■

Practice 8: Graph the following linear equation and check.

$$2x + y = 3$$

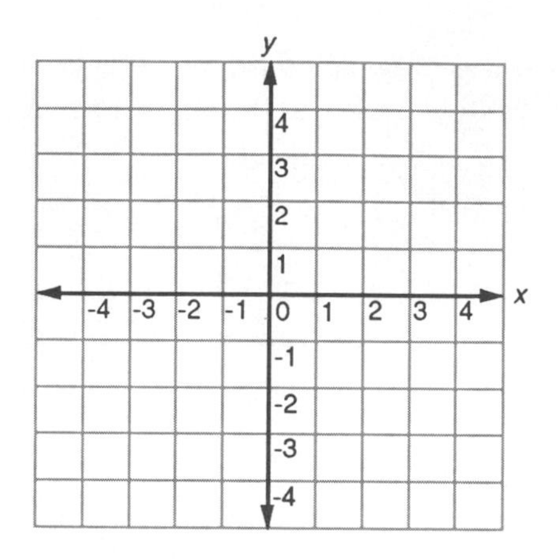

Practice 9: Graph the following linear equation and check.

$$-5x + y = 7$$

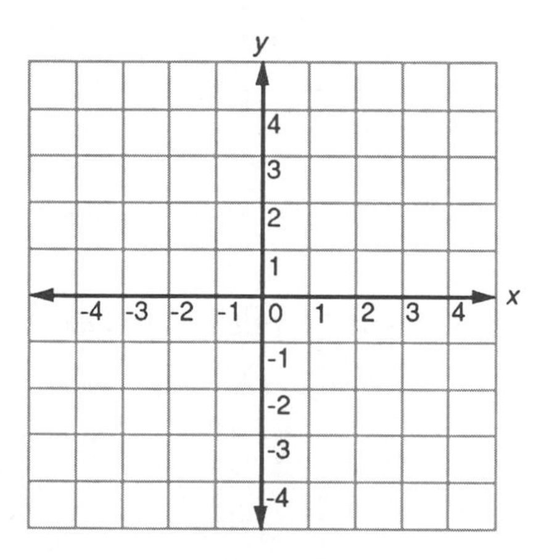

Practice 10: Graph the following linear equation and check.

$$-2x + 3y = 5$$

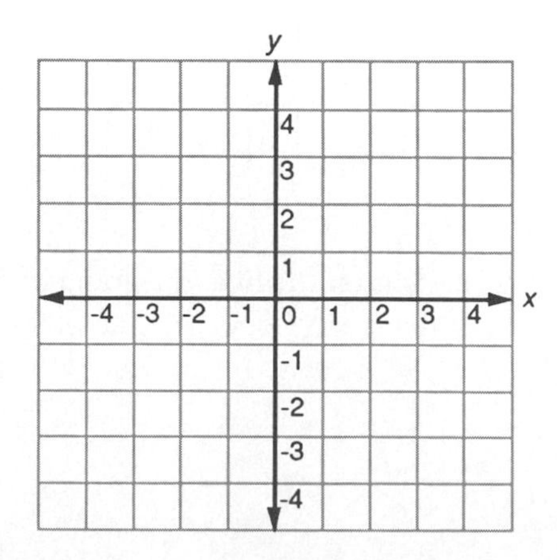

Practice 11: Graph the following linear equation and check.

$$-3x - 2y = 5$$

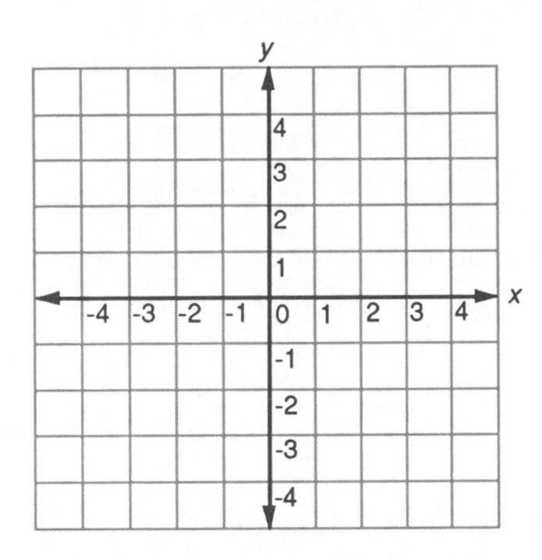

Example 3 **Graph the following linear equation.**

$$2y = 6$$

Step 1: Write the equation in the form $ax + by = c$.

$$0x + 2y = 6$$

Step 2: Find two ordered pairs of numbers that make the equation true.

a. Pick any number for x and substitute it in the equation. Let $x = 3$ and substitute 3 for x in the equation.

$$0x + 2y = 6$$
$$0(3) + 2y = 6$$
$$0 + 2y = 6$$
$$2y = 6$$

b. Solve the resulting equation for y.

$$2y = 6$$
$$y = 3$$

c. Check to see if (3,3) satisfies the original equation (optional).

$$0x + 2y = 6$$
$$0(3) + 2(3) \stackrel{?}{=} 6$$
$$0 + 6 \stackrel{?}{=} 6$$
$$6 = 6$$

Thus, (3,3) is an ordered pair of numbers that satisfies the equation. To find a second ordered pair we repeat Steps a and b.

d. Pick a number other than 3 for x and substitute it in the equation. Let $x = -2$ and substitute it for x in the equation.

$$0x + 2y = 6$$
$$0(-2) + 2y = 6$$
$$0 + 2y = 6$$
$$2y = 6$$

e. Solve the resulting equation for y.

$$2y = 6$$
$$y = 3$$

f. Check to see if $(-2,3)$ satisfies the original equation (optional).

$$0x + 2y = 6$$
$$0(-2) + 2(3) \stackrel{?}{=} 6$$
$$0 + 6 \stackrel{?}{=} 6$$
$$6 = 6$$

Thus, $(-2,3)$ is a second ordered pair of numbers that satisfies the equation.

We have two ordered pairs, (3,3) and (−2,3) that have the same number for y.

That's correct, Charlie. Since the coefficient of x is 0, it makes no difference what number we substitute for x; the resulting equation will always be $2y = 6$ and y will always equal 3.

<u>Step 3:</u> Locate (3,3) and $(-2,3)$ in the following rectangular coordinate system.

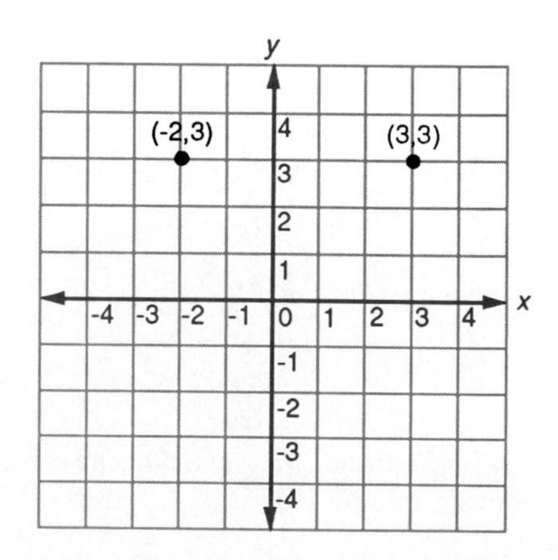

Step 4: Draw the line through the two points located by (3,3) and (−2,3). This line is the graph of $0x + 2y = 6$ or $2y = 6$.

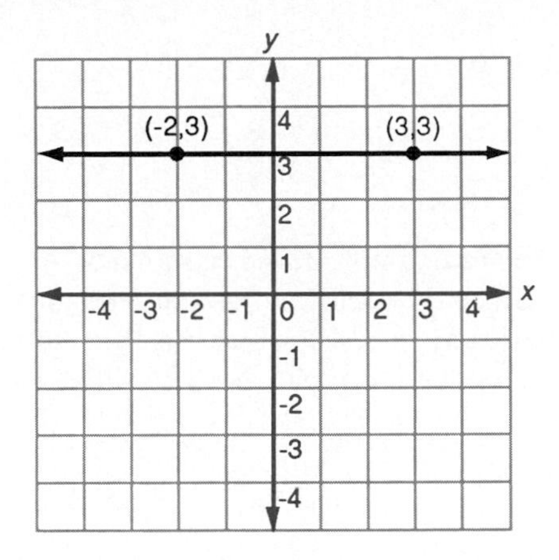

The graph of $2y = 6$ is a horizontal line. In fact, the graph of any equation of the form $by = c$ is a **horizontal line.** ■

Is there a shorter way to graph equations like 2y = 6?

Yes, Charlie. Let's look at the next example.

Example 4 **Graph the following linear equation.**

$$5y = -10$$

Step 1: Solve the equation for y.

$$5y = -10$$
$$y = -2$$

Step 2: Pick two numbers for x and pair them with −2.

Let $x = -3$ and 2. Now, (−3,−2) and (2,−2) are two ordered pairs of numbers that satisfy the equation; that is,

$$0x + 5y = -10 \qquad\qquad 0x + 5y = -10$$
$$0(-3) + 5(-2) \stackrel{?}{=} -10 \qquad\qquad 0(2) + 5(-2) \stackrel{?}{=} -10$$
$$0 + (-10) \stackrel{?}{=} -10 \qquad\qquad 0 + (-10) \stackrel{?}{=} -10$$
$$-10 = -10 \qquad\qquad -10 = -10$$

Wait a minute! How did you do that?

Well, Charlie, just as in Example 3, the equation $5y = -10$ is really the equation $0x + 5y = -10$. Since the coefficient of x is zero, any number that you select for x gives you the equation $5y = -10$, which has the solution $y = -2$. Therefore, it does not matter what numbers you pick for x, y will always equal -2.

Step 3: Locate $(-3,-2)$ and $(2,-2)$ in the following rectangular coordinate system.

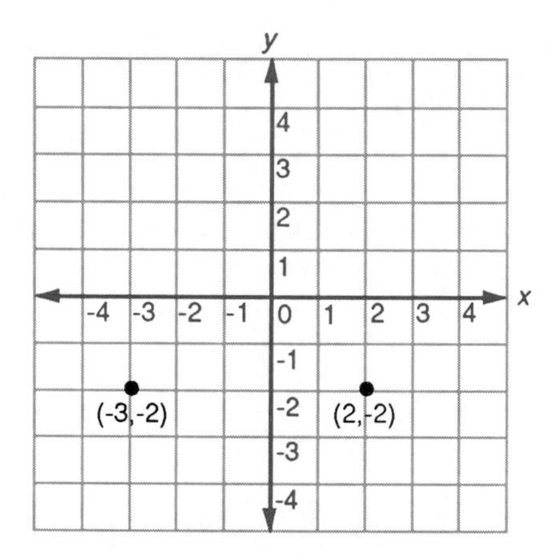

Step 4: Draw a line through the two points located by $(-3,-2)$ and $(2,-2)$. This line is the graph of $5y = -10$.

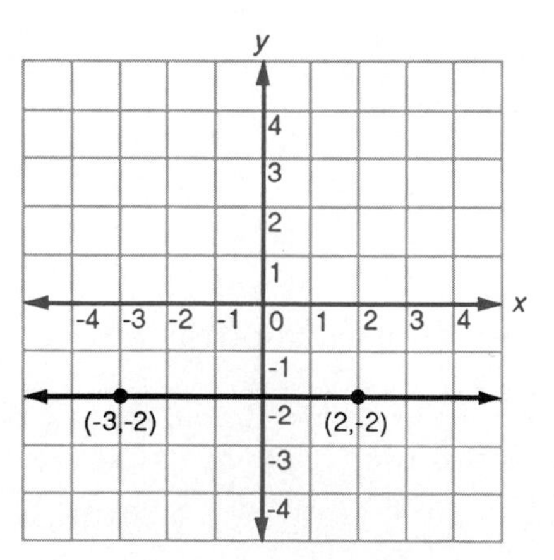

■

Practice 12: Graph the following linear equation.

$$5y = 15$$

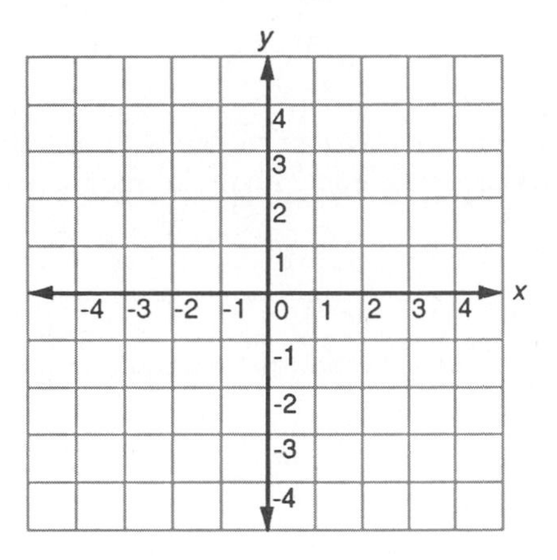

Practice 13: Graph the following linear equation.

$$-5y = 20$$

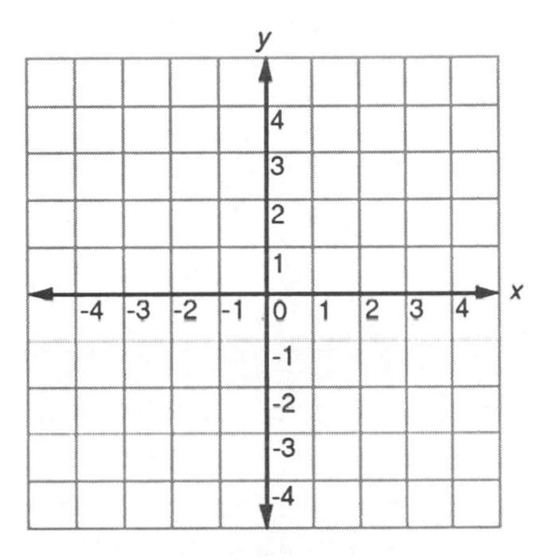

Practice 14: Graph the following linear equation.

$$3y = -9$$

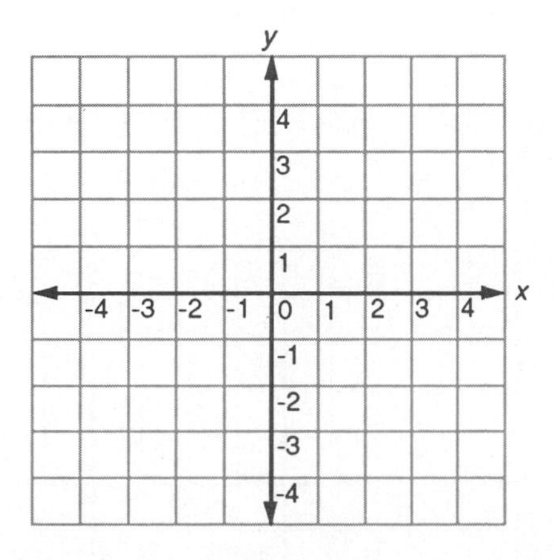

Example 5 **Graph the following linear equation.**

$$3x = 12$$

Step 1: Solve the equation for x.

$$3x = 12$$

$$x = 4$$

Step 2: Pick two numbers for y and pair them with 4. Let $y = 3$ and 1. Now, (4,3) and (4,1) are two ordered pairs of numbers that satisfy the equation.

I didn't understand that step!

Well, Charlie, the equation $3x = 12$ is really the equation $3x + 0y = 12$. Since the coefficient of y is zero, any number you select for y gives you the equation $3x = 12$, which has the solution $x = 4$. Therefore, it does not matter what numbers you pick for y, x will always equal 4.

Step 3: Locate (4,3) and (4,1) in the following rectangular coordinate system.

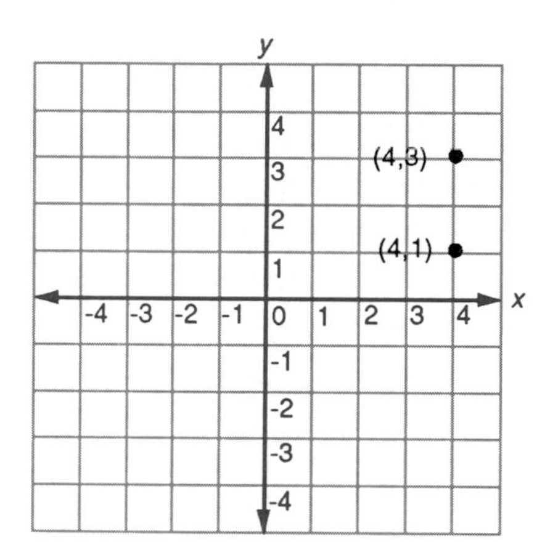

Step 4: Draw a line through the two points located by (4,3) and (4,1). This line is the graph of $3x = 12$.

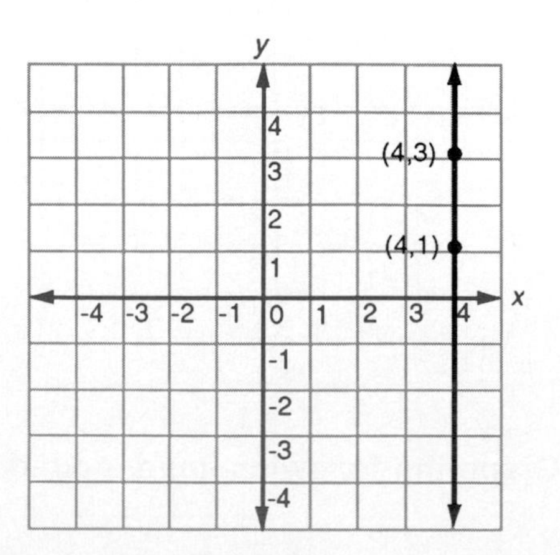

The graph of $3x = 12$ is a vertical line. In fact, the graph of any linear equation of the form $ax = c$ is a **vertical line.** ■

Practice 15: Graph the following linear equation.

$$3x = 6$$

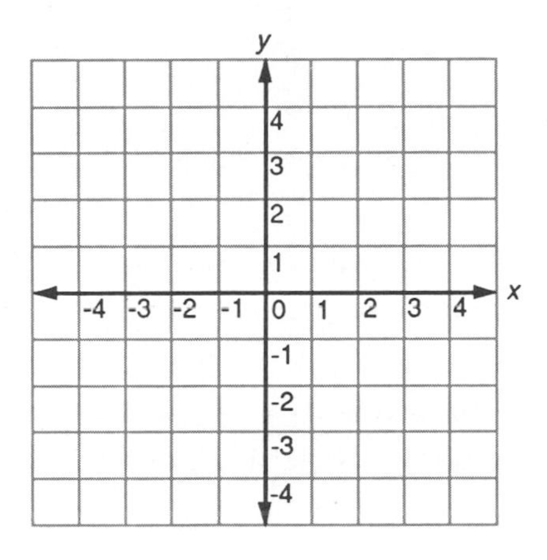

Practice 16: Graph the following linear equation.

$$5x = -10$$

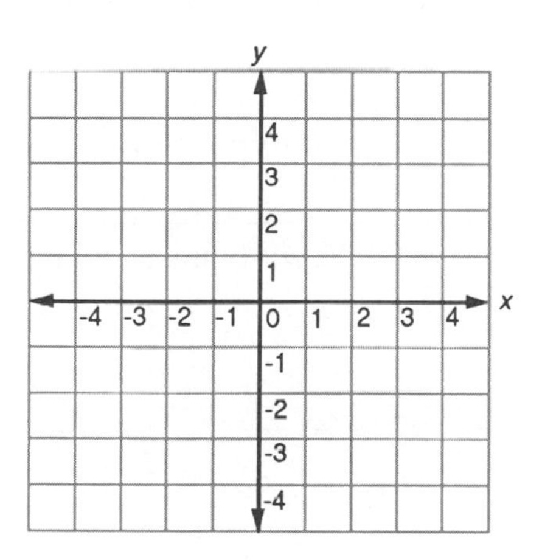

Practice 17: Graph the following linear equation.

$$-6x = 18$$

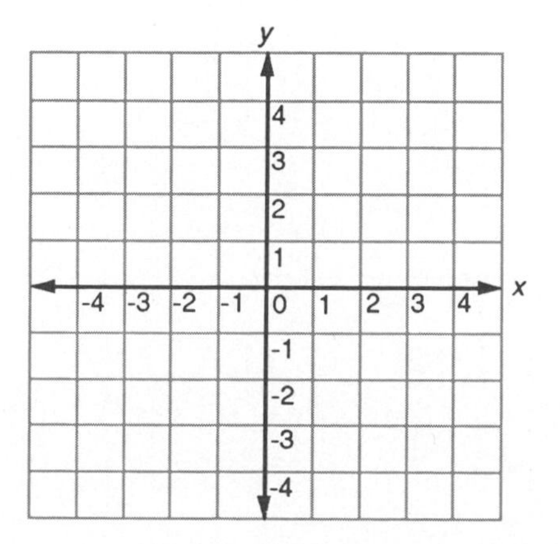

EXERCISE 10.1

Graph the following linear equations and check.

A

1. $x + y = 3$

2. $2x + y = 5$

3. $3x + y = -4$

4. $-2x + y = 1$

5. $-5x + y = -2$

6. $-4x + y = -5$

7. $7x + y = -6$

8. $3x + y = 9$

9. $-8x + y = -4$

10. $-5x + y = -10$

B

11. $2x + 3y = 0$

12. $2x + 4y = 6$

13. $-3x + 3y = 9$

14. $2y = 12$

15. $3y = -9$

16. $3x = 15$

17. $-5x = 15$

18. $-3x + 4y = 12$

19. $-5x + 2y = 3$

20. $2x - 5y = 7$

10.2 GRAPHING LINEAR EQUATIONS BY THE SLOPE-INTERCEPT METHOD

Some interesting things happen when we solve a linear equation of the form $ax + by = c$ for the variable y and then draw the graph. Let's look at an example.

Example 1 **Graph the following linear equation.**

$$-3x + 2y = 2$$

Step 1: Solve $-3x + 2y = 2$ for y.

$$-3x + 2y = 2$$

$$2y = 3x + 2$$

$$y = \frac{(3x + 2)}{2}$$

$$y = \frac{3}{2}x + 1$$

Step 2: Find two ordered pairs that make the equation true.

a. Let $x = 0$. Substitute 0 for x in the equation and find the value of y.

$$y = \frac{3}{2}x + 1$$

$$y = \frac{3}{2}(0) + 1$$

$$y = 0 + 1$$

$$y = 1$$

Thus, (0,1) locates a point on the line.

b. Let $x = 2$. Substitute 2 for x in the equation and find the value of y.

$$y = \frac{3}{2}x + 1$$

$$y = \frac{3}{2}(2) + 1$$

$$y = 3 + 1$$

$$y = 4$$

Thus, (2,4) locates a point on the line.

Step 3: Locate (0,1) and (2,4) in the rectangular coordinate system, then draw the line.

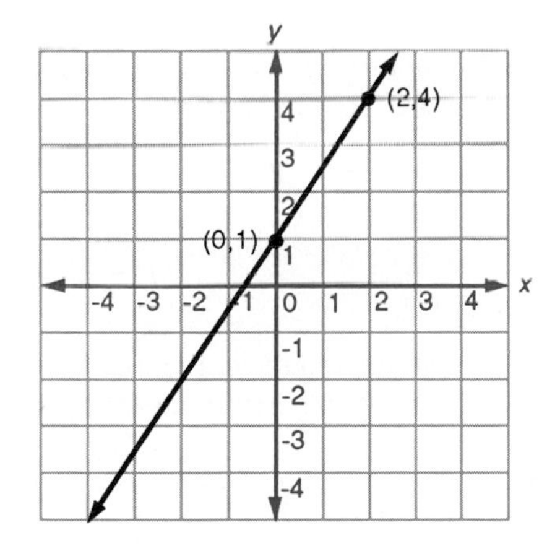

■

What's so interesting about this example?

Well, Charlie, the **y-intercept** of a line is the point where the line crosses the y-axis. When you look at the above line, you will notice that it crosses the y-axis at (0,1)—that is, when $y = 1$. Now, 1 is the constant term in the equation $y = \frac{3}{2}x + 1$, and it is also the y-coordinate of the y-intercept. Next, when you go from (0,1) to (2,4), you go up 3 units and to the right 2 units; that is, the line has a rise of 3 units and a run of 2 units. Thus, the slope of the line is $\frac{3}{2}$; which is the coefficient of x in the equation $y = \frac{3}{2}x + 1$.

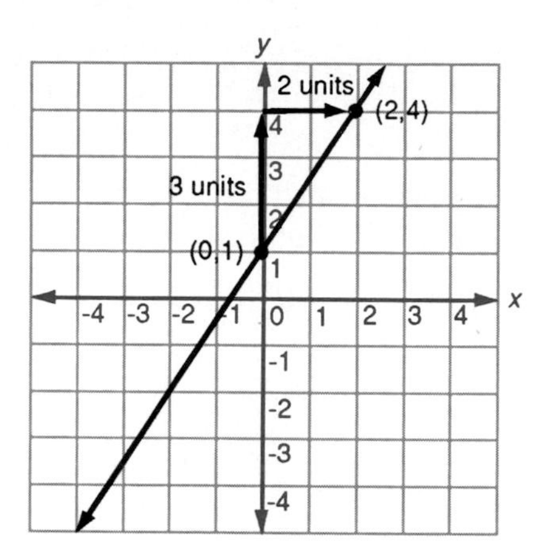

In general, when you solve a linear equation for y and write the equation in the form $y = mx + b$, the equation is said to be written in **slope-intercept form.** The constant term, b, gives us $(0,b)$, the **y-intercept** of the line, and the coefficient of x, m, is the **slope** of the line.

I have two questions. First, how did you know the y-intercept was (0,b)?

Well, Charlie, we know that any point on the y-axis has an x-coordinate of 0. So, to find the y-intercept of a line (where it crosses the y-axis) we substitute 0 for x in the equation $y = mx + b$.

$$y = mx + b$$
$$y = m(0) + b$$
$$y = 0 + b$$
$$y = b$$

Thus, the y-intercept of the line is $(0,b)$.

Okay, Charlie, what is your second question?

What is slope?

Well, Charlie, **slope** is one way of measuring the incline or steepness of a line. The slope of a line is the ratio of the rise of the line (change in vertical distance) to the run of the line (change in horizontal distance); that is,

$$\text{Slope} = \frac{\text{Rise}}{\text{Run}} = \frac{\text{Change in vertical distance}}{\text{Change in horizontal distance}}$$

Let's look at an example of how we can graph a linear equation using the slope-intercept method.

Example 2 **Graph the following linear equation using the slope-intercept method.**

$$-2x + y = -3$$

Step 1: Solve for y and write the equation in the form $y = mx + b$.

$$-2x + y = -3$$
$$y = -3 + 2x$$
$$y = 2x - 3$$
$$y = 2x + (-3)$$

Step 2: Find the y-intercept. Since the constant term of $y = 2x + (-3)$ is -3, the line crosses the y-axis at $(0, -3)$. Thus, $(0, -3)$, the y-intercept, locates one point on the line.

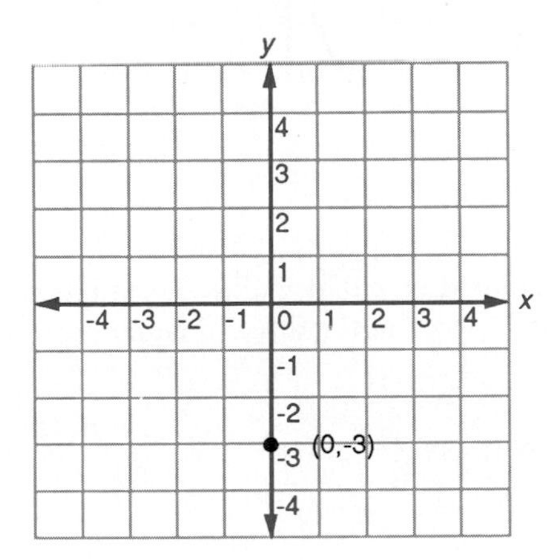

Step 3: Use the slope of the line to find a second point on the line. Since the coefficient of x is 2 in the equation $y = 2x + (-3)$, the slope of the line is 2. A slope of $2 = \frac{2}{1}$ tells us that for every 2 units we go up, we go to the right 1 unit; thus, we can find a second point on the line by starting at $(0,-3)$ and going up 2 units to -1 and to the right 1 unit to 1. Therefore, $(1,-1)$ locates a second point on the line.

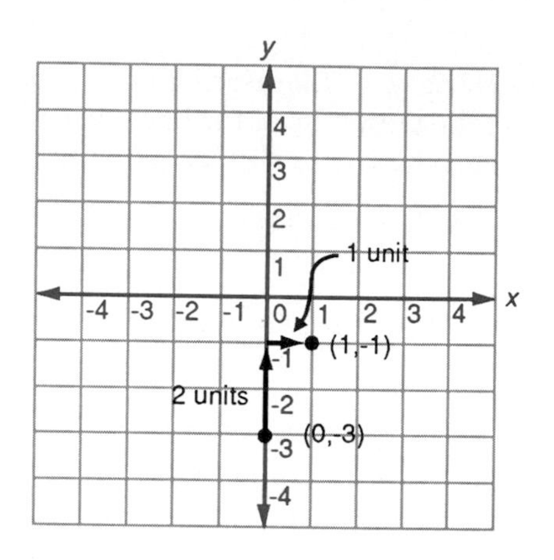

Step 4: Draw the line through the two points located by $(0,-3)$ and $(1,-1)$.

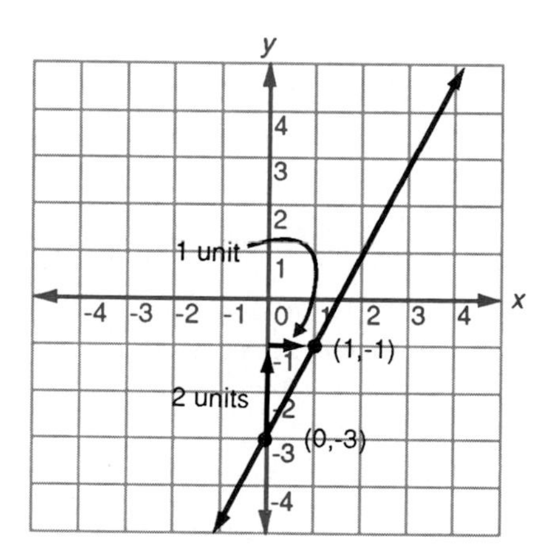

Check: Find a third ordered pair that makes the equation true and check to see if the point is on the line.

Step 1: Find a third ordered pair that makes the equation true. Let $x = 3$. Now substitute 3 for x in the original equation and solve for y.

$$-2x + y = -3$$
$$-2(3) + y = -3$$
$$-6 + y = -3$$
$$y = -3 + 6$$
$$y = 3$$

Now, (3,3) is an ordered pair that satisfies $-2x + y = -3$.

Step 2: Check to see if (3,3) locates a point on the line.

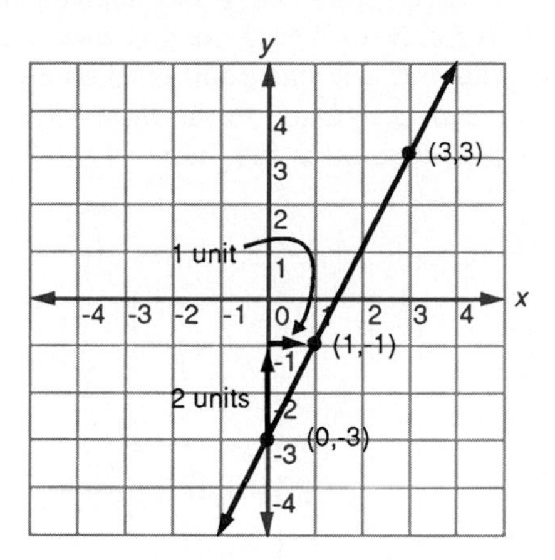

Now, (3,3) locates a point on the line and the check is complete. ■

Example 3 **Graph the following linear equation using the slope-intercept method.**

$$4x + 6y = 12$$

Step 1: Solve for y and write the equation in the form $y = mx + b$.

$$4x + 6y = 12$$

$$6y = -4x + 12$$

$$y = \frac{(-4x + 12)}{6}$$

$$y = \frac{-4}{6}x + \frac{12}{6}$$

$$y = \frac{-2}{3}x + 2$$

Step 2: Find the y-intercept. Since the constant term of the equation $y = \frac{-2}{3}x + 2$ is 2, the line crosses the y-axis at (0,2). Thus, (0,2), the y-intercept, locates one point on the line.

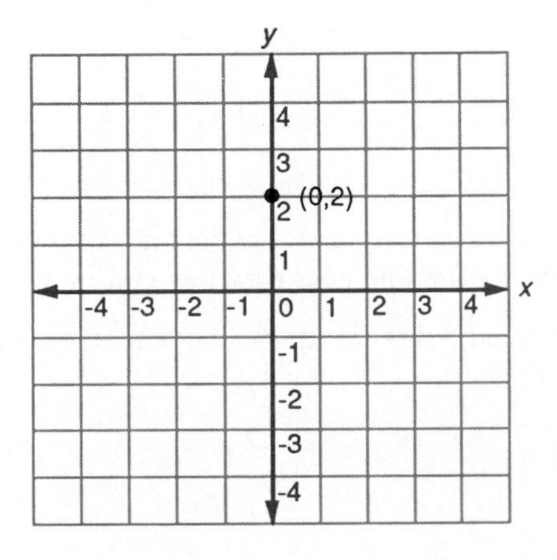

Step 3: Use the slope of the line to find a second point on the line. Since the coefficient of x is $\frac{-2}{3}$($y = \frac{-2}{3}x + 2$), the slope of the line is $\frac{-2}{3}$. A slope of $\frac{-2}{3}$ tells us that for every 2 units we go *down,* we also go to the right 3 units; thus, we can find a second point on the line by starting at (0,2) and going down 2 units to 0 and to the right 3 units to 3. Therefore, (3,0) locates a second point on the line.

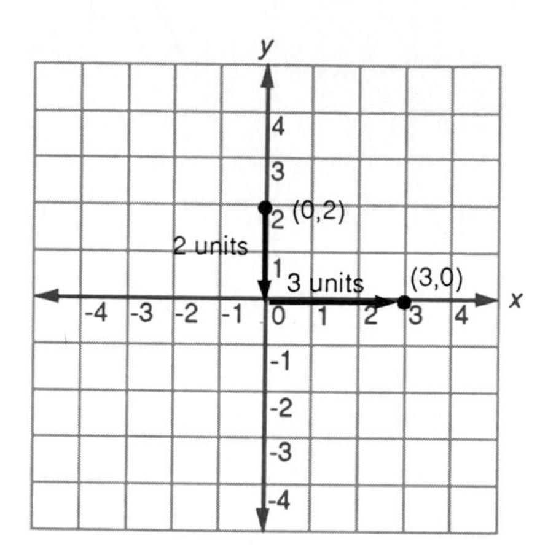

Step 4: Draw the line through the two points located by (0,2) and (3,0).

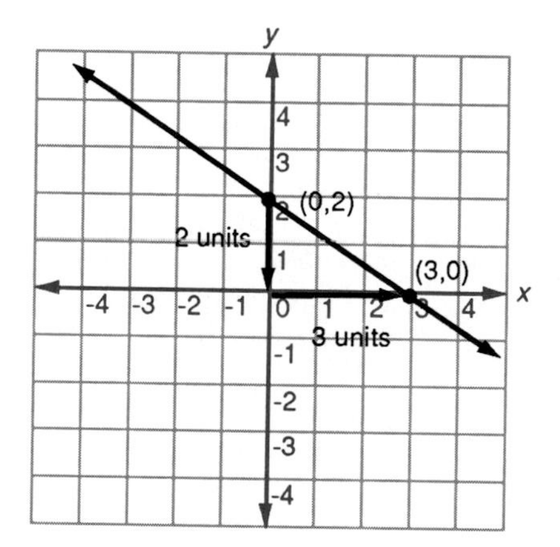

What does a negative slope mean?

Well, Charlie, a *positive slope* means the line goes up as you move to the right and a *negative slope* means the line goes down as you move to the right.

Check: Find a third ordered pair that makes the equation true and check to see if the point is on the line.

A N S W E R

Step 1: Find a third ordered pair that makes the equation true. Let $x = -3$. Now substitute -3 for x in the equation and solve for y.

$$4x + 6y = 12$$
$$4(-3) + 6y = 12$$
$$-12 + 6y = 12$$
$$6y = 12 + 12$$
$$6y = 24$$
$$y = 4$$

Now, $(-3,4)$ is an ordered pair that satisfies $4x + 6y = 12$.

Step 2: Check to see if $(-3,4)$ locates a point on the line.

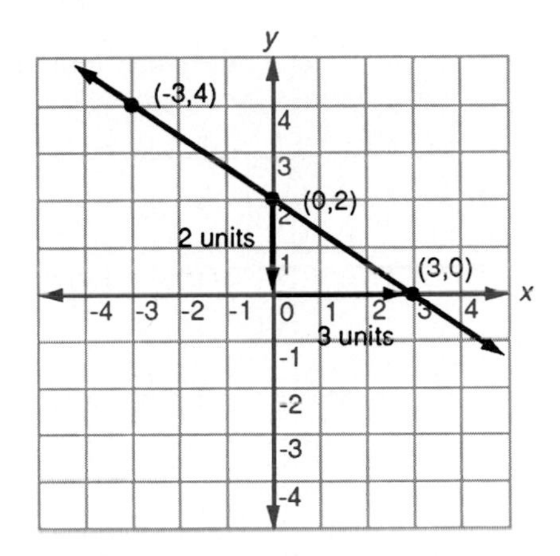

Now, $(-3,4)$ locates a point on the line and the check is complete. ■

Practice 1: Graph $-2x + y = -1$ using the slope-intercept method and check.

1. 

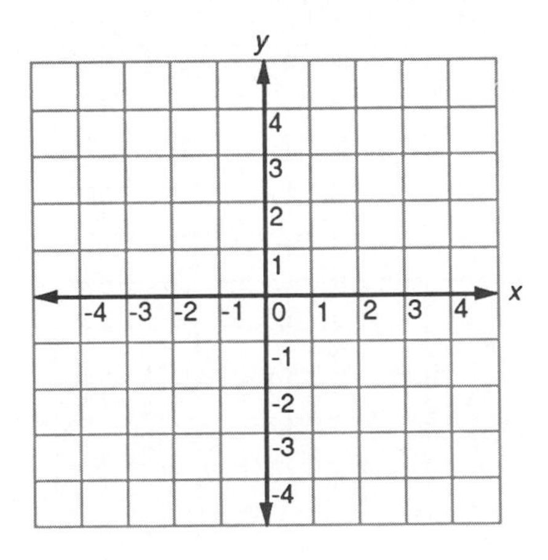

A N S W E R

2. ____________________

Practice 2: Graph $2x + y = 1$ using the slope-intercept method and check.

3. ____________________

Practice 3: Graph $-2x + 6y = 12$ using the slope-intercept method and check.

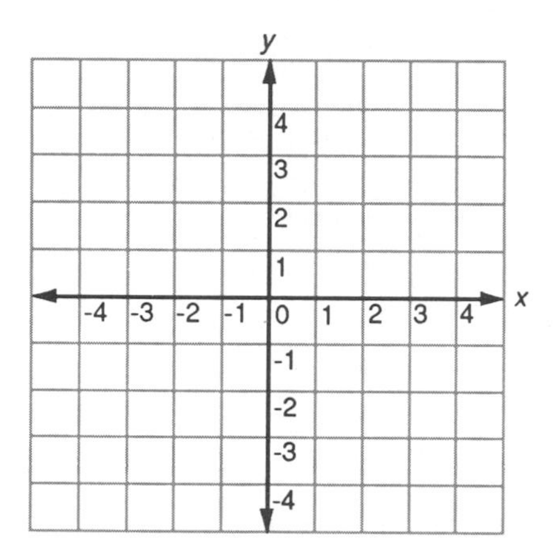

4. ____________________

Practice 4: Graph $5x + 4y = 16$ using the slope-intercept method and check.

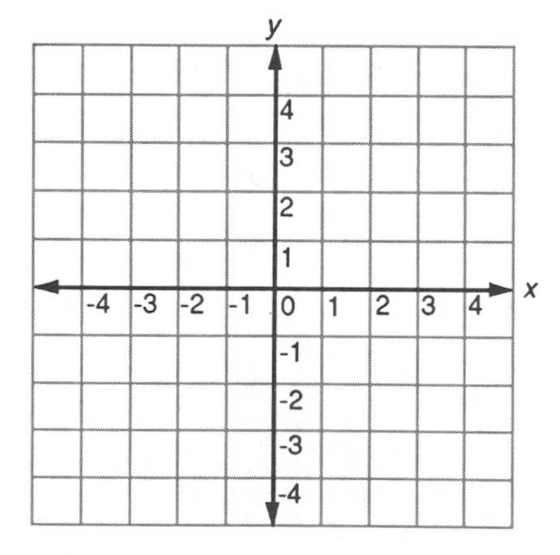

EXERCISE 10.2

Graph the following linear equations using the slope-intercept method and check.

A

1. $5x + y = 8$
2. $7x + y = -5$
3. $-3x + y = 6$
4. $-8x + y = -7$
5. $-2x + 5y = 30$
6. $-5x + 4y = 20$
7. $-6x + 7y = -98$
8. $-8x + 6y = -42$
9. $4x + 5y = 35$
10. $8x + 10y = -40$

B

11. $5x - 4y = 32$
12. $8x - 12y = 36$
13. $9x - 12y = -60$
14. $12x - 15y = -45$
15. $-15x - 25y = 50$
16. $-18x - 24y = 48$
17. $-7x - 21y = -63$
18. $-11x - 33y = -99$
19. $22y = 44$
20. $5y = -30$

10.3 GRAPHING QUADRATIC EQUATIONS

In this section, we will graph quadratic equations in two variables of the form

$$y = ax^2 + bx + c$$

where a, b, and c are real numbers and $a \neq 0$. For example,

$$y = 3x^2 + 5x + 6$$

$$y = -2x^2 + 7$$

$$y = 5x^2$$

are quadratic equations of the form $y = ax^2 + bx + c$.

The graph of a quadratic equation of the form $y = ax^2 + bx + c$ is a U-shaped curve called a **parabola**. These parabolas will have one of the following two forms.

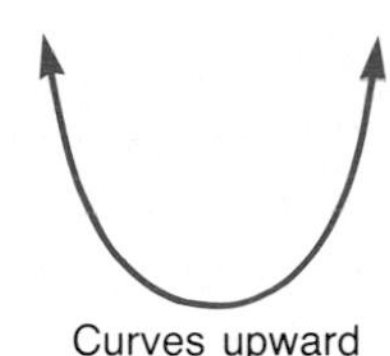

Curves upward Curves downward

Think of a parabola as having an umbrella shape.

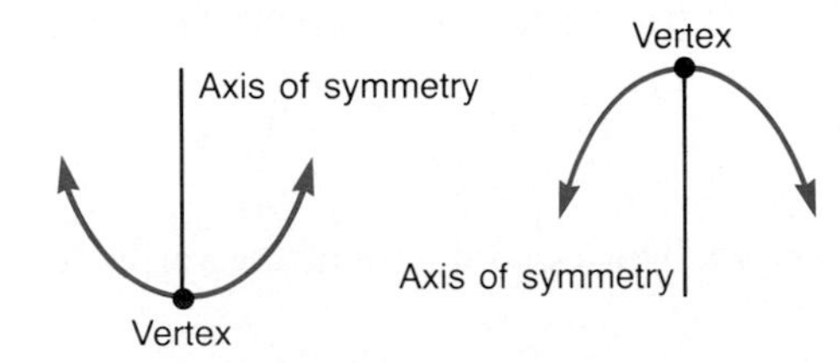

The handle of the umbrella represents the **axis of symmetry** of the parabola. The axis of symmetry divides the parabola into two mirror images. Where the handle joins the umbrella is the **vertex** of the parabola. The vertex of a parabola in this section is the lowest or highest point on the parabola.

To draw a parabola, locate several points whose ordered pairs satisfy the quadratic equation. Then draw a smooth curve through the points.

How many points will we have to locate?

Well, Charlie, you should locate at least five points on the parabola before drawing it.

Will any five points be okay?

No, Charlie. You should first locate the vertex of the parabola and then at least two points on each side of the vertex.

How do I find the vertex of the parabola?

Well, Charlie, the x-coordinate of the vertex of the parabola whose equation is

$$y = ax^2 + bx + c$$

is $\frac{-b}{2a}$. Arriving at this fact involves completing the square, a procedure used when we developed the quadratic formula. You may want to look this up in the library as a special project. Examples 1 and 2 will show you how to use $\frac{-b}{2a}$ to locate the vertex of a parabola.

Example 1 **Graph the following quadratic equation.**

$$y = x^2 - 2x - 3$$

Step 1: Locate the vertex. The x-coordinate of the vertex is

$$\frac{-b}{2a} = \frac{-(-2)}{2(1)}$$

$$= \frac{2}{2}$$

$$= 1$$

The y-coordinate of the vertex is found by substituting 1 for x in the original equation and solving for y.

$$y = x^2 - 2x - 3$$
$$y = (1)^2 - 2(1) - 3$$
$$y = 1 - 2 - 3$$
$$y = 1 - 5$$
$$y = -4$$

Thus, the vertex is located at $(1,-4)$.

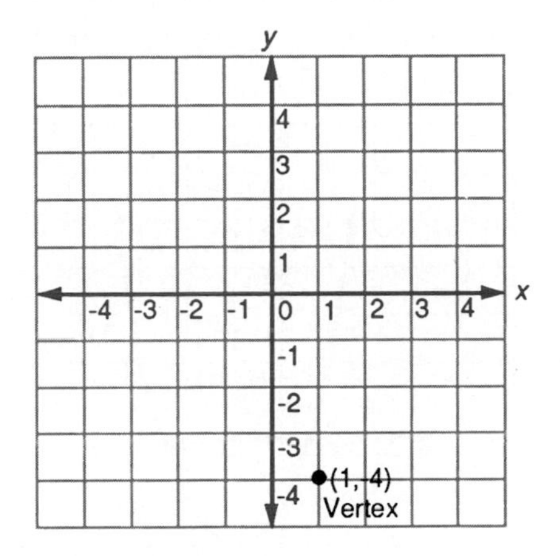

Step 2: Locate two points on the parabola on each side of the vertex.

a. Find two points on the left of the vertex. Let $x = 0$, then

$$y = x^2 - 2x - 3$$
$$y = (0)^2 - 2(0) - 3$$
$$y = 0 - 0 - 3$$
$$y = -3$$

Thus, $(0,-3)$ locates one point on the parabola to the left of the vertex. Let $x = -1$, then

$$y = x^2 - 2x - 3$$
$$y = (-1)^2 - 2(-1) - 3$$
$$y = 1 + 2 - 3$$
$$y = 0$$

Thus, $(-1,0)$ locates a second point on the parabola to the left of the vertex.

b. Find two points on the right of the vertex. Let $x = 2$, then

$$y = x^2 - 2x - 3$$
$$y = (2)^2 - 2(2) - 3$$
$$y = 4 - 4 - 3$$
$$y = -3$$

Thus, $(2,-3)$ locates one point on the parabola to the right of the vertex. Let $x = 3$, then

$$y = x^2 - 2x - 3$$
$$y = (3)^2 - 2(3) - 3$$
$$y = 9 - 6 - 3$$
$$y = 0$$

Thus, $(3,0)$ locates a second point on the parabola to the right of the vertex.

Step 3: Draw a smooth curve through the four points located by $(0,-3)$, $(-1,0)$, $(2,-3)$, $(3,0)$, and the vertex located by $(1,-4)$.

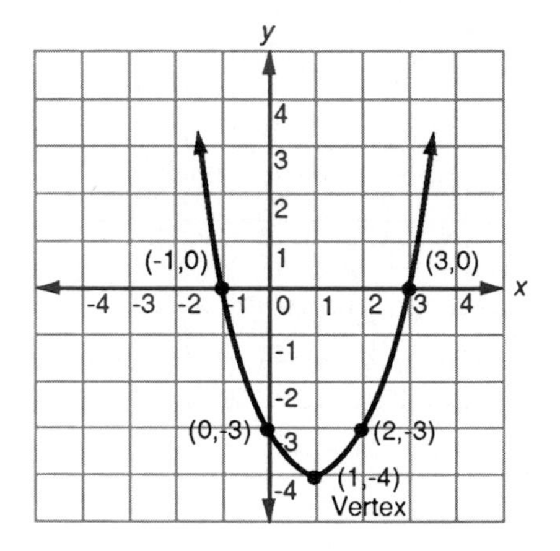

■

Is there a check to find if we have drawn the right parabola?

Well, Charlie, if you are in doubt, find an additional point on each side of the vertex whose ordered pair satisfies the quadratic equation. Locate these two points in the rectangular coordinate system and see if they lie on the parabola.

Example 2 **Graph the following quadratic equation.**

$$y = -2x^2 - 4x + 2$$

Step 1: Locate the vertex. The x-coordinate of the vertex is

$$\frac{-b}{2a} = \frac{-(-4)}{2(-2)}$$
$$= \frac{4}{-4}$$
$$= -1$$

The y-coordinate of the vertex is found by substituting -1 for x in the original equation and solving for y.

$$y = -2x^2 - 4x + 2$$
$$y = -2(-1)^2 - 4(-1) + 2$$
$$y = -2(1) + 4 + 2$$
$$y = -2 + 6$$
$$y = 4$$

Thus, the vertex is located at $(-1,4)$.

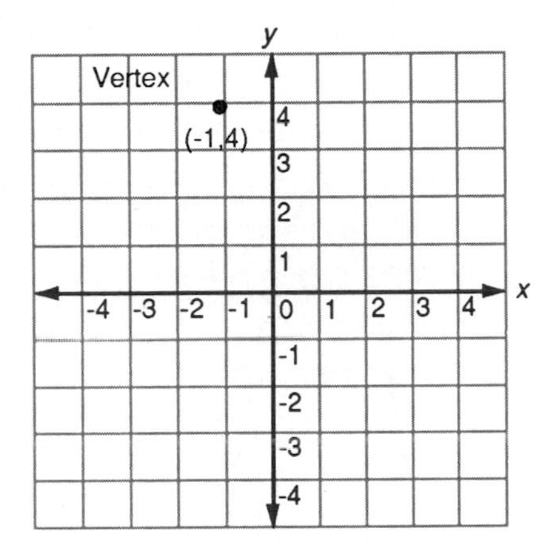

Step 2: Locate two points on the parabola on each side of the vertex.

a. Find two points on the left of the vertex. Let $x = -2$, then

$$y = -2x^2 - 4x + 2$$
$$y = -2(-2)^2 - 4(-2) + 2$$
$$y = -2(4) + 8 + 2$$
$$y = -8 + 8 + 2$$
$$y = 2$$

Thus, $(-2,2)$ locates one point on the parabola to the left of the vertex.
Let $x = -3$, then

$$y = -2x^2 - 4x + 2$$
$$y = -2(-3)^2 - 4(-3) + 2$$
$$y = -2(9) + 12 + 2$$
$$y = -18 + 12 + 2$$
$$y = -4$$

Thus, $(-3,-4)$ locates a second point on the parabola to the left of the vertex.

b. Find two points on the right of the vertex. Let $x = 0$, then

$$y = -2x^2 - 4x + 2$$
$$y = -2(0)^2 - 4(0) + 2$$
$$y = 0 - 0 + 2$$
$$y = 2$$

Thus, (0,2) locates one point on the parabola to the right of the vertex. Let $x = 1$, then

$$y = -2x^2 - 4x + 2$$
$$y = -2(1)^2 - 4(1) + 2$$
$$y = -2 - 4 + 2$$
$$y = -4$$

Thus, $(1,-4)$ locates a second point on the parabola to the right of the vertex.

Step 3: Draw a smooth curve through the four points located by $(-2,2)$, $(-3,-4)$, $(0,2)$, $(1,-4)$, and the vertex located by $(-1,4)$.

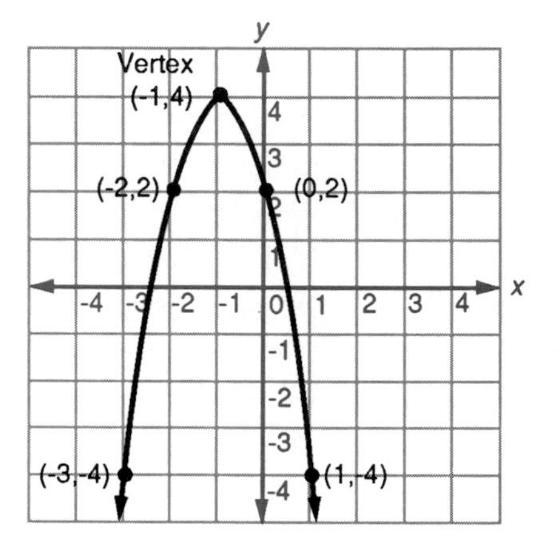

■

Can we tell whether a parabola is curved upward or downward without graphing it?

Yes, Charlie, **when the coefficient of x^2 is positive, the parabola curves upward** (as in Example 1); **when the coefficient of x^2 is negative, the parabola curves downward** (as in Example 2).

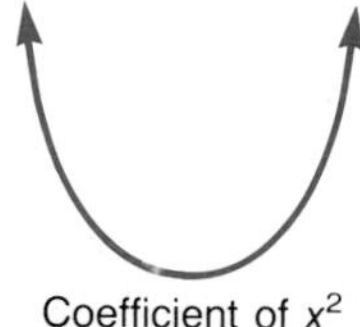

Coefficient of x^2 positive.

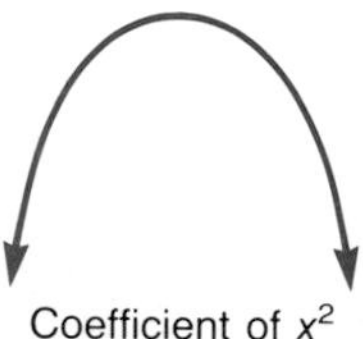

Coefficient of x^2 negative.

Practice 1: Graph the following quadratic equation.

$$y = x^2$$

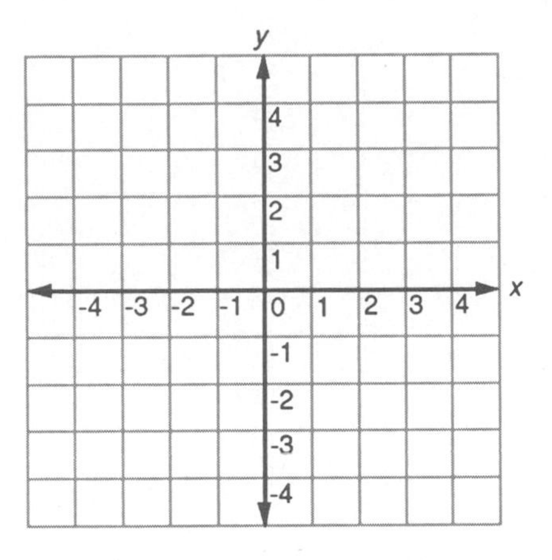

Practice 2: Graph the following quadratic equation.

$$y = x^2 - 2x + 1$$

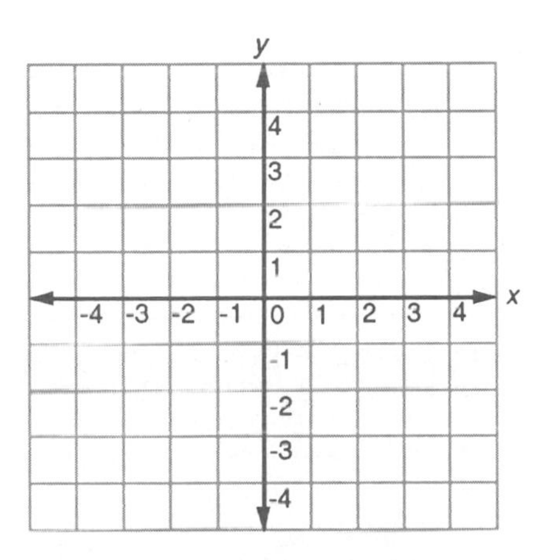

Practice 3: Graph the following quadratic equation.

$$y = -2x^2 + 3$$

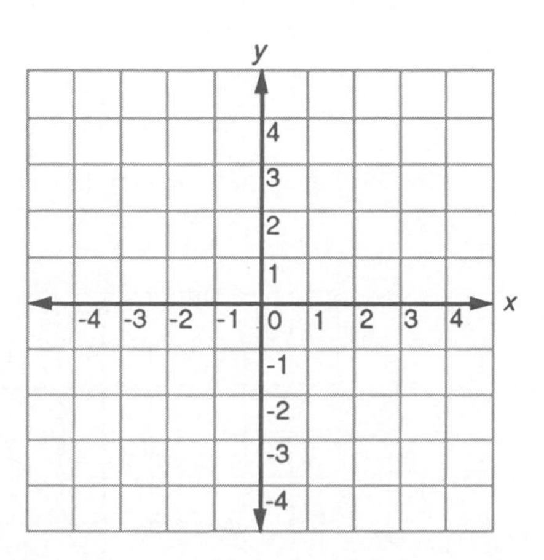

Practice 4: Graph the following quadratic equation.

$$y = -2x^2 + 4x + 1$$

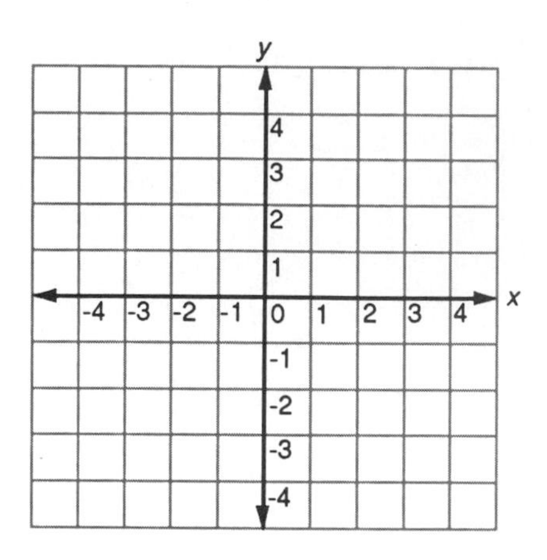

EXERCISE 10.3

Graph the following quadratic equations.

A

1. $y = 2x^2 - 4$
2. $y = -x^2$
3. $y = -3x^2 + 6$
4. $y = 5x^2 - 10$
5. $y = 3x^2$
6. $y = x^2 - 2x + 5$
7. $y = x^2 + 6x - 7$
8. $y = x^2 + 4x - 9$
9. $y = -x^2 + 4x - 5$
10. $y = -x^2 + 6x - 8$

B

11. $y = 5x^2 + 10x - 5$
12. $y = 2x^2 - 8x + 8$
13. $y = -3x^2 + 12x - 4$
14. $y = -2x^2 - 12x - 20$
15. $y = 6x^2 - 18x - 3$
16. $y = 4x^2 - 12x + 1$
17. $y = -5x^2 + 20x - 10$
18. $y = -4x^2 + 16x - 9$
19. $y = 7x^2 - 14x - 3$
20. $y = -2x^2 + 20x - 40$

USEFUL INFORMATION

1. You locate **points in the rectangular coordinate system** using ordered pairs of the form (x,y). If x is a positive number, then (x,y) locates a point x units to the right of the y-axis and $(-x,y)$ locates a point x units to the left of the y-axis. If y is a positive number, then (x,y) locates a point y units up from the x-axis and $(x,-y)$ locates a point y units down from the x-axis.
2. A **linear equation in two variables** is an equation that can be written in the form $ax + by = c$, where a and b are not both equal to zero.
3. The **graph of a linear equation** is a line and it takes two points to determine a line.
4. To locate two points on a line, you find two ordered pairs which satisfy the equation of the line. For example, using $x + y = 3$, $(1,2)$ and $(0,3)$ satisfy the equation since $1 + 2 = 3$ and $0 + 3 = 3$. Thus, these two ordered pairs locate two points on the line.
5. The graph of a linear equation of the form $x = a$ is a **vertical line** through $(a,0)$. For example, the graph of $x = -4$ is a vertical line through $(-4,0)$.
6. The graph of a linear equation of the form $y = b$ is a **horizontal line** through $(0,b)$. For example, the graph of $y = 2$ is a horizontal line through $(0,2)$.
7. The **slope** (steepness of the line) is the ratio of the rise of the line (change in the vertical distance) to the run of the line (change in the horizontal distance); that is,

$$\text{Slope} = \frac{\text{Rise}}{\text{Run}} = \frac{\text{Change in vertical distance}}{\text{Change in horizontal distance}}$$

For example, the following line has a slope of $\frac{3}{2}$.

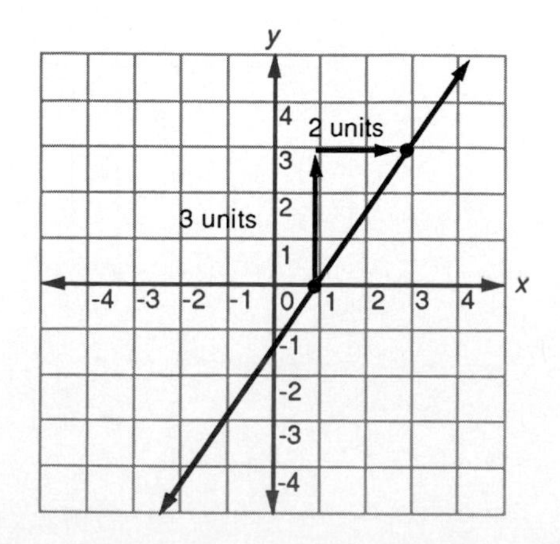

8. The **y-intercept** is the point where the line crosses the y-axis.

9. Using the slope and y-intercept, you have a second method of graphing a linear equation. Write the equation in **slope-intercept form,** $y = mx + b$. Using the slope, m, and the y-intercept, $(0,b)$, you can graph the equation. See Examples 2 and 3 in Section 10.2.

10. The **graph of a quadratic equation** of the form $y = ax^2 + bx + c$ is a U-shaped curve called a **parabola.** These parabolas have one of the following two forms.

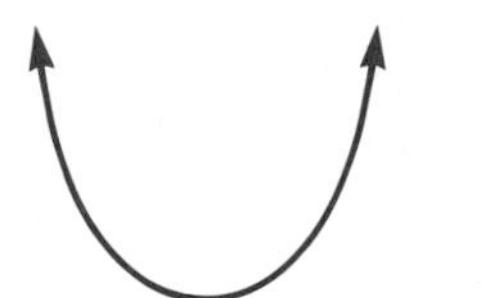

When a is positive, the parabola curves upward.

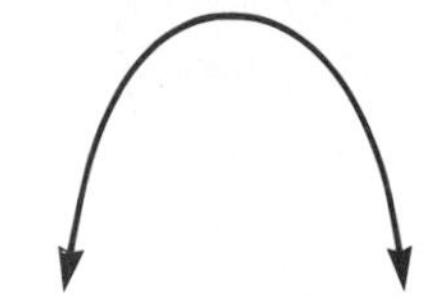

When a is negative, the parabola curves downward.

11. The **vertex** of the parabola in this chapter is the lowest or highest point on the parabola. The x-coordinate of the vertex of the parabola whose equation is $y = ax^2 + bx + c$ is $\frac{-b}{2a}$.

12. When graphing a quadratic equation, you need to locate at least five points—the vertex and two points on each side of the vertex.

REVIEW PROBLEMS

Section 10.1

Graph the following linear equations and check.

1. $x + y = 1$
2. $x + y = -2$
3. $5x + y = -2$
4. $3x - y = -1$
5. $-x + 2y = 7$
6. $-2x + y = -3$
7. $4x + 2y = 6$
8. $5x + 3y = -2$
9. $-2x - 5y = 3$
10. $-4x - 6y = -2$

Section 10.2

Graph the following linear equations using the slope-intercept method and check.

1. $y = 2x + 3$
2. $y = 3x - 1$
3. $y = -4x - 2$
4. $y = -2x + 7$
5. $3x + y = 4$
6. $2x + y = -1$
7. $4x + 2y = 6$
8. $-3x + 2y = 10$
9. $-5x + 3y = -3$
10. $-3x + 4y = -8$

Section 10.3

Graph the following quadratic equations.

1. $y = -2x^2$
2. $y = -3x^2$
3. $y = 4x^2$
4. $y = 5x^2$
5. $y = 2x^2 + 4$
6. $y = -4x^2 + 8$
7. $y = -3x^2 - 2x + 3$
8. $y = 4x^2 - 4x - 3$
9. $y = 4x^2 + 8x - 7$
10. $y = -5x^2 + 10x + 2$

ANSWERS TO PRACTICE PROBLEMS

Section 10.1

1.–7.

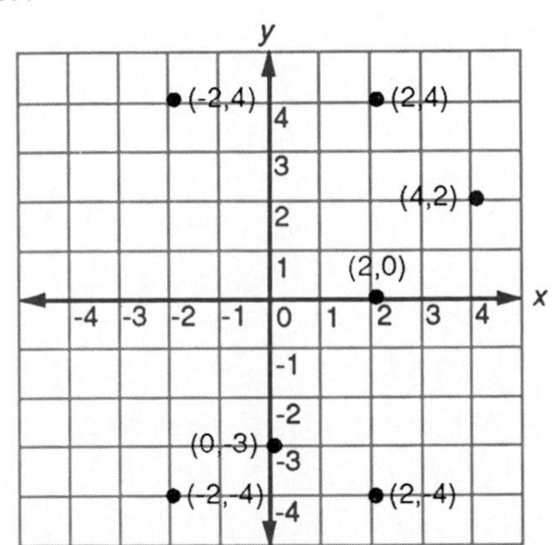

8.

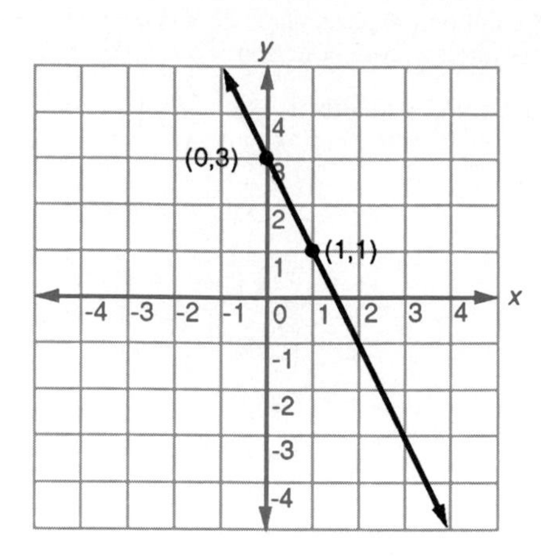
y
x
(0,3)
(1,1)
-4 -3 -2 -1 0 1 2 3 4
4 3 2 1 -1 -2 -3 -4

9.

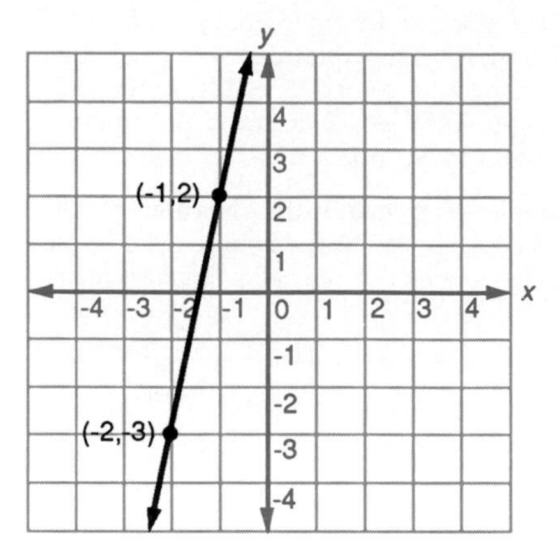
y
x
(-1,2)
(-2,-3)
-4 -3 -2 -1 0 1 2 3 4
4 3 2 1 -1 -2 -3 -4

10.

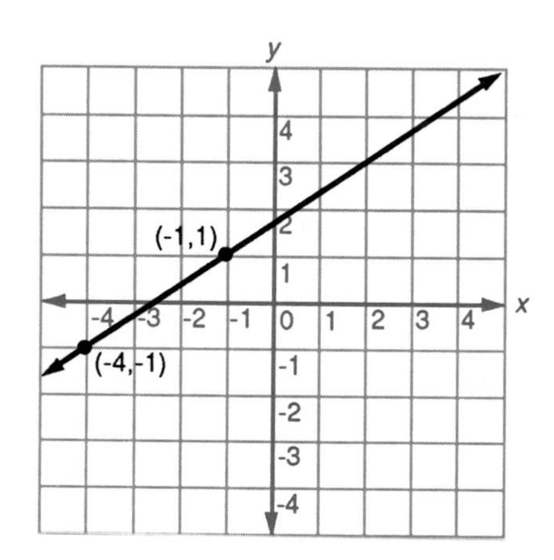
y
x
(-1,1)
(-4,-1)
-4 -3 -2 -1 0 1 2 3 4
4 3 2 1 -1 -2 -3 -4

11.

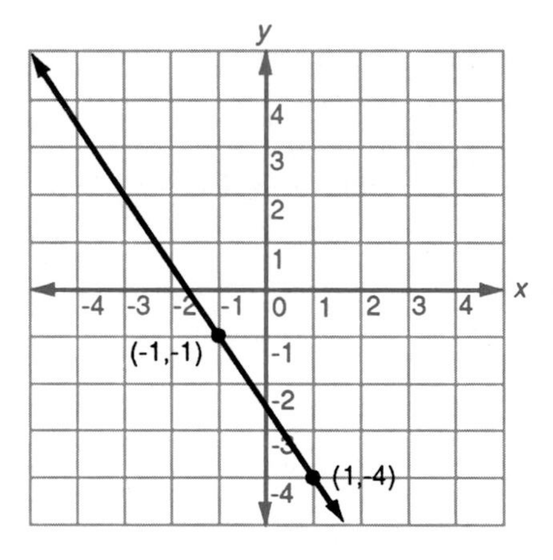
y
x
(-1,-1)
(1,-4)
-4 -3 -2 -1 0 1 2 3 4
4 3 2 1 -1 -2 -3 -4

12.

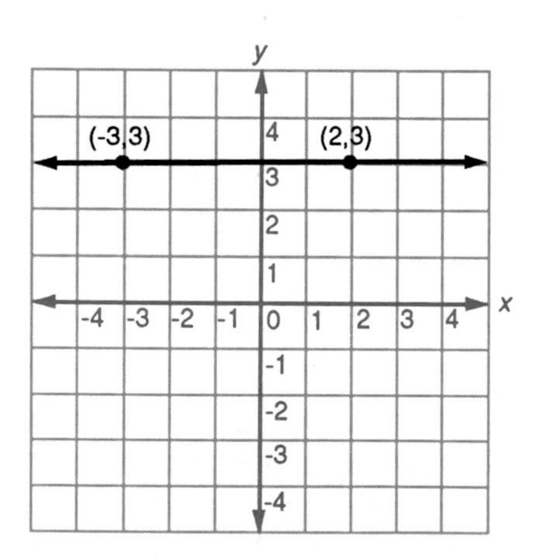
y
x
(-3,3)
(2,3)
-4 -3 -2 -1 0 1 2 3 4
4 3 2 1 -1 -2 -3 -4

13.

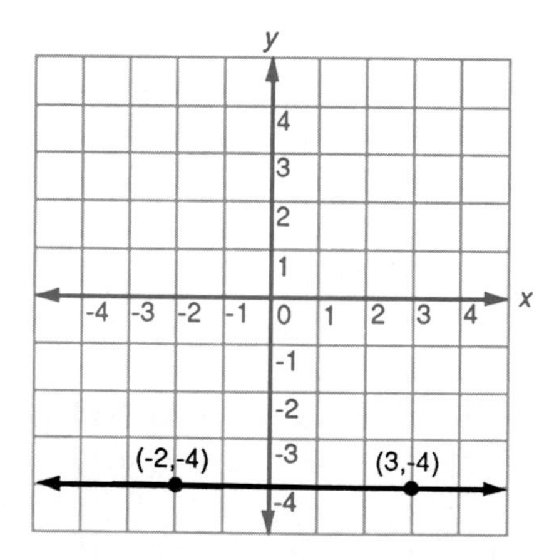
y
x
(-2,-4)
(3,-4)
-4 -3 -2 -1 0 1 2 3 4
4 3 2 1 -1 -2 -3 -4

14.

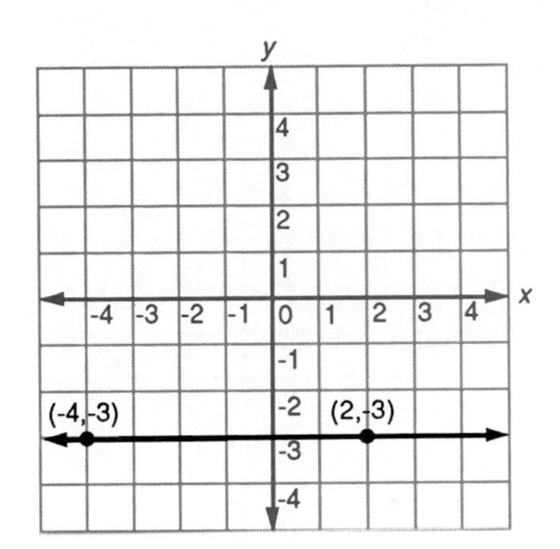

15.

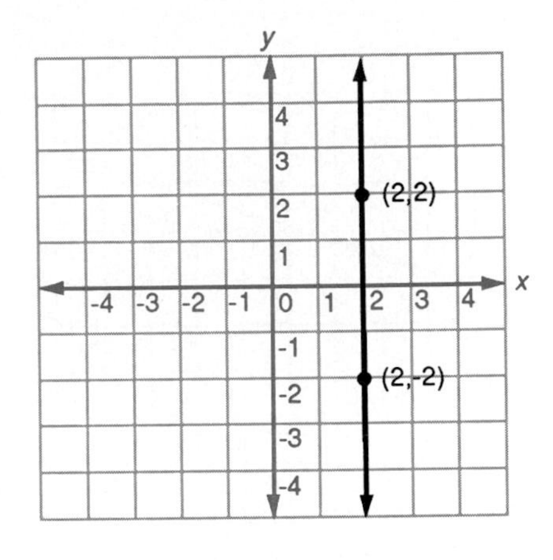

16.

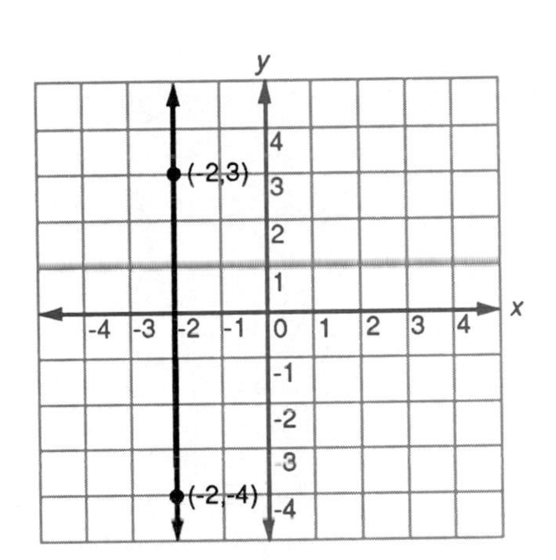

17.

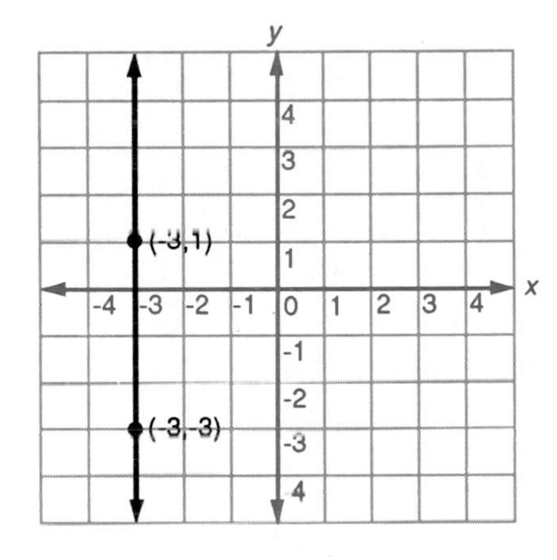

Section 10.2

1. $y = 2x - 1$

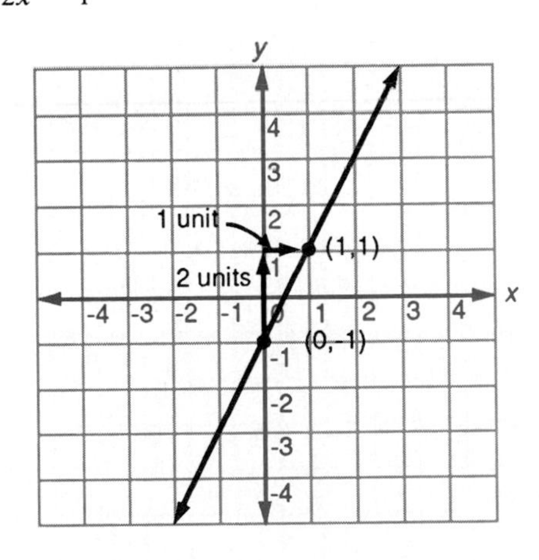

2. $y = -2x + 1$

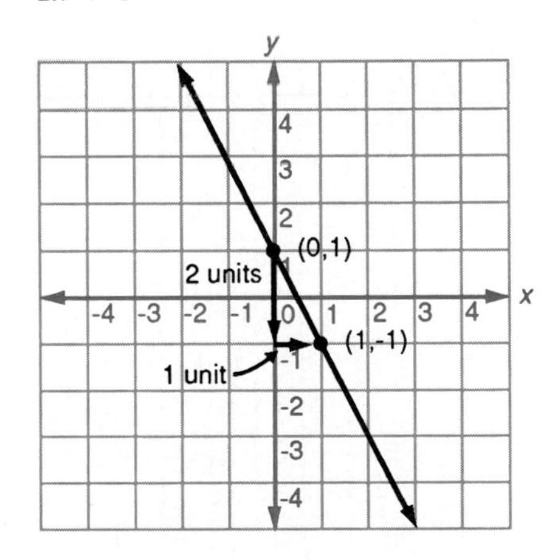

3. $y = \frac{1}{3}x + 2$

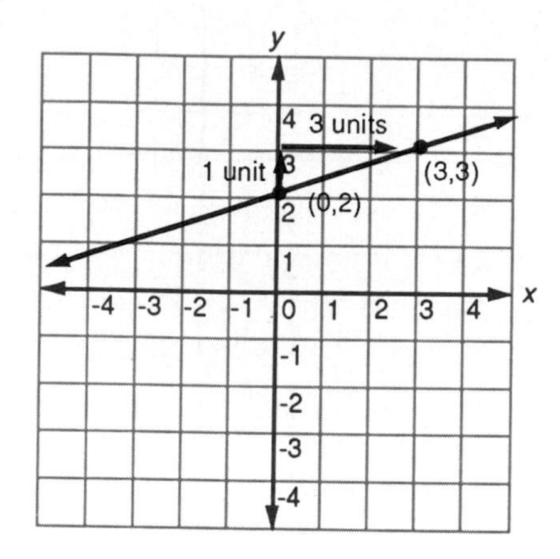

4. $y = -\frac{5}{4}x + 4$

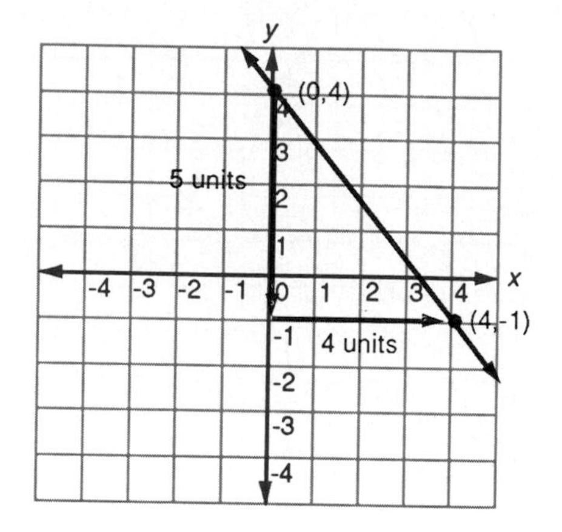

Section 10.3

1.

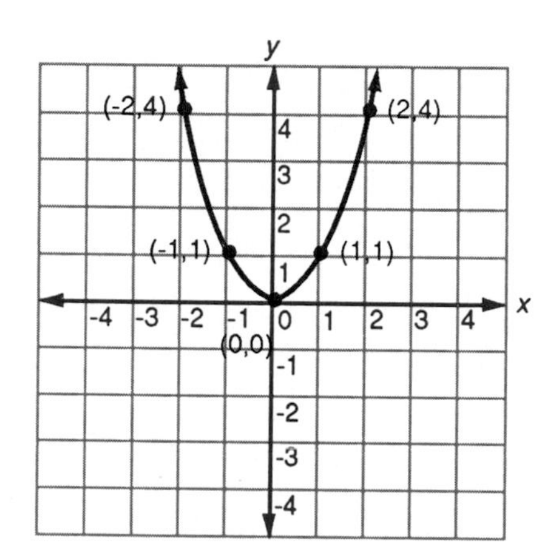

2.

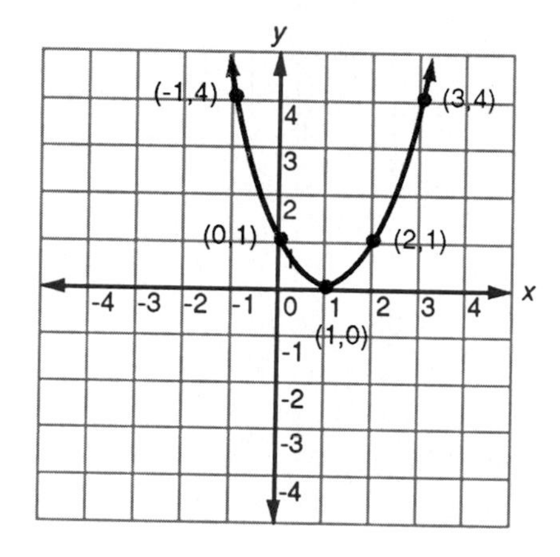

3.

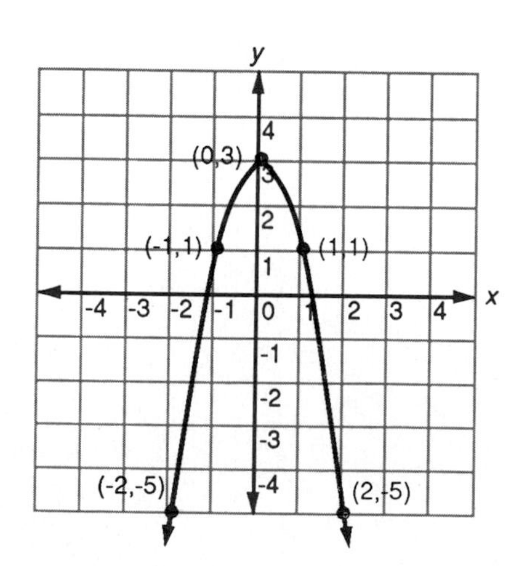

4.

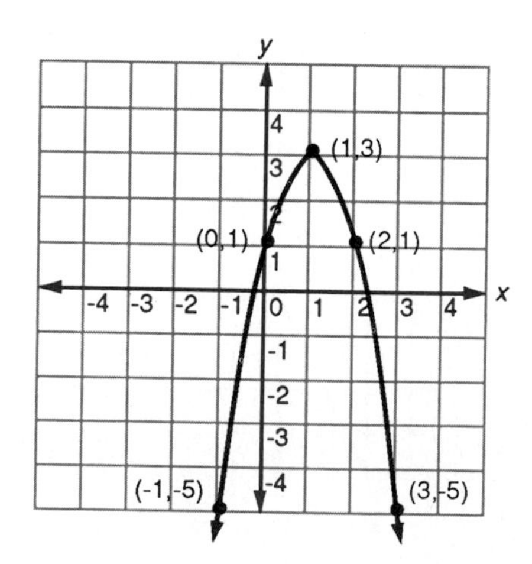

C H A P T E R 11

Solving Systems of Equations

The objective of this chapter is:

1. **To be able to solve certain systems of two equations in two unknowns (variables).**

When working practical problems in the real world, we sometimes have to solve two equations with two unknowns in each equation. This is called solving a **system of equations.** The **solution of a system of equations** is the common solution of the equations in the system.

11.1 SOLVING A SYSTEM OF TWO LINEAR EQUATIONS BY SUBSTITUTION

In Chapter 10, you learned that the graph of a linear equation is a line. Thus, the graph of a system of two linear equations is two lines which can have one point in common, no point in common, or all points in common.

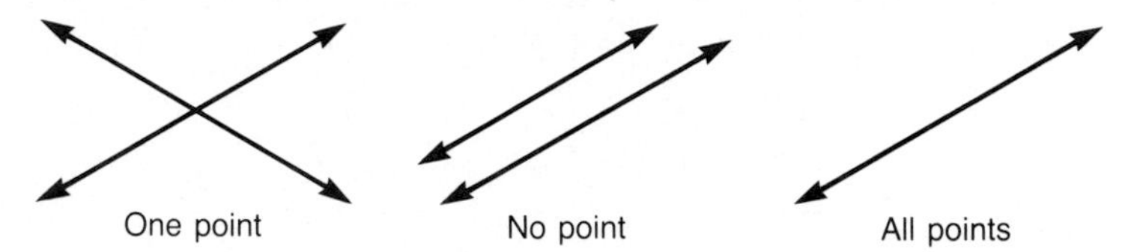

I don't understand how two lines can have all points in common.

Well, Charlie, remember that two equivalent equations have the same solution. Thus, any two equivalent linear equations will graph the same line. Let's look at some examples that explain the three cases.

Example 1 **Graph the following system of linear equations.**

$$2x + y = 5$$
$$-5x + 2y = -8$$

Step 1: Graph $2x + y = 5$.

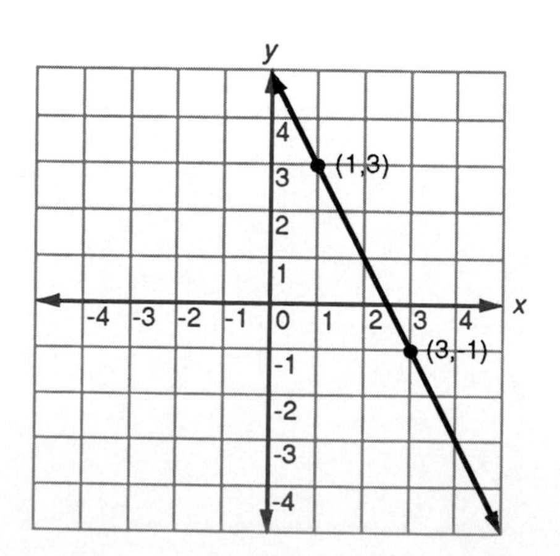

Step 2: Graph $-5x + 2y = -8$ on the same coordinate system.

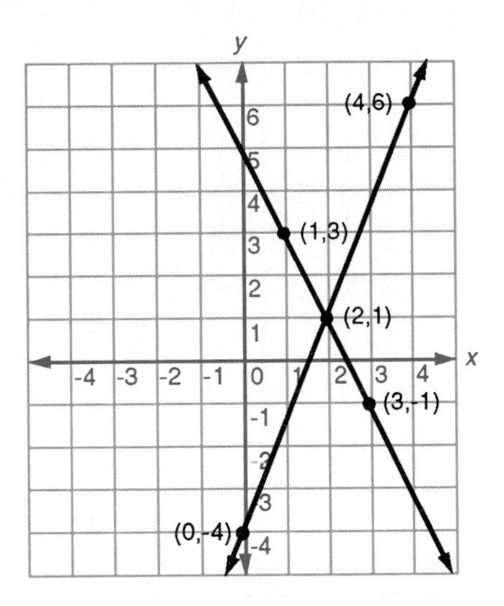

Looking at the graphs of the two equations, the two lines intersect at (2,1). Thus, the two equations have one common solution, $x = 2$ and $y = 1$.

Note: Your selections of x and thus your ordered pairs may be different from the example, but you should get the same lines and point of intersection.

Check: Check to see if (2,1) satisfies the original equations.

$$2x + y = 5 \qquad -5x + 2y = -8$$
$$2(2) + 1 \stackrel{?}{=} 5 \qquad -5(2) + 2(1) \stackrel{?}{=} -8$$
$$4 + 1 \stackrel{?}{=} 5 \qquad -10 + 2 \stackrel{?}{=} -8$$
$$5 = 5 \qquad -8 = -8$$

The check verifies that (2,1) is the solution of the system of equations. ■

Example 2 **Graph the following system of equations.**

$$3x + y = 7$$
$$6x + 2y = 8$$

Step 1: Graph $3x + y = 7$.

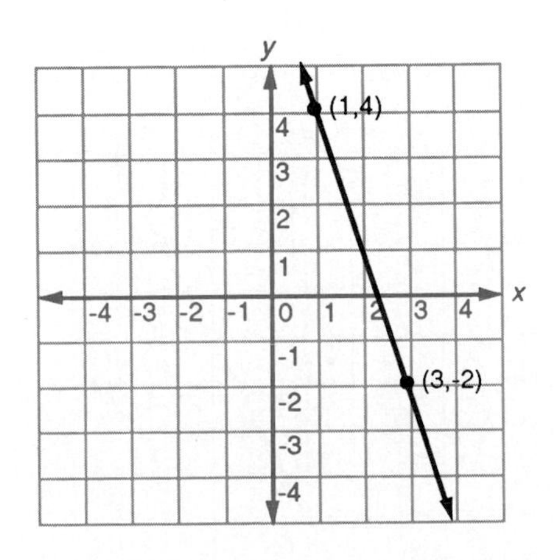

Step 2: Graph $6x + 2y = 8$ on the same coordinate system.

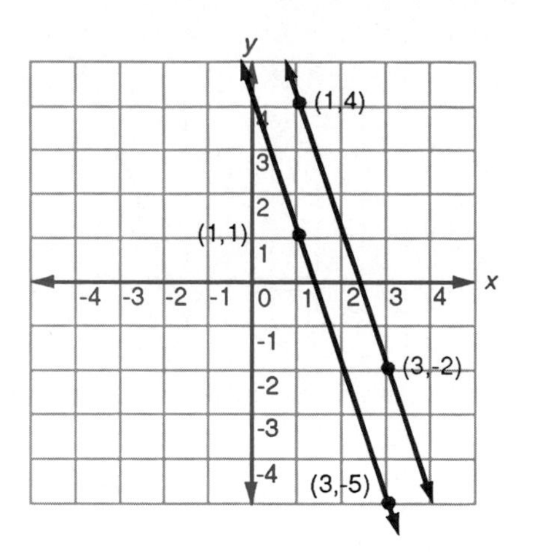

Looking at the graphs of the two equations, the two lines do not intersect. Thus, the two equations do not have a common solution; that is, there is no ordered pair (x,y) which satisfies both equations. ■

Example 3 **Graph the following system of equations.**

$$-2x + y = -3$$
$$-4x + 2y = -6$$

Step 1: Graph $-2x + y = -3$.

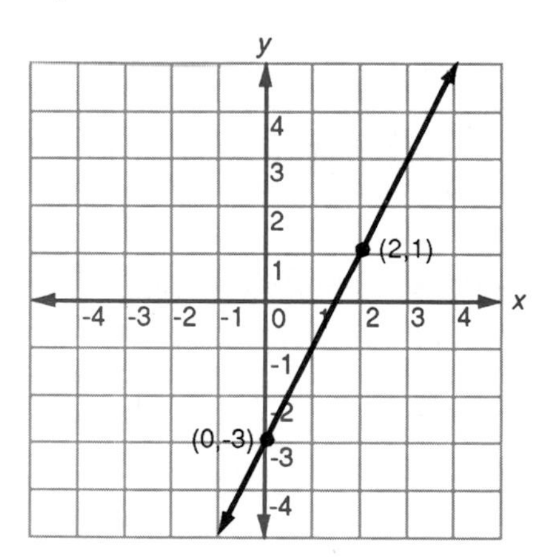

Step 2: Graph $-4x + 2y = -6$ on the same coordinate system.

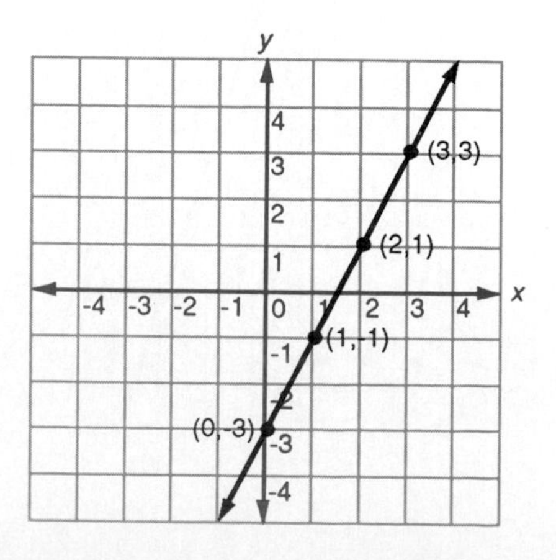

Looking at the graphs of the two equations, we have the same line. Thus, the two equations have all solutions in common. For example, $(0,-3)$, $(1,-1)$, $(2,1)$, and $(3,3)$ satisfy both equations and locate points on the same line. ■

So, we can find the solution of a system of linear equations by graphing the equations and locating the point of intersection.

That's correct, Charlie. If a system of equations has only one solution, then you can find this solution by locating the point of intersection of the graphs (lines). Otherwise, the graphs will show that the system does not have a solution (parallel lines) or has many solutions (the same line). You can solve a system of linear equations by graphing; however, graphing is time-consuming and does not always give a completely accurate answer. Therefore, we usually solve systems of two linear equations in two variables by other methods. One method is by eliminating one of the variables through substitution. When one of the variables is eliminated, you have a linear equation in one variable which you learned to solve in Chapter 4.

Let's look at an example of how to solve a system of linear equations by the method of substitution.

Example 4 **Solve the following system of linear equations by substitution.**

$$\text{Equation 1: } 2x + y = 4$$

$$\text{Equation 2: } 3x - y = 6$$

Step 1: Select one of the equations and solve for x or y. Let's use Equation 1 and solve for y.

$2x + y = 4$ **Equation 1.**

$y = -2x + 4$ **Solve for y.**

Does it matter which equation and variable we select?

No, Charlie. But whichever equation and variable we select in Step 1, we must substitute for that variable in the other equation in Step 2.

Step 2: Substitute $-2x + 4$ for y in Equation 2.

$3x - y = 6$ **Equation 2.**

$3x - (\mathbf{-2x + 4}) = 6$ **Substitute $-2x + 4$ for y.**

Now we have a linear equation with only one variable.

That's correct, Charlie. We now solve this equation for x.

Step 3: Solve the resulting linear equation.

$3x - (-2x + 4) = 6$	**Equation from Step 2.**
$3x + 2x - 4 = 6$	**Remove parentheses.**
$5x - 4 = 6$	**Combine like terms.**
$5x = 6 + 4$	**Add 4 to both sides.**
$5x = 10$	**Simplify.**
$x = 2$	**Divide both sides by 5.**

So, the solution is 2.

Well, Charlie, part of the solution is 2. Remember that we have two variables in each equation, x and y; thus, we must find the value of y.

Step 4: Substitute 2 for x in the equation $y = -2x + 4$ and find the value of y.

$y = -2x + 4$	**Equation from Step 1.**
$y = -2(\mathbf{2}) + 4$	**Substitute 2 for x.**
$y = -4 + 4$	**Perform the multiplication.**
$y = 0$	**Perform the addition.**

Check: Substitute 2 for x and 0 for y in the original equations.

Equation 1	Equation 2
$2x + y = 4$	$3x - y = 6$
$2(\mathbf{2}) + \mathbf{0} \stackrel{?}{=} 4$	$3(\mathbf{2}) - \mathbf{0} \stackrel{?}{=} 6$
$4 + 0 \stackrel{?}{=} 4$	$6 - 0 \stackrel{?}{=} 6$
$4 = 4$	$6 = 6$

Now, $x = 2$ and $y = 0$ satisfy the original equations; thus, (2,0) is the solution of the system of equations. ■

I have some questions. First, when finding the value of y in Step 4, why did we substitute in the equation from Step 1 and not in one of the original equations?

Well, Charlie, in Step 1 we solved Equation 1, $2x + y = 4$, for y, getting $y = -2x + 4$. Now, it is easier to substitute for x and find the value of y in an equation that has already been solved for y.

Why did we check the solution in both equations?

The solution of a system of equations must satisfy all equations in the system.

My last question is how do we decide which equation to select and which variable to solve for in Step 1?

Well, Charlie, as stated earlier you can select either equation and solve for either variable. However, we usually select the equation that contains a variable with a coefficient of 1 and solve for that variable. If none of the variables has a coefficient of 1, then we simply pick Equation 1 and solve for y. Let's look at some examples.

Example 5 **Select one of the following equations and solve for one of the variables.**

Equation 1: $5x + y = 3$

Equation 2: $3x - 5y = 7$

Solution: Select Equation 1 and solve for y.

$$5x + y = 3$$

$$y = -5x + 3$$

■

So, if one of the variables has a coefficient of 1, then we select that equation and solve for that variable.

That's correct, Charlie.

Example 6 **Select one of the following equations and solve for one of the variables.**

Equation 1: $3x + 4y = 8$

Equation 2: $x - 7y = 12$

Solution: Select Equation 2 and solve for x.

$$x - 7y = 12$$
$$x = 7y + 12$$

■

We selected Equation 2 and solved for x *because the coefficient of* x *was 1.*

That's correct, Charlie.

Example 7 **Select one of the following equations and solve for one of the variables.**

Equation 1: $-6x + 3y = 9$

Equation 2: $12x - 4y = 8$

Solution: Select Equation 1 and solve for y.

$$-6x + 3y = 9$$
$$3y = 6x + 9$$
$$y = 2x + 3$$

■

How did we know to select Equation 1 and solve for y*?*

Well, Charlie, this is one of those cases when there is no clear choice as to which equation and variable to select. In a case such as this, we usually select Equation 1 and solve for y.

Example 8 **Solve the following system of linear equations by substitution.**

Equation 1: $5x + 6y = 7$

Equation 2: $2x + 12y = 6$

Step 1: Select one of the equations and solve for x or y. Let's use Equation 1 and solve for y.

$5x + 6y = 7$	**Equation 1.**
$6y = 7 - 5x$	**Subtract $5x$ from both sides.**
$y = \frac{7 - 5x}{6}$	**Divide both sides by 6.**

Step 2: Substitute $\frac{7-5x}{6}$ for y in Equation 2 and solve for x.

$2x + 12y = 6$	**Equation 2.**
$2x + 12\left(\frac{7 - 5x}{6}\right) = 6$	**Substitute $\frac{7-5x}{6}$ for y.**
$2x + 2(7 - 5x) = 6$	**Simplify.**
$2x + 14 - 10x = 6$	**Remove parentheses.**
$14 - 8x = 6$	**Combine like terms.**
$-8x = 6 - 14$	**Subtract 14 from both sides.**
$-8x = -8$	**Simplify.**
$x = 1$	**Divide both sides by -8.**

Step 3: Substitute 1 for x in $y = \frac{7-5x}{6}$ and find y.

$y = \frac{7 - 5x}{6}$	**Equation from Step 1.**
$y = \frac{7 - 5(\mathbf{1})}{6}$	**Substitute 1 for x.**
$y = \frac{7 - 5}{6}$	**Perform the multiplication.**
$y = \frac{2}{6}$	**Perform the subtraction.**
$y = \frac{1}{3}$	**Reduce.**

Check: Substitute 1 for x and $\frac{1}{3}$ for y in the original equations.

Equation 1	Equation 2
$5x + 6y = 7$	$2x + 12y = 6$
$5(\mathbf{1}) + 6\left(\frac{\mathbf{1}}{\mathbf{3}}\right) \stackrel{?}{=} 7$	$2(\mathbf{1}) + 12\left(\frac{\mathbf{1}}{\mathbf{3}}\right) \stackrel{?}{=} 6$
$5 + 2 \stackrel{?}{=} 7$	$2 + 4 \stackrel{?}{=} 6$
$7 = 7$	$6 = 6$

Now, $x = 1$ and $y = \frac{1}{3}$ satisfy the original equations; thus, $(1, \frac{1}{3})$ is the solution of the system of equations. ■

A N S W E R

1. ____________________

Practice 1: Solve the following system of linear equations by substitution and check your answer.

$$-5x + y = 6$$
$$2x + 3y = 1$$

2. ____________________

Practice 2: Solve the following system of linear equations by substitution and check your answer.

$$x + 3y = 8$$
$$2x - 4y = 6$$

3. ____________________

Practice 3: Solve the following system of linear equations by substitution and check your answer.

$$7x + 2y = 3$$
$$5x + 2y = -1$$

Practice 4: Solve the following system of linear equations by substitution and check your answer.

$$-9x + 4y = 5$$
$$-3x + 8y = -5$$

ANSWER

4. ______________

Now, let's look at an example of a system of linear equations that does not have a solution.

Example 9 **Solve the following system of linear equations by substitution.**

Equation 1: $3x + 4y = 1$

Equation 2: $9x + 12y = 12$

Step 1: Select one of the equations and solve for x or y. Let's use Equation 1 and solve for y.

$3x + 4y = 1$	**Equation 1.**
$4y = 1 - 3x$	**Subtract $3x$ from both sides.**
$y = \dfrac{1 - 3x}{4}$	**Divide both sides by 4.**

Step 2: Substitute $\frac{1-3x}{4}$ for y in Equation 2 and solve for x.

$9x + 12y = 12$	**Equation 2.**
$9x + 12\left(\dfrac{1 - 3x}{4}\right) = 12$	**Substitute $\frac{1-3x}{4}$ for y.**
$9x + 3(1 - 3x) = 12$	**Simplify.**
$9x + 3 - 9x = 12$	**Remove parentheses.**
$3 = 12$	**Combine like terms.**

■

What happened? Three does not equal 12.

That's true, Charlie. This means that it is not possible for the two original equations to have a common solution. Thus, this system of equations does *not* have a solution.

So, anytime we get a statement that is not true, such as $3 = 12$ when solving a system of equations, we know the system does not have a solution.

That's correct, Charlie. If you want to check your work, you can graph the two equations. When the graph is two parallel lines—two lines that do not intersect—you know the two equations do not have a common solution. Now, let's look at an example of a system of linear equations that has many solutions.

Example 10 **Solve the following system of linear equations by substitution.**

Equation 1: $-2x + 5y = -3$

Equation 2: $-4x + 10y = -6$

Step 1: Select one of the equations and solve for x or y. Let's use Equation 1 and solve for y.

$$-2x + 5y = -3 \qquad \textbf{Equation 1.}$$

$$5y = -3 + 2x \qquad \textbf{Add } 2x \textbf{ to both sides.}$$

$$y = \frac{-3 + 2x}{5} \qquad \textbf{Divide both sides by 5.}$$

Step 2: Substitute $\frac{-3 + 2x}{5}$ for y in Equation 2 and solve for x.

$$-4x + 10y = -6 \qquad \textbf{Equation 2.}$$

$$-4x + 10\left(\frac{-3 + 2x}{5}\right) = -6 \qquad \textbf{Substitute } \frac{-3 + 2x}{5} \textbf{ for } y.$$

$$-4x + 2(-3 + 2x) = -6 \qquad \textbf{Simplify.}$$

$$-4x - 6 + 4x = -6 \qquad \textbf{Remove parentheses.}$$

$$-6 = -6 \qquad \textbf{Combine like terms.}$$

■

What happened? Everybody knows $-6 = -6$.

That's true, Charlie. This means that every solution of one equation is also a solution of the other equation. Thus, the system of equations has *many* solutions.

ANSWER

So, anytime we get a statement that is always true, such as $-6 = -6$ when solving a system of equations, we know the system has many solutions.

That's true, Charlie. If you want to check your work, you can graph the two equations. When the graph is the same line, you know the two equations have all solutions in common.

Practice 5: Solve the following system of linear equations by substitution.

$$-3x + 4y = 2$$
$$-6x + 8y = -5$$

5. ____________

Practice 6: Solve the following system of linear equations by substitution.

$$4x + 2y = 5$$
$$-16x + 8y = 20$$

6. ____________

Practice 7: Solve the following system of linear equations by substitution.

$$-14x + 22y = 15$$
$$-7x + 11y = 25$$

7. ____________

ANSWER

8. ______________

Practice 8: Solve the following system of linear equations by substitution.

$$-24x + 28y = 16$$
$$-6x + 7y = 4$$

Now, let's summarize how to solve a system of two linear equations by the substitution method.

To solve a system of two linear equations by substitution:

1. Select one of the equations and solve for one of the variables in terms of the other.
2. Substitute the solution from Step 1 in the other equation to get a linear equation in one variable and solve the equation.
   a. If you get a statement that is not true, such as $3 = 5$, then the system of equations has no solution.
   b. If you get a statement that is always true, such as $0 = 0$, then the system of equations has many solutions.
   c. If you get a statement such as $x = 5$ or $y = 7$, then the system of equations has one solution and you proceed to Step 3.
3. Substitute the solution from Step 2 in the equation from Step 1 to find the value of the second variable.
4. Check your answer by substituting it in both of the original equations (optional).

EXERCISE 11.1

Solve the following system of linear equations by substitution.

A

1. $y = x - 2$
   $5x + 2y = 10$

2. $y = 2x + 5$
   $-3x + 7y = 2$

3. $x + 5y = -2$
   $4x + 8y = 10$

4. $6x + y = 8$
   $12x + 3y = 6$

5. $-7x + y = 3$
$10x - 2y = -2$

6. $-x + 5y = 4$
$-5x + 3y = -2$

7. $3x + y = 2$
$6x + 2y = -5$

8. $-6x + y = -3$
$36x - 6y = 18$

9. $-3x - y = 5$
$15x - 5y = -5$

10. $-8x - y = -4$
$-20x - 2y = 0$

11. $x + 7y = -3$
$2x + 6y = 10$

12. $-11x + y = 8$
$24x - 3y = -15$

13. $x - 7y = 0$
$5x - 35y = 0$

14. $7x + y = -5$
$-21x - 3y = 7$

B

15. $9x + 3y = 12$
    $3x + 4y = 7$

16. $-10x + 5y = 15$
    $-4x + 3y = -11$

17. $3x + 2y = 6$
    $5x + 2y = 4$

18. $-5x + 3y = 7$
    $-12x + 6y = 8$

19. $-4x + 5y = -3$
    $6x - 10y = -6$

20. $-15x + 12y = 5$
    $20x - 11y = 5$

11.2 SOLVING A SYSTEM OF TWO LINEAR EQUATIONS BY ADDITION

Another method of solving a system of two linear equations in two variables makes use of the addition principle to eliminate one of the variables. Some systems of equations are easier to solve using the addition principle.

The addition principle: If equal quantities are added to equal quantities, the sums are equal. That is,

if	$a = b$
and	$c = d$
then	$a + c = b + d$

For example,

if $x = 3$

and $y = 2$

then $x + y = 3 + 2$

or $x + y = 5$

Now, let's look at an example of how to solve a system of linear equations using the addition principle.

Example 1 **Solve the following system of linear equations using the addition principle.**

Equation 1: $5x + y = 7$

Equation 2: $2x - y = 0$

Step 1: Make the x or y term in one equation the opposite of the corresponding term in the other equation. The y term in Equation 2 is the opposite of the y term in Equation 1.

$$5x + y = 7$$
$$2x - y = 0$$

Why do we want the x *or* y *terms in the two equations to be opposites of each other?*

Well, Charlie, when we add the two equations in Step 2, we want to eliminate one of the variables. If we have variable terms that are opposites, then when we add the two equations we eliminate that variable.

Step 2: Add the two equations using the addition principle.

$$\begin{array}{r} 5x + y = 7 \\ \underline{2x - y = 0} \\ 7x \qquad = 7 \end{array}$$

Oh! I see why we want variable terms to be opposites. The y *term disappeared when we added.*

A N S W E R

That's correct, Charlie. The resulting equation contains only the variable x.

Step 3: Solve the resulting linear equation.

$7x = 7$	**Equation from Step 2.**
$x = 1$	**Divide both sides by 7.**

Step 4: Substitute 1 for x in one of the original equations and solve for y. Let's use Equation 1.

$5x + y = 7$	**Equation 1.**
$5(\mathbf{1}) + y = 7$	**Substitute 1 for x.**
$5 + y = 7$	**Perform the multiplication.**
$y = 7 - 5$	**Subtract 5 from both sides.**
$y = 2$	**Simplify.**

Check: Substitute 1 for x and 2 for y in the original equations.

Equation 1	Equation 2
$5x + y = 7$	$2x - y = 0$
$5(\mathbf{1}) + \mathbf{2} \stackrel{?}{=} 7$	$2(\mathbf{1}) - \mathbf{2} \stackrel{?}{=} 0$
$5 + 2 \stackrel{?}{=} 7$	$2 - 2 \stackrel{?}{=} 0$
$7 = 7$	$0 = 0$

Now, $x = 1$ and $y = 2$ satisfy the original equations; thus, (1,2) is the solution of the system of equations. ■

1. 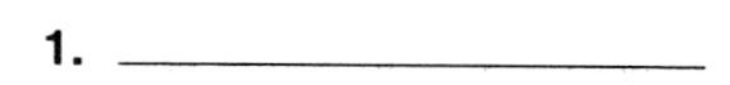

Practice 1: Solve the following system of linear equations using the addition principle.

$$5x + 2y = 3$$
$$5x - 2y = 7$$

2. ____________________

Practice 2: Solve the following system of linear equations using the addition principle.

$$-3x + 7y = 8$$
$$3x - 2y = 2$$

Example 2 **Solve the following system of linear equations using the addition principle.**

Equation 1: $3x + 2y = -12$

Equation 2: $x + 3y = 3$

Step 1: Make the x or y term in one equation the opposite of the corresponding term in the other equation. Let's make the x term in Equation 2 the opposite of the x term in Equation 1.

Why did we decide to make the x *term in Equation 2 the opposite of the* x *term in Equation 1?*

Well, Charlie, the coefficient of x in Equation 2 is 1. Now, when we multiply Equation 2 by -3, the x terms will be opposites. Multiply the second equation by -3.

$x + 3y = 3$	**Equation 2.**
$-3(x + 3y) = -3(3)$	**Multiply both sides by −3.**
$-3x - 9y = -9$	**Simplify.**

This gives us the following two equations.

$$3x + 2y = -12$$
$$-3x - 9y = -9$$

Now, when we add the equations, the x variable will be eliminated.

Step 2: Add the two equations from Step 1 using the addition principle.

$$\begin{array}{r} -3x + 2y = -12 \\ \underline{-3x - 9y = -\ \ 9} \\ -7y = -21 \end{array}$$

I see. We always want one of the variables to disappear when we add the two equations.

That's correct, Charlie.

Step 3: Solve the resulting linear equation.

$-7y = -21$	**Equation from Step 2.**
$y = 3$	**Divide both sides by −7.**

Step 4: Substitute 3 for y in one of the original equations and solve for x. Let's use Equation 2.

$x + 3y = 3$	**Equation 2.**
$x + 3(3) = 3$	**Substitute 3 for y.**
$x + 9 = 3$	**Perform the multiplication.**
$x = 3 - 9$	**Subtract 9 from both sides.**
$x = -6$	**Simplify.**

How do we decide which original equation to use when we substitute?

Well, Charlie, we can use either of the original equations; however, we try to select the equation that will be easier to solve for the remaining variable. We selected the second equation this time because the coefficient of x was 1, thus, making it easier to solve for x.

Check: Substitute -6 for x and 3 for y in the original equations.

Equation 1	Equation 2
$3x + 2y = -12$	$x + 3y = 3$
$3(\mathbf{-6}) + 2(\mathbf{3}) \stackrel{?}{=} -12$	$\mathbf{-6} + 3(\mathbf{3}) \stackrel{?}{=} 3$
$-18 + 6 \stackrel{?}{=} -12$	$-6 + 9 \stackrel{?}{=} 3$
$-12 = -12$	$3 = 3$

Now, $x = -6$ and $y = 3$ satisfy the original equations; thus, $(-6,3)$ is the solution of the system of equations. ■

Example 3 **Solve the following system of linear equations using the addition principle.**

$$2y = 3x + 5$$
$$4x - 6y = 10$$

Step 1: Write the first equation in the form $ax + by = c$.

Equation 1: $-3x + 2y = 5$

Equation 2: $4x - 6y = 10$

Step 2: Make the x or y term in one equation the opposite of the corresponding term in the other equation. Let's make the y term in Equation 1 the opposite of the y term in Equation 2.

Why did we decide to make the y *term in Equation 1 the opposite of the* y *term in Equation 2?*

Well, Charlie, the coefficient of y in Equation 2 is a multiple of the coefficient of y in Equation 1. Now, when we multiply Equation 1 by 3, we will have the y terms opposites. Multiply Equation 1 by 3.

$-3x + 2y = 5$	**Equation 1.**
$3(-3x + 2y) = 3(5)$	**Multiply both sides by 3.**
$-9x + 6y = 15$	**Perform the multiplications.**

This gives us the following two equations.

$$-9x + 6y = 15$$
$$4x - 6y = 10$$

Now, when we add the equations the y variable will be eliminated.

Step 3: Add the two equations from Step 2 using the addition principle.

$$\begin{array}{r} -9x + 6y = 15 \\ 4x - 6y = 10 \\ \hline -5x \quad\quad = 25 \end{array}$$

Step 4: Solve the resulting linear equation.

$-5x = 25$	**Equation from Step 3.**
$x = -5$	**Divide both sides by −5.**

Step 5: Substitute -5 for x in one of the original equations and solve for y. Let's use Equation 1.

$-3x + 2y = 5$	**Equation 1.**
$-3(-5) + 2y = 5$	**Substitute −5 for *x*.**
$15 + 2y = 5$	**Perform the multiplication.**
$2y = 5 - 15$	**Subtract 15 from both sides.**
$2y = -10$	**Simplify.**
$y = -5$	**Divide both sides by 2.**

Check: Substitute -5 for x and -5 for y in the original equations.

Equation 1	Equation 2
$2y = 3x + 5$	$4x - 6y = 10$
$2(\mathbf{-5}) \stackrel{?}{=} 3(\mathbf{-5}) + 5$	$4(\mathbf{-5}) - 6(\mathbf{-5}) \stackrel{?}{=} 10$
$-10 \stackrel{?}{=} -15 + 5$	$-20 + 30 \stackrel{?}{=} 10$
$-10 = -10$	$10 = 10$

Now, $x = -5$ and $y = -5$ satisfy the original equations; thus, $(-5,-5)$ is the solution of the system of equations. ■

A N S W E R

3. ______________

Practice 3: Solve the following system of linear equations using the addition principle.

$$-3x + y = -8$$
$$8x - 2y = 6$$

4. ______________

Practice 4: Solve the following system of linear equations using the addition principle.

$$5x - 15y = 10$$
$$x - 4y = 0$$

5. ______________

Practice 5: Solve the following system of linear equations using the addition principle.

$$3x + 2y = 6$$
$$5x - 4y = 12$$

6. ______________

Practice 6: Solve the following system of linear equations using the addition principle.

$$7x + 3y = 6$$
$$6x + 9y = 3$$

Example 4 **Solve the following system of linear equations using the addition principle.**

Equation 1: $-5x + 3y = -4$

Equation 2: $4x - 2y = 6$

Step 1: Make the x or y term in one equation the opposite of the corresponding term in the other equation. Let's make the y terms opposites.

Why did we decide to make the y *terms opposites?*

Well, Charlie, neither the x term nor the y term in Equation 1 is a multiple of the corresponding term in Equation 2 and neither the x term nor the y term in Equation 2 is a multiple of the corresponding term in Equation 1. Thus, there is no simple way to make the x or y terms opposites. We will have to go through a process similar to that of finding the least common denominator of two fractions. We picked the y terms in this case simply because the coefficients of y were smaller than the coefficients of x.

Multiply Equation 1 by 2.

$-5x + 3y = -4$	**Equation 1.**
$2(-5x + 3y) = 2(-4)$	**Multiply both sides by 2.**
$-10x + 6y = -8$	**Perform the multiplications.**

Multiply Equation 2 by 3.

$4x - 2y = 6$	**Equation 2.**
$3(4x - 2y) = 3(6)$	**Multiply both sides by 3.**
$12x - 6y = 18$	**Perform the multiplications.**

We now have the following two equations.

$$-10x + 6y = -8$$

$$12x - 6y = 18$$

We had to multiply the first equation by 2 and the second equation by 3 in order to make the y *terms opposites.*

Yes, Charlie. We often have to multiply the two equations by different numbers in order to get the x or y terms opposite.

A N S W E R

Step 2: Add the two equations from Step 1 using the addition principle.

$$\begin{array}{rcr} -10x + 6y &=& -\ 8 \\ 12x - 6y &=& 18 \\ \hline 2x \qquad &=& 10 \end{array}$$

Step 3: Solve the resulting linear equation.

$2x = 10$ **Equation from Step 2.**

$x = 5$ **Divide both sides by 2.**

Step 4: Substitute 5 for x in one of the original equations and solve for y. Let's use Equation 1.

$-5x + 3y = -4$ **Equation 1.**

$-5(5) + 3y = -4$ **Substitute 5 for x.**

$-25 + 3y = -4$ **Perform the multiplication.**

$3y = -4 + 25$ **Add 25 to both sides.**

$3y = 21$ **Simplify.**

$y = 7$ **Divide both sides by 3.**

Check: Substitute 5 for x and 7 for y in the original equations.

Equation 1	Equation 2
$-5x + 3y = -4$	$4x - 2y = 6$
$-5(\mathbf{5}) + 3(\mathbf{7}) \stackrel{?}{=} -4$	$4(\mathbf{5}) - 2(\mathbf{7}) \stackrel{?}{=} 6$
$-25 + 21 \stackrel{?}{=} -4$	$20 - 14 \stackrel{?}{=} 6$
$-4 = -4$	$6 = 6$

Now, $x = 5$ and $y = 7$ satisfy the original equations; thus, (5,7) is the solution of the system of equations. ■

7. ______________

Practice 7: Solve the following system of linear equations using the addition principle.

$$2x - 5y = 7$$
$$-3x + 6y = 3$$

Practice 8: Solve the following system of linear equations using the addition principle.

$$4x - 3y = -1$$
$$-7x + 5y = 2$$

ANSWER

8. ______________

Example 5 **Solve the following system of linear equations using the addition principle.**

Equation 1: $4x + 12y = 7$

Equation 2: $6x + 18y = 5$

Step 1: Make the x or y term in one equation the opposite of the corresponding term in the other equation. Let's make the x terms opposites. Multiply Equation 1 by 3.

$4x + 12y = 7$	**Equation 1.**
$3(4x + 12y) = 3(7)$	**Multiply both sides by 3.**
$12x + 36y = 21$	**Perform the mulitplications.**

Multiply Equation 2 by -2.

$6x + 18y = 5$	**Equation 2.**
$-2(6x + 18y) = -2(5)$	**Multiply both sides by -2.**
$-12x - 36y = -10$	**Perform the multiplications.**

We now have the following two equations.

$$12x + 36y = 21$$
$$-12x - 36y = -10$$

How did we know which numbers to multiply the equations by?

Well, Charlie, we wanted to make the x terms opposites—that is, we wanted to get the two x terms equal except for their signs. The coefficient of x in Equation 1 is 4 and the coefficient of x in Equation 2 is 6. The method used to find which number to multiply Equation 1 by and which number to multiply Equation 2 by is the same method used to find the least common denominator of two fractions. Let's review this method.
Completely factor 4 and 6.

$$4 = 2 \times 2 \qquad 6 = 2 \times 3$$

In order for two numbers to be equal, they must contain exactly the same factors when completely factored. Since 4 does not have a factor of 3, we multiply 4 by 3. Since 6 only has one factor of 2, we multiply 6 by 2.

$$4 = 2 \times 2 \qquad\qquad 6 = 2 \times 3$$

$$4 \times 3 = 2 \times 2 \times 3 \qquad\qquad 2 \times 6 = 2 \times 2 \times 3$$

$$12 = 2 \times 2 \times 3 \qquad\qquad 12 = 2 \times 2 \times 3$$

Thus, we decided to multiply Equation 1 by 3 and Equation 2 by -2 so the x terms would be opposites.

Step 2: Add the two equations from Step 1 using the addition principle.

$$\begin{array}{r} 12x + 36y = 21 \\ -12x - 36y = -10 \\ \hline 0 = 11 \end{array}$$

I remember from the last section that anytime we get a statement that is not true, such as 0 = 11, the system does not have a solution.

That's correct, Charlie. This system has no solution. ■

Example 6 **Solve the following system of linear equations using the addition principle.**

$$8x = 4y + 16$$

$$10x - 5y = 20$$

Step 1: Write the first equation in the form $ax + by = c$.

Equation 1: $8x - 4y = 16$

Equation 2: $10x - 5y = 20$

Step 2: Make the x or y term in one equation the opposite of the corresponding term in the other equation. Let's make the y terms opposites. Multiply Equation 1 by 5.

$8x - 4y = 16$ **Equation 1.**

$5(8x - 4y) = 5(16)$ **Multiply both sides by 5.**

$40x - 20y = 80$ **Perform the multiplications.**

ANSWER

Multiply Equation 2 by -4.

$$10x - 5y = 20 \quad \textbf{Equation 2.}$$

$$-4(10x - 5y) = -4(20) \quad \textbf{Multiply both sides by } \mathbf{-4.}$$

$$-40x + 20y = -80 \quad \textbf{Perform the multiplications.}$$

We now have the following two equations.

$$40x - 20y = 80$$

$$-40x + 20y = -80$$

Step 3: Add the two equations from Step 2 using the addition principle.

$$\begin{array}{r} 40x - 20y = 80 \\ -40x + 20y = -80 \\ \hline 0 = 0 \end{array}$$

If we get a statement that is always true, such as 0 = 0, does that mean the system has many solutions?

Yes, Charlie. Every solution of one equation is also a solution of the other equation. Thus, the system of equations has many solutions. ■

Practice 9: Solve the following system of linear equations using the addition principle.

$$4x - 6y = -1$$

$$-8x + 12y = 2$$

9. ____________

Practice 10: Solve the following system of linear equations using the addition principle.

$$6x - 3y = 5$$

$$12x - 6y = 2$$

10. ____________

A N S W E R

11. ____________

Practice 11: Solve the following system of linear equations using the addition principle.

$$6x - 4y = 8$$
$$-15x + 10y = -10$$

12. ____________

Practice 12: Solve the following system of linear equations using the addition principle.

$$6x - 9y = -15$$
$$8x - 12y = -20$$

Now, let's summarize how to solve a system of two linear equations by the addition method.

To solve a system of two linear equations by addition:

1. Write the equations in the form of $ax + by = c$.
2. Make the x or y term in one equation the opposite of the corresponding term in the other equation.
3. Eliminate one of the variables by adding the two equations.
4. Solve the resulting linear equation in one variable.
   a. If you get a statement that is not true, such as $3 = 5$, then the system of equations has no solution.
   b. If you get a statement that is always true, such as $0 = 0$, then the system of equations has many solutions.
   c. If you get a statement such as $x = 5$ or $y = 7$, then the system of equations has one solution and you proceed to Step 5.
5. Substitute the solution from Step 4 in one of the original equations and solve for the other variable.
6. Check your answer by substituting it in both of the original equations (optional).

EXERCISE 11.2

Solve the following systems of linear equations using the addition principle.

1. $x + y = 2$
$x - y = 0$

2. $3x + y = -1$
$2x - y = -4$

3. $3x - 2y = 4$
$4x + 2y = 10$

4. $5x + 3y = 7$
$x - 3y = 5$

5. $-2x + 5y = 3$
$-6x + 15y = 7$

6. $7x - 5y = 0$
$14x - 10y = 0$

7. $2x + 3y = 2$
$3x - y = 3$

8. $4x - 5y = 6$
$2x + 10y = -2$

9. $-6x + 20y = 8$
$3x - 10y = -4$

10. $-4x - 5y = 5$
$8x + 10y = -6$

11. $3x + 5y = -3$
$x + 2y = 2$

12. $4x + 7y = -10$
$y = -3x + 1$

13. $x - 3y = -1$
$5x = 2y + 21$

14. $12x - 3y = 3$
$2x - y = -5$

B

15. $5y = -2x - 2$
$6x + 2y = 20$

16. $5x - 7y = 8$
$4x - 14y = -2$

17. $4x - 6y = -8$
$6x - 5y = 4$

18. $8x - 12y = 5$
$8y = 5x - 2$

19. $15x + 11y = 0$
$10x = -12y - 14$

20. $-12x + 16y = 6$
$18x - 24y = -9$

11.3 SOLVING A SYSTEM OF ONE LINEAR EQUATION AND ONE QUADRATIC EQUATION

In this section, you will learn how to solve a system of one linear equation and one quadratic equation of the form $y = ax^2 + bx + c$.

In Chapter 10, you learned that the graph of a linear equation is a line and the graph of a quadratic equation of the form $y = ax^2 + bx + c$ is a parabola. Thus, the graph of a system of one linear equation and one quadratic equation of the form $y = ax^2 + bx + c$ is a line and a parabola. A line and a parabola can have two points in common, one point in common, or no point in common.

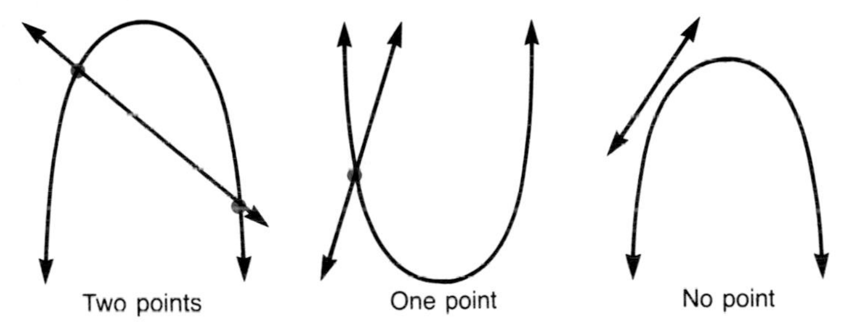

Example 1 **Graph the following system of equations.**

$$y = 2x + 2$$
$$y = x^2 + 4x - 1$$

<u>*Step 1:*</u> Graph $y = x^2 + 4x - 1$.

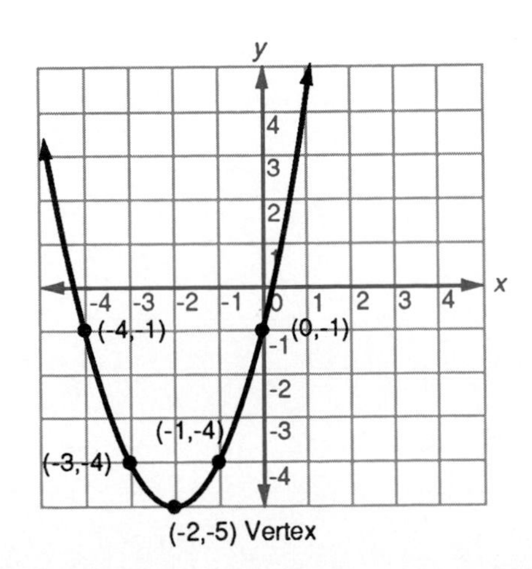

Step 2: Graph $y = 2x + 2$ on the same coordinate system.

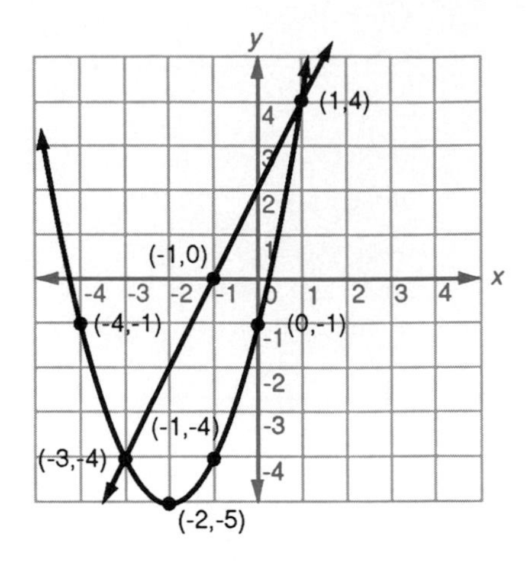

Looking at the graph of the two equations, the line and parabola intersect at two points located by (1,4) and $(-3,-4)$. Thus, the two equations have two common solutions, $x = 1$ and $y = 4$; $x = -3$ and $y = -4$. ■

Example 2 **Graph the following system of equations.**

$$y = -2x - 2$$
$$y = x^2 - 1$$

Step 1: Graph $y = x^2 - 1$.

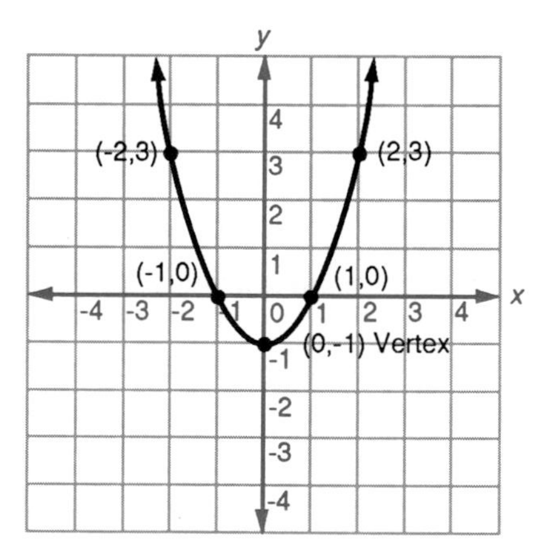

Step 2: Graph $y = -2x - 2$ on the same coordinate system.

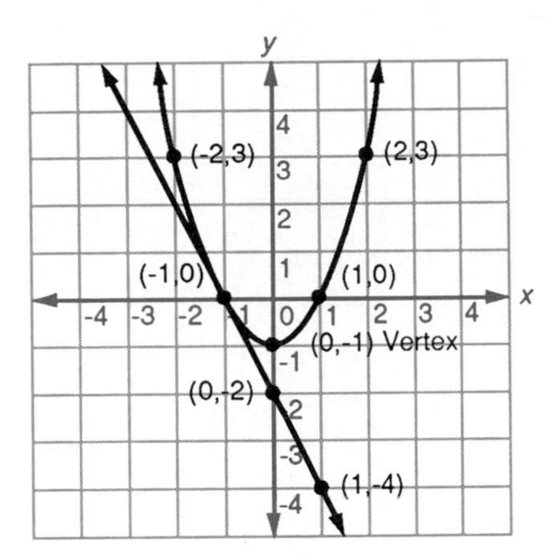

Looking at the graph of the two equations, the line and parabola intersect at the point located by $(-1,0)$. Thus, the two equations have only one common solution, $x = -1$ and $y = 0$. ■

Example 3 **Graph the following system of equations.**

$$y = 2x + 4$$
$$y = -2x^2 + 3$$

Step 1: Graph $y = -2x^2 + 3$.

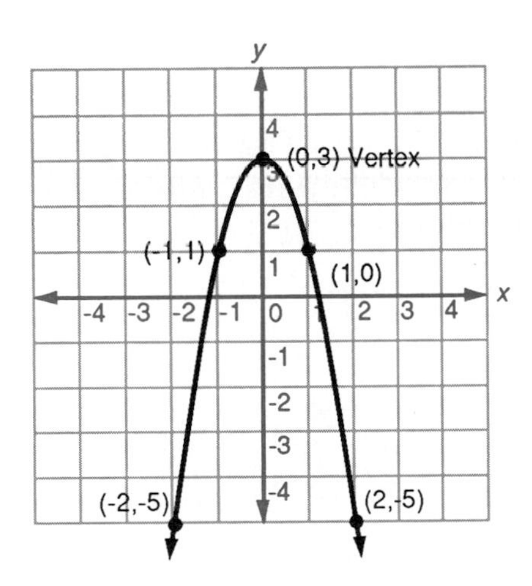

Step 2: Graph $y = 2x + 4$ on the same coordinate system.

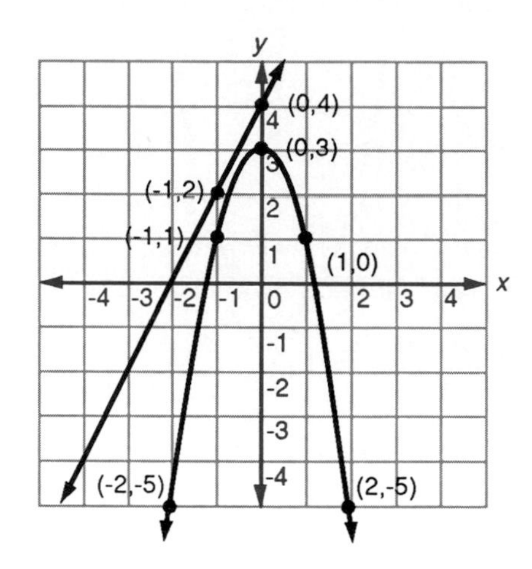

Looking at the graph of the two equations, the line and parabola do not intersect. Thus, the two equations do not have a common solution. ■

So, we can find the solution of a system of one linear and one quadratic equation by graphing the equations and locating the points of intersection.

That's true, Charlie. But just like a system of two linear equations, graphing does not always give an accurate answer. Besides, it takes too much time. Therefore, we will solve a system of one linear and one quadratic equation by substitution. Let's look at an example.

Example 4 **Solve the following system of equations by substitution.**

$$y - x = -1$$
$$y = x^2 + 2x - 3$$

Step 1: Solve the linear equation for y.

$y - x = -1$ **Original linear equation.**

$y = x - 1$ **Add x to both sides.**

Step 2: Substitute $x - 1$ for y in the quadratic equation.

$y = x^2 + 2x - 3$ **Original quadratic equation.**

$x - 1 = x^2 + 2x - 3$ **Substitute $x - 1$ for y.**

Why did we substitute x $-$ *1 for* y*?*

Well, Charlie, when we substitute $x - 1$ for y, the original quadratic equation in two variables becomes a quadratic equation in one variable. You can solve this quadratic equation using the techniques of Chapter 9.

Step 3: Solve the resulting quadratic equation.

$x - 1 = x^2 + 2x - 3$	**Equation from Step 2.**
$0 = x^2 + 2x - 3 - (x - 1)$	**Subtract $x - 1$ from both sides.**
$0 = x^2 + 2x - 3 - x + 1$	**Remove parentheses.**
$0 = x^2 + x - 2$	**Combine like terms.**
$0 = (x + 2)(x - 1)$	**Factor.**
$x + 2 = 0$ or $x - 1 = 0$	**Zero product rule.**

Solve the two linear equations.

$$x + 2 = 0 \qquad x - 1 = 0$$
$$x = -2 \qquad x = 1$$

Step 4: Substitute -2 and 1 for x in the linear equation $y = x - 1$ from Step 1.

Substitute -2 for x.	Substitute 1 for x.
$y = x - 1$	$y = x - 1$
$y = -2 - 1$	$y = 1 - 1$
$y = -3$	$y = 0$

Thus, we have $x = -2$ and $y = -3$; $x = 1$ and $y = 0$.

Why did you substitute -2 and 1 for x *in* y = x $-$ *1 and not in one of the original equations?*

The equation $y = x - 1$ had already been solved for y in terms of x, thus making the substitutions easier.

Check: Substitute $x = -2$ and $y = -3$, and $x = 1$ and $y = 0$ in the original equations.

Substitute $x = -2$ and $y = -3$ in the original equations.

$$\begin{aligned} y - x &= -1 \\ -3 - (-2) &\stackrel{?}{=} -1 \\ -3 + 2 &\stackrel{?}{=} -1 \\ -1 &= -1 \end{aligned} \qquad \begin{aligned} y &= x^2 + 2x - 3 \\ -3 &\stackrel{?}{=} (-2)^2 + 2(-2) - 3 \\ -3 &\stackrel{?}{=} 4 - 4 - 3 \\ -3 &= -3 \end{aligned}$$

Now, $x = -2$ and $y = -3$ satisfy the original equations; thus, $(-2,-3)$ is one solution of the system of equations.
Substitute $x = 1$ and $y = 0$ in the original equations.

$$\begin{aligned} y - x &= -1 \\ 0 - 1 &\stackrel{?}{=} -1 \\ -1 &= -1 \end{aligned} \qquad \begin{aligned} y &= x^2 + 2x - 3 \\ 0 &\stackrel{?}{=} (1)^2 + 2(1) - 3 \\ 0 &\stackrel{?}{=} 1 + 2 - 3 \\ 0 &= 0 \end{aligned}$$

Now, $x = 1$ and $y = 0$ satisfy the original equations; thus, $(1,0)$ is a second solution of the system of equations. ■

Example 5 **Solve the following system of equations by substitution.**

$$10x + 2y = 6$$
$$y = 10x^2 - 6x + 1$$

Step 1: Solve the linear equation for y.

$10x + 2y = 6$	**Original linear equation.**
$2y = -10x + 6$	**Subtract $10x$ from both sides.**
$y = \dfrac{-10x + 6}{2}$	**Divide both sides by 2.**
$y = -5x + 3$	**Perform the division.**

Do we always solve the linear equation for y?

Usually, Charlie, since the quadratic equation is of the form $y = ax^2 + bx + c$, it is easier to substitute for y than x in the quadratic equation.

Step 2: Substitute $-5x + 3$ for y in the quadratic equation.

$y = 10x^2 - 6x + 1$	**Original quadratic equation.**
$-5x + 3 = 10x^2 - 6x + 1$	**Substitute $-5x + 3$ for y.**

Step 3: Solve the resulting quadratic equation.

$-5x + 3 = 10x^2 - 6x + 1$	**Equation from Step 2.**
$0 = 10x^2 - 6x + 1 - (-5x + 3)$	**Subtract $-5x + 3$ from both sides.**
$0 = 10x^2 - 6x + 1 + 5x - 3$	**Remove parentheses.**
$0 = 10x^2 - x - 2$	**Combine like terms.**
$0 = (2x - 1)(5x + 2)$	**Factor.**
$2x - 1 = 0$ or $5x + 2 = 0$	**Zero product rule.**

Solve the two linear equations.

$$2x - 1 = 0 \qquad 5x + 2 = 0$$
$$2x = 1 \qquad 5x = -2$$
$$x = \frac{1}{2} \qquad x = \frac{-2}{5}$$

Step 4: Substitute $\frac{1}{2}$ and $\frac{-2}{5}$ for x in the linear equation $y = -5x + 3$ from Step 1.

Substitute $\frac{1}{2}$ for x.

$$y = -5x + 3$$
$$y = -5\left(\frac{1}{2}\right) + 3$$
$$y = \frac{-5}{2} + \frac{6}{2}$$
$$y = \frac{1}{2}$$

Substitute $\frac{-2}{5}$ for x.

$$y = -5x + 3$$
$$y = -5\left(\frac{-2}{5}\right) + 3$$
$$y = 2 + 3$$
$$y = 5$$

Thus, we have $x = \frac{1}{2}$ and $y = \frac{1}{2}$; $x = \frac{-2}{5}$ and $y = 5$.

Check: Substitute $x = \frac{1}{2}$ and $y = \frac{1}{2}$, and $x = \frac{-2}{5}$ and $y = 5$ in the original equations.

Substitute $x = \frac{1}{2}$ and $y = \frac{1}{2}$ in the original equations.

$$10x + 2y = 6$$
$$10\left(\frac{1}{2}\right) + 2\left(\frac{1}{2}\right) \stackrel{?}{=} 6$$
$$5 + 1 \stackrel{?}{=} 6$$
$$6 = 6$$

$$y = 10x^2 - 6x + 1$$
$$\frac{1}{2} \stackrel{?}{=} 10\left(\frac{1}{2}\right)^2 - 6\left(\frac{1}{2}\right) + 1$$
$$\frac{1}{2} \stackrel{?}{=} 10\left(\frac{1}{4}\right) - 3 + 1$$
$$\frac{1}{2} \stackrel{?}{=} 2\frac{1}{2} - 3 + 1$$
$$\frac{1}{2} \stackrel{?}{=} 2\frac{1}{2} - 2$$
$$\frac{1}{2} = \frac{1}{2}$$

Now, $x = \frac{1}{2}$ and $y = \frac{1}{2}$ satisfy the original equations; thus, $(\frac{1}{2}, \frac{1}{2})$ is one solution of the system of equations.

ANSWER

Substitute $x = -\frac{2}{5}$ and $y = 5$ in the original equations.

$$10x + 2y = 6 \qquad y = 10x^2 - 6x + 1$$

$$10\left(\frac{-2}{5}\right) + 2(5) \stackrel{?}{=} 6 \qquad 5 \stackrel{?}{=} 10\left(\frac{-2}{5}\right)^2 - 6\left(\frac{-2}{5}\right) + 1$$

$$-4 + 10 \stackrel{?}{=} 6 \qquad 5 \stackrel{?}{=} 10\left(\frac{4}{25}\right) + \frac{12}{5} + 1$$

$$6 = 6 \qquad 5 \stackrel{?}{=} \frac{8}{5} + \frac{12}{5} + 1$$

$$5 \stackrel{?}{=} \frac{20}{5} + 1$$

$$5 \stackrel{?}{=} 4 + 1$$

$$5 = 5$$

Now, $x = \frac{-2}{5}$ and $y = 5$ satisfy the original equations; thus, $(\frac{-2}{5}, 5)$ is a second solution of the system of equations. ■

Now, let's summarize how to solve a system of one linear and one quadratic equation by the substitution method.

To solve a system of one linear and one quadratic equation by substitution:

1. Solve the linear equation for y in terms of x.
2. Substitute the solution of the linear equation for y in the quadratic equation.
3. Solve the quadratic equation in one variable to find the value(s) of x.
   a. If the quadratic equation in one variable has no solution, then the system of equations has no solution.
   b. If the quadratic equation in one variable has one solution, then the system of equations has one solution.
   c. If the quadratic equation in one variable has two solutions, then the system of equations has two solutions.
4. Substitute the solution(s) of the quadratic equation for x in the linear equation from Step 1 to find the value(s) of y.
5. Check your answer(s) by substituting the values of x and y in both of the original equations (optional).

1. ______________________

Practice 1: Solve the following system of equations by substitution.

$$x + y = 5$$

$$y = x^2 - 2x - 1$$

Practice 2: Solve the following system of equations by substitution.

$$y = 2x + 5$$
$$y = x^2 - x - 5$$

ANSWER

2. ____________________

Practice 3: Solve the following system of equations by substitution.

$$y = 3x$$
$$y = 4x^2 - x + 1$$

3. ____________________

Practice 4: Solve the following system of equations by substitution.

$$-3x + y = 4$$
$$y = 6x^2 - 2x - 2$$

4. ____________________

EXERCISE 11.3

Solve the following systems of equations by substitution.

1. $y = 2x$
$y = x^2 - 3x + 6$

2. $y = -x$
$y = x^2 - 2$

3. $y = 2x - 1$
$y = x^2$

4. $y = x - 5$
$y = x^2 - 2x - 9$

5. $x + y = 3$
$y = x^2 + 7x + 10$

6. $3x + y = 1$
$y = x^2 - 2x - 19$

7. $-2x + y = 3$
$y = x^2 + 2x + 7$

8. $-5x + y = 1$
$y = x^2 + x + 4$

9. $5x - y = 7$
$y = x^2 + 8x - 17$

10. $-2x - y = -3$
$y = x^2 - 12x + 28$

B

11. $-3x + y = -1$
$y = 2x^2 + 2x - 4$

12. $-4x + 2y = 6$
$y = 10x^2 + x$

13. $-6x + 3y = 12$
$y = 15x^2 - 19x + 10$

14. $-10x - 5y = 15$
$y = 2x^2 + 11x - 10$

15. $-20x + 10y = -30$
$y = 25x^2 - 18x + 1$

11.4 STRATEGIES FOR SOLVING APPLICATIONS

Sometimes it is easier to solve a word problem with two unknowns by using a system of two linear equations.

Example 1 **There are 28 students in a math class. The number of juniors is 24 less than three times the number of seniors. How many juniors and seniors are in the class?**

Solution: Let x = number of juniors.

Let y = number of seniors.

Since there are 28 students in the class we have our first equation.

Equation 1: $x + y = 28$

Since the number of juniors is 24 less than three times the number of seniors, we have to add 24 to the number of juniors in order to get three times the number of seniors.

Equation 2: $x + 24 = 3y$

We now have our system of equations.

Equation 1: $x + y = 28$

Equation 2: $x + 24 = 3y$

Let's solve this system using the substitution method.

Step 1: Solve Equation 1 for x.

$$x + y = 28$$
$$x = 28 - y$$

Step 2: Substitute $28 - y$ for x in Equation 2 and solve the resulting equation for y.

$$x + 24 = 3y$$
$$(28 - y) + 24 = 3y$$
$$28 - y + 24 = 3y$$
$$-y + 52 = 3y$$
$$52 = 3y + y$$
$$52 = 4y$$
$$13 = y$$
$$y = 13$$

Step 3: Substitute 13 for y in the equation $x = 28 - y$ and find the value of x.

$$x = 28 - y$$
$$x = 28 - 13$$
$$x = 15$$

Thus, there are 15 juniors and 13 seniors. ■

Example 2 **A collection of quarters and dimes contains 44 coins and has a total value of \$6.50. How many dimes and quarters are in the collection?**

Solution: Let x = number of dimes.

Let y = number of quarters.

Since there 44 coins, we have our first equation.

$$\text{Equation 1: } x + y = 44$$

The value of the dimes is $.10x$ and the value of the quarters is $.25y$. Since the total value is 6.50 we have our second equation.

$$\text{Equation 2: } .10x + .25y = 6.50$$

We now have our system of equations.

$$\text{Equation 1: } x + y = 44$$
$$\text{Equation 2: } .10x + .25y = 6.50$$

Let's solve this system using the substitution method.

Step 1: Solve Equation 1 for x.

$$x + y = 44$$
$$x = 44 - y$$

Step 2: Substitute $44 - y$ for x in Equation 2 and solve the resulting equation for y.

$$.10x + .25y = 6.50$$
$$.10(44 - y) + .25y = 6.50$$
$$4.4 - .10y + .25y = 6.50$$
$$4.4 + .15y = 6.50$$
$$.15y = 6.50 - 4.4$$
$$.15y = 2.10$$
$$y = 14$$

Step 3: Substitute 14 for y in the equation $x = 44 - y$ and find the value of x.

$$x = 44 - y$$
$$x = 44 - 14$$
$$x = 30$$

Thus, there are 30 dimes and 14 quarters. ■

Example 3 **A rectangular field is 15 feet longer than it is wide. If the length and width are each increased by 5 feet, then the new area exceeds the old area by 550 square feet. Find the length and width of the original field.**

Solution: Let L = length of the original field.

Let W = width of the original field.

Then $L + 5$ is the length of the new field and $W + 5$ is the width of the new field.

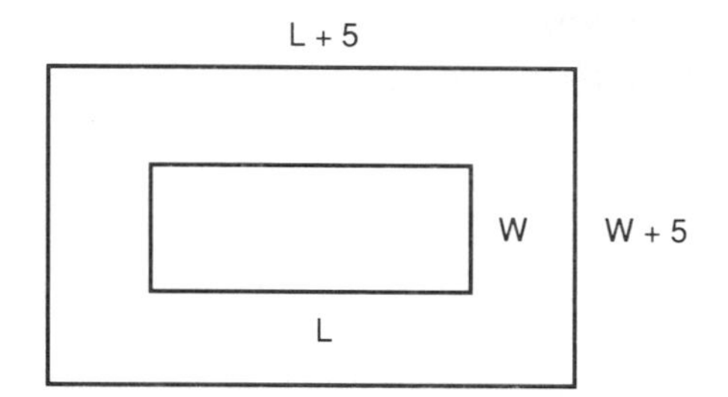

Since the length of the original field is 15 feet longer than the width, we have to add 15 to the width in order to get the length.

Equation 1: $L = W + 15$

The area of a rectangle is length times width, so the area of the original field is LW and the area of the new field is $(L + 5)(W + 5)$. Since the new area exceeds (is more than) the old area by 550 square feet, we have to add 550 to the old area in order to get the new area.

Equation 2: $LW + 550 = (L + 5)(W + 5)$

We now have our system of equations.

Equation 1: $L = W + 15$

Equation 2: $LW + 550 = (L + 5)(W + 5)$

We will solve this system using the substitution method.

Step 1: From Equation 1, we substitute $W + 15$ for L in Equation 2 and solve the resulting equation for W.

$$LW + 550 = (L + 5)(W + 5)$$
$$(W + 15)W + 550 = [(W + 15) + 5](W + 5)$$
$$(W + 15)W + 550 = (W + 20)(W + 5)$$
$$W^2 + 15W + 550 = W^2 + 25W + 100$$
$$W^2 + 15W + 550 - W^2 = 25W + 100$$
$$15W + 550 = 25W + 100$$
$$15W + 550 - 25W = 100$$
$$-10W + 550 = 100$$
$$-10W = 100 - 550$$
$$-10W = -450$$
$$W = 45$$

Step 2: Substitute 45 for W in the equation $L = W + 15$ and find the value of L.

$$L = W + 15$$
$$L = 45 + 15$$
$$L = 60$$

Thus, the original field is 45 feet wide and 60 feet long. ■

Example 4 **A boat whose speed in still water is 20 miles per hour can travel 12 miles downstream in the same time it takes to travel 8 miles upstream. Find the rate of the water current.**

Solution: Let r = rate of the water current. Then $20 + r$ = rate of the boat downstream and $20 - r$ = rate of the boat upstream. Let t = the time the boat goes upstream which equals the time the boat goes downstream. We know that rate multiplied by time equals distance ($r \times t = d$); thus, we can set up our two equations.

Equation 1: $(20 + r)t = 12$

Equation 2: $(20 - r)t = 8$

Step 1: Remove parentheses.

$$20t + rt = 12$$
$$20t - rt = 8$$

Step 2: Add the two equations using the addition principle and solve the resulting linear equation.

$$\begin{array}{r}20t + rt = 12\\ 20t - rt = 8\\ \hline 40t = 20\\ t = .5\end{array}$$

Step 3: Substitute .5 for t in $20t + rt = 12$ and solve for r.

$$20t + rt = 12$$
$$20(.5) + r(.5) = 12$$
$$10 + .5r = 12$$
$$.5r = 12 - 10$$
$$.5r = 2$$
$$r = 4$$

Thus, the rate of the water current is 4 miles per hour. ■

Example 5 **An airplane flying with a tail wind flew 750 miles between two cities in three hours. The return trip with an equal head wind took five hours. Find the rate of the plane in calm air and the rate of the wind.**

Solution: Let x = rate of plane in calm air.

Let y = rate of the wind.

Now, $x + y$ is the rate of the plane with a tail wind and $x - y$ is the rate of the plane with a head wind.
Since rate multiplied by time equals distance ($r \times t = d$), we can set up our two equations.

$$(x + y)3 = 750$$
$$(x - y)5 = 750$$

Step 1: Remove parentheses.

$$\text{Equation 1: } 3x + 3y = 750$$
$$\text{Equation 2: } 5x - 5y = 750$$

Step 2: Make the y terms opposites.

Multiply Equation 1 by 5.

$$3x + 3y = 750$$
$$5(3x + 3y) = 5(750)$$
$$15x + 15y = 3750$$

Multiply Equation 2 by 3.

$$5x - 5y = 750$$
$$3(5x - 5y) = 3(750)$$
$$15x - 15y = 2250$$

We now have the following two equations.

$$15x + 15y = 3750$$
$$15x - 15y = 2250$$

Step 3: Add the equations from Step 2 using the addition principle and solve for x.

$$\begin{array}{r} 15x + 15y = 3750 \\ \underline{15x - 15y = 2250} \\ 30x \qquad\quad = 6000 \\ x = 200 \end{array}$$

Step 4: Substitute 200 for x in $3x + 3y = 750$ and solve for y.

$$3x + 3y = 750$$
$$3(200) + 3y = 750$$
$$600 + 3y = 750$$
$$3y = 750 - 600$$
$$3y = 150$$
$$y = 50$$

Thus, the rate of the plane in calm air is 200 miles per hour and the rate of the wind is 50 miles per hour. ■

APPLICATION EXERCISES

1. The sum of two numbers is 40 and the difference of the two numbers is 10. Find the numbers.
2. Sue earns \$5.50 per hour and Mary earns \$6.50 per hour. Together they worked a total of 45 hours one week and earned a total of \$266.50. How many hours did each work?
3. Three dozen eggs and five loaves of bread cost \$6.10; four dozen eggs and two loaves of bread cost \$5.10. Find the prices of a dozen eggs and a loaf of bread.
4. Adult tickets for a concert cost \$5.00 each and student tickets cost \$4.00. Receipts from ticket sales totaled \$875. If 210 tickets were sold in all, how many were sold at each rate?
5. A barge can travel 96 kilometers downstream in four hours, but it requires six hours to travel the same distance upstream. Find the rate of the current and the rate of the barge in still water.
6. The sum of two numbers is 105. One number is four times as large as the other. What are the two numbers?
7. Jimmy has 31 nickels and dimes in his pocket worth a total of \$2.20. How many coins of each kind does he have?
8. A rectangle has a perimeter of 104 feet. The difference between the length and width of the rectangle is 14 feet. Find the length and width of the rectangle.
9. In 3 hours an airplane flies 630 miles with a tail wind. Returning against the same wind, the airplane takes $3\frac{1}{2}$ hours to fly the 630 miles. Find the speed of the wind and the speed of the plane in still air.
10. A canoeist paddling with the current can travel 15 miles in three hours. Against the current it takes five hours to travel the same distance. Find the rate of the current and the rate of the canoeist in still water.

USEFUL INFORMATION

1. The graph of **a system of two linear equations** is two lines which can have one point in common, no point in common, or all points in common (one line). This means that a system of two linear equations can have one solution, no solution, or many solutions.
2. Finding the solution of a system of linear equations by graphing takes longer and is not as accurate as the substitution or addition method.
3. If you get a statement that is not true, such as $5 = 6$, when solving a system of two linear equations, then **the system does not have a solution.** This means that the graph of the system is two parallel lines.
4. If you get a statement that is always true, such as $2 = 2$, when solving a system of two linear equations, then **the system has many solutions.** This means that the two equations in the system are equivalent and the graph is one line.
5. **The addition principle:** If $a = b$ and $c = d$ then $a + c = b + d$.
6. The graph of **a system of one linear equation and one quadratic equation** of the form $y = ax^2 + bx + c$ is a line and a parabola. A line and a parabola can have two points in common, one point in common, or no point in common. This means that a system of one linear and one quadratic equation can have two solutions, one solution, or no solution.

7. To solve a system of two linear equations by substitution:

a. Select one of the equations and solve for one of the variables in terms of the other.
b. Substitute the solution from Step a in the other equation to get a linear equation in one variable and solve the equation.
   i. If you get a statement that is not true, such as $3 = 5$, then the system of equations has no solution.
   ii. If you get a statement that is always true, such as $0 = 0$, then the system of equations has many solutions.
   iii. If you get a statement such as $x = 5$ or $y = 7$, then the system of equations has one solution and you proceed to Step c.
c. Substitute the solution from Step b in the equation from Step a to find the value of the second variable.
d. Check your answer by substituting it in both of the original equations (optional).

8. To solve a system of two linear equations by addition:

a. Write the equations in the form $ax + by = c$.
b. Make the x or y term in one equation the opposite of the corresponding term in the other equation.
c. Eliminate one of the variables by adding the two equations.
d. Solve the resulting linear equation in one variable.
   i. If you get a statement that is not true, such as $3 = 5$, then the system of equations has no solution.
   ii. If you get a statement that is always true, such as $0 = 0$, then the system of equations has many solutions.
   iii. If you get a statement such as $x = 5$ or $y = 7$, then the system of equations has one solution and you proceed to Step e.
e. Substitute the solution from Step d in one of the original equations and solve for the other variable.
f. Check your answer by substituting it in both of the original equations (optional).

9. To solve a system of one linear and one quadratic equation by substitution:

a. Solve the linear equation for y in terms of x.
b. Substitute the solution of the linear equation for y in the quadratic equation.
c. Solve the quadratic equation in one variable to find the value(s) of x.
   i. If the quadratic equation in one variable has no solution, then the system of equations has no solution.
   ii. If the quadratic equation in one variable has one solution, then the system of equations has one solution.
   iii. If the quadratic equation in one variable has two solutions, then the system of equations has two solutions.
d. Substitute the solution(s) of the quadratic equation for x in the linear equation from Step a to find the value(s) of y.
e. Check your answer(s) by substituting the values of x and y in both of the original equations (optional).

REVIEW PROBLEMS

Section 11.1

Solve the following systems of linear equations by substitution.

1. $y = -5x$
$2x + y = 3$

2. $y = 7x$
$-3x + y = 12$

3. $-2x + y = 3$
$3x - 2y = 2$

4. $-5x + y = 1$
$7x + 2y = -15$

5. $3x + y = 5$
$6x + 2y = 4$

6. $x - 3y = 7$
$-2x + 3y = 4$

7. $x + 5y = 1$
$-2x - 6y = 10$

8. $5x + y = 3$
$10x + 2y = 6$

9. $-6x + y = 7$
$-18x + 3y = 21$

10. $6x + 2y = 10$
$7x + 4y = 0$

Section 11.2

Solve the following systems of linear equations using the addition principle.

1. $2x + 3y = 5$
$-2x + 2y = 10$

2. $5x - 7y = 0$
$2x + 7y = 14$

3. $3x - y = 5$
$2x + 2y = 6$

4. $4x - 3y = 2$
$-8x + 6y = -4$

5. $3x - 4y = 2$
$-6x + 8y = 5$

6. $5x - 2y = 3$
$4x + 3y = 7$

7. $-8x + 11y = 13$
$6x - 5y = 0$

8. $4x + 7y = -5$
$6x + 8y = 10$

9. $18x - 24y = 15$
$12x - 16y = 10$

10. $12x - 10y = 10$
$18x - 5y = -5$

Section 11.3

Solve the following systems of equations by substitution.

1. $y = -4x$
$y = x^2 - 3x - 2$

2. $y = 9x$
$y = x^2 + 5x + 3$

3. $-2x + y = 1$
$y = x^2 - 7x + 15$

4. $-5x + y = -2$
$y = x^2 + 6x - 14$

5. $x + y = 3$
$y = x^2 - 5x + 7$

6. $4x + y = 1$
$y = x^2 - 4x + 10$

7. $2x - y = 5$
$y = 6x^2 + 15x - 13$

8. $3x - y = 0$
$y = 2x^2 - 2x + 2$

9. $5x + y = 0$
$y = 6x^2 - 6x - 2$

10. $-6x + 2y = 8$
$y = 5x^2 - 21x - 1$

ANSWERS TO PRACTICE PROBLEMS

Section 11.1

1. $(-1,1)$
2. $(5,1)$
3. $\left(2,-\frac{11}{2}\right)$
4. $(-1,-1)$
5. No solution
6. Many solutions
7. No solution
8. Many solutions

Section 11.2

1. $(1,-1)$
2. $(2,2)$
3. $(-5,-23)$
4. $(8,2)$
5. $\left(\frac{24}{11},\frac{-3}{11}\right)$
6. $\left(1,\frac{-1}{3}\right)$
7. $(-19,-9)$
8. $(-1,-1)$
9. Many solutions
10. No solution
11. No solution
12. Many solutions

Section 11.3

1. $(-2,7)$ and $(3,2)$
2. $(-2,1)$ and $(5,15)$
3. $\left(\frac{1}{2},\frac{3}{2}\right)$
4. $\left(\frac{-2}{3},2\right)$ and $\left(\frac{3}{2},\frac{17}{2}\right)$

CHAPTER 12

Inequalities

The objectives of this chapter are:

1. To be able to solve linear inequalities in one variable.
2. To be able to graph linear inequalities in one and two variables.

A N S W E R

In order for Sam to lose weight, Sam must eat *less than* 1600 calories per day. Sam's telephone bill must be *less than* \$50 per month in order for Sam to stay within budget. In order for Sam to buy a new stereo system this year, Sam must save *more than* \$30 a month. These statements, called inequalities, occur quite often in everyday life.

12.1 THE SYMBOLS OF INEQUALITY

For two real numbers a and b, a is said to be **greater than** b and we write $a > b$ if $a - b$ is a positive number. If $a - b$ is a negative number, then a is said to be **less than** b and we write $a < b$.

Example 1 $8 > 2$, read 8 is greater than 2, since $8 - 2 = 6$ and 6 is a positive number. ■

Example 2 $8 < 10$, read 8 is less than 10, since $8 - 10 = -2$ and -2 is a negative number. ■

Example 3 $5 > -7$, read 5 is greater than -7, since $5 - (-7) = 5 + 7 = 12$ and 12 is a positive number. ■

Example 4 $-6 < -2$, read -6 is less than -2, since $-6 - (-2) = -6 + 2 = -4$ and -4 is a negative number. ■

1. True ____ False ____ Practice 1: $9 > 2$
2. True ____ False ____ Practice 2: $7 > 6$
3. True ____ False ____ Practice 3: $8 > 12$
4. True ____ False ____ Practice 4: $-6 > -5$
5. True ____ False ____ Practice 5: $5 > -9$
6. True ____ False ____ Practice 6: $-15 > 5$
7. True ____ False ____ Practice 7: $-10 > -20$
8. True ____ False ____ Practice 8: $4 < 7$
9. True ____ False ____ Practice 9: $9 < 3$
10. True ____ False ____ Practice 10: $16 < 23$
11. True ____ False ____ Practice 11: $-5 < -3$
12. True ____ False ____ Practice 12: $-8 < -17$

If $a < b$, then a is to the left of b on the number line.

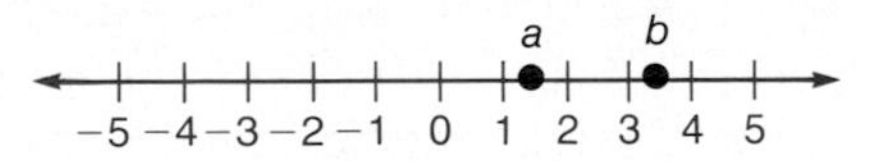

If $a > b$, then a is to the right of b on the number line.

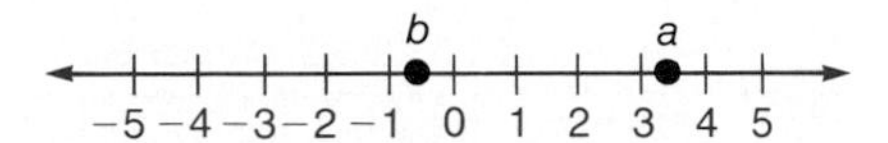

Then all negative numbers are less than zero and all positive numbers are greater than zero.

That's correct, Charlie. You made a very good observation. You should also observe that a negative number is always less than a positive number.

Practice 13: Rework Practices 1 through 12 using the number line.

Sam has \$4 to spend for lunch; thus, the price of Sam's lunch must be less than or equal to \$4. For a birthday present, Sam receives a \$20 gift certificate for a sweater; thus, the price of the sweater must be less than or equal to \$20.

To write such statements using algebraic symbols, we combine the less than symbol, $<$, with the equal symbol, $=$, and form the less than or equal to symbol $\leq$.

For two real numbers a and b, a is said to be **less than or equal to** b and we write $a \leq b$ if $a < b$ or $a = b$.

For two real numbers a and b, a is said to be **greater than or equal to** b and we write $a \geq b$ if $a > b$ or $a = b$.

Example 5 $5 \leq 7$, read 5 is less than or equal to 7, since $5 < 7$. ■

Example 6 $-12 \leq -12$, read -12 is less than or equal to -12, since $-12 = -12$. ■

Example 7 $-3 \geq -7$, read -3 is greater than or equal to -7, since $-3 > -7$. ■

Example 8 $4 \geq 4$, read 4 is greater than or equal to 4, since $4 = 4$. ■

Now, let's summarize the symbols of inequality.

Definition of $>$ and $<$:
For two real numbers a and b, a is said to be **greater than** b and we write $a > b$ if $a - b$ is a positive number
For two real numbers a and b, a is said to be **less than** b and we write $a < b$ if $a - b$ is a negative number

Definition of $\geq$ and $\leq$:
For two real numbers a and b, a is said to be **greater than or equal to** b and we write $a \geq b$ if $a > b$ or $a = b$
For two real numbers a and b, a is said to be **less than or equal to** b and we write $a \leq b$ if $a < b$ or $a = b$

Statements of the form $a < b$, $a > b$, $a \leq b$, and $a \geq b$ are called **inequalities.** Inequalities are read in the following manner.

$a > b$	***a* is greater than *b***
$a < b$	***a* is less than *b***
$a \geq b$	***a* is greater than or equal to *b***
$a \leq b$	***a* is less than or equal to *b***

There are two ways to compare 3 and 5. We can say that 3 is less than 5 and write $3 < 5$, or we can say that 5 is greater than 3 and write $5 > 3$.

Example 9 **Fill in the blanks with $<$ or $>$.**

7 ___ 11 or 11 ___ 7

Solution: $7 < 11$ or $11 > 7$ ∎

Note: The **direction of an inequality** is the way that the inequality symbol points. The inequality symbol should always point from the larger number to the smaller number.

So, the inequality symbol is like an arrow pointing from the larger number to the smaller number.

That's correct, Charlie.

Practice 14: Fill in the blanks with $<$ or $>$.

15 ____ 25 or 25 ____ 15

Practice 15: Fill in the blanks with $<$ or $>$.

12 ____ 4 or 4 ____ 12

Practice 16: Fill in the blanks with $<$ or $>$.

-5 ____ -8 or -8 ____ -5

Practice 17: Fill in the blanks with $<$ or $>$.

3 ____ -7 or -7 ____ 3

Now, let's look at inequalities that contain a variable, such as $x > 2$. A **solution** to this inequality is any value of x that makes the inequality true. There are many real numbers greater than 2; for example, $2\frac{1}{2}$, 3, 4, 4.2, and 5 are all greater than 2. Thus, there are many solutions to this inequality. Let's graph the solutions of $x > 2$ using the number line.

The "(" at 2 indicates that 2 *is not part* of the solution.

How would we graph it, if 2 were part of the solution?

Well, Charlie, if 2 were part of the solution, then we would have the inequality $x \geq 2$. Let's graph the solution.

x ≥ 2
−5 −4 −3 −2 −1 0 1 2 3 4 5

The "[" at 2 indicates that 2 *is part* of the solution. When graphing an inequality using the number line, use "[" when the inequality is $\geq$; use "]" when the inequality is $\leq$; use "(" when the inequality is $>$; use ")" when the inequality is $<$.

Example 10 **Graph the solutions of $x < -1$ using the number line and give six numbers that satisfy the inequality.**

Solution:

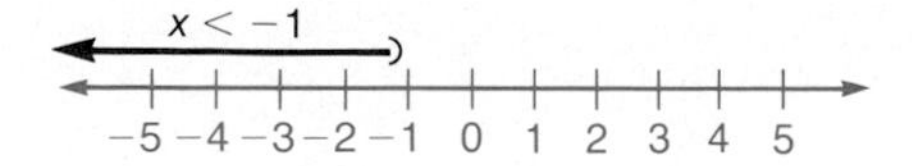

ANSWER

Six values of x that satisfy $x < -1$ are $-1\frac{1}{3}$, -2, -2.7, -2.85, -3, and $-3\frac{1}{5}$. There are many other numbers that also satisfy $x < -1$. ■

Example 11 **Graph the solutions to $x \leq -1$ using the number line and give four numbers that satisfy the inequality.**

Solution:

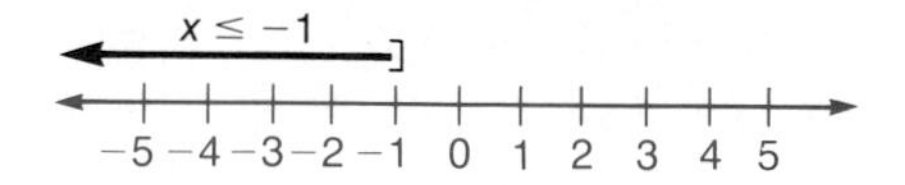

Four values of x that satisfy $x \leq -1$ are -1, -5, -12 and -23. Again, there are many other numbers that also satisfy the inequality. ■

18. ______________

Practice 18: Graph the solutions to $x > 3$ using the number line and give three numbers that satisfy the inequality.

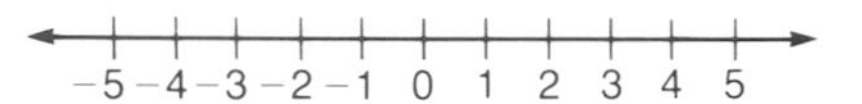

19. ______________

Practice 19: Graph the solutions to $x \geq 3$ using the number line and give five numbers that satisfy the inequality.

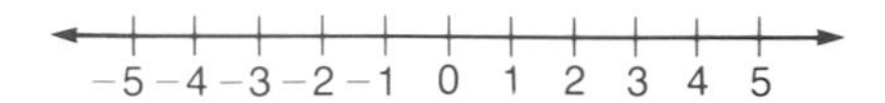

20. ______________

Practice 20: Graph the solutions to $x < 4$ using the number line and give seven numbers that satisfy the inequality.

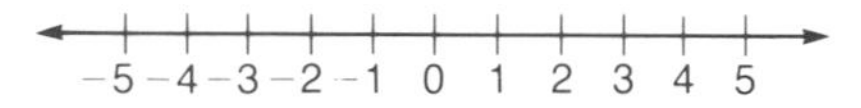

21. ______________

Practice 21: Graph the solutions to $x \geq 4$ using the number line and give three numbers that satisfy the inequality.

22. ______________

Practice 22: Graph the solutions to $x \leq -3$ using the number line and give eight numbers that satisfy the inequality.

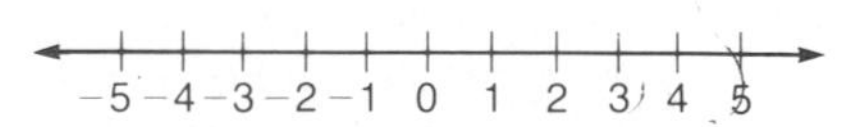

23. ______________

Practice 23: Graph the solutions to $x < -3$ using the number line and give five numbers that satisfy the inequality.

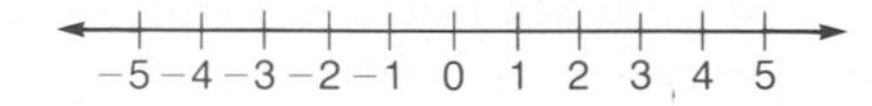

Example 12 **Graph the solutions to $-3 < x$ using the number line and give six numbers that satisfy the inequality.**

Solution:

$-3 < x$

$-5\ -4\ -3\ -2\ -1\ \ 0\ \ 1\ \ 2\ \ 3\ \ 4\ \ 5$

Six values of x that satisfy the inequality are -2, -1.7, $-1\frac{1}{4}$, -1, 0, and 4. There are many other numbers that also satisfy $-3 < x$. ■

Example 13 **Graph the solutions to $x > -3$ using the number line and give six numbers that satisfy the inequality.**

Solution:

$x > -3$

$-5\ -4\ -3\ -2\ -1\ \ 0\ \ 1\ \ 2\ \ 3\ \ 4\ \ 5$

Six values of x that satisfy the inequality are 2, 1.7, $1\frac{1}{4}$, 1, 0, and 4. There are many other numbers that also satisfy $x > -3$. ■

We have the same graphs for Examples 12 and 13.

That's correct, Charlie. The inequality $-3 < x$ is equivalent to the inequality $x > -3$. In both cases the inequality symbol points from x to -3 indicating that x is the larger quantity and -3 is the smaller quantity.

Example 14 **If $x < 7$ fill in the blank with $<$ or $>$.**

$$7 \_\_\_\_ x$$

Solution: The inequality symbol must point from the larger quantity to the smaller quantity; that is, the inequality symbol must point from 7 to x. Thus, we write $7 > x$. ■

Practice 24: If $x < 3$ fill in the blank with $<$ or $>$.

$$3 \_\_\_\_ x$$

Practice 25: If $5 > x$ fill in the blank with $<$ or $>$.

$$x \_\_\_\_ 5$$

Practice 26: If $-7 \leq x$ fill in the blank with $\leq$ or $\geq$.

$$x \_\_\_\_ -7$$

Practice 27: If $x > -4$ fill in the blank with $<$ or $>$.

$$-4 \underline{\quad} x$$

Practice 28: If $x \leq 1$ fill in the blank with $\leq$ or $\geq$.

$$1 \underline{\quad} x$$

EXERCISE 12.1

Graph the solutions to the following inequalities using the number line and give five values that satisfy each inequality.

1. $x > -4$

2. $x > 0$

3. $x < -2$

4. $x \geq 1$

5. $x \leq 5$

6. $x \geq -4$

7. $x < 7$

8. $x < 5$

9. $x > 2\frac{1}{2}$

10. $x \geq -5$

11. $x \leq -1\frac{1}{2}$

12. $x \geq 0$

12.2 SOLVING LINEAR INEQUALITIES USING ADDITION

A linear inequality looks very much like a linear equation except that instead of the equals symbol ($=$) it has an inequality symbol ($>$, $\geq$, $<$, $\leq$). Solving linear inequalities is similar to solving linear equations. Just as with equations, two inequalities are **equivalent** if they have exactly the same solution(s). We will first look at inequalities that can be solved using the addition and subtraction property of inequalities.

Addition and subtraction property of inequalities: You can change an inequality to an equivalent one (one that has the same solutions) by adding or subtracting the *same* quantity on both sides of the inequality.

For example,

$$\text{if } a < b \text{ then } a + c < b + c$$

$$\text{if } a < b \text{ then } a - c < b - c$$

We have **solved a linear inequality** when the variable is isolated on one side with a coefficient of 1 and a number is on the other side.

So, to solve a linear inequality we isolate the variable on one side of the inequality just as we isolated the variable on one side of an equation when solving linear equations.

That's correct, Charlie.

Example 1 **Solve the following linear inequality and graph the solutions using the number line.**

$$x + 5 > 7$$

Solution: Subtract 5 from both sides of the inequality in order to isolate x.

$x + 5 > 7$	**Original inequality.**
$x + 5 - 5 > 7 - 5$	**Subtract 5 from both sides.**
$x + 0 > 2$	**Perform the subtractions.**
$x > 2$	**The addition property of zero.**

Check: Substitute a few numbers greater than 2 for x in the original inequality, $x + 5 > 7$.

Substitute 3 for x.	Substitute 6 for x.
$x + 5 > 7$	$x + 5 > 7$
$3 + 5 \overset{?}{>} 7$	$6 + 5 \overset{?}{>} 7$
$8 > 7$	$11 > 7$

Thus, 3 and 6 are two solutions of $x + 5 > 7$. Since every number greater than 2 is a solution to $x + 5 > 7$, it is not possible to check all the solutions. However, checking a few solutions will often show an error when solving the inequality. ■

Example 2 **Solve the following linear inequality and graph the solutions using the number line.**

$$-5 \geq x - 8$$

Solution: Add 8 to both sides of the inequality in order to isolate x.

$-5 \geq x - 8$	**Original inequality.**
$-5 + 8 \geq x - 8 + 8$	**Add 8 to both sides.**
$3 \geq x + 0$	**Perform the additions.**
$3 \geq x$	**The addition property of zero.**

Check: Substitute a few numbers that 3 is greater than or equal to for x in the original inequality, $-5 \geq x - 8$.

Substitute 3 for x.	Substitute 1 for x.
$-5 \geq x - 8$	$-5 \geq x - 8$
$-5 \overset{?}{\geq} 3 - 8$	$-5 \overset{?}{\geq} 1 - 8$
$-5 \geq -5$	$-5 \geq -7$

Thus, 3 and 1 are two solutions of $-5 \geq x - 8$. Since every number that 3 is greater than or equal to is a solution to $-5 \geq x - 8$, it is not possible to check all the solutions. However, checking a few solutions will often show an error when solving the inequality. ■

A N S W E R

1. ____________________

Practice 1: Solve the following linear inequality and graph the solutions using the number line.

$$x - 12 < -8$$

−5 −4 −3 −2 −1 0 1 2 3 4 5

2. ____________________

Practice 2: Solve the following linear inequality and graph the solutions using the number line.

$$7 > x + 2$$

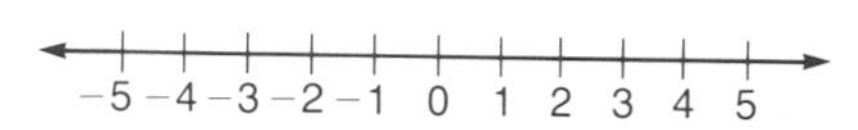

3. ____________________

Practice 3: Solve the following linear inequality and graph the solutions using the number line.

$$x + 5 \leq 8$$

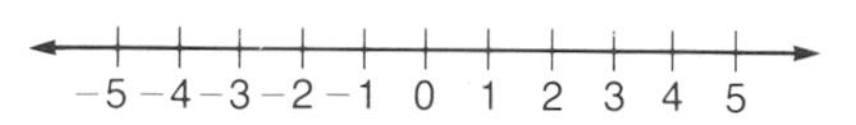

4. ____________________

Practice 4: Solve the following linear inequality and graph the solutions using the number line.

$$-1 \geq x - 4$$

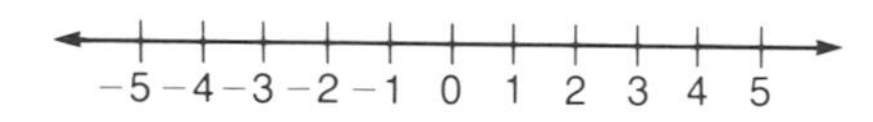

EXERCISE 12.2

Solve the following inequalities and graph the solutions using the number line.

1. $x + 12 > 15$

2. $x - 7 < -11$

3. $17 < x + 15$

4. $x - 2 \geq -4$

5. $-2 \geq x + 3$

6. $x - 24 \leq -30$

7. $5 \leq -7 + x$

8. $17 > x + 5$

9. $x - 22 < -17$

10. $-13 < x - 12$

11. $x + 7 \geq 5$

12. $6 > x + 8$

12.3 SOLVING LINEAR INEQUALITIES USING MULTIPLICATION

In this section, we will look at linear inequalities that can be solved using the multiplication and division property of inequalities.

> **Multiplication and division property of inequalities:** You can change an inequality to an equivalent one by multiplying or dividing both sides of the inequality by the *same* nonzero quantity and giving the resulting inequality the proper direction.
> **Case 1:** If both sides of an inequality are multiplied or divided by the same *positive* number, *the direction of the inequality is not changed.*
> **Case 2:** If both sides of an inequality are multiplied or divided by the same *negative* number, *the direction of the inequality is reversed.*

Example 1

Case 1: $3 < 5$

$2(3) < 2(5)$ **Multiply both sides by 2.**

$6 < 10$ **Now, 6 is less than 10 and the direction of the inequality did not change.**

In the original inequality, the smaller number, 3, is on the left and the larger number, 5, is on the right. Thus, the direction of the inequality is from right to left. Remember that the inequality symbol always points from the larger number to the smaller number. After multiplying both sides by 2, the smaller number is still on the left and the larger number is still on the right. Thus, the direction of the inequality did not change.

Case 2: $3 < 5$

$-2(3) > -2(5)$ **Multiply both sides by −2.**

$-6 > -10$ **Now, −6 is greater than −10 and the direction of the inequality changed.**

In the original inequality, the smaller number, 3, is on the left and the larger number, 5, is on the right. Thus, the direction of the inequality is from right to left. After multiplying both sides by −2, the smaller number is on the right and the larger number is on the left. Thus, the direction of the inequality changed. ■

I will have to be sure to change the direction of the inequality when I multiply or divide both sides by a negative number.

That's true, Charlie. The mistake that is made most often when solving inequalities is that students forget to change the direction of an inequality when they multiply or divide both sides by a negative number. Except for this, solving a linear inequality is just like solving a linear equation.

Example 2 **Solve the following linear inequality and graph the solutions using the number line.**

$$5x > 20$$

Solution: Divide both sides of the inequality by 5 in order to isolate x.

$5x > 20$	**Original inequality.**
$\dfrac{5x}{5} > \dfrac{20}{5}$	**Divide both sides by 5.**
$x > 4$	**Perform the divisions.**

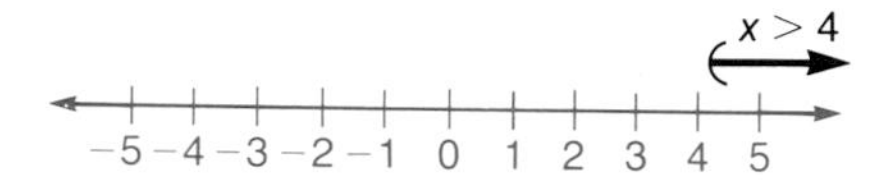

Check: Substitute a few numbers greater than 4 for x in the original inequality, $5x > 20$.

Substitute 5 for x.	Substitute 8 for x.
$5x > 20$	$5x > 20$
$5(5) \overset{?}{>} 20$	$5(8) \overset{?}{>} 20$
$25 > 20$	$40 > 20$

Thus, 5 and 8 are two solutions of $5x > 20$. Checking a few solutions will often show an error when solving the inequality. ∎

Example 3 **Solve the following linear inequality and graph the solutions using the number line.**

$$12 > -3x$$

Solution: Divide both sides of the inequality by -3 in order to isolate x.

$12 > -3x$	**Original inequality.**
$\dfrac{12}{-3} < \dfrac{-3x}{-3}$	**Divide both sides by -3 and change the direction of the inequality.**
$-4 < x$	**Perform the divisions.**

−4 < x
−5 −4 −3 −2 −1 0 1 2 3 4 5

A N S W E R

Check: Substitute a few numbers that -4 is less than for x in the original inequality, $12 > -3x$.

Substitute -3 for x.	Substitute -1 for x.
$12 > -3x$	$12 > -3x$
$12 \overset{?}{>} -3(-3)$	$12 \overset{?}{>} -3(-1)$
$12 > 9$	$12 > 3$

Thus, -3 and -1 are two solutions of $12 > -3x$. ■

Example 4 **Solve the following linear inequality and graph the solutions using the number line.**

$$-2x \leq -6$$

Solution: Divide both sides of the inequality by -2 in order to isolate x.

$-2x \leq -6$	Original inequality.
$\dfrac{-2x}{-2} \geq \dfrac{-6}{-2}$	Divide both sides by -2 and change the direction of the inequality.
$x \geq 3$	Perform the divisions.

Check: Substitute a few numbers greater than or equal to 3 for x in the original inequality, $-2x \leq -6$.

Substitute 3 for x.	Substitute 4 for x.
$-2x \leq -6$	$-2x \leq -6$
$-2(3) \overset{?}{\leq} -6$	$-2(4) \overset{?}{\leq} -6$
$-6 \leq -6$	$-8 \leq -6$

Thus, 3 and 4 are two solutions of $-2x \leq -6$. Since every number greater than or equal to 3 is a solution of $-2x \leq -6$, it is not possible to check all the solutions. ■

Practice 1: Solve the following linear inequality and graph the solutions using the number line.

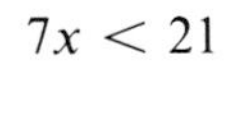

$$7x < 21$$

1. ____________

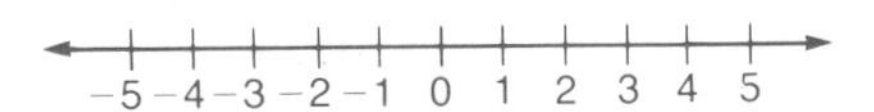

A N S W E R

2. ______________________

Practice 2: Solve the following linear inequality and graph the solutions using the number line.

$$9x \geq 18$$

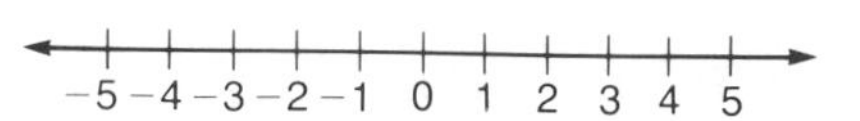

3. ______________________

Practice 3: Solve the following linear inequality and graph the solutions using the number line.

$$-11x < 22$$

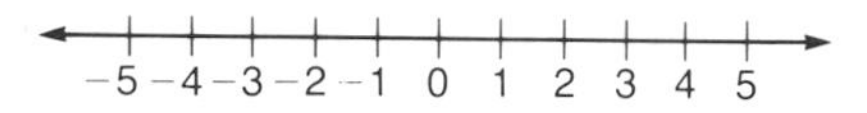

4. ______________________

Practice 4: Solve the following linear inequality and graph the solutions using the number line.

$$-5x \leq 15$$

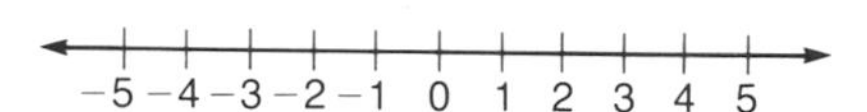

5. ______________________

Practice 5: Solve the following linear inequality and graph the solutions using the number line.

$$-7x \geq -28$$

−5 −4 −3 −2 −1 0 1 2 3 4 5

EXERCISE 12.3

Solve the following inequalities and graph the solutions using the number line.

1. $8x > 8$

2. $9x \leq -36$

3. $27x > -54$

4. $34 \geq 17x$

5. $-38 < 19x$

6. $-12x \geq 36$

7. $-11x \leq -33$

8. $-48 > -12x$

9. $-51 < 17x$

10. $27x \leq -81$

11. $5 > -2x$

12. $-8 < -5x$

12.4 SOLVING LINEAR INEQUALITIES USING ADDITION AND MULTIPLICATION

To solve many inequalities, it is necessary to use both the addition and multiplication properties of inequalities. In this section, we will look at inequalities of this type.

Example 1 **Solve the following linear inequality and graph the solutions using the number line.**

$$2x + 3 < 7$$

Step 1: Subtract 3 from both sides of the inequality in order to isolate $2x$.

$2x + 3 < 7$	**Original inequality.**
$2x + 3 - 3 < 7 - 3$	**Subtract 3 from both sides.**
$2x < 4$	**Perform the subtractions.**

Step 2: Divide both sides of the resulting inequality by 2 in order to isolate x.

$2x < 4$ **Inequality from Step 1.**

$\frac{2x}{2} < \frac{4}{2}$ **Divide both sides by 2.**

$x < 2$ **Perform the divisions.**

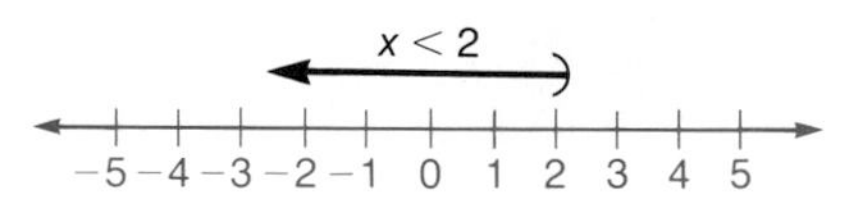

Check: Substitute a few numbers less than 2 for x in the original inequality, $2x + 3 < 7$.

Substitute 1 for x.	Substitute 0 for x.
$2x + 3 < 7$	$2x + 3 < 7$
$2(1) + 3 \overset{?}{<} 7$	$2(0) + 3 \overset{?}{<} 7$
$2 + 3 \overset{?}{<} 7$	$0 + 3 \overset{?}{<} 7$
$5 < 7$	$3 < 7$

Thus, 1 and 0 are two solutions of $2x + 3 < 7$. Since every number less than 2 is a solution of $2x + 3 < 7$, it is not possible to check all the solutions. ■

Example 2 **Solve the following linear inequality and graph the solutions using the number line.**

$$-14 > -3x - 2$$

Step 1: Add 2 to both sides of the inequality in order to isolate $-3x$.

$-14 > -3x - 2$ **Original inequality.**

$-14 + 2 > -3x - 2 + 2$ **Add 2 to both sides.**

$-12 > -3x$ **Perform the additions.**

Step 2: Divide both sides of the resulting inequality by -3 in order to isolate x.

$-12 > -3x$ **Inequality from Step 1.**

$\frac{-12}{-3} < \frac{-3x}{-3}$ **Divide both sides by −3 and change the direction of the inequality.**

$4 < x$ **Perform the divisions.**

Check: Substitute a few numbers that 4 is less than for x in the original inequality, $-14 > -3x - 2$.

Substitute 5 for x.	Substitute 7 for x.
$-14 > -3x - 2$	$-14 > -3x - 2$
$-14 \overset{?}{>} -3(5) - 2$	$-14 \overset{?}{>} -3(7) - 2$
$-14 \overset{?}{>} -15 - 2$	$-14 \overset{?}{>} -21 - 2$
$-14 > -17$	$-14 > -23$

Thus, 5 and 7 are two solutions of $-14 > -3x - 2$. ■

Practice 1: Solve the following linear inequality and graph the solutions using the number line.

$$5x - 2 > 8$$

ANSWER

1. ____________

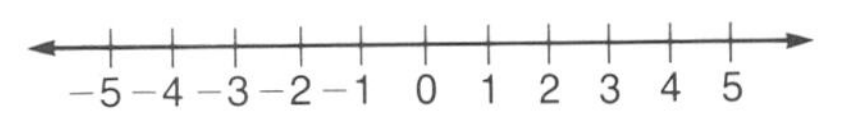

Practice 2: Solve the following linear inequality and graph the solutions using the number line.

2. ____________

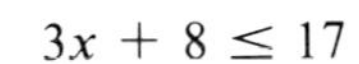

$$3x + 8 \leq 17$$

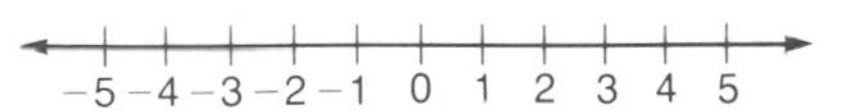

Practice 3: Solve the following linear inequality and graph the solutions using the number line.

3. ____________

$$-4x + 3 \geq 15$$

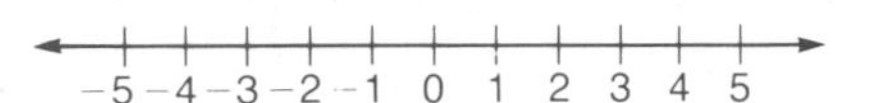

A N S W E R

4. ____________

Practice 4: Solve the following linear inequality and graph the solutions using the number line.

$$-22 > -8x - 6$$

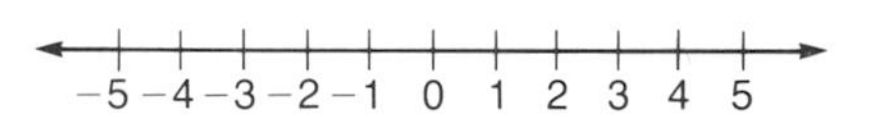

Example 3 **Solve the following linear inequality and graph the solutions using the number line.**

$$-5x + 3 \le 2x - 11$$

Step 1: Subtract 3 from both sides of the inequality.

$-5x + 3 \le 2x - 11$	**Original inequality.**
$-5x + 3 - 3 \le 2x - 11 - 3$	**Subtract 3 from both sides.**
$-5x \le 2x - 14$	**Perform the subtractions.**

Step 2: Subtract $2x$ from both sides of the resulting inequality.

$-5x \le 2x - 14$	**Inequality from Step 1.**
$-5x - 2x \le 2x - 14 - 2x$	**Subtract $2x$ from both sides.**
$-7x \le -14$	**Perform the subtractions.**

Step 3: Divide both sides of the resulting inequality by -7.

$-7x \le -14$	**Inequality from Step 2.**
$\frac{-7x}{-7} \ge \frac{-14}{-7}$	**Divide both sides by −7 and change the direction of the inequality.**
$x \ge 2$	**Perform the divisions.**

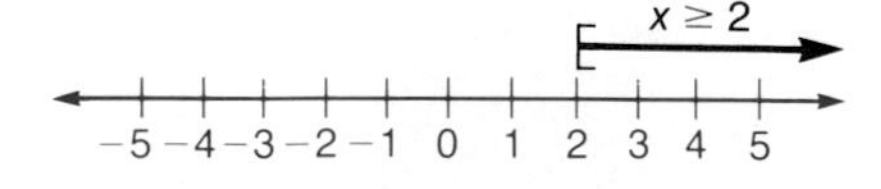

Check: Substitute a few numbers greater than or equal to 2 for x in the original inequality, $-5x + 3 \le 2x - 11$.

Substitute 2 for x.	Substitute 3 for x.
$-5x + 3 \le 2x - 11$	$-5x + 3 \le 2x - 11$
$-5(2) + 3 \overset{?}{\le} 2(2) - 11$	$-5(3) + 3 \overset{?}{\le} 2(3) - 11$
$-10 + 3 \overset{?}{\le} 4 - 11$	$-15 + 3 \overset{?}{\le} 6 - 11$
$-7 \le -7$	$-12 \le -5$

Thus, 2 and 3 are two solutions of $-5x + 3 \le 2x - 11$. ■

A N S W E R

Practice 5: Solve the following linear inequality and graph the solutions using the number line.

$$-2x + 7 > -3x + 5$$

5. ______________

Practice 6: Solve the following linear inequality and graph the solutions using the number line.

$$5x - 8 \geq -3x + 16$$

6. ______________

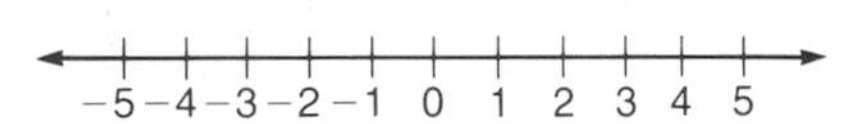

Practice 7: Solve the following linear inequality and graph the solutions using the number line.

$$-6x + 15 < -2x + 3$$

7. ______________

−5 −4 −3 −2 −1 0 1 2 3 4 5

A N S W E R

8. ________________

Practice 8: Solve the following linear inequality and graph the solutions using the number line.

$$-12x + 3 \leq -5x - 11$$

EXERCISE 12.4

Solve the following inequalities and graph the solutions using the number line.

A

1. $2x + 3 < 5$

2. $5x - 2 > 8$

3. $3x - 5 > 10$

4. $7x + 8 < -13$

5. $12 \geq 3x - 6$

6. $-15 < 5x - 20$

7. $-2x + 8 < 4$

8. $-7x - 15 \geq 6$

9. $14 \leq -6x - 4$

10. $-23 > -4x - 3$

B

11. $5x + 7 < 3x - 5$

12. $2x - 8 < -3x + 2$

13. $-6x - 4 \geq -8x + 2$

14. $3x + 2 \leq -2x - 13$

15. $5x - 8 > -3x + 16$

16. $-4x + 7 > 2x - 5$

17. $3x + 12 < 5x + 8$

18. $-7x - 8 \geq -3x + 8$

19. $-12x - 15 \leq 3x + 30$

20. $7x - 11 < 10x + 4$

12.5 GRAPHING LINEAR INEQUALITIES IN TWO VARIABLES

The graph of a linear equation separates a rectangular coordinate system into three regions: the line itself and the region on each side of the line. For example, the graph of $y = 2x - 1$ separates the rectangular coordinate system into the following three regions.

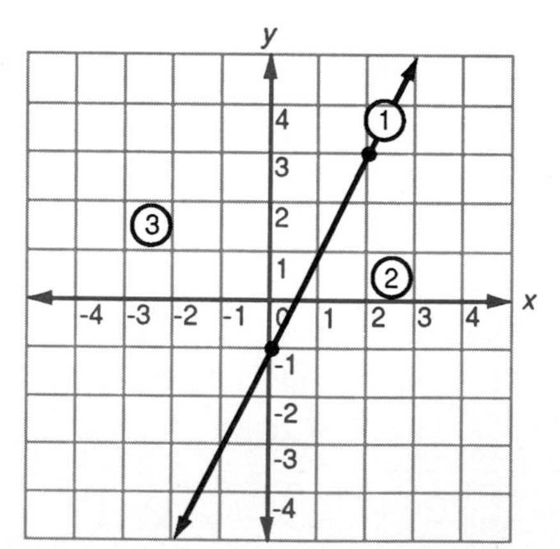

To graph a linear inequality, replace the inequality symbol with an equals symbol and draw the graph of the linear equation.

a. The graph of an inequality $>$ or $<$ is one of the two regions on either side of the line.

b. The graph of an inequality $\geq$ or $\leq$ is the line itself and one of the two regions on either side of the line.

For example, the graph of $y < 2x - 1$ is Region 2 in the above figure and the graph of $y \leq 2x - 1$ is Region 1 (the line) and Region 2.

So, the graph of a linear equation is a line and the graph of a linear inequality is an area.

That's correct, Charlie.

Example 1 **Graph the following linear inequality.**

$$y \leq 3x - 5$$

Step 1: Graph the linear equation $y = 3x - 5$.

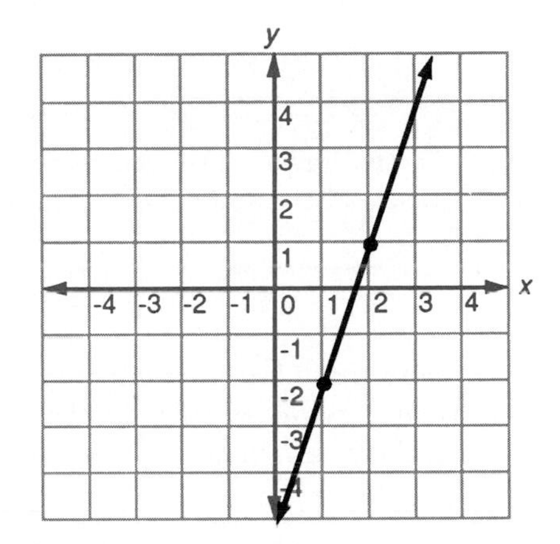

Step 2: Pick an ordered pair that locates a point on one side of the line and substitute its values in $y \leq 3x - 5$. Let's use the ordered pair (3,0). Now, substitute 3 for x and 0 for y in the original inequality.

$$y \leq 3x - 5$$
$$0 \overset{?}{\leq} 3(3) - 5$$
$$0 \overset{?}{\leq} 9 - 5$$
$$0 \leq 4$$

Since 0 is less than 4, (3,0) satisfies $y \leq 3x - 5$ and (3,0) locates a point on the graph of $y \leq 3x - 5$. Therefore, the graph of $y \leq 3x - 5$ consists of the line and the region below and to the right of the line since (3,0) is in that region.

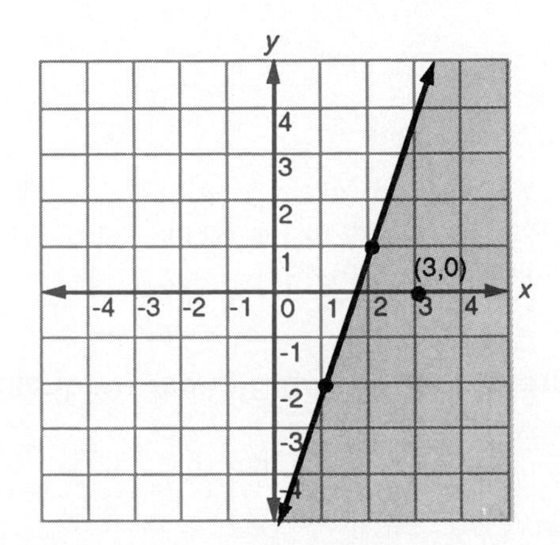

■

I have some questions about Example 1.

Okay, Charlie, what are they?

Why did you change the linear inequality to a linear equation and then graph the equation?

Well, Charlie, the graph of the linear equation separates the rectangular coordinate system into three regions. The graph of the inequality will consist of one or more of these three regions.

After you graphed the linear equation, you picked an ordered pair that located a point on one side of the line and then found if it satisfied the inequality. Why?

Well, Charlie, we know that one of the regions on either side of the line will be part of the graph of the inequality; but, we don't know which region. If the ordered pair that we pick satisfies the inequality, then the region containing the point located by this ordered pair is part of the graph of the inequality. If the ordered pair that we pick does not satisfy the inequality, then the region on the other side of the line will be part of the graph of the inequality.

My last question is, how did you know that the line was part of the graph of the inequality?

The inequality $y \leq 3x - 5$ means $y = 3x - 5$ or $y < 3x - 5$. Now, $y = 3x - 5$ gives us the line as part of the graph. Notice that the line was drawn solid to show that it is part of the graph.

Example 2 **Graph the following linear inequality.**

$$2x + 3y > 6$$

Step 1: Graph the linear equation $2x + 3y = 6$ drawing a dotted line.

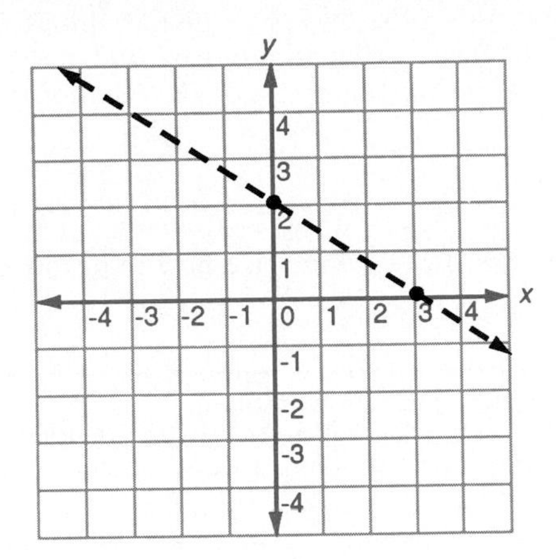

Why did we draw a dotted line?

Well, Charlie, we drew a dotted line because the line itself is not part of the graph of $2x + 3y > 6$.

Step 2: Pick an ordered pair that locates a point on one side of the dotted line and substitute its values in $2x + 3y > 6$. Let's use the ordered pair (0,1). Now, substitute 0 for x and 1 for y in $2x + 3y > 6$.

$$2x + 3y > 6$$
$$2(0) + 3(1) \overset{?}{>} 6$$
$$0 + 3 \overset{?}{>} 6$$
$$3 \not> 6$$

Since 3 is not greater than 6, (0,1) does not satisfy $2x + 3y > 6$ and (0,1) does not locate a point on the graph of $2x + 3y > 6$. Therefore, the graph of $2x + 3y > 6$ consists of the region above and to the right of the line—the region that does not include the point located by (0,1).

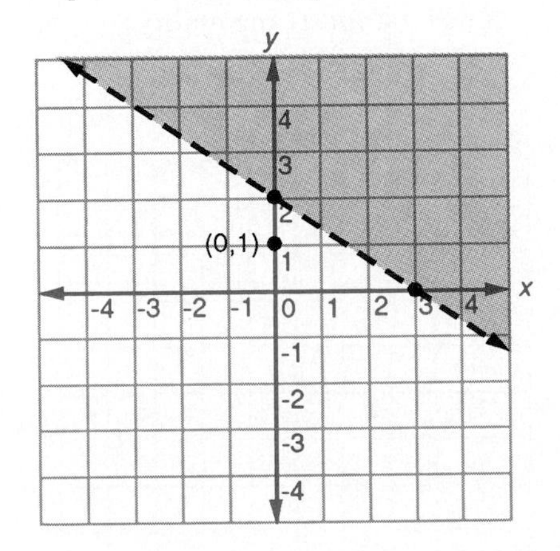

■

So a solid line indicates that the line is part of the graph of the inequality and a dotted line indicates that the line is not part of the graph of the inequality.

That's correct, Charlie. Let's summarize how to graph a linear inequality.

To graph a linear inequality in two variables:

1. Treat the inequality as if it were an equation and graph the equation.
   a. Draw a dotted line if the inequality is $>$ or $<$.
   b. Draw a solid line if the inequality is $\geq$ or $\leq$.
2. Select an ordered pair that locates a point on one side of the line and substitute its values in the original inequality for x and y.
   a. If the values satisfy the inequality, then the region containing the point is part of the graph of the inequality.
   b. If the values do not satisfy the inequality, then the region on the other side of the line is part of the graph of the inequality.
3. If the inequality is $\geq$ or $\leq$, then the graph is the solid line from Step 1 and the region found in Step 2.
4. If the inequality is $>$ or $<$, then the graph is just the region found in Step 2.

Practice 1: Graph the following linear inequality.

$$y > x + 3$$

Practice 2: Graph the following linear inequality.

$$y \leq 2x - 5$$

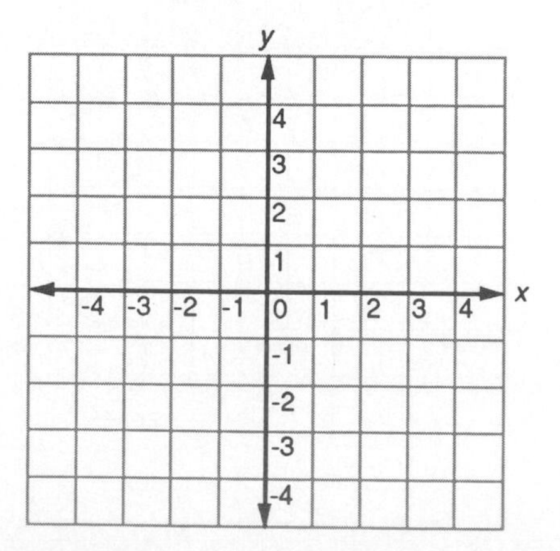

Practice 3: Graph the following linear inequality.

$$3x - 2y \geq 4$$

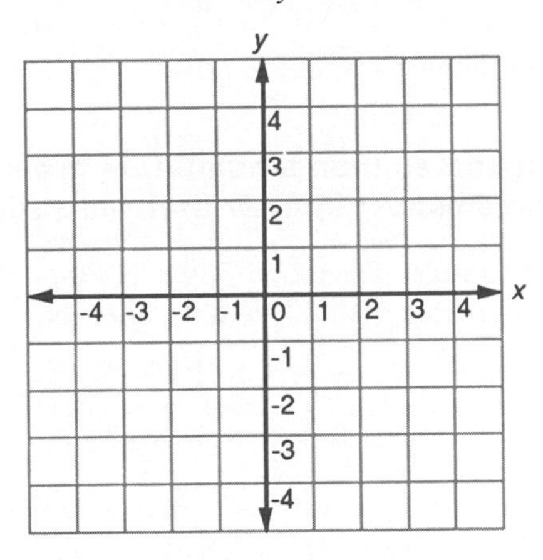

Practice 4: Graph the following linear inequality.

$$-2x + 5y > 10$$

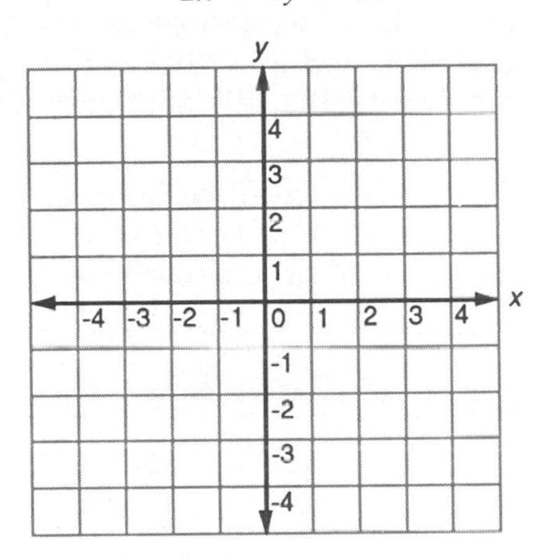

EXERCISE 12.5

Graph the following linear inequalities.

A

1. $y < 3x$

2. $y \geq -3x$

3. $y > -2x$

4. $y \geq 5x$

5. $y \leq 2x - 3$

6. $y \geq -x + 4$

7. $y \geq -4x + 5$

8. $y < -5x + 8$

9. $y < \frac{1}{2}x + 2$

10. $y > \frac{3}{2}x - 5$

B

11. $3x + 2y < 7$

12. $2x + 5y > 3$

13. $-2x + 4y \geq -4$

14. $-5x + 2y \leq 4$

15. $5x - 3y < -8$

12.6 STRATEGIES FOR SOLVING APPLICATIONS

Read the following examples carefully.

Example 1 **Eight less than a number is greater than ten. Find the smallest integer that will satisfy the inequality.**

Solution: Let $x =$ the number. Then $x - 8$ is 8 less than the number. Since 8 less than the number is greater than 10 we have our inequality.

$$x - 8 > 10$$

$$x - 8 + 8 > 10 + 8$$

$$x > 18$$

The smallest integer that satisfies $x > 18$ is 19. ■

Example 2 **Carl must have at least 450 points out of a possible 500 points on five tests to receive an A in mathematics. His scores on the first four tests were 94, 87, 77, and 95. What is the lowest score Carl can make on the fifth test and still receive an A for the course?**

Solution: Let $x =$ score on the fifth test. Then $94 + 87 + 77 + 95 + x =$ total points on the five tests. In order for Carl to receive an A in the course, this total must be greater than or equal to 450. Thus, we have our inequality.

$$94 + 87 + 77 + 95 + x \geq 450$$

$$353 + x \geq 450$$

$$353 - 353 + x \geq 450 - 353$$

$$x \geq 97$$

Therefore, Carl must make at least 97 on the fifth test in order to receive an A in the course. ■

Example 3 **A rectangular pen can only be 10 feet wide, but it can be as long as desired. Find the minimum length of the pen so that the area is greater than or equal to 200 square feet.**

Solution: We know area $=$ length $\times$ width and that the width is 10 feet. Let x represent the length of the pen. Then the area of the pen is $10x$. Since the area is greater than or equal to 200 square feet we have our inequality.

$$10x \geq 200$$

$$x \geq 20.$$

Hence the minimum length of the pen is 20 feet. ■

Example 4 **Company A rents cars for \$15 a day and 10 cents for every mile driven. Company B rents cars for \$35 a day with unlimited mileage. You want to rent a car for one week. What is the maximum number of miles you can drive a Company A car if it is to cost you less than a Company B car?**

Solution: Let x = number of miles driven. The cost of a Company A car for one week (7 days) is $(7 \times 15) + .10x$ since the car rents for $15 a day and 10 cents a mile. The cost of a Company B car for one week (7 days) is $7 \times 35 = 245$. We want the cost of a Company A car to be less than the cost of a Company B car. This gives us our inequality.

$$(7 \times 15) + .10x < 245$$
$$105 + .10x < 245$$
$$.10x < 245 - 105$$
$$.10x < 140$$
$$x < 1400$$

Therefore, 1399 miles is the maximum number of miles you can drive a Company A car and have it cost less than a Company B car. ■

Example 5 **An appliance dealer will make a profit on the sale of a television set if the cost of the new set is less than 70% of the selling price. What minimum selling price will enable the dealer to make a profit on a television set that costs the dealer $345?**

Solution: Let x = selling price of the television set. Then, 70% of the selling price is $.70x$. Since the dealer's cost, $345, has to be less than 70% of the selling price of the set we have our inequality.

$$345 < .70x$$
$$492.857 < x$$

Therefore, the minimum selling price on which the dealer would make a profit is $492.86. ■

APPLICATION EXERCISES

1. Marie received grades of 63, 73, 94, and 80 on four tests in a mathematics course. What grade on the fifth test will enable her to receive a minimum of 400 points on the five tests?

2. Five times a number plus 7 is less than -8. Find the largest number that satisfies this condition.

3. A rectangular pen can only be 8 feet wide but can be as long as desired. Find the minimum length of the pen so that the area is greater than 104 square feet.

4. Company A rents cars for $12 a day and 12 cents for every mile driven. Company B rents cars for $8 a day and 20 cents per mile driven. You want to rent a car for one week. What is the maximum number of miles you can drive a Company B car if it is to cost you less than a Company A car?

5. The base of a triangle is 10 inches and the height is x inches. Find the maximum height (integer) of the triangle so that the area is less than 80 square inches.

6. Three times a number is less than 20 minus the number. Find the largest integer that satisfies this condition.

7. An appliance dealer will make a profit on the sale of a refrigerator if the cost of the new refrigerator is less than 88% of the selling price. What minimum selling price will enable the dealer to make a profit on a refrigerator that costs the dealer $540?

8. Regulations require that grade A hamburger not contain more than 20% fat. Find the maximum amount of fat that a grocer can mix with 200 pounds of lean meat if he wishes to meet the 20% regulation for grade A hamburger.

USEFUL INFORMATION

1. The symbol "$>$" means **greater than** and the symbol "$<$" means **less than.**
2. For any two real numbers a and b, **$a > b$ if $a - b$ is a positive number.** For example, $11 > 7$ since $11 - 7 = 4$ is a positive number.
3. For any two real numbers a and b, **$a < b$ if $a - b$ is a negative number.** For example, $1 < 3$ since $1 - 3 = -2$ is a negative number.
4. The **direction of an inequality** is the way that the inequality symbol points. The inequality symbol always points from the larger number to the smaller number.
5. For any two real numbers a and b, **$a \geq b$ if $a = b$ or $a > b$** and **$a \leq b$ if $a = b$ or $a < b$.**
6. When graphing an inequality using the number line, use "[" when the inequality is $\geq$; use "]" when the inequality is $\leq$; use "(" when the inequality is $>$; use ")" when the inequality is $<$.
7. A linear inequality looks very much like a linear equation except that instead of the equals symbol ($=$) it has an inequality symbol ($<$, $\leq$, $>$, or $\geq$).
8. You can **change an inequality to an equivalent one** by adding or subtracting the same quantity on both sides of the inequality. For example, if $a < b$, then $a + c < b + c$ and $a - c < b - c$.
9. You have **solved a linear inequality** when the variable with a coefficient of 1 is isolated on one side with a number on the other side.
10. It is usually not possible to check all the solutions of an inequality; however, checking a few solutions will often show an error when solving the inequality.
11. You can **change an inequality to an equivalent one** by multiplying or dividing both sides of the inequality by the same nonzero quantity and giving the resulting inequality the proper direction.

    Case 1: If both sides of an inequality are multiplied or divided by the same positive number, the direction of the inequality is not changed.

    Case 2: If both sides of an inequality are multiplied or divided by the same negative number, the direction of the inequality is reversed.
12. When solving an inequality using both the addition and multiplication properties, it is usually easier to use the addition property first, then the multiplication property.
13. The graph of a linear equation in two variables separates the rectangular coordinate system into three regions: the line and the region on each side of the line.
    a. The graph of an inequality $>$ or $<$ is one of the two regions on either side of the line.
    b. The graph of an inequality $\geq$ or $\leq$ is the line and one of the two regions on either side of the line.

REVIEW PROBLEMS

Section 12.1

Graph the solutions to the following inequalities using the number line and give three values that satisfy each inequality.

1. $x < 8$
2. $x > 6$
3. $x \leq -1$
4. $x \geq -5$
5. $x > -3.5$
6. $x < 4\frac{1}{2}$
7. $x \geq -8$
8. $x \leq -5$
9. $x < 12$
10. $x > 8$

Section 12.2

Solve the following inequalities and graph the solutions using the number line.

1. $x + 2 > -3$
2. $x + 4 > 5$
3. $x - 1 < 11$
4. $x - 5 < 8$
5. $7 \geq x - 3$
6. $-4 \leq x - 3$

7. $-11 \leq x + 6$

8. $9 > x + 16$

9. $x - 22 < -18$

10. $x - 34 > -30$

Section 12.3

Solve the following inequalities and graph the solutions using the number line.

1. $5x > 30$

2. $7x < 21$

3. $15x \leq 45$

4. $20x \geq 60$

5. $39 < 13x$

6. $42 > 7x$

7. $-18x \geq 36$

8. $-9x < -45$

9. $-15 < -5x$

10. $-32 > -8x$

11. $-12x \leq 48$

12. $-64 \leq -16x$

Section 12.4

Solve the following inequalities and graph the solutions using the number line.

A

1. $2x - 7 > 3$

2. $4x + 8 < -4$

3. $8x + 26 \leq -6$

4. $12x - 32 > 16$

5. $16 > 6x - 8$

6. $14 < -4x - 6$

7. $-19 \geq -5x + 1$

8. $-7x + 3 \geq -11$

9. $-5x - 9 < -24$

10. $3 > -11x - 19$

B

11. $7x + 2 < 3x - 6$

12. $9x - 12 > 6x + 3$

13. $-11x + 7 \leq 4x - 8$

14. $2x - 5 < 7x + 10$

15. $-3x + 8 > -5x + 12$

16. $2x - 18 \geq 5x - 12$

17. $11x + 23 \geq 14x - 3$

18. $-21x + 15 < -18x - 3$

19. $23x - 31 < 17x + 5$

20. $37x - 52 > 25x + 8$

Section 12.5

Graph the following linear inequalities.

1. $y > -6x$

2. $y \geq 8x$

3. $y \leq 2x - 7$

4. $y < -4x + 5$

5. $y > -3x - 2$

6. $x + y < -3$

7. $3x + 2y < -5$

8. $-5x + 4y \leq -3$

9. $7x - 5y > 8$

10. $-3x + 7y > -5$

ANSWERS TO PRACTICE PROBLEMS

Section 12.1

1. true
2. true
3. false
4. false
5. true
6. false
7. true
8. true
9. false
10. true
11. true
12. false
14. $15 < 25$ or $25 > 15$
15. $12 > 4$ or $4 < 12$
16. $-5 > -8$ or $-8 < -5$
17. $3 > -7$ or $-7 < 3$
18. $4, 4\frac{1}{2}, 6$

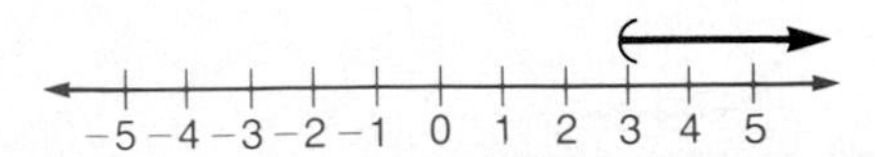

19. $3, 3.2, 7, 7.5, 9$

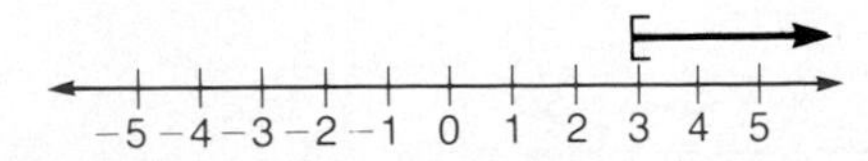

20. 3, 2.5, 2, 1, .5, -2, -5

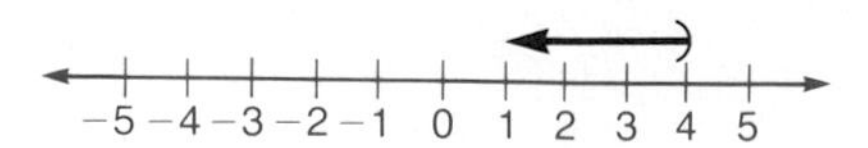

21. 4, 5, 7

22. -3, -3.5, $-3\frac{3}{4}$, -4, -5, -6, -8, -20

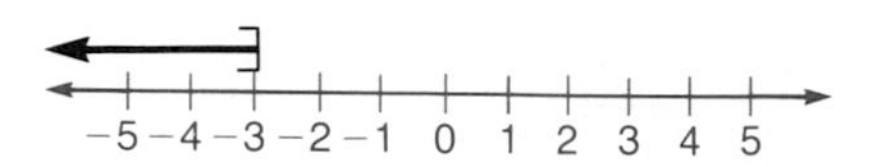

23. -3.5, -4, -8, -25, -50

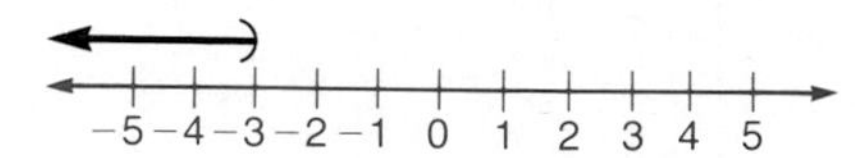

24. $3 > x$

25. $x < 5$

26. $x \geq -7$

27. $-4 < x$

28. $1 \geq x$

Section 12.2

1. $x < 4$

2. $5 > x$ or $x < 5$

3. $x \leq 3$

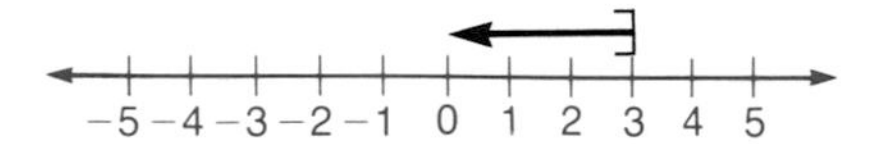

4. $3 \geq x$ or $x \leq 3$

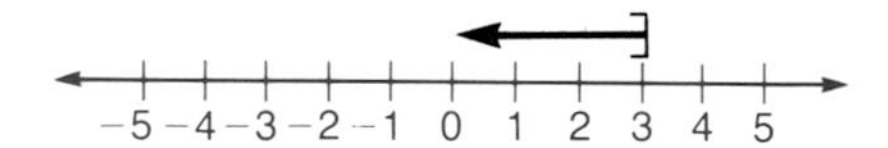

Section 12.3

1. $x < 3$

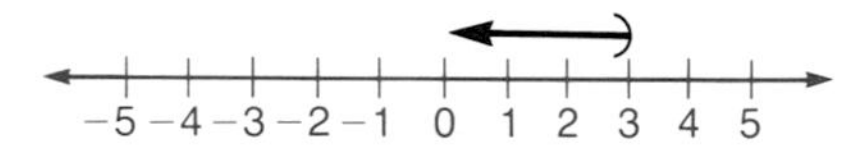

2. $x \geq 2$

3. $x > -2$

4. $x \geq -3$

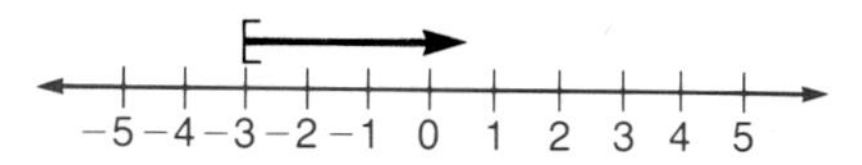

5. $x \leq 4$

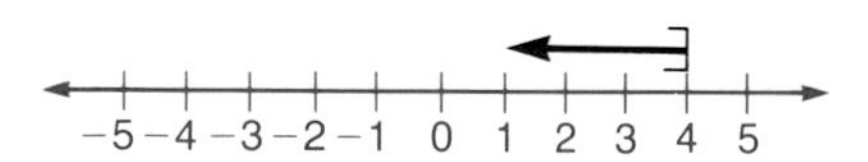

Section 12.4

1. $x > 2$

2. $x \leq 3$

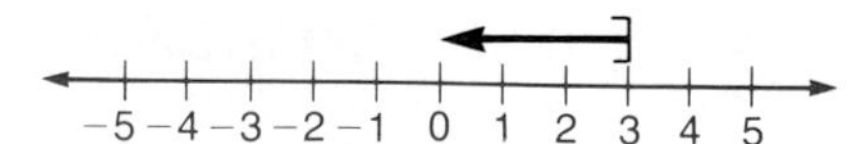

3. $x \leq -3$

4. $2 < x$ or $x > 2$

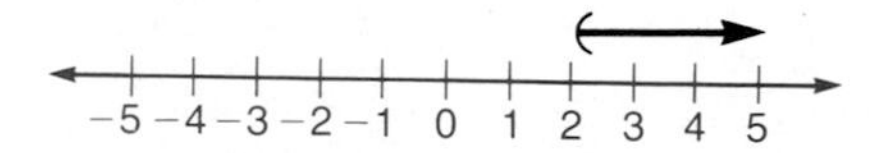

5. $x > -2$

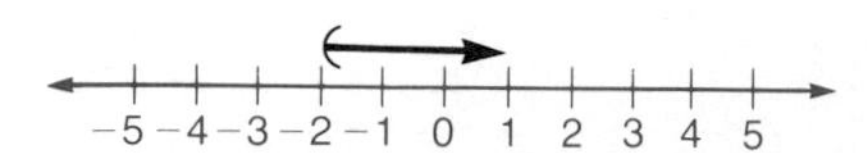

6. $x \geq 3$

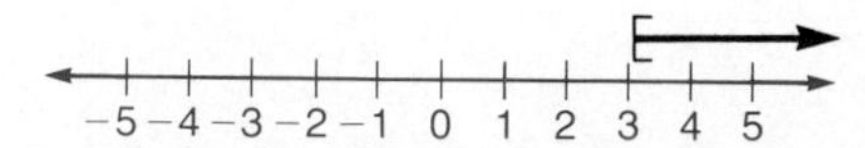

7. $x > 3$

8. $x \geq 2$

Section 12.5

1.

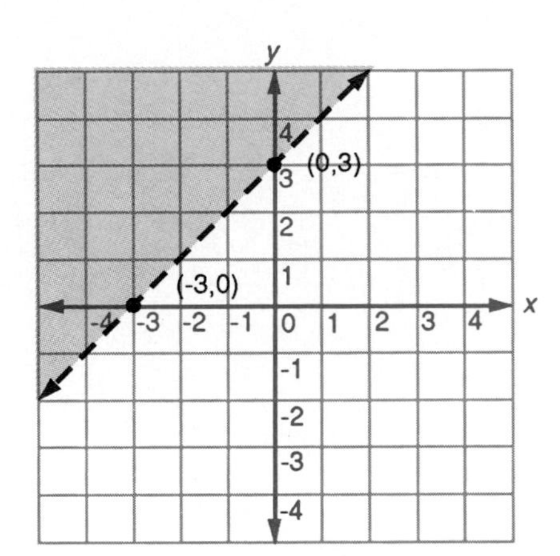

2.

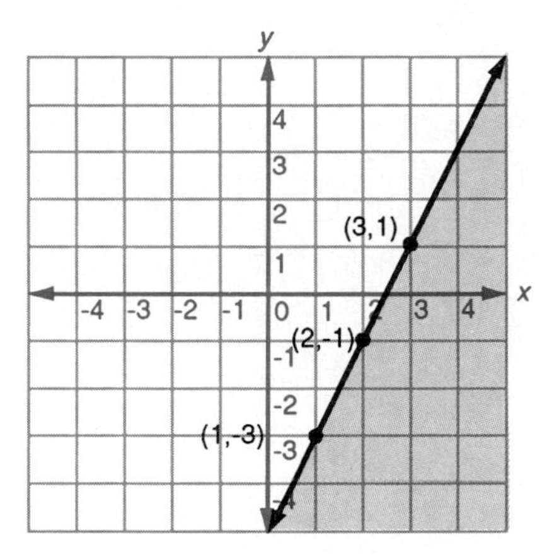

3.

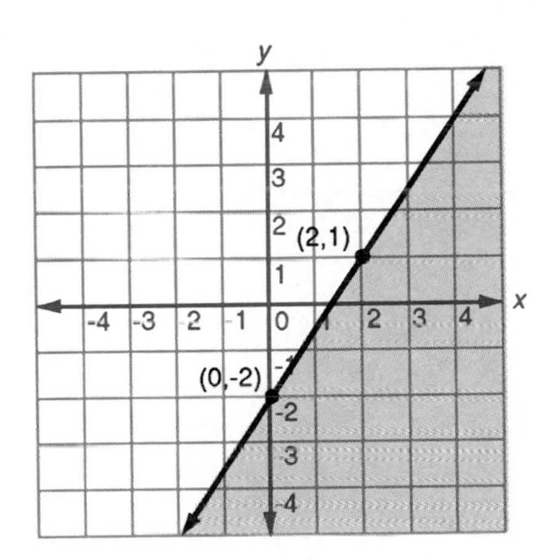

4.

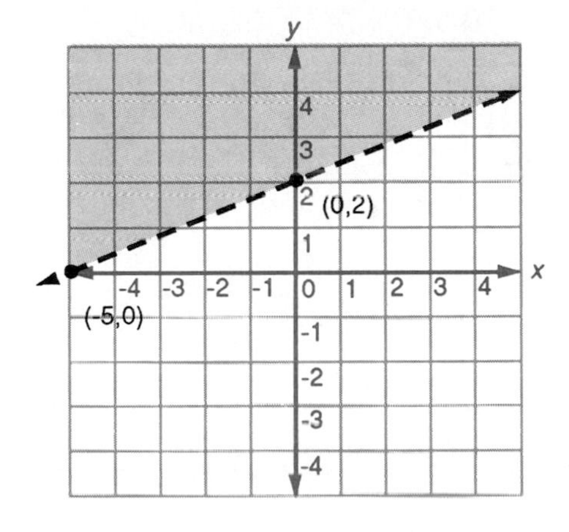

A P P E N D I X A

Computing Square Roots

To find the square root of a positive number that is not a perfect square, we will use the divide and average method for estimating square roots.

Example 1 **Find the square root of 30.**

Step 1: Find the perfect squares that are immediately below and above 30.

$$5 \times 5 = 25$$

$$6 \times 6 = 36$$

Thus, $\sqrt{30}$ is between 5 and 6.

Step 2: Estimate $\sqrt{30}$ to one decimal place.

Since 30 is approximately half way between 25 and 36 we estimate $\sqrt{30}$ to be 5.5.

Step 3: Divide 30 by 5.5, taking the answer to two decimal places.

$$\frac{30}{5.5} = 5.45 \text{ to two decimal places}$$

Step 4: Average 5.5 and 5.45 to two decimal places to find the second estimate.

$$\frac{5.5 + 5.45}{2} = \frac{10.95}{2} = 5.48$$

Step 5: Divide 30 by 5.48 to three decimal places.

$$\frac{30}{5.48} = 5.474 \text{ to three decimal places}$$

Step 6: Average 5.48 and 5.474 to three decimal places to find the third estimate.

$$\frac{5.48 + 5.474}{2} = \frac{10.954}{2} = 5.477$$

Now, rounding 5.477 to two decimal places, we get 5.48. Thus, $\sqrt{30}$ equals 5.48 to two decimal places. In order to get an answer accurate to three or more decimal places, you simply continue the process. ■

A N S W E R

1. ____________________

Practice 1: Find the square root of 19 to two decimal places using the divide and average method.

2. ____________________

Practice 2: Find the square root of 23 to two decimal places using the divide and average method.

EXERCISES

Find the square root to two decimal places using the divide and average method.

1. $\sqrt{2}$

2. $\sqrt{12}$

3. $\sqrt{70}$

4. $\sqrt{85}$

5. $\sqrt{32}$

ANSWERS TO PRACTICE PROBLEMS

1. 4.36

2. 4.80

A P P E N D I X B

Square Root Table

Table of Squares and Square Roots

n	n^2	$\sqrt{n}$	$\sqrt{10n}$	n	n^2	$\sqrt{n}$	$\sqrt{10n}$
1	1	1.000	3.162	51	2601	7.141	22.583
2	4	1.414	4.472	52	2704	7.211	22.804
3	9	1.732	5.477	53	2809	7.280	23.002
4	16	2.000	6.325	54	2916	7.348	23.238
5	25	2.236	7.071	55	3025	7.416	23.452
6	36	2.449	7.746	56	3136	7.483	23.664
7	49	2.646	8.367	57	3249	7.550	23.875
8	64	2.828	8.944	58	3364	7.616	24.083
9	81	3.000	9.487	59	3481	7.681	24.290
10	100	3.162	10.000	60	3600	7.746	24.495
11	121	3.317	10.488	61	3721	7.810	24.698
12	144	3.464	10.954	62	3844	7.874	24.900
13	169	3.606	11.402	63	3969	7.937	25.100
14	196	3.742	11.832	64	4096	8.000	25.298
15	225	3.873	12.247	65	4225	8.062	25.495
16	256	4.000	12.649	66	4356	8.124	25.690
17	289	4.123	13.038	67	4489	8.185	25.884
18	324	4.243	13.416	68	4624	8.246	26.077
19	361	4.359	13.784	69	4761	8.307	26.268
20	400	4.472	14.142	70	4900	8.367	26.458
21	441	4.583	14.491	71	5041	8.426	26.646
22	484	4.690	14.832	72	5184	8.485	26.833
23	529	4.796	15.166	73	5329	8.544	27.019
24	576	4.899	15.492	74	5476	8.602	27.203
25	625	5.000	15.811	75	5625	8.660	27.386
26	676	5.099	16.125	76	5776	8.718	27.568
27	729	5.196	16.432	77	5929	8.775	27.749
28	784	5.292	16.733	78	6084	8.832	27.928
29	841	5.385	17.029	79	6241	8.888	28.107
30	900	5.477	17.321	80	6400	8.944	28.284
31	961	5.568	17.607	81	6561	9.000	28.460
32	1024	5.657	17.889	82	6724	9.055	28.636
33	1089	5.745	18.166	83	6889	9.110	28.810
34	1156	5.831	18.439	84	7056	9.165	28.983
35	1225	5.916	18.708	85	7225	9.220	29.155
36	1296	6.000	18.974	86	7396	9.274	29.326
37	1369	6.083	19.235	87	7569	9.327	29.496
38	1444	6.164	19.494	88	7744	9.381	29.665
39	1521	6.245	19.748	89	7921	9.434	29.833
40	1600	6.325	20.000	90	8100	9.487	30.000
41	1681	6.403	20.248	91	8281	9.539	30.166
42	1764	6.481	20.494	92	8464	9.592	30.332
43	1849	6.557	20.736	93	8649	9.644	30.496
44	1936	6.633	20.976	94	8836	9.695	30.659
45	2025	6.708	21.213	95	9025	9.747	30.822
46	2116	6.782	21.448	96	9216	9.798	30.984
47	2209	6.856	21.679	97	9409	9.849	31.145
48	2304	6.928	21.909	98	9604	9.899	31.305
49	2401	7.000	22.136	99	9801	9.950	31.464
50	2500	7.071	22.361	100	10000	10.000	31.623

A P P E N D I X C

Answers to Odd-Numbered Exercises

Chapter 1

Exercise 1.1

1. $3(10) + 2$ **3.** $3(10) + 9$ **5.** $9(10) + 3$
7. $6(10) + 7$ **9.** $4(10) + 0$ **11.** $3(10) + 3$
13. $2(10^2) + 3(10) + 5$ **15.** $7(10^2) + 4(10) + 9$
17. $3(10^2) + 6(10) + 0$ **19.** $4(10^2) + 4(10) + 8$
21. $3(10^2) + 8(10) + 6$ **23.** $8(10^2) + 7(10) + 0$
25. $3(10^3) + 6(10^2) + 0(10) + 9$
27. $8(10^3) + 3(10^2) + 0(10) + 0$
29. $8(10^3) + 0(10^2) + 8(10) + 8$
31. $5(10^3) + 8(10^2) + 0(10) + 0$
33. $3(10^4) + 3(10^3) + 0(10^2) + 1(10) + 0$
35. $8(10^4) + 8(10^3) + 8(10^2) + 8(10) + 8$

Exercise 1.2

1. $4(10) + 7 = 47$ **3.** $5(10) + 9 = 59$
5. $2(10) + 5 = 25$ **7.** $9(10) + 8 = 98$
9. $1(10^2) + 5(10) + 9 = 159$
11. $1(10^2) + 4(10) + 3 = 143$
13. $1(10^2) + 3(10) + 7 = 137$
15. $1(10^2) + 7(10) + 6 = 176$
17. $4(10^2) + 4(10) + 7 = 447$
19. $4(10^2) + 2(10) + 6 = 426$
21. $9(10^2) + 7(10) + 3 = 973$
23. $1(10^3) + 1(10^2) + 2(10) + 0 = 1120$
25. $8(10^2) + 4(10) + 1 = 841$
27. $1(10^3) + 3(10^2) + 7(10) + 0 = 1370$
29. $1(10^3) + 2(10^2) + 0(10) + 5 = 1205$
31. $6(10^3) + 7(10^2) + 8(10) + 6 = 6786$
33. $1(10^4) + 2(10^3) + 8(10^2) + 0(10) + 2 = 12802$
35. $1(10^4) + 3(10^3) + 1(10^2) + 0(10) + 8 = 13108$
37. $7(10^4) + 6(10^3) + 1(10^2) + 9(10) + 2 = 76192$
39. $1(10^5) + 3(10^4) + 5(10^3) + 7(10^2) + 8(10) + 7 = 135787$

Applications/Section 1.2

1. 48 feet **3.** 290 feet **5.** 160 feet

Exercise 1.3

1. $2(10) + 1 = 21$ **3.** $4(10) + 1 = 41$
5. $3(10) + 3 = 33$ **7.** $3(10) + 5 = 35$
9. $3(10) + 1 = 31$ **11.** 6 **13.** $1(10) + 7 = 17$
15. $1(10) + 7 = 17$ **17.** $5(10^2) + 3(10) + 3 = 533$
19. $5(10^2) + 3(10) + 4 = 534$
21. $2(10^2) + 7(10) + 3 = 273$
23. $2(10^2) + 6(10) + 9 = 269$
25. $1(10^2) + 6(10) + 5 = 165$
27. $3(10^2) + 0(10) + 1 = 301$ **29.** $7(10) + 8 = 78$
31. $1(10^3) + 9(10^2) + 5(10) + 4 = 1954$
33. $1(10^3) + 1(10^2) + 5(10) + 5 = 1155$
35. $1(10^3) + 9(10^2) + 9(10) + 6 = 1996$
37. $2(10^3) + 2(10^2) + 9(10) + 7 = 2297$
39. $2(10^3) + 4(10^2) + 0(10) + 1 = 2401$

Applications/Section 1.3

1. \$207 **3.** 86 points **5.** \$8445

Exercise 1.4

1. $8(10) + 0 = 80$ **3.** $1(10^2) + 1(10) + 2 = 112$
5. $7(10) + 5$ **7.** $2(10^2) + 2(10) + 4 = 224$
9. $4(10^2) + 8(10) + 4 = 484$
11. $2(10^2) + 3(10) + 4 = 234$
13. $6(10^2) + 0(10) + 0 = 600$
15. $2(10^3) + 2(10^2) + 6(10) + 8 = 2268$
17. $3(10^3) + 8(10^2) + 1(10) + 9 = 3819$
19. $9(10^3) + 2(10^2) + 1(10) + 5 = 9215$
21. 24750 **23.** 45500 **25.** 51585 **27.** 31929
29. 65910 **31.** 24920 **33.** 56252 **35.** 87912
37. 21632 **39.** 85932 **41.** 140625 **43.** 319859
45. 431235 **47.** 708300 **49.** 2618826

Applications/Section 1.4

1. 2304 square feet **3.** \$690 **5.** \$360,815

Exercise 1.5

1. $1(10) + 4 = 14$ **3.** 9 **5.** 6
7. $2(10) + 1 + R3 = 21\ R3$ **9.** $1(10) + 7 = 17$
11. $2(10) + 1 = 21$ **13.** $4(10) + 2 = 42$
15. $2(10) + 2 = 22$ **17.** $5(10) + 7 = 57$
19. $7(10) + 9 = 79$ **21.** 3 **23.** 3
25. $2(10) + 1 = 21$ **27.** $3(10) + 1 = 31$
29. $4(10) + 3 = 43$ **31.** $2(10) + 1 = 21$
33. 14 R5 **35.** 13 **37.** 8 **39.** 4 R20 **41.** 29
43. 93 **45.** 216 **47.** 247 **49.** 32 **51.** 82

Applications/Section 1.5

1. \$5 **3.** 32 cents and 34 cents; 3 liter **5.** \$714

Application Exercises/Section 1.6

1. 3632 **3.** 54,259 pounds **5.** \$1842 **7.** \$16,000
9. \$9528 **11.** \$4536 **13.** \$152,800 **15.** \$13,866,840
17. \$832 **19.** \$584

Review Problems/Section 1.1

1. $1(10) + 7$ **3.** $4(10) + 7$ **5.** $6(10) + 4$
7. $7(10) + 1$ **9.** $4(10) + 8$
11. $3(10^2) + 7(10) + 2$ **13.** $7(10^2) + 0(10) + 0$
15. $2(10^2) + 2(10) + 2$ **17.** $8(10^2) + 3(10) + 9$
19. $1(10^2) + 0(10) + 1$
21. $4(10^3) + 7(10^2) + 3(10) + 4$
23. $8(10^3) + 0(10^2) + 3(10) + 7$
25. $5(10^3) + 0(10^2) + 0(10) + 2$
27. $1(10^3) + 0(10^2) + 0(10) + 0$
29. $5(10^4) + 3(10^3) + 0(10^2) + 3(10) + 0$

Review Problems/Section 1.2

1. $2(10) + 7 = 27$ **3.** $5(10) + 7 = 57$
5. $1(10^2) + 2(10) + 9 = 129$
7. $1(10^2) + 6(10) + 5 = 165$
9. $1(10^2) + 7(10) + 0 = 170$
11. $4(10^2) + 5(10) + 7 = 457$
13. $4(10^2) + 2(10) + 1 = 421$
15. $1(10^3) + 1(10^2) + 2(10) + 2 = 1122$
17. $1(10^3) + 1(10^2) + 3(10) + 1 = 1131$
19. $1(10^3) + 2(10^2) + 1(10) + 0 = 1210$
21. $5(10^3) + 5(10^2) + 1(10) + 9 = 5519$
23. $6(10^3) + 3(10^2) + 8(10) + 5 = 6385$
25. $1(10^4) + 5(10^3) + 6(10^2) + 6(10) + 2 = 15662$

Review Problems/Section 1.3

1. 8 **3.** $4(10) + 3 = 43$ **5.** $2(10) + 2 = 22$
7. 6 **9.** $3(10) + 3 = 33$
11. $1(10^2) + 2(10) + 7 = 127$
13. $2(10^2) + 0(10) + 7 = 207$
15. $1(10^2) + 7(10) + 7 = 177$
17. $5(10^2) + 0(10) + 7 = 507$
19. $2(10^2) + 1(10) + 3 = 213$
21. $2(10^3) + 0(10^2) + 1(10) + 2 = 2012$
23. $4(10^3) + 3(10^2) + 5(10) + 9 = 4359$ **25.** 1

Review Problems/Section 1.4

1. $7(10) + 8 = 78$ **3.** $8(10) + 8 = 88$
5. $2(10^2) + 6(10) + 0 = 260$
7. $1(10^3) + 0(10^2) + 1(10) + 5 = 1015$
9. $2(10^3) + 5(10^2) + 0(10) + 8 = 2508$ **11.** 6600
13. 33768 **15.** 40401 **17.** 23079 **19.** 76632
21. 41303 **23.** 105600 **25.** 810000
27. 1467416 **29.** 3452988

Review Problems/Section 1.5

1. 7 **3.** 8 **5.** 3 R3 **7.** $2(10) + 8 = 28$
9. $5(10) + 8 = 58$ **11.** 3 **13.** 6
15. $1(10) + 1 = 11$ **17.** $8(10) + 1 = 81$
19. $2(10) + 2 = 22$ **21.** 81 **23.** 24 **25.** 5 R10
27. 114 **29.** 46

Chapter 2

Exercise 2.1

1. −1 **3.** −26 **5.** 1 **7.** 23 **9.** −9
11. −5 **13.** 35 **15.** −58 **17.** −69
19. −25 **21.** −1 **23.** 5.05 **25.** −3.2
27. −.2 **29.** −5.74

Applications/Section 2.1

1. −15 degrees **3.** −12% **5.** −\$289 **7.** −5

Exercise 2.2

1. 14 **3.** −28 **5.** −16 **7.** −1 **9.** 38
11. 10 **13.** −40 **15.** −4 **17.** 0 **19.** 75
21. −20 **23.** 38 **25.** 10 **27.** 28.08
29. −1.67 **31.** 10.5

Applications/Section 2.2

1. −\$3980

Exercise 2.3

1. 56 **3.** −50 **5.** −24 **7.** 96 **9.** 168
11. −20 **13.** −220 **15.** 48 **17.** 720
19. −400 **21.** 56 **23.** −8.68 **25.** −102
27. −92.415 **29.** 10.142 **31.** −30.6

Applications/Section 2.3

1. −85 points

Exercise 2.4

1. −6 **3.** 9 **5.** −14 **7.** 14 **9.** 27
11. −29 **13.** −3 **15.** 4.4 **17.** 5.8 **19.** 3.2

Applications/Section 2.4

1. −13 points

Exercise 2.5/Part 1

1. 0 3. $\frac{-1}{2}$ 5. -1 7. $\frac{-5}{6}$ 9. $\frac{-1}{7}$
11. $\frac{1}{4}$ 13. $\frac{1}{2}$ 15. $\frac{11}{60}$ 17. $\frac{2}{5}$ 19. $\frac{-17}{42}$

Exercise 2.5/Part 2

1. $-\frac{1}{3}$ 3. $\frac{3}{16}$ 5. $-\frac{1}{4}$ 7. $\frac{1}{3}$ 9. $-\frac{2}{3}$
11. $-\frac{7}{25}$ 13. $\frac{2}{5}$ 15. $-\frac{1}{5}$

Exercise 2.5/Part 3

1. $\frac{-2}{3}$ 3. $\frac{2}{3}$ 5. $\frac{-7}{2}$ 7. $-\frac{3}{2}$ 9. $\frac{4}{15}$
11. $-\frac{5}{6}$ 13. $\frac{9}{20}$ 15. $-\frac{8}{27}$

Exercise 2.5/Part 4

1. $-1\frac{1}{4}$ 3. $2\frac{3}{4}$ 5. $-5\frac{7}{10}$ 7. $-2\frac{5}{6}$
9. $-2\frac{11}{15}$

Exercise 2.5/Part 5

1. $-19\frac{1}{4}$ 3. 13 5. -2 7. 6 9. $-5\frac{1}{4}$

Exercise 2.5/Part 6

1. 1 3. -2 5. $-\frac{5}{6}$ 7. $\frac{1}{2}$ 9. $\frac{1}{2}$

Application Exercises/Section 2.6

1. -1 3. -15 degrees 5. $125,000

Review Problems/Part 1

The examples given in the following problems are one of many examples that will meet the given requirement. Your answer will probably differ from the one given.
1. $8 + 4 = 12$ 3. $8 + (-5) = 3$ 5. $6 - 2 = 4$
7. $3 - 7 = -4$ 9. $5 - (-7) = 12$
11. $2 \times 5 = 10$ 13. $3(-2) = -6$ 15. $\frac{-12}{-6} = 2$

Review Problems/Section 2.1

1. 1 3. 4 5. -8 7. -95 9. -53
11. -4 13. -11 15. 2 17. -15.8

Review Problems/Section 2.2

1. 8 3. 7 5. -15 7. 6 9. 23 11. 1
13. -5.2 15. -9.78

Review Problems/Section 2.3

1. -12 3. -56 5. 49 7. 90 9. -88
11. 14 13. -28 15. 300 17. -105
19. -22.4 21. -8.5 23. -4.42 25. -90.1

Review Problems/Section 2.4

1. -3 3. -9 5. 3 7. -7 9. 3 11. 3
13. -1 15. -3.2 17. 7.5 19. -16

Review Problems/Section 2.5/Part 1

1. $\frac{1}{2}$ 3. $-\frac{1}{3}$ 5. $\frac{1}{2}$ 7. 0 9. $-\frac{26}{27}$

Review Problems/Section 2.5/Part 2

1. $-\frac{2}{5}$ 3. 2 5. $\frac{3}{4}$ 7. $\frac{4}{3}$ 9. $-\frac{1}{10}$

Review Problems/Section 2.5/Part 3

1. $-2\frac{3}{8}$ 3. $-6\frac{7}{10}$ 5. $-\frac{11}{12}$ 7. $-1\frac{15}{16}$
9. $3\frac{7}{22}$

Review Problems/Section 2.5/Part 4

1. $-9\frac{1}{3}$ 3. $-2\frac{2}{9}$ 5. 24 7. $-15\frac{3}{4}$
9. $1\frac{1}{14}$

Review Problems/Section 2.5/Part 5

1. $\frac{1}{2}$ 3. $-\frac{1}{3}$ 5. -1 7. $-1\frac{4}{5}$ 9. $-3\frac{1}{2}$

Chapter 3

Exercise 3.1.1

1. $9x$ **3.** $14y$ **5.** $20x$ **7.** $12x + 2y$
9. $10a + 8b$ **11.** $-2a$ **13.** $-20y$
15. $-15a + 10b$ **17.** $8x - 3xy$ **19.** $10x^2 - 3x$
21. $-6x^2 - 3x - 17$ **23.** $3xy^2 + 2xy + 23x$

Exercise 3.1.2

1. $7x$ **3.** $7x + 3$ **5.** $6y + 5$ **7.** $7x + 3$
9. $8x - 6$ **11.** $8x^2 + 7x + 7$ **13.** $8x^2 + 5x - 2$
15. $9x^2 - 2x$ **17.** $13x^2 + 13x - 5$
19. $30x^2 - 2x - 30$ **21.** $3y^2 + 3y + 4$
23. $-2y^2 - 2$ **25.** $8x^2 - 5x + 9$
27. $5x^2 - 3x + 3$

Exercise 3.2.1

1. $5x$ **3.** $10a$ **5.** $10x$ **7.** $-6a - 11b$
9. $-13x + 7y$ **11.** $-4x$ **13.** $-13x^2 + 5xy$
15. $-33abc + 22ab - 12bc$

Exercise 3.2.2

1. $3a$ **3.** 0 **5.** $x + 5$ **7.** $x + 5$ **9.** $5x + 5$
11. $3x^2 + 4x + 2$ **13.** $x + 5$ **15.** $3x^2 + 3x + 2$
17. $2x^2 - 3x - 4$ **19.** $-3x^2 + 4x - 6$
21. $2x^2 + 4x + 2$ **23.** $4x^2 + 22x + 12$
25. $8x^2 - x + 2$ **27.** $3a - 15b + 5c$
29. $8a - 6b - 2c$ **31.** $3y^2 - y - 7$
33. $2x^2 - 5x - 1$ **35.** $12a + 11b - 8c$

Exercise 3.3

1. $6x + 2$ **3.** $35a - 56$ **5.** $-24b + 12$
7. $40x^2 - 20x + 10$ **9.** $60x^2 - 12x + 48$
11. $35x^2 - 42x + 7$ **13.** $12y^2 - 16y - 3$
15. $27x^2 + 48x - 35$ **17.** $63x^2 - 41x + 6$
19. $60b^2 - 41b + 3$ **21.** $6a^2 + ab - b^2$
23. $20a^2 - 23ab + 6b^2$ **25.** $49x^2 - 14xy - 15y^2$
27. $6x^3 + 13x^2 + 17x + 20$
29. $3x^3 - 13x^2 + 27x - 20$
31. $40x^3 - 7x^2 + 42x + 9$
33. $30b^3 - 22b^2 - 66b + 56$
35. $3x^4 - 4x^3 - 4x^2 - 8x - 3$

Exercise 3.4

1. $2x + 1$ **3.** $3a + 2$ **5.** $x^2 - 3x + 9$
7. $x + 1$ **9.** $2x + 5$ **11.** $2x - 6$ **13.** $3x - 4$
15. $3x - 5$ **17.** $4x - 7$ **19.** $7x - 2$
21. $3a - 5 + \frac{14}{a + 4}$ **23.** $x + 3$
25. $3y + 9 + \frac{35}{2y - 7}$ **27.** $7x + 6$

Application Exercises/Section 3.5

1. $5 + x$ **3.** $x - 11$ **5.** $x + 3$ **7.** $8x$
9. $\frac{x}{-2}$ **11.** $16 - 2x$ **13.** $x + 5$ **15.** $3x$
17. $x + 7$ **19.** $.20c$ **21.** $x^2 - 4$

Review Problems/Section 3.1

1. $-5xy - 6x$ **3.** $-13x + 2y$ **5.** $10x - 20$
7. $28x - 32$ **9.** $9z - 30$ **11.** $4z^2 + 9z + 6$
13. $5y^2 - 8y + 4$ **15.** $9y^2 + 9y - 2$
17. $20x^2 - 10x - 5$ **19.** $4x^2 + 3x$

Review Problems/Section 3.2

1. $7x + y$ **3.** $24a - 5ab$ **5.** $a + 3$ **7.** $8a - 22$
9. $-7y + 1$ **11.** $2x^2 + 3x + 2$ **13.** $4x^2 + x + 3$
15. $4a^2 - 2a + 5$ **17.** $x^2 - 3x - 3$
19. $6y^2 - 30y + 25$ **21.** $9x^2 + 9x - 8$
23. $-16y^2 - y - 2$ **25.** $7x^5 - x^4 + 5x^3 + 5x^2 + 4$

Review Problems/Section 3.3

1. $8x - 8$ **3.** $-10a^2 + 35a - 15$
5. $2x^2 + 9x + 9$ **7.** $x^2 + 11x + 30$
9. $12b^2 + 9b - 30$ **11.** $6x^3 + 7x^2 + 8x + 3$
13. $30x^3 + 57x^2 + 31x + 14$
15. $12y^3 - y^2 - 38y + 24$
17. $10x^3 - 56x^2 + 56x + 32$
19. $18b^3 + 30b^2 - 60b + 12$
21. $-24x^3 + 57x^2 + 66x - 63$
23. $2x^4 - 4x^3 + 3x^2 - 2x + 1$

Review Problems/Section 3.4

1. $3x + 1$ **3.** $5a - 2$ **5.** $a + 1$ **7.** $x + 5$
9. $5x + 8$ **11.** $b - 2$ **13.** $3x - 2$ **15.** $4x + 6$
17. $7x - 1$ **19.** $4x - 2 + \frac{1}{2x + 3}$
21. $2a + 1 - \frac{9}{a + 5}$ **23.** $3x - 2 + \frac{9}{7x + 2}$
25. $4x - 2 + \frac{1}{2x + 1}$

Chapter 4

Exercise 4.0

1. yes **3.** yes **5.** no **7.** yes **9.** no **11.** yes
13. no **15.** ycs **17.** no

Exercise 4.1

1. 2 **3.** 4 **5.** 16 **7.** 19 **9.** -3 **11.** -10
13. 3 **15.** 27 **17.** 15.5 **19.** $8\frac{1}{2}$ **21.** $10\frac{1}{3}$
23. 3.7 **25.** $20\frac{4}{5}$

Exercise 4.2

1. 5 **3.** -6 **5.** 16 **7.** 5 **9.** -4 **11.** 21
13. 0 **15.** 12 **17.** 6 **19.** 21 **21.** -28
23. $5\frac{1}{2}$ **25.** -8.5 **27.** $2\frac{1}{2}$ **29.** $-7\frac{1}{2}$ **31.** 9
33. 21 **35.** -20 **37.** -16 **39.** 10 **41.** $\frac{5}{6}$

Exercise 4.3

1. 7 **3.** 7 **5.** -2 **7.** -6 **9.** 6 **11.** -4
13. -7 **15.** -3 **17.** 5 **19.** $-\frac{4}{5}$ **21.** 10
23. 9.2 **25.** -12 **27.** 10 **29.** 60

Exercise 4.4

1. -2 **3.** -9 **5.** 10 **7.** 4 **9.** 7 **11.** -3
13. 5 **15.** 6 **17.** -2 **19.** 5

Exercise 4.5

1. 16 **3.** 5 **5.** -1 **7.** -1.9 **9.** 2 **11.** 1
13. -3 **15.** -3 **17.** $\frac{1}{2}$ **19.** 3.6

Exercise 4.6

1. $6x + 3$ **3.** $10a + 15$ **5.** $30b - 54$
7. $-16y - 56$ **9.** $-2x - 6$ **11.** $-2a - 1$
13. $6x + 2$ **15.** $12x - 28$ **17.** $3y - 5$
19. $9x - 21$ **21.** $4t - 2$ **23.** $13x - 15$
25. $3b + 2$ **27.** $2y - 6$ **29.** $-5t + 15$
31. $2x + 6$ **33.** 0 **35.** 2 **37.** $13a + 12$
39. $14y - 37$ **41.** $-17b - 17$ **43.** $2x - 22$
45. $-16a - 1$ **47.** $2y + 2$ **49.** $25x - 5$
51. $-29x - 10$

Exercise 4.7

1. -3 **3.** 3 **5.** -7 **7.** -1 **9.** -1
11. 6 **13.** 1 **15.** -2 **17.** 1 **19.** -5

Application Exercises/Section 4.8

1. $7 + x = 15$; 8 **3.** $x - 11 = 48$; 59
5. $6x = x + 15$; 3 **7.** $2x + 4x = 72$; 12
9. $2w + 2(w + 5) = 70$; width $= 15$ inches, length $= 20$ inches **11.** $x + (x + 4.50) = 10$; \$2.75, \$7.25
13. $x + (x + 8) = 36$; 14 inches, 22 inches
15. $x + (x + 11) = 39$; Mary is 14 years old, Lynn is 25 years old **17.** $55t = 330$; 6 hours **19.** $x = \frac{12}{20}$; 60%

Review Problems/Section 4.1

1. 2 **3.** -14 **5.** 3 **7.** -7 **9.** -1
11. 10.8 **13.** 2 **15.** $\frac{1}{2}$

Review Problems/Section 4.2

1. 4 **3.** -4 **5.** -6 **7.** -5 **9.** 15 **11.** 12
13. $\frac{-3}{4}$ **15.** $17\frac{1}{2}$ **17.** 6.2 **19.** 2.4 **21.** 18.7
23. 3 **25.** 15.6 **27.** $\frac{-4}{5}$ **29.** $\frac{-3}{5}$

Review Problems/Section 4.3

1. 4 **3.** -5 **5.** -3 **7.** 4 **9.** 5 **11.** 1.6
13. 4 **15.** 4 **17.** -20 **19.** 2 **21.** -5.5

Review Problems/Section 4.5

1. 9 **3.** 1 **5.** -18 **7.** 2 **9.** -29

Review Problems/Section 4.6

1. $15a + 21$ **3.** $18t - 10$ **5.** $-6x - 30$
7. $-6b - 12$ **9.** $-20x + 28$ **11.** $10a + 15$
13. $-y - 2$ **15.** $-6y - 10$ **17.** $-4x - 1$
19. $x - 2$ **21.** $11x + 3$ **23.** $7t - 30$

Review Problems/Section 4.7

1. 3 **3.** 1 **5.** 6 **7.** 8 **9.** -13 **11.** 13
13. 3 **15.** -1

Chapter 5

Exercise 5.1/Part 1

1. 7 **3.** 2 **5.** 2 **7.** 5 **9.** 2 **11.** $a + b$
13. 11 **15.** 0

Exercise 5.1/Part 2

1. a^3 **3.** a^4b^4 **5.** $x^3y^2z^4$ **7.** y^4 **9.** x^5
11. b^9 **13.** $6a^6$ **15.** x^4y **17.** a^5b^6 **19.** x^2y^8
21. $6x^3y^3 + 10x^4y^2$ **23.** $32a^4b^9 - 28a^7b^6$ **25.** x^4
27. a^8 **29.** $-10a^3b^4$ **31.** a^6y^3
33. $6a^5b^5c - 15a^3b^4c^5$ **35.** $35x^5y^4 - 40x^5y^8 + 35xy^3$
37. x^8y^{12} **39.** $25a^6$ **41.** a^8b^9 **43.** $-15a^{10}b^3$
45. $a^{12}b^{18}$ **47.** $36x^{10}y^{10}$ **49.** $1024t^{30}$

Exercise 5.2/Part 1

1. $m + n$ **3.** nm **5.** n **7.** n **9.** $m + (-n)$
11. n **13.** 5 **15.** -2 **17.** 5 **19.** 0;1 **21.** 4
23. 1 **25.** 8 **27.** -27

Exercise 5.2/Part 2

1. x **3.** 1 **5.** $\frac{1}{x^2y^3}$ **7.** $\frac{1}{x^2}$ **9.** $16a^4$
11. $\frac{4x^2}{y^6}$ **13.** $\frac{x^4}{y^6}$ **15.** $\frac{25a^2}{b^4}$ **17.** a^{10} **19.** $\frac{1}{2x}$
21. $\frac{16x^4}{9y^6}$

Exercise 5.3

1. $\frac{a^2b^2 + 2ab + 1}{b^2}$ **3.** $\frac{x^2 + 4x + 4}{x^2}$ **5.** $\frac{xy}{y - x}$
7. $(2x + y)^2 = (2x + y)(2x + y) = 4x^2 + 4xy + y^2$

Review Problems/Section 5.1

1. x^7 **3.** $-10a^2$ **5.** $-9a^3$ **7.** x^{12} **9.** $4a^6$
11. $6x^3 + 8x^2$ **13.** $-12x^5 + 12x^4$ **15.** x^4y^3
17. $20a^3b^3$ **19.** $8x^9y^3$ **21.** a^6b^8 **23.** $4x^6$
25. $9a^4b^6$ **27.** $15x^5y^3 - 20x^7y^5$
29. $-6x^7y^3 + 10x^9y^2$ **31.** 1 **33.** $12a^7b^9$
35. $-a^3\,b^9\,c^6$

Review Problems/Section 5.2

1. $\frac{1}{a^2}$ **3.** $\frac{1}{x^5}$ **5.** a **7.** $\frac{1}{y}$ **9.** $\frac{2y^3}{x^4}$ **11.** $\frac{1}{a^2}$
13. $\frac{5}{a^5}$ **15.** $\frac{1}{b^4}$ **17.** $\frac{4x^2}{9y^4}$ **19.** $\frac{-8x^6}{27y^6}$ **21.** y^6
23. $\frac{1}{a^2b^2}$ **25.** $\frac{9a^4b^6}{4c^2}$ **27.** $\frac{1}{9x^6}$ **29.** $\frac{x^3}{2}$

Review Problems/Section 5.3

1. $\frac{(x^2 + 2x + 1)}{x^2}$ **3.** $\frac{a^2 - 2a + 1}{b^2}$
5. $\frac{1}{(a^4 - 2a^2b + b^2)}$

Chapter 6

Exercise 6.1

1. $x^2 - 4x + 4$ **3.** $4x^2 + 4x + 1$
5. $4x^2 - 12x + 9$ **7.** $9x^2 - 4$ **9.** $25x^2 - 10x + 1$
11. $16a^2 + 56a + 49$ **13.** $4a^2 - 49$
15. $x^2 + 5x + 6$ **17.** $a^2 - a - 6$ **19.** $4c^2 - 25$
21. $y^2 - 10y - 24$ **23.** $10x^2 - 39x - 27$
25. $7x^2 - 32x - 15$ **27.** $4a^2 - 2a - 12$
29. $a^2 - 4b^2$ **31.** $a^2 + 6ab + 9b^2$ **33.** $4x^2 - 9y^2$
35. $9x^2 + 24xy + 16y^2$ **37.** $2a^2 + 7ab + 3b^2$
39. $12x^2 + xy - 6y^2$

Exercise 6.2

1. $4(x + 2)$ **3.** $8(y + 3)$ **5.** $5(2x - 3)$
7. $19(y - 2)$ **9.** $9(3a - 4)$ **11.** $8y(y - 2)$
13. $-x(x^2 + x - 1)$ **15.** no common factors
17. $5a^2(a^2 + 5)$ **19.** $8x(2x^2 + 6x + 3)$
21. $xy(xy + 1)$ **23.** $9ay^2(3a^2 + 5)$
25. $45x^2(2x^2 - x + 4)$ **27.** $-5a^2bc(3b^2 + 7a^2c^2)$
29. $7x^2y^2(4x^2y^2 + 2xy + 5)$

Exercise 6.3

1. $(x + 2)(x - 2)$ **3.** $(x + 5)(x - 5)$
5. $(4y + 1)(4y - 1)$ **7.** $(8b + 5)(8b - 5)$
9. $(9a + 11)(9a - 11)$ **11.** $(x^2 + 15)(x^2 - 15)$
13. $(t^3 + 3)(t^3 - 3)$ **15.** $(2x^5 + 3)(2x^5 - 3)$
17. $(xy + 6)(xy - 6)$ **19.** $(9x^2 + 13y^3)\,(9x^2 - 13y^3)$

Exercise 6.4

1. $(x + 4)(x + 2)$ **3.** $(x + 5)(x - 1)$
5. $(x - 3)(x - 1)$ **7.** $(t - 7)(t - 5)$
9. $(x + 5)(x - 3)$ **11.** $(x + 3)(x + 3)$
13. $(a - 6)(a - 6)$ **15.** $(2x + 3)(x - 1)$
17. $(3x - 5)(2x - 1)$ **19.** $(2t + 1)(t - 5)$
21. $(2y - 5)(y - 5)$ **23.** $(3x + 1)(3x + 1)$
25. $(2x - 5)(2x - 5)$

Exercise 6.5

1. $2(x + 2)(x + 1)$ **3.** $3(a - 5)(a + 4)$
5. $2(x - 7)(x - 2)$ **7.** $12(x^2 - 5)$
9. $8(3x + 1)(3x - 1)$ **11.** $4t(t + 6)(t - 2)$
13. $2a(a + 3)(a + 3)$ **15.** $3x(2x^2 + 3x - 3)$
17. $2x(x + 5)(x - 3)$ **19.** $3a(a - 2)(a + 1)$
21. $x^3(x + 2)(x + 2)$ **23.** $5a^2(a - 1)(a - 1)$

Exercise 6.6

1. $-2,0$ **3.** 5 **5.** $-3,5$ **7.** 2,9 **9.** $-5,5$
11. $-4,4$ **13.** 7,9 **15.** $-10,-4$ **17.** $-1,\frac{7}{5}$
19. $-\frac{5}{8},\frac{5}{8}$ **21.** $\frac{1}{2},6$ **23.** $-\frac{1}{2},\frac{3}{7}$
25. $-\frac{7}{8},\frac{3}{2}$

Application Exercises/Section 6.7

1. 4 **3.** 8,9
5. base = 4 inches and height = 10 inches **7.** 14,16
9. 5 **11.** 10 inches **13.** base = 12 inches
15. 180 feet

Review Problems/Section 6.1

1. $x^2 - 2x + 1$ **3.** $4y^2 + 12y + 9$ **5.** $y^2 - 25$
7. $3t^2 - 11t - 4$ **9.** $25a^2 + 110a + 121$
11. $16t^2 + 22t - 3$ **13.** $14p^2 + 51p + 45$
15. $a^2 - 4ab + 4b^2$ **17.** $42a^2 - 11ab - 3b^2$
19. $49a^2 - 36b^2$

Review Problems/Section 6.2

1. $12(x+2)$ **3.** $3(5a-4)$ **5.** $8(4y-5)$
7. $7(5x+7)$ **9.** $21(3b-2)$ **11.** $9a(a-2)$
13. $3(x^4-5)$ **15.** $2(3t^2-6t+2)$
17. $7x(x^2-2x+1)$ **19.** $-3x^2(x^2+2)$
21. $5xy(3x^2y^2+5xy+8)$ **23.** $5ab(6a-2b-3)$
25. $-13ab^2(2a^2b^2-3a+4)$

Review Problems/Section 6.3

1. $(x+8)(x-8)$ **3.** $(3b+4)(3b-4)$
5. $(4x+7)(4x-7)$ **7.** $(3a+13)(3a-13)$
9. $(8x+11)(8x-11)$ **11.** $(t+m)(t-m)$
13. $(3y+1)(3y-1)$ **15.** $(x^5+1)(x^5-1)$
17. $(t^2+7)(t^2-7)$ **19.** $(a^3+b^2)(a^3-b^2)$

Review Problems/Section 6.4

1. $(x+5)(x+1)$ **3.** $(x+6)(x+5)$
5. $(y+7)(y-3)$ **7.** $(x-15)(x+2)$
9. $(a-16)(a+3)$ **11.** $(2x+1)(2x+1)$
13. $(2x+3)(2x+3)$ **15.** $(2a-3)(3a+1)$
17. $(4x+3)(x+3)$ **19.** $(3t-4)(4t+3)$
21. $(3x-7)(4x+5)$ **23.** $(2b+3)(b-1)$
25. $(3y-10)(2y+7)$

Review Problems/Section 6.5

1. $2(3x+1)(3x+1)$ **3.** $6(2x+5)(x-2)$
5. $15(3x-2)(2x-1)$ **7.** $6(x^2+25)$
9. $3(4x+7)(4x-7)$ **11.** $2x^2(3x-5)(x-1)$
13. $5x(2x^2-7x-5)$ **15.** $4x(x+2)(x-2)$
17. $2y(3y+1)(y+2)$ **19.** $x^3(4x+1)(4x-1)$

Review Problems/Section 6.6

1. $-3,5$ **3.** $4,6$ **5.** $\frac{-1}{3},\frac{1}{2}$ **7.** $\frac{3}{2},\frac{5}{3}$
9. $\frac{-1}{4},\frac{5}{2}$ **11.** $-4,-2$ **13.** $-6,1$
15. $5,6$ **17.** $-3,3$ **19.** $\frac{-1}{2},3$

Chapter 7

Exercise 7.1

1. $\frac{1}{3}$ **3.** $\frac{3ab}{4}$ **5.** $\frac{2y^2}{3x^3z^2}$ **7.** $\frac{a}{6}$
9. $\frac{1}{a-3}$ **11.** $\frac{2y+3}{2y-3}$ **13.** $a+b$ **15.** $a-5b$
17. $\frac{y-6}{y-1}$ **19.** $\frac{m+4}{2m+3}$ **21.** $\frac{x+5}{x+4}$ **23.** -1

Exercise 7.2

1. $\frac{3m}{4}$ **3.** $\frac{2}{x^2y^3}$ **5.** $\frac{-3x^2}{8y}$ **7.** $\frac{6}{a+b}$ **9.** 1
11. $\frac{2a}{3}$ **13.** z^2 **15.** 1 **17.** $\frac{x^2+x}{3x-3}$
19. $\frac{y+1}{y-3}$ **21.** $\frac{2x+8}{x-2}$ **23.** $\frac{x^2-13x+42}{x^2+9x+14}$

Exercise 7.3

1. $\frac{1}{2xy}$ **3.** $\frac{3}{x+2}$ **5.** 3 **7.** $\frac{y^2+y}{y-1}$
9. $\frac{5z-10}{z-1}$ **11.** $\frac{1}{x}$ **13.** $\frac{18}{r^2}$ **15.** -1
17. $\frac{y+3}{y+4}$ **19.** $\frac{a^2+2a}{a^2+5a+24}$ **21.** $\frac{a+3}{a+4}$

Exercise 7.4/Part 1

1. $\frac{5x}{x+1}$ **3.** $\frac{x^2y+1}{x^3}$
5. $\frac{3m^2+5m^2n+8n}{4m^3n^3}$ **7.** $\frac{m-2}{m^2-9}$
9. $\frac{x^2}{x^2-1}$ **11.** $\frac{1}{y^2-9y+18}$

Exercise 7.4/Part 2

1. $\frac{x^2-4y}{6xy}$ **3.** $\frac{1}{3ab}$ **5.** $\frac{6-2x}{x^2}$ **7.** $\frac{4-a}{4a-4b}$
9. $\frac{3}{(m+1)(m-1)(m+2)}$
11. $\frac{6x}{(2x-1)(2x+1)(x+5)}$

Exercise 7.5

1. $\frac{1}{2}$ **3.** 24 **5.** 1 **7.** 1 **9.** -8 **11.** 0
13. $4,6$ **15.** $\frac{-3}{8}$ **17.** $\frac{13}{4}$ **19.** $\frac{-8}{3}$

Application Exercises/Section 7.6

1. $\frac{12}{9}$ **3.** 5 **5.** 2 **7.** $1\frac{1}{5}$ hours
9. $1\frac{3}{5}$ inches **11.** 18 cars **13.** 15 gallons
15. 24 pounds **17.** 15 mph

Review Problems/Section 7.1

1. $\frac{5}{y}$ 3. $\frac{1}{5}$ 5. $\frac{1}{x+2}$ 7. $n+1$ 9. $\frac{z}{8}$
11. $\frac{m+1}{m+3}$ 13. $\frac{n+2}{2n+1}$ 15. $\frac{x+y}{x-y}$ 17. $\frac{2x-4}{x-5}$
19. $\frac{3x-2}{5x-1}$

Review Problems/Section 7.2

1. $\frac{3}{x}$ 3. $\frac{3w+1}{5w-2}$ 5. $\frac{30}{7}$ 7. $\frac{9x+12}{2x+6}$
9. $\frac{2x+1}{2}$ 11. 1 13. $\frac{3x^2-2x-8}{12x^2+16x+4}$
15. $\frac{42x^2+27x+3}{3x^2+7x+2}$ 17. $\frac{10}{3x^2}$

Review Problems/Section 7.3

1. $\frac{6}{7}$ 3. $\frac{x}{x-1}$ 5. $\frac{x-1}{x(y+1)}$
7. $\frac{x+2}{6x+9}$ 9. $\frac{z-2}{z+2}$

Review Problems/Section 7.4/Part 1

1. $\frac{7}{x}$ 3. $\frac{y+x}{xy}$ 5. $\frac{37}{4x+8}$
7. $\frac{4x+1}{(x-2)(x-2)(x+1)}$ 9. $\frac{a^2+6a-9}{(a+3)(a+3)(a-3)}$

Review Problems/Section 7.4/Part 2

1. $\frac{3}{8x}$ 3. 3 5. $\frac{-x+4}{x^2-1}$
7. $\frac{x^2+4x-6}{(x+2)(x-2)(x+7)}$ 9. $\frac{8}{x^2-16}$

Review Problems/Section 7.5

1. $\frac{3}{2}$ 3. 3 5. 3 7. $-2,1$ 9. -1

Chapter 8

Exercise 8.1/Part 1

1. $-5,5$ 3. $-15,15$ 5. $-25,25$
7. $-14,14$

Exercise 8.1/Part 2

1. 2 3. -8 5. -9 7. -30 9. .1

Exercise 8.1/Part 3

1. $\sqrt{8}$ 3. $\sqrt{5}$ 5. $\sqrt{y}$ 7. $\sqrt{y^3}$ 9. $\sqrt{(a-b)}$

Exercise 8.1/Part 4

1. $8^{1/2}$ 3. $(x^2)^{1/2}$ 5. $(x-1)^{1/2}$ 7. $(5y^2)^{1/2}$
9. $(x^3)^{1/2}$

Exercise 8.1/Part 5

1. x 3. $5x$ 5. $13x^2$ 7. $3xy^2$ 9. $7x^4$

Exercise 8.1/Part 6

1. 16 3. $16x^2$ 5. x^2 7. $2x$ 9. 81

Exercise 8.2

1. 5 3. $3\sqrt{10}$ 5. 8 7. $4\sqrt{10}$ 9. $2x\sqrt{x}$
11. $3x^2\sqrt{2x}$ 13. $5y\sqrt{5xy}$ 15. $3y^2\sqrt{3x}$ 17. $8a^3b^2$
19. $4x^2y^4$ 21. $4m^2n^4\sqrt{2mn}$

Exercise 8.3

1. $\sqrt{6}$ 3. 6 5. $3\sqrt{2}$ 7. $3\sqrt{5}$ 9. $b\sqrt{a}$
11. $3a\sqrt{2}$ 13. $3\sqrt{5}-3$ 15. $x\sqrt{2}-3\sqrt{x}$
17. $2a\sqrt{b}+ab\sqrt{2}$ 19. $x-25$
21. $a-2\sqrt{a}+1$ 23. $y+\sqrt{y}-6$
25. $9x-y$

Exercise 8.4

1. $\sqrt{\frac{6}{2}}$ 3. $\sqrt{\frac{30}{12}}$ 5. $2\sqrt{\frac{x}{x}}$
7. $4\sqrt{\frac{2x}{x}}$ 9. $2\sqrt{\frac{17}{17}}$ 11. $\sqrt{\frac{2x}{2x}}$
13. $\frac{1}{x}$ 15. $x\sqrt{\frac{3y}{y}}$ 17. $\sqrt{\frac{a(b+1)}{b+1}}$
19. $2x^2\sqrt{\frac{2x}{y}}$

Exercise 8.5

1. 5 3. $\sqrt{5}$ 5. $\sqrt{2x}$ 7. $x\sqrt{\frac{3y}{3y}}$
9. $\frac{2+\sqrt{2}}{2}$ 11. $\frac{5\sqrt{3}+6\sqrt{5}}{15}$ 13. $2-\sqrt{2}$
15. $3\sqrt{2}-3$ 17. $5\sqrt{2}-2\sqrt{10}$ 19. $\frac{3\sqrt{x}-3}{x-1}$

21. $\frac{2a - a\sqrt{a}}{4 - a}$ **23.** $3 - 2\sqrt{2}$
25. $2 - \sqrt{3}$ **27.** $\frac{3y + 10\sqrt{y} + 3}{y - 9}$

Exercise 8.6

1. $4\sqrt{2}$ **3.** $5\sqrt{5}$ **5.** $2\sqrt{2}$ **7.** $8\sqrt{5}$
9. $(a - a^2)\sqrt{a}$ **11.** $(3 - 2a)\sqrt{a}$ **13.** $19\sqrt{x}$
15. $(x - 3)\sqrt{y}$ **17.** $(1 + x - \sqrt{y})\sqrt{x}$ **19.** 0

Exercise 8.7

1. 1 **3.** 1 **5.** 3 **7.** $\frac{5}{8}$ **9.** 4 **11.** 2
13. 3 **15.** no solution **17.** 0,4 **19.** 5
21. -1

Application Exercises/Section 8.8

1. 5 inches **3.** 50 inches **5.** 40 feet
7. 5 and 15 **9.** 21

Review Problems/Section 8.1/Part 1

1. $\sqrt{6}$ **3.** $\sqrt{a}$ **5.** $\sqrt{2x}$ **7.** $5\sqrt{a}$
9. $\sqrt{x - 1}$

Review Problems/Section 8.1/Part 2

1. $5^{1/2}$ **3.** $(2a)^{1/2}$ **5.** $\left(\frac{y}{5}\right)^{1/2}$ **7.** $(a^2 + b^2)^{1/2}$
9. $\left(\frac{y - 2}{3}\right)^{1/2}$

Review Problems/Section 8.1/Part 3

1. 4 **3.** x^2 **5.** $2y$ **7.** 5 **9.** a^6b^6

Review Problems/Section 8.2

1. 6 **3.** $2\sqrt{2}$ **5.** $6\sqrt{2}$ **7.** $x^2y\sqrt{y}$ **9.** $4b\sqrt{2a}$
11. $9xy\sqrt{x}$

Review Problems/Section 8.3

1. $3\sqrt{7}$ **3.** $3\sqrt{10}$ **5.** ab **7.** y^2
9. $x - \sqrt{x}$ **11.** $x - 1$

Review Problems/Section 8.4

1. $\sqrt{\frac{3}{2}}$ **3.** $\sqrt{\frac{3}{3}}$ **5.** $\sqrt{\frac{5y}{y}}$ **7.** $\sqrt{\frac{3}{3}}$
9. $\frac{\sqrt{x - 1}}{x - 1}$ **11.** $\frac{3\sqrt{3x}}{2y}$

Review Problems/Section 8.5

1. $\frac{\sqrt{6} + \sqrt{2}}{2}$ **3.** $\frac{y + \sqrt{y}}{y}$ **5.** $4\sqrt{2} - 4$
7. $\sqrt{3} - \sqrt{2}$ **9.** $\frac{x + x\sqrt{x}}{1 - x}$ **11.** $\sqrt{a} + \sqrt{b}$

Review Problems/Section 8.6

1. $3\sqrt{5}$ **3.** $9\sqrt{2}$ **5.** $(2 - a)\sqrt{a}$
7. $(x^2 + x + 1)\sqrt{x}$ **9.** $(a - 2)\sqrt{b}$
11. $3\sqrt{y - 1}$

Review Problems/Section 8.7

1. -1 **3.** 9 **5.** 7 **7.** 48 **9.** 4 **11.** 7

Chapter 9

Exercise 9.1

1. $-4,4$ **3.** $-3,3$ **5.** $-2,2$ **7.** $-5,5$
9. $-3,1$ **11.** $-5\sqrt{2},5\sqrt{2}$ **13.** $-5,5$
15. $-\sqrt{10},\sqrt{10}$ **17.** $-4,4$ **19.** $-\frac{1}{2},\frac{1}{2}$

Exercise 9.2

1. $\frac{-1}{2},1$ **3.** $0,\frac{3}{5}$ **5.** $-3,2$ **7.** $-2,-1$
9. $-5,7$ **11.** $\frac{-1}{2},3$ **13.** 0,6 **15.** 4
17. $-2,3$ **19.** $\frac{-5}{3},3$

Exercise 9.3

1. 2,3 **3.** $-3,7$ **5.** $\frac{1 \pm \sqrt{17}}{4}$ **7.** no solution
9. $-4 \pm \sqrt{13}$ **11.** $3 \pm \sqrt{17}$ **13.** $\frac{-1 \pm \sqrt{85}}{14}$
15. no solution **17.** $\frac{-2 \pm \sqrt{10}}{2}$ **19.** $\frac{1 \pm \sqrt{11}}{2}$

Application Exercises/Section 9.4

1. -6 or 6 **3.** 10 and 12
5. length = 14 cm and width = 10 cm
7. $\frac{-2+\sqrt{46}}{2}$ cm and $\frac{2+\sqrt{46}}{2}$ cm
9. 100 cm² and 225 cm²

Review Problems/Section 9.1

1. $-5,5$ **3.** $-3\sqrt{2},3\sqrt{2}$ **5.** $\pm 3\sqrt{2}$ **7.** $-3,3$
9. $\frac{-1}{3},\frac{1}{3}$

Review Problems/Section 9.2

1. $-1,6$ **3.** $-7,1$ **5.** $\frac{-1}{2},2$ **7.** $\frac{1}{6},3$
9. $-\frac{7}{2},0$

Review Problems/Section 9.3

1. $-5,4$ **3.** $1,9$ **5.** $1 \pm \sqrt{3}$ **7.** $\frac{-1 \pm \sqrt{10}}{3}$
9. no solution

Chapter 10

Exercise 10.1

1.

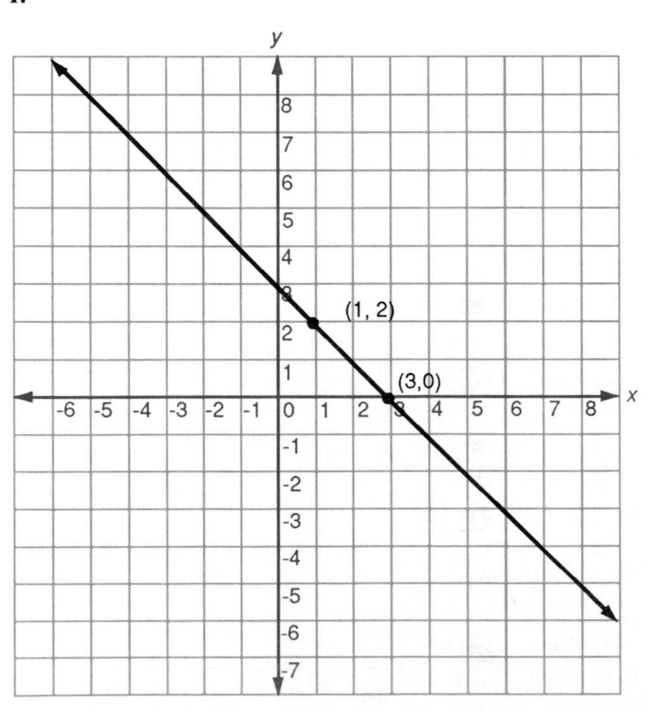

3.

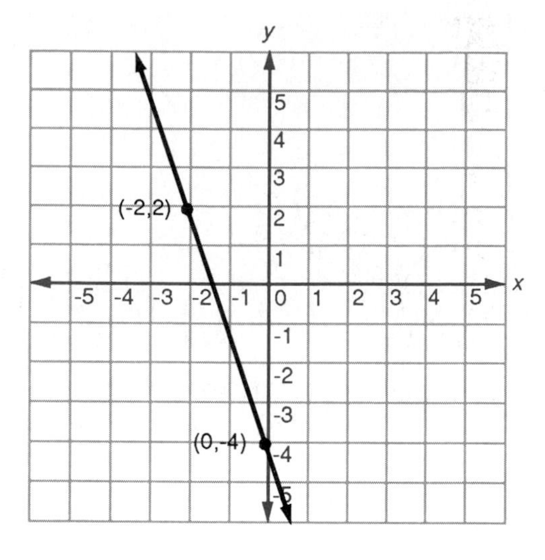

5.

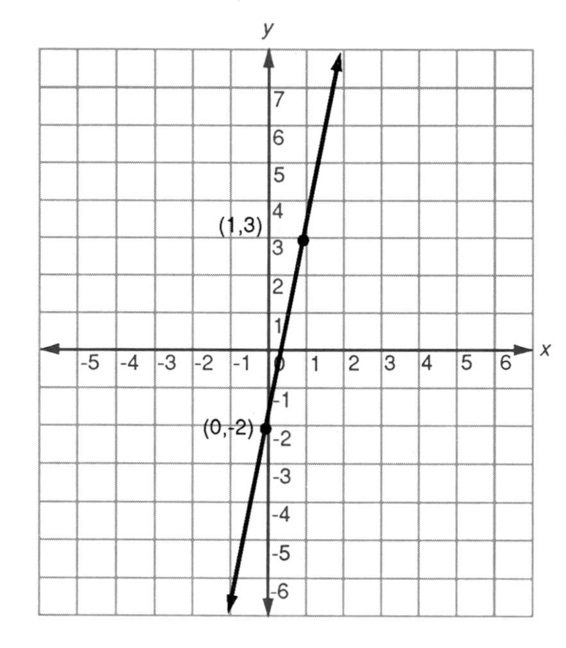

7.

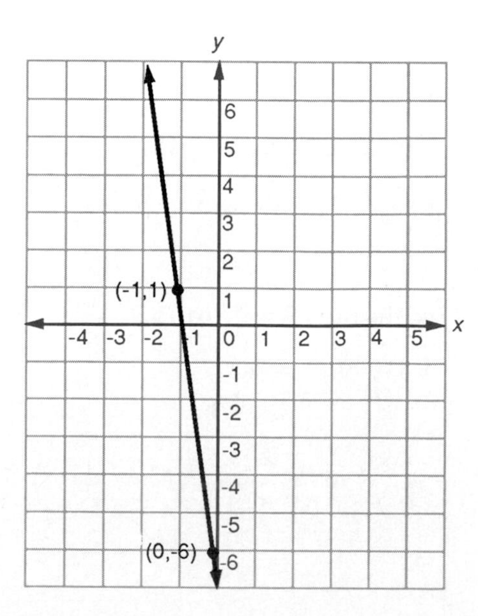

9.

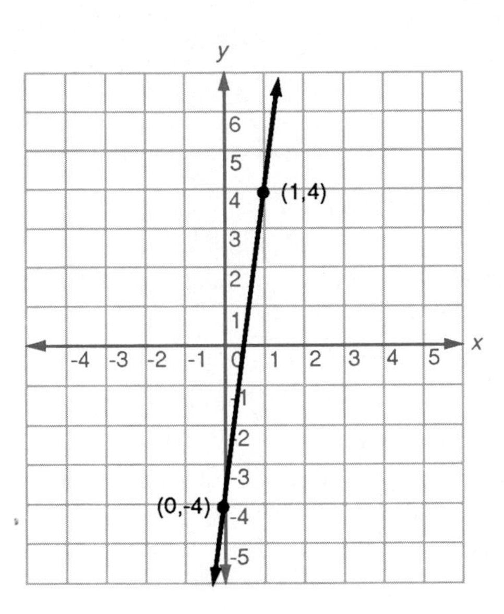

11.

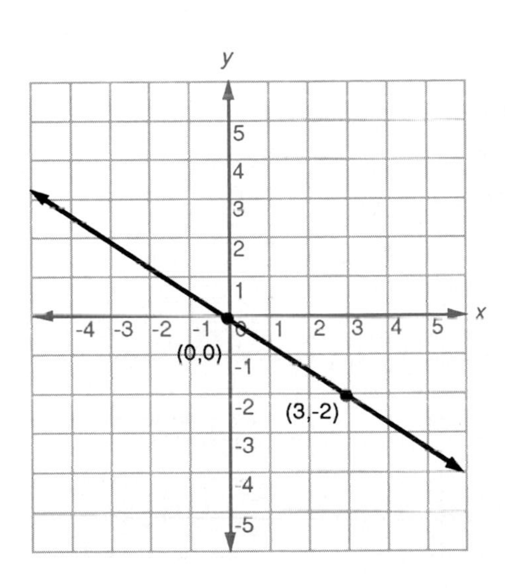

13.

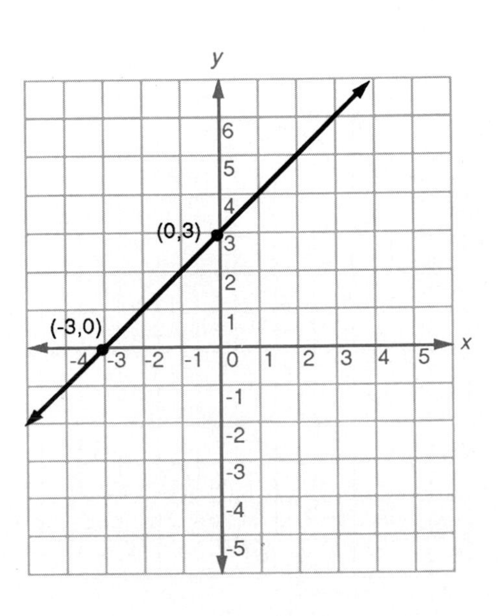

15.

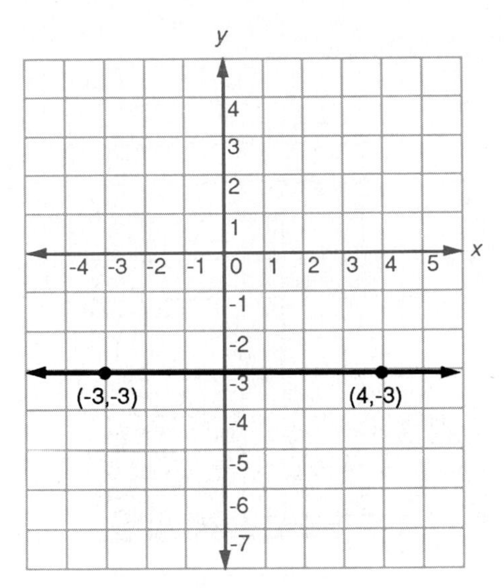

17.

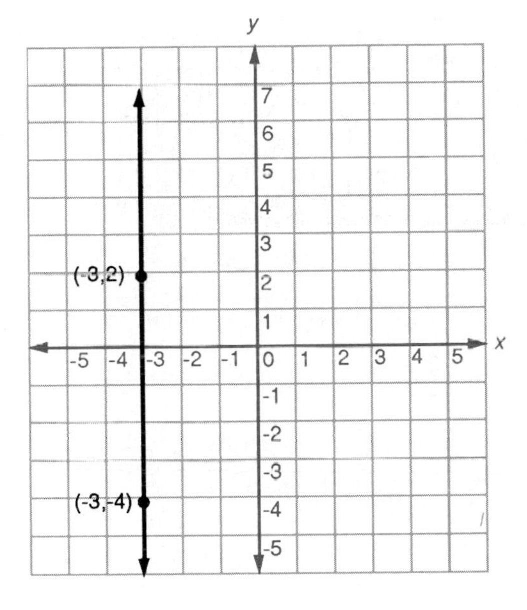

19.

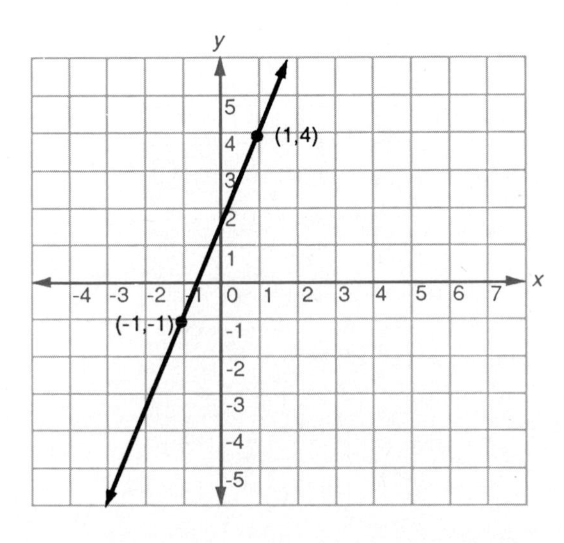

Exercise 10.2

1. $y = -5x + 8$

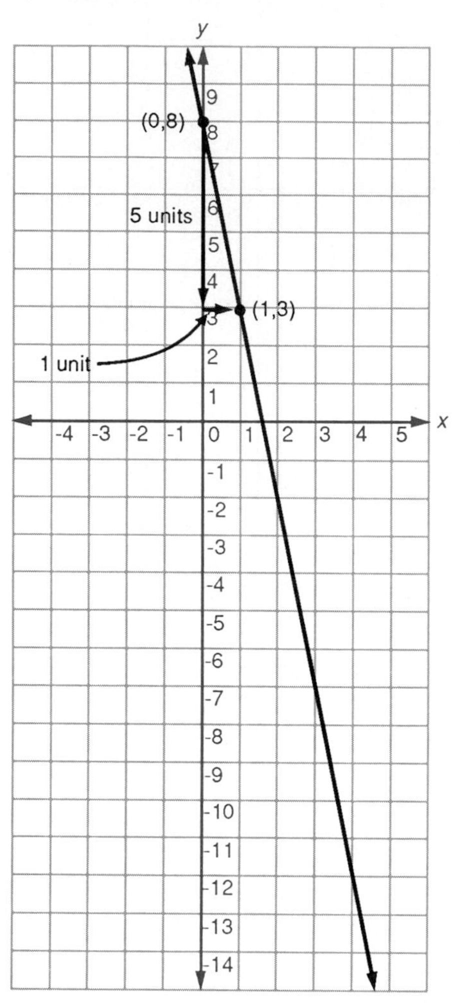

3. $y = 3x + 6$

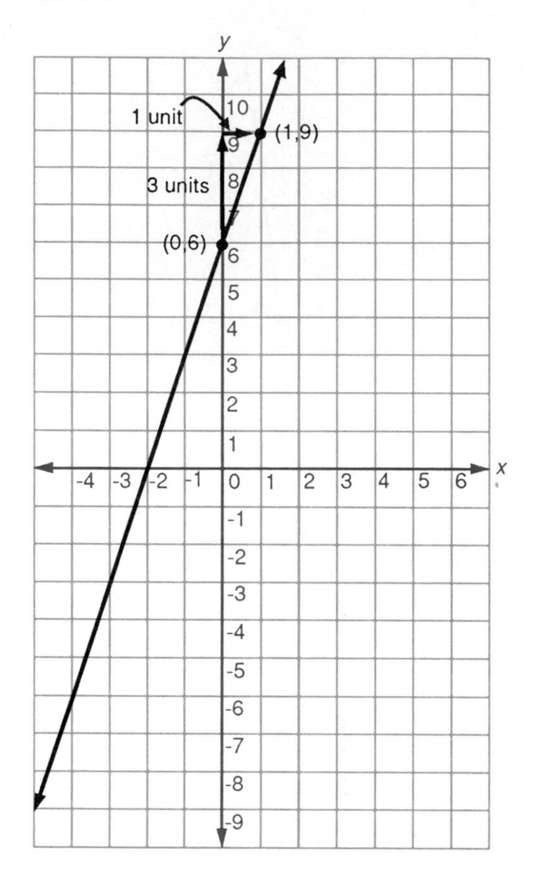

5. $y = \frac{2}{5}x + 6$

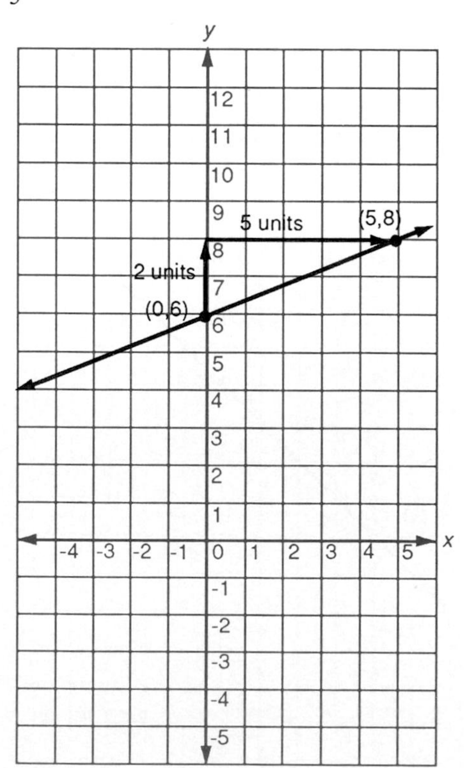

7. $y = \frac{6}{7}x - 14$

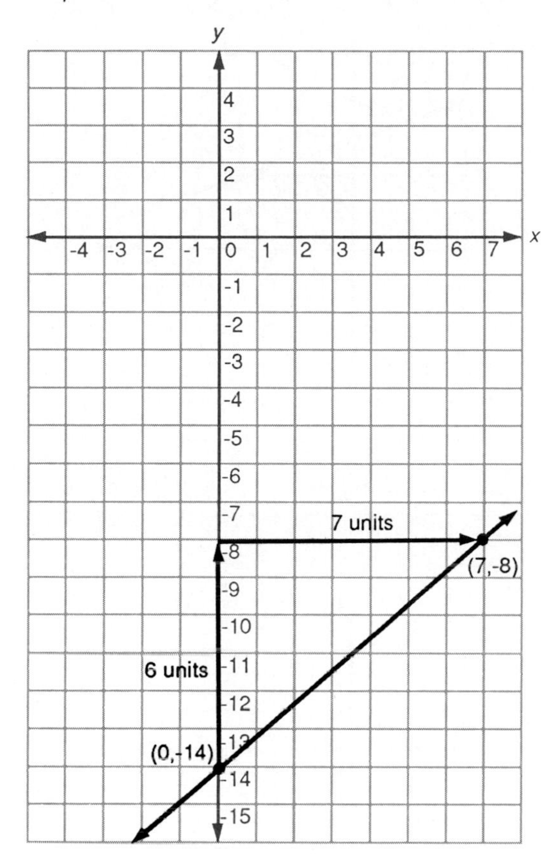

9. $y = -\frac{4}{5}x + 7$

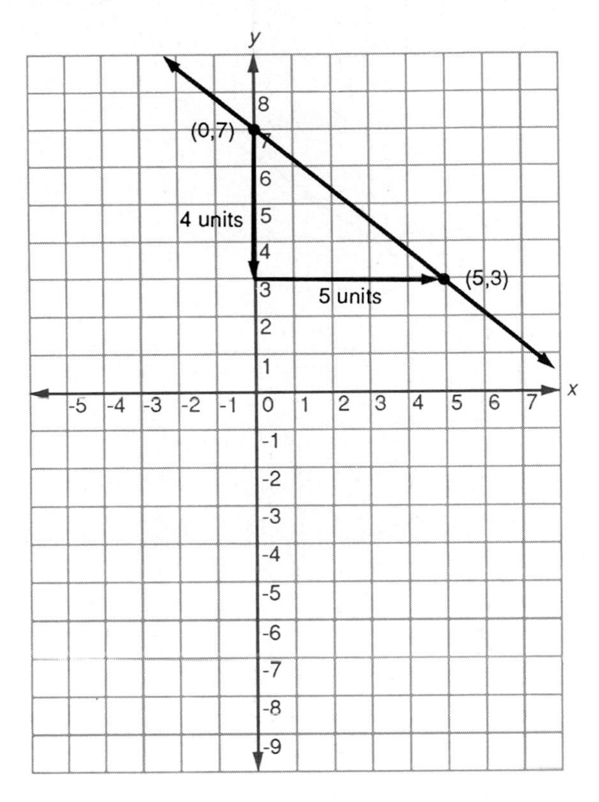

11. $y = \frac{5}{4}x - 8$

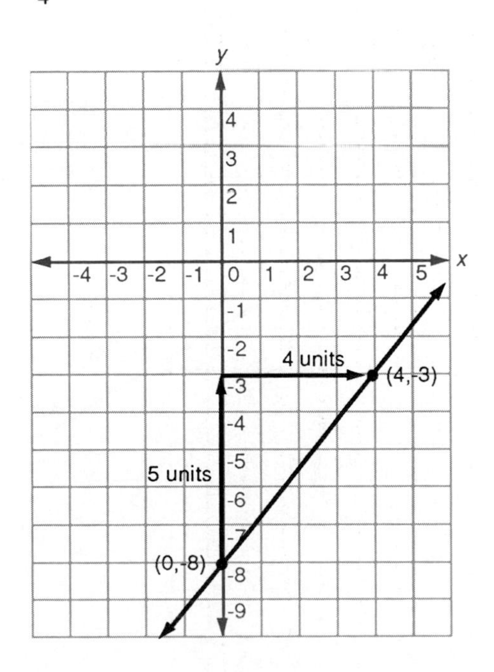

13. $y = \frac{3}{4}x + 5$

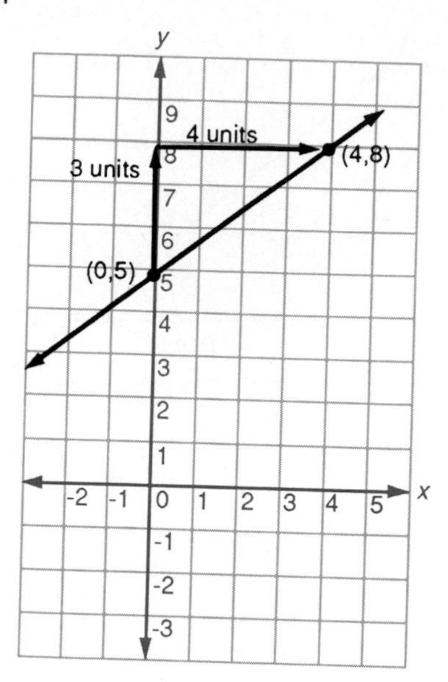

15. $y = \frac{-3}{5}x - 2$

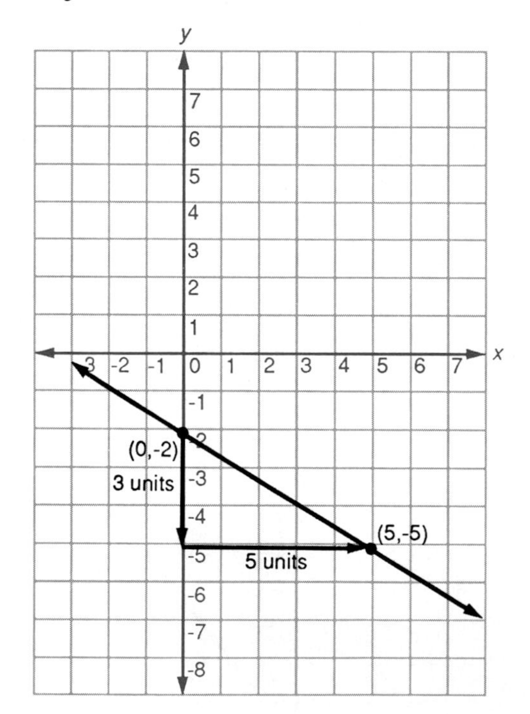

17. $y = \frac{-1}{3}x + 3$

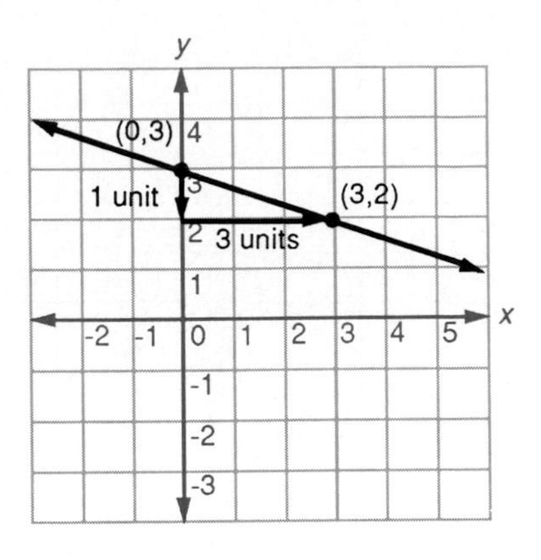

19. $y = 2y = 0x + 2y = \frac{0}{1}x + 2$

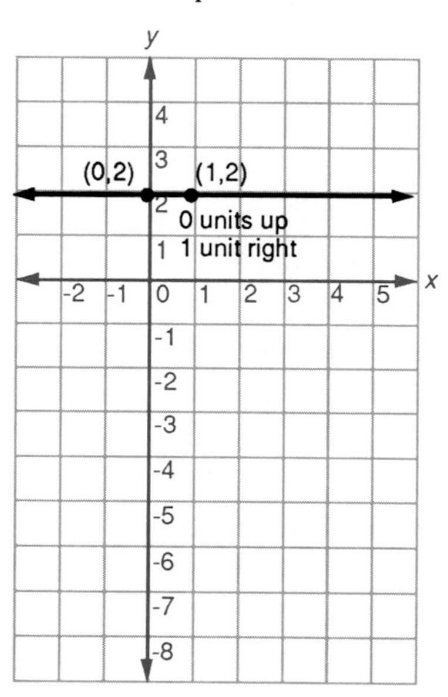

Exercise 10.3

1.

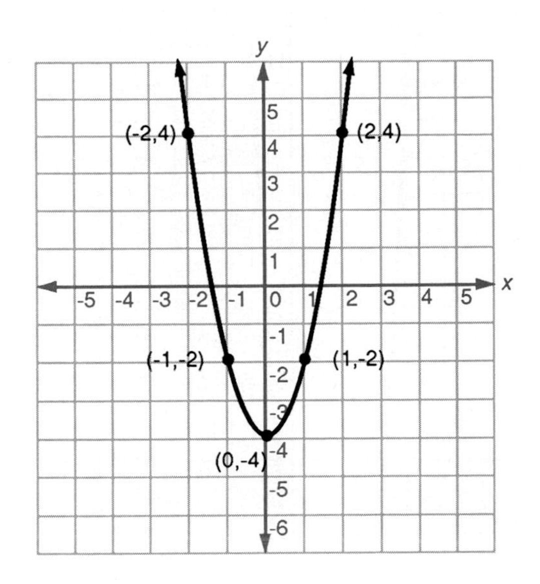

3.

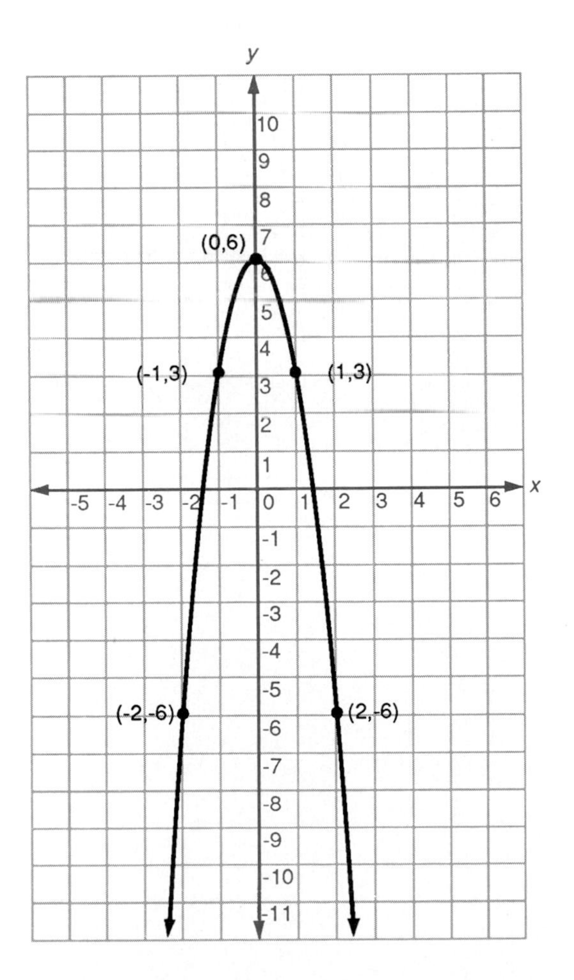

5.

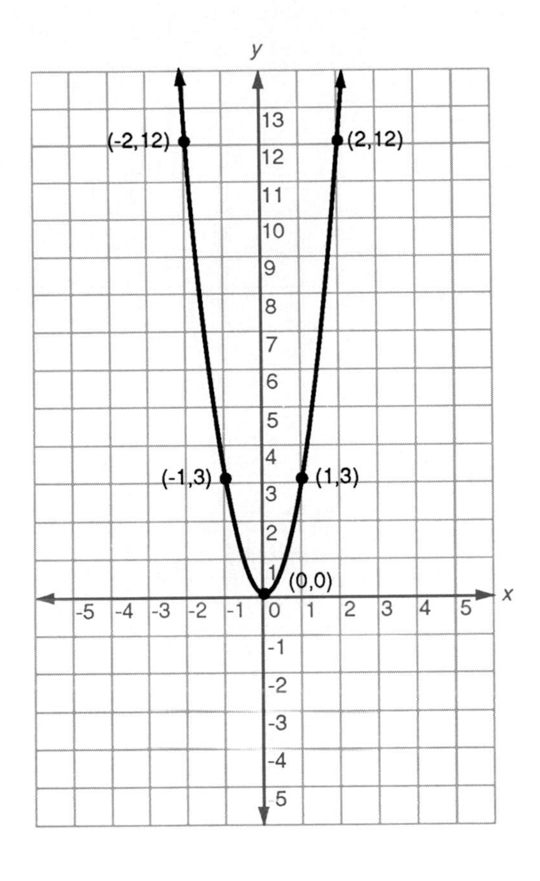

7.

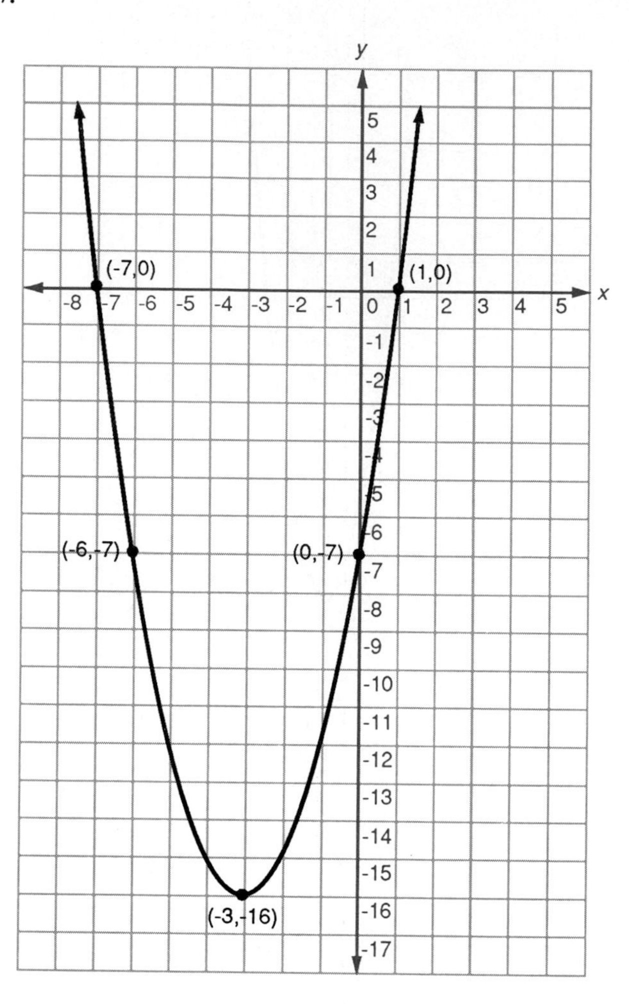

11.

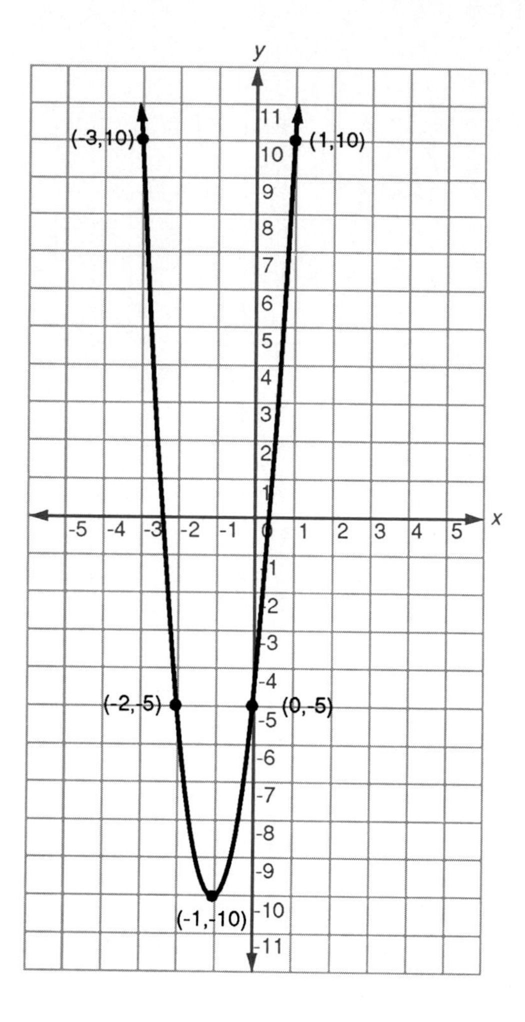

9.

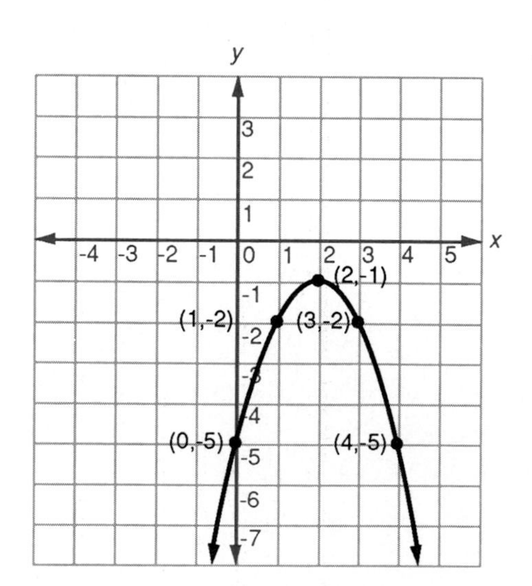

13.

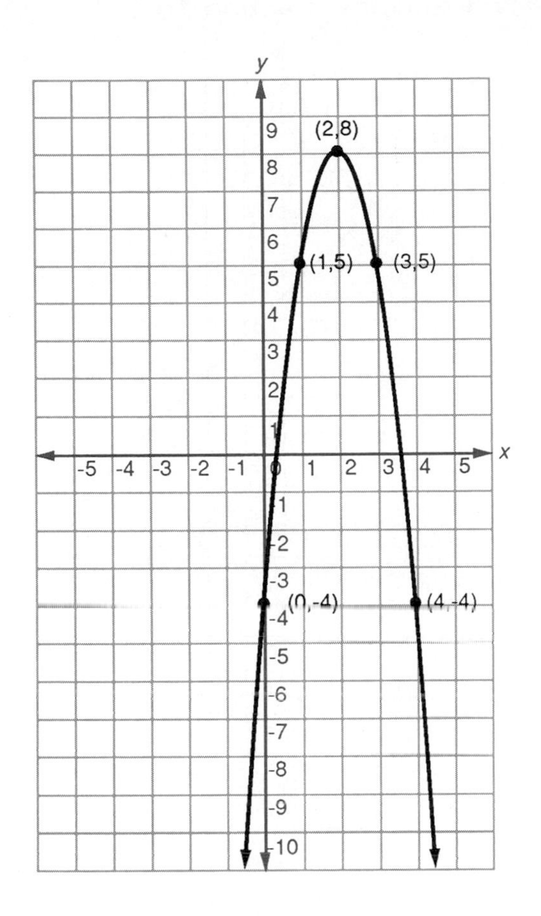

15.

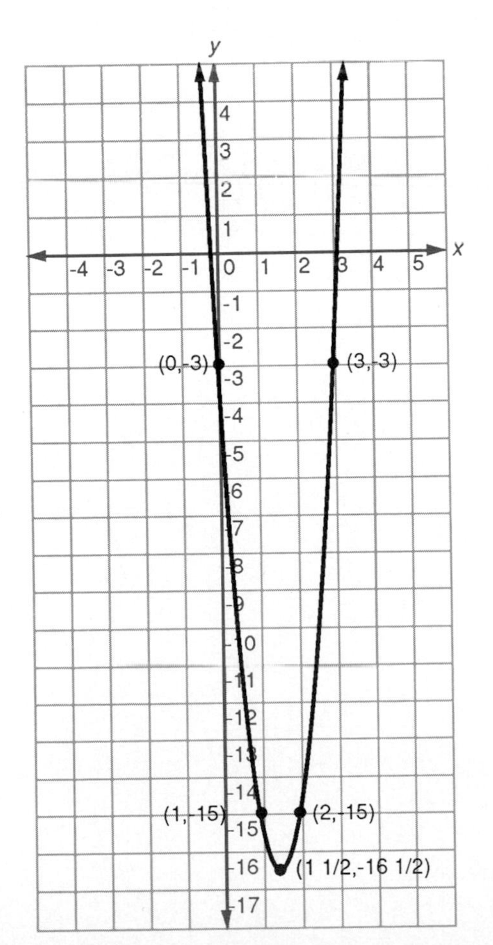

17.

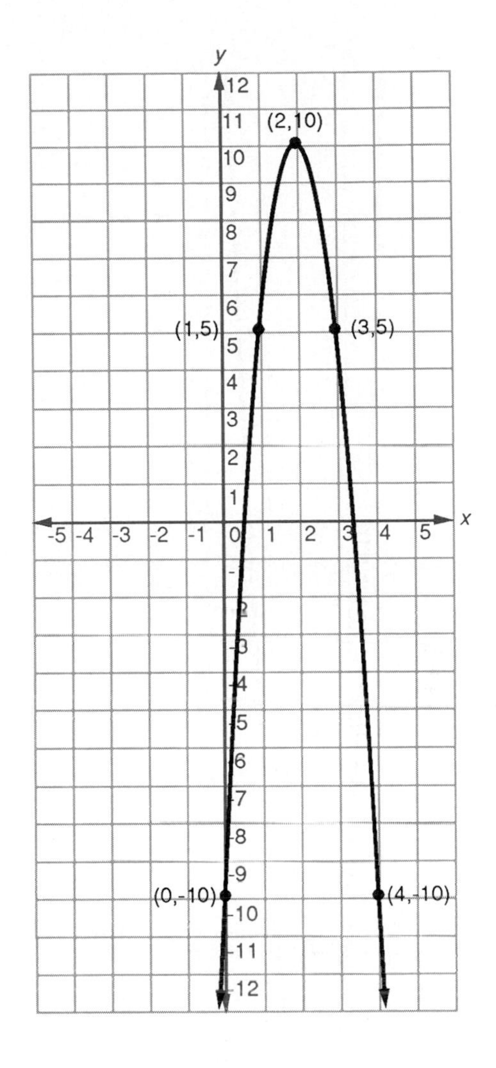

19.

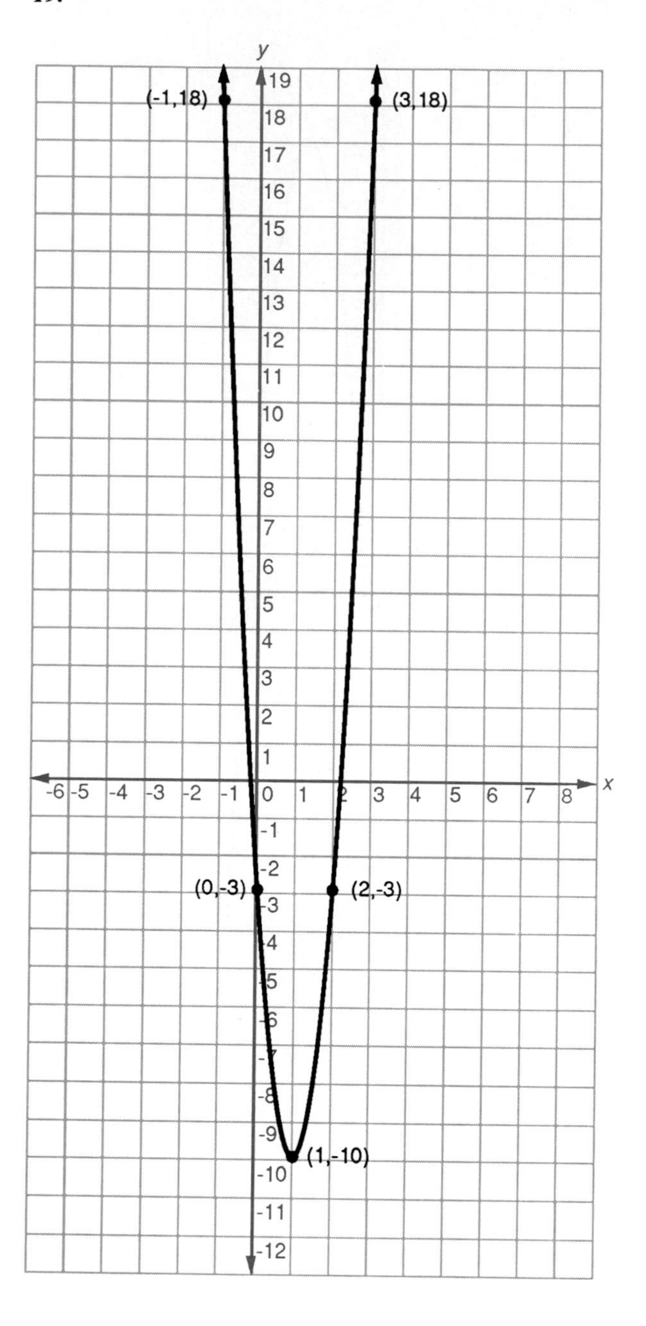

Review Problems/Section 10.1

1.

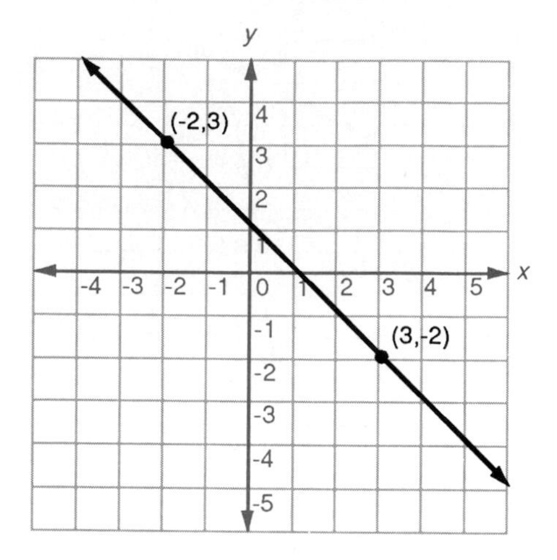

3.

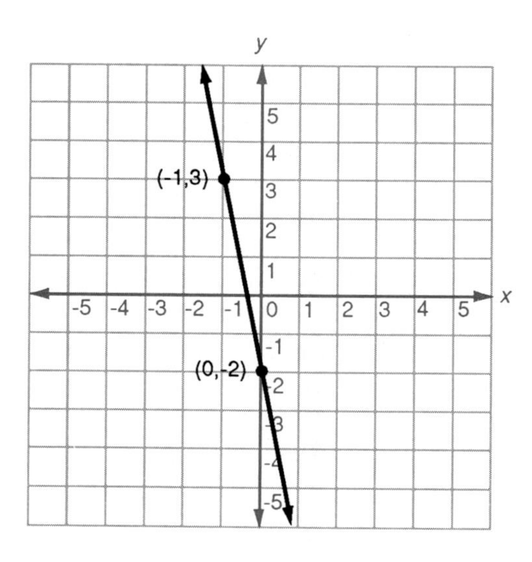

5.

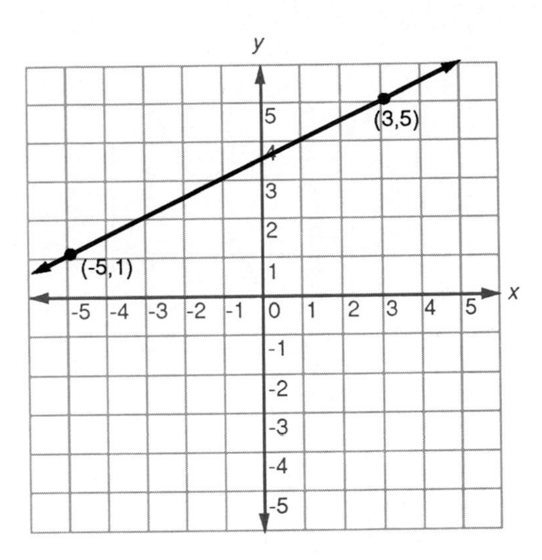

7.

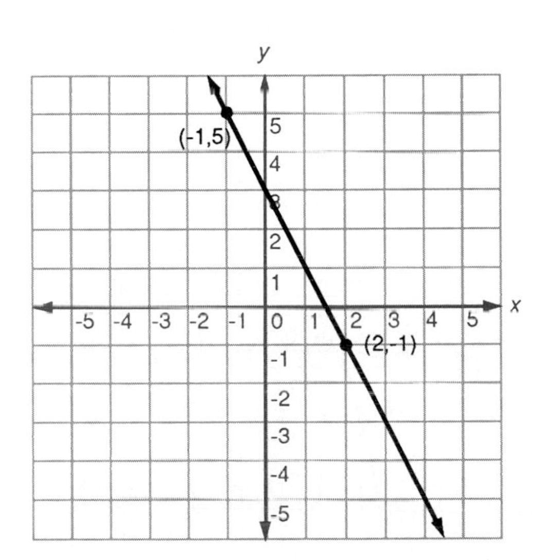

9.

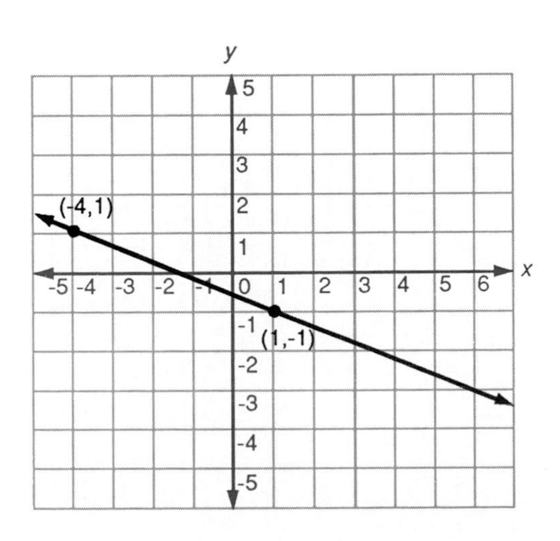

Review Problems/Section 10.2

1. $y = 2x + 3$

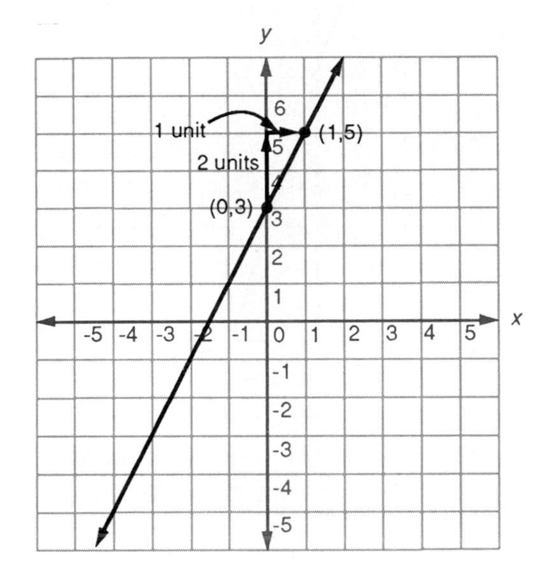

3. $y = -4x - 2$

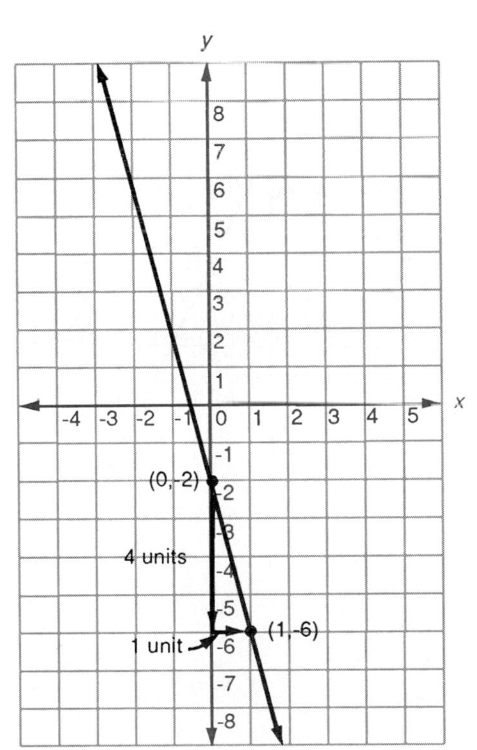

5. $y = -3x + 4$

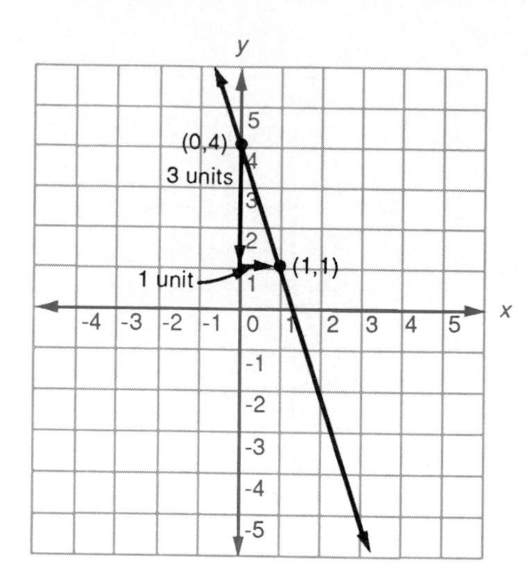

7. $y = -2x + 3$

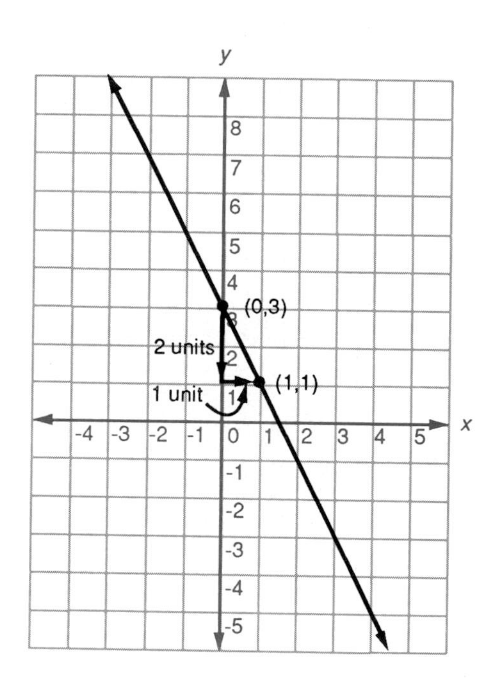

9. $y = \frac{5}{3}x - 1$

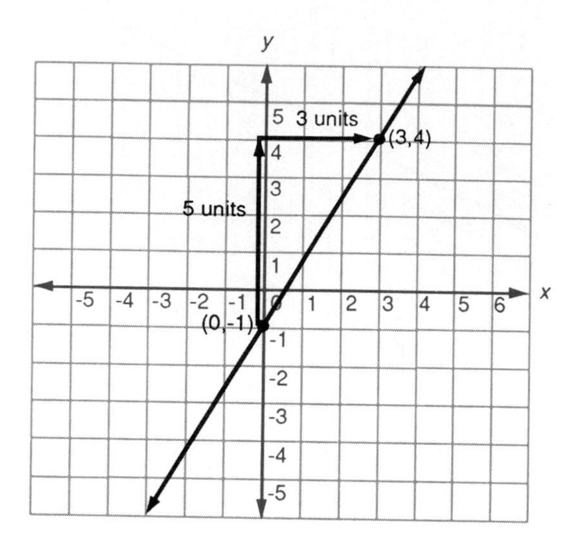

Review Problems/Section 10.3

1.

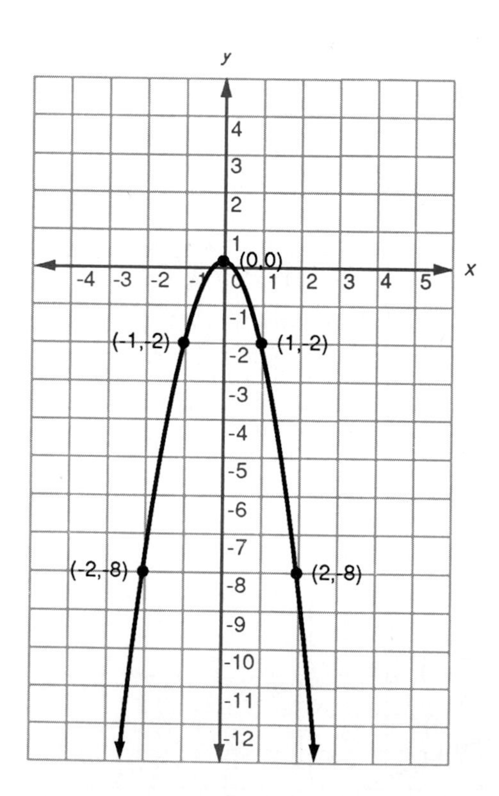

3.

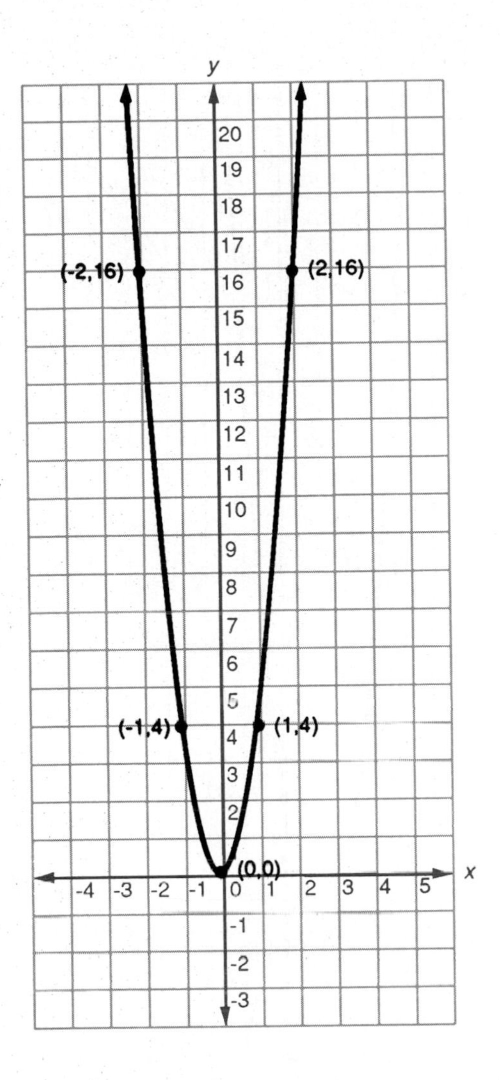
y
x
(-2,16)
(2,16)
(-1,4)
(1,4)
(0,0)
20
19
18
17
16
15
14
13
12
11
10
9
8
7
6
5
4
3
2
-1
-2
-3
-4
-3
-2
-1
0
1
2
3
4
5

5.

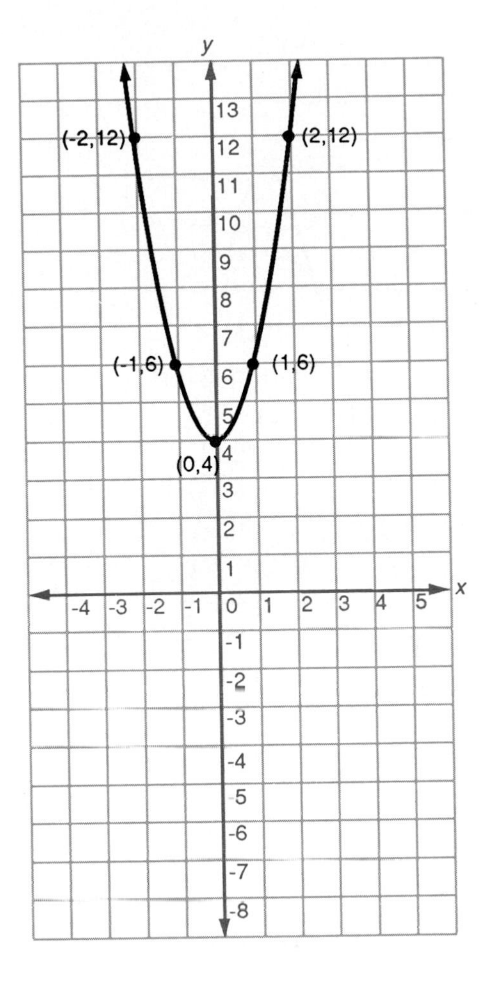
y
x
(-2,12)
(2,12)
(-1,6)
(1,6)
(0,4)
13
12
11
10
9
8
7
6
5
4
3
2
1
-4
-3
-2
-1
0
1
2
3
4
5
-1
-2
-3
-4
-5
-6
-7
-8

7.

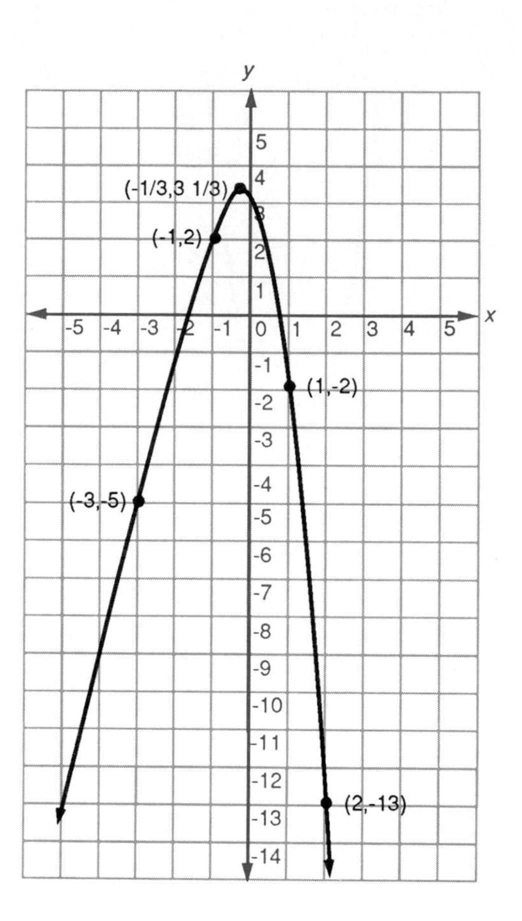

9.

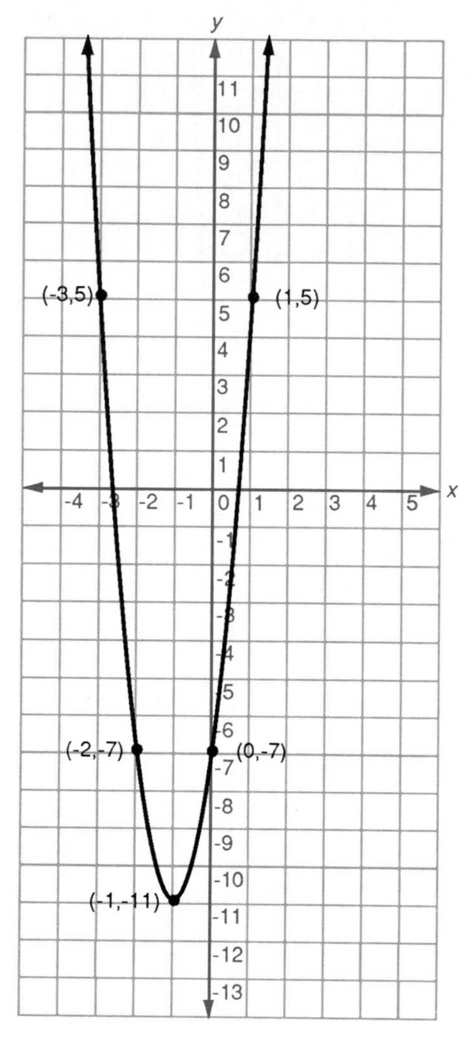

Chapter 11

Exercise 11.1

1. (2,0) **3.** $\left(\frac{11}{2}, -\frac{3}{2}\right)$ **5.** $(-1,-4)$
7. no solution **9.** $(-1,-2)$ **11.** $(11,-2)$
13. many solutions **15.** (1,1) **17.** $\left(-1,\frac{9}{2}\right)$
19. $\left(6,\frac{21}{5}\right)$

Exercise 11.2

1. (1,1) **3.** (2,1) **5.** no solution **7.** (1,0)
9. many solutions **11.** $(-16,9)$ **13.** (5,2)
15. $(4,-2)$ **17.** (4,4) **19.** $(2.2,-3)$

Exercise 11.3

1. (2,4) and (3,6) **3.** (1,1) **5.** (−1,4) and (−7,10)
7. no solution **9.** (−5,−32) and (2,3)
11. $\left(\frac{3}{2}, \frac{7}{2}\right)$ and (−1,−4) **13.** (1,6) and $\left(\frac{2}{5}, 4\frac{4}{5}\right)$
15. $\left(\frac{2}{5}, -\frac{11}{5}\right)$

Application Exercises/Section 11.4

1. 25 and 15
3. A dozen eggs costs \$.95. A loaf of bread costs \$.65.
5. barge: 20 km per hour; current: 4 km per hour
7. 18 nickels and 13 dimes
9. wind: 15 mph; plane: 195 mph

Review Problems/Section 11.1

1. (−1,5) **3.** (−8,−13) **5.** no solution
7. (−14,3) **9.** many solutions

Review Problems/Section 11.2

1. (−2,3) **3.** (2,1) **5.** no solution
7. $\left(\frac{5}{2}, 3\right)$ **9.** many solutions

Review Problems/Section 11.3

1. (−2,8) and (1,−4) **3.** (2,5) and (7,15) **5.** (2,1)
7. $\left(\frac{1}{2}, -4\right)$ and $\left(-2\frac{2}{3}, -10\frac{1}{3}\right)$
9. $\left(\frac{2}{3}, -\frac{10}{13}\right)$ and $\left(-\frac{1}{2}, \frac{5}{2}\right)$

Chapter 12

Exercise 12.1

1. −3,0,5,7,9

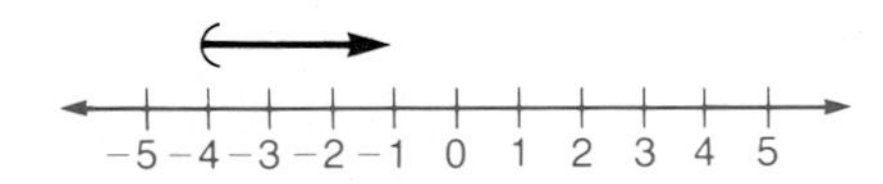

3. −3,−5,−7,−8,−9

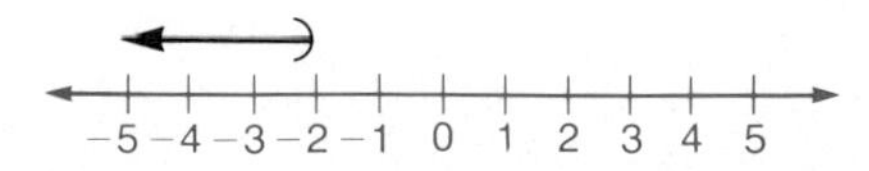

5. 5,4,2,0,−1

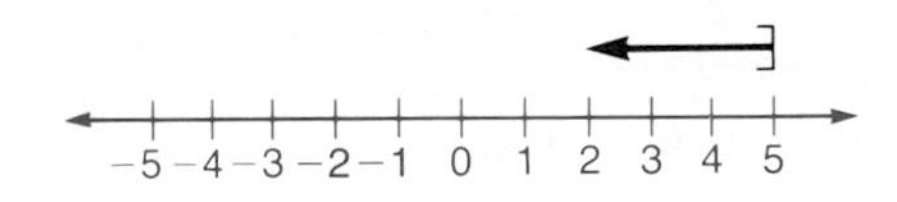

7. 6,5,4,3,2

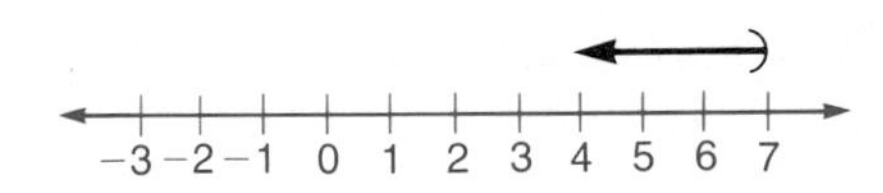

9. 3,5,6,10,15

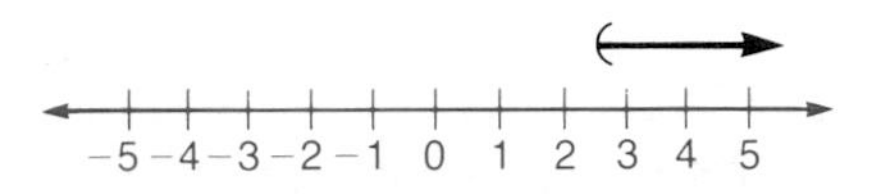

11. −2,−3,−7,−11,−24

Exercise 12.2

1. $x > 3$

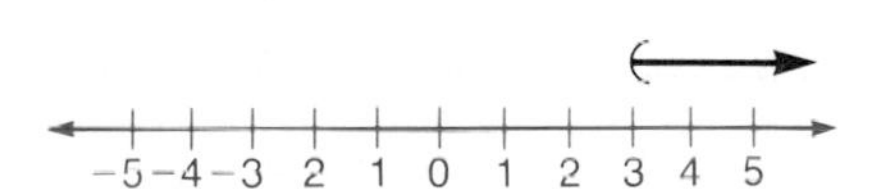

3. $2 < x$

5. $-5 \geq x$

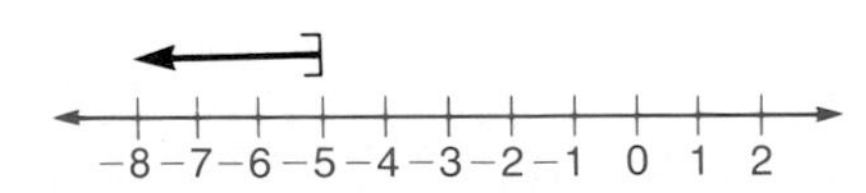

7. $12 \leq x$

9. $x < 5$

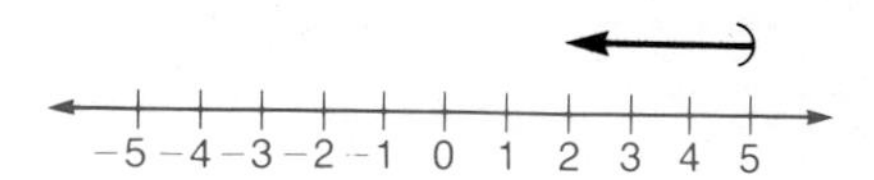

11. $x \geq -2$

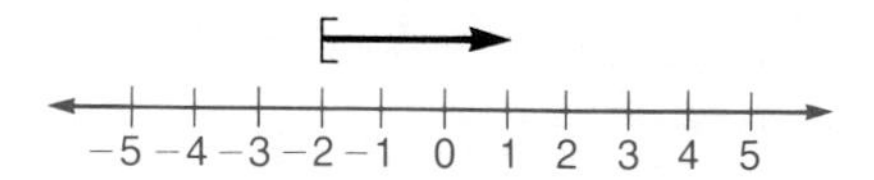

Exercise 12.3

1. $x > 1$

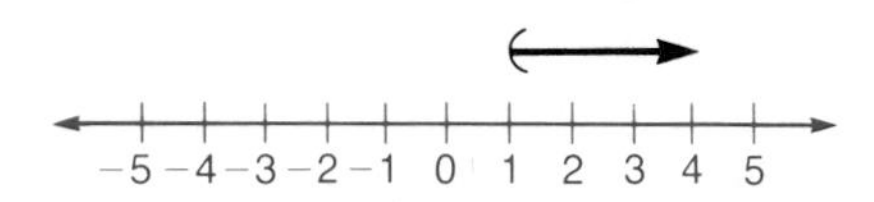

3. $x > -2$

5. $-2 < x$

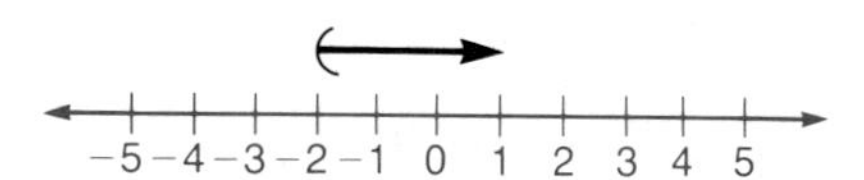

7. $x \geq 3$

9. $-3 < x$

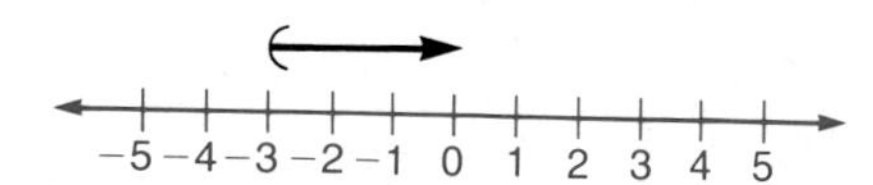

11. $-2.5 < x$

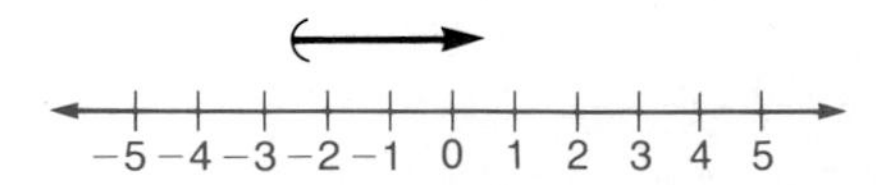

Exercise 12.4

1. $x < 1$

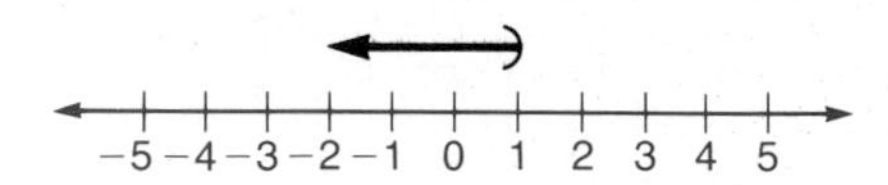

3. $x > 5$

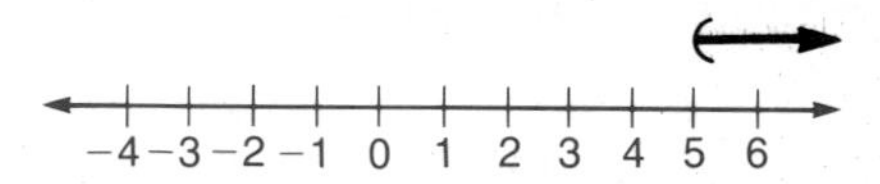

5. $6 \geq x$

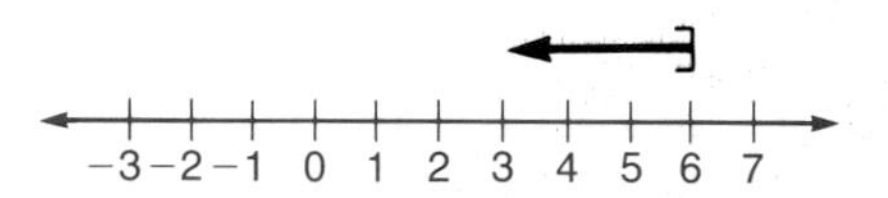

7. $x > 2$

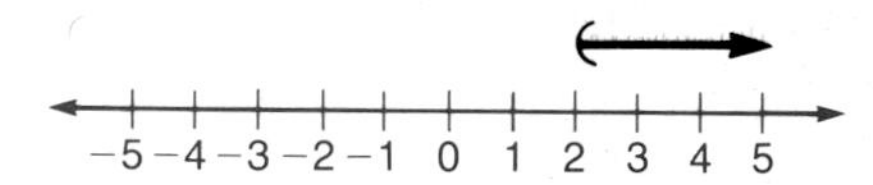

9. $-3 \geq x$

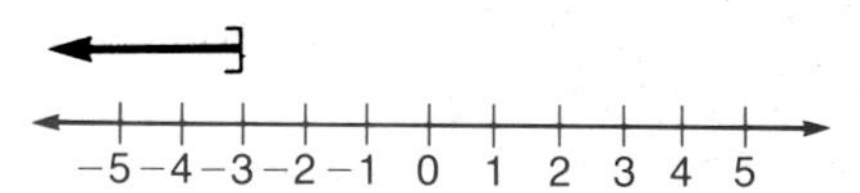

11. $x < -6$

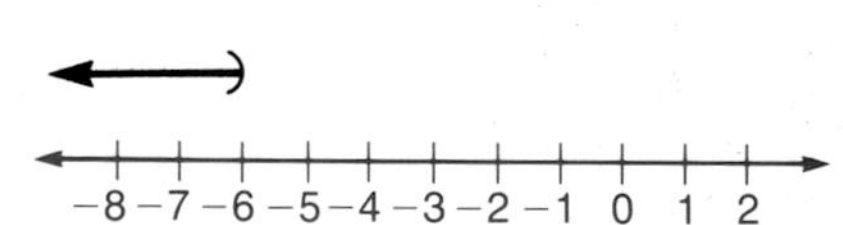

13. $x \geq 3$

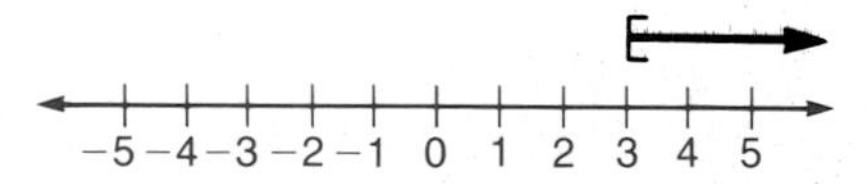

15. $x > 3$

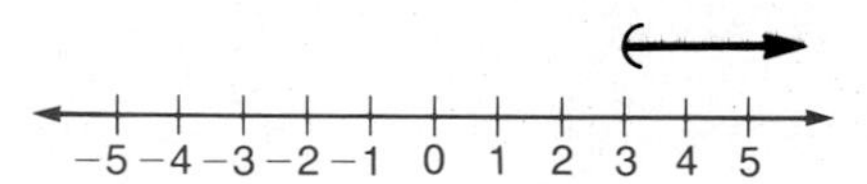

17. $x > 2$

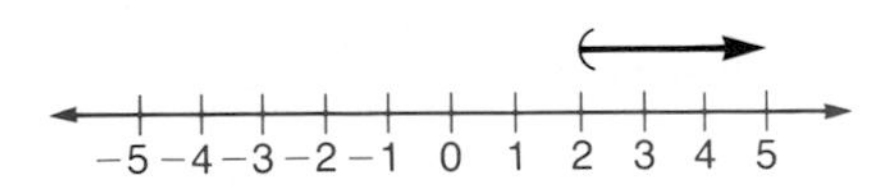

19. $x \geq -3$

Exercise 12.5

1.

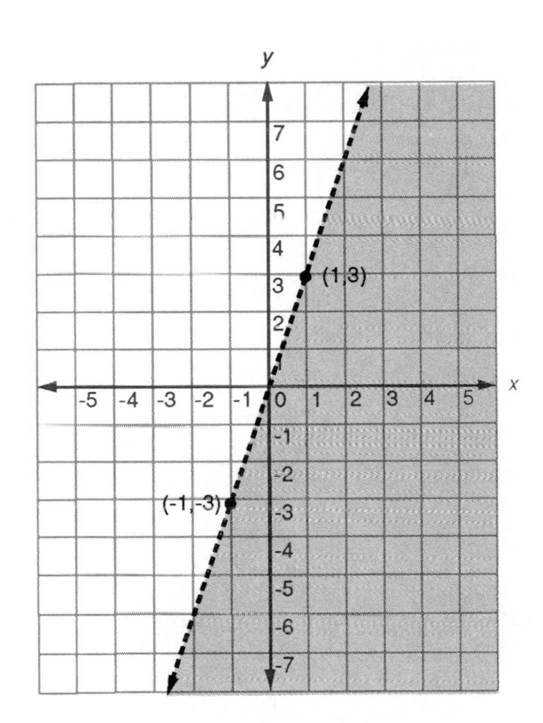

3.

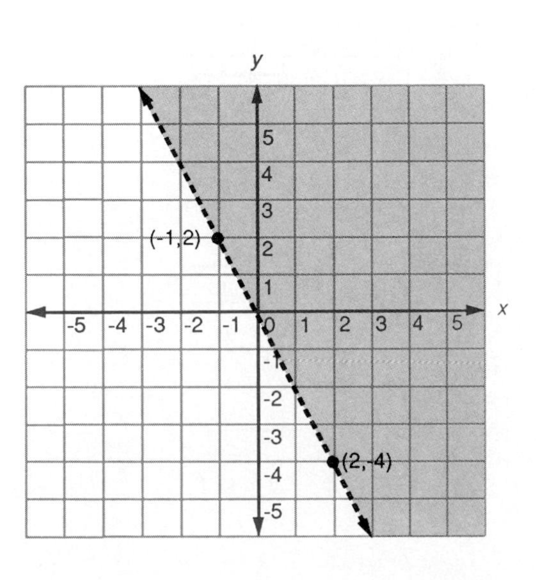

5.

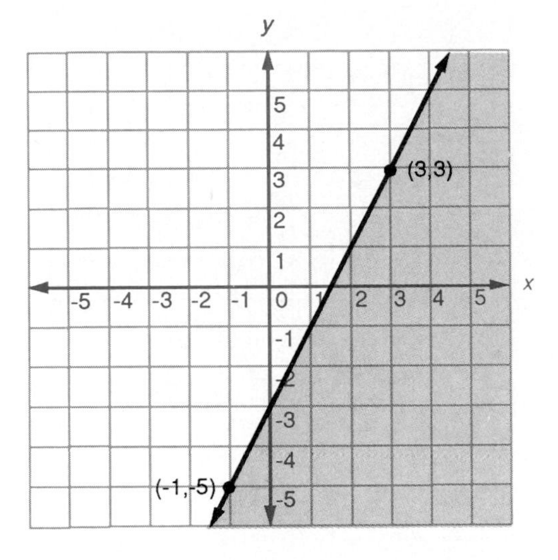

7.

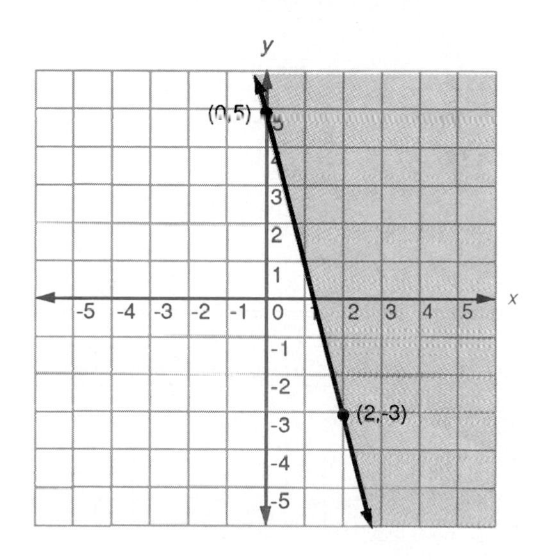

9.

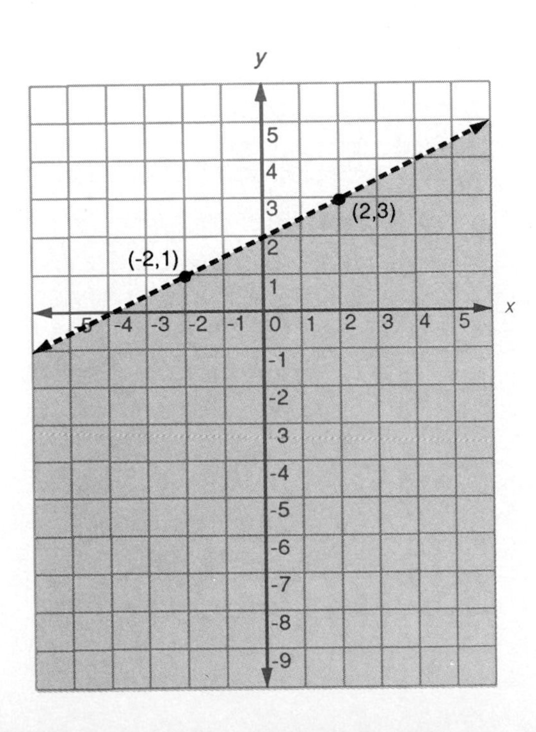

11.

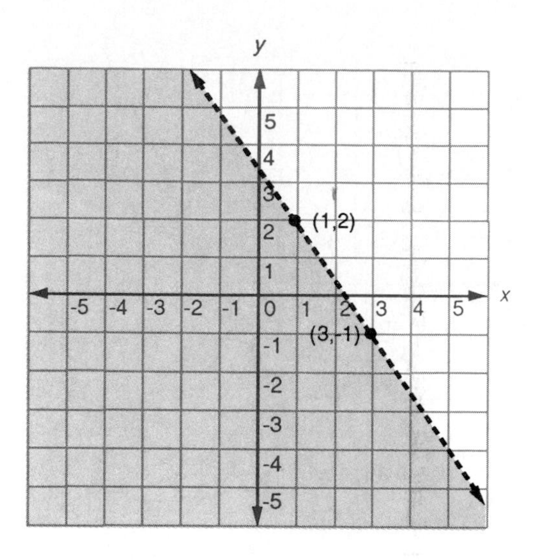

13.

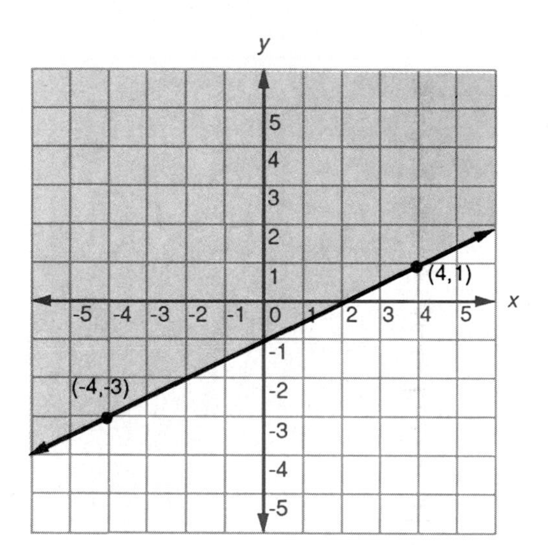

15.

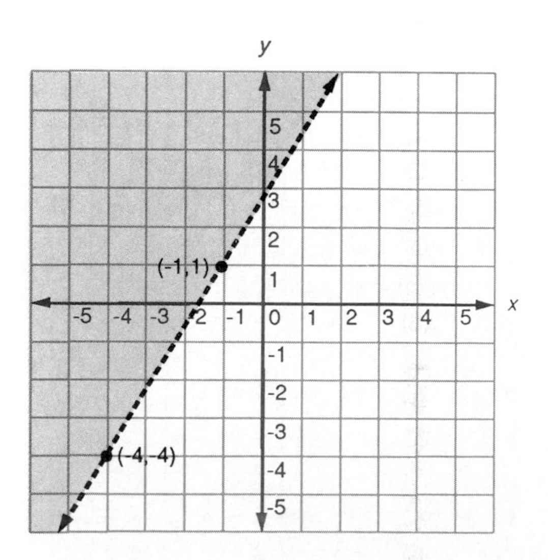

Review Problems/Section 12.1

1. 7,6,5

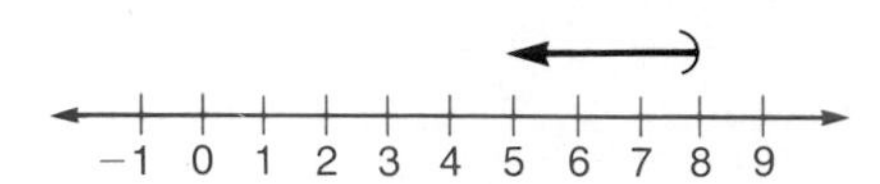

3. $-2, -3, -4$

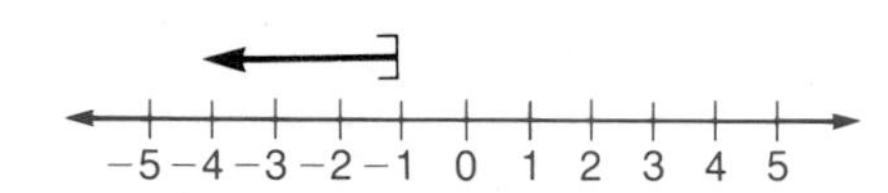

5. $-3, -2, -1$

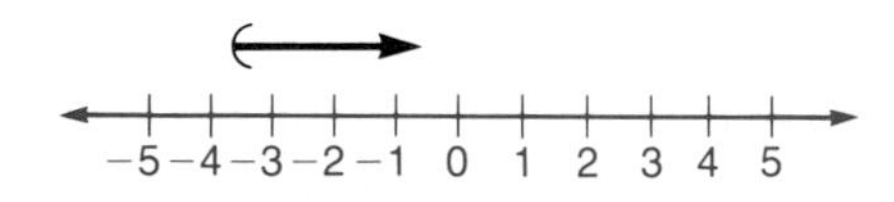

7. $-7, -6, -5$

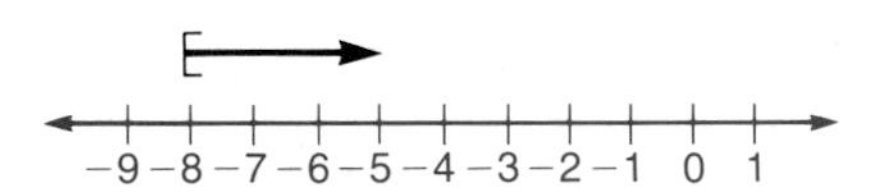

9. 11,10,9

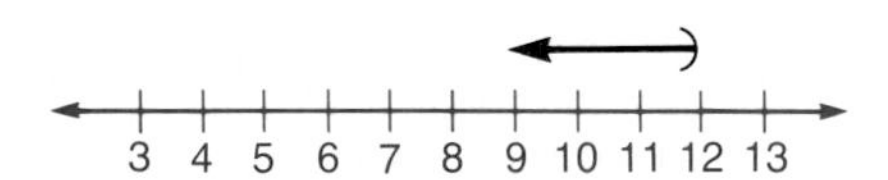

Review Problems/Section 12.2

1. $x > -5$

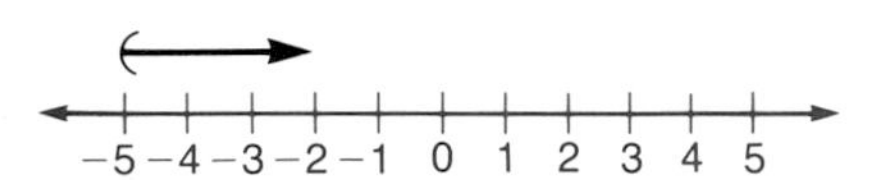

3. $x < 12$

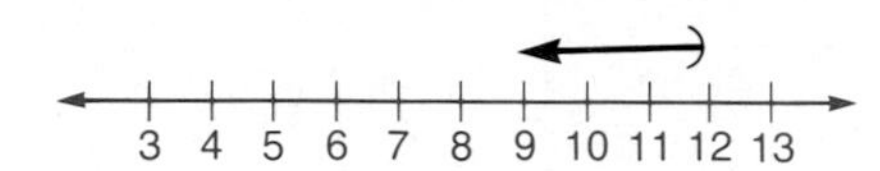

5. $10 \geq x$

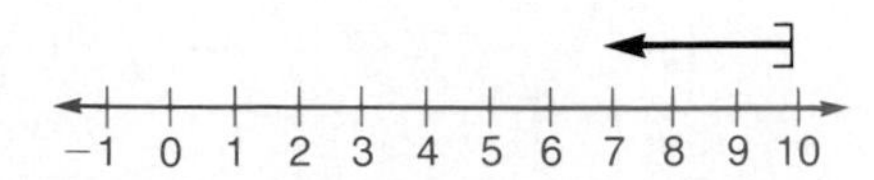

7. $-17 \leq x$

9. $x < 4$

Review Problems/Section 12.3

1. $x > 6$

3. $x \leq 3$

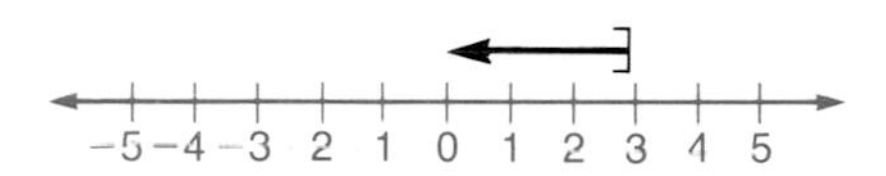

5. $3 < x$

7. $x \leq -2$

9. $3 > x$

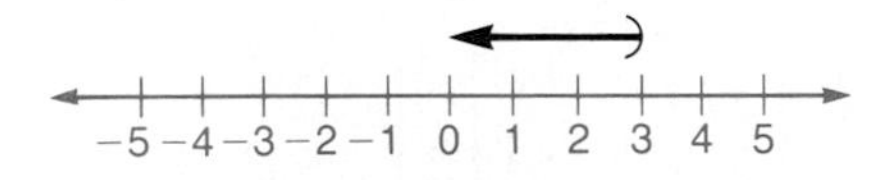

11. $x \geq -4$

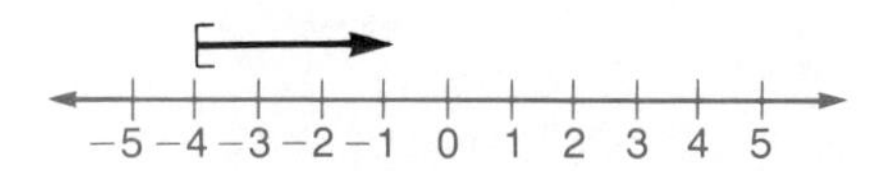

Review Problems/Section 12.4

1. $x > 5$

3. $x \leq -4$

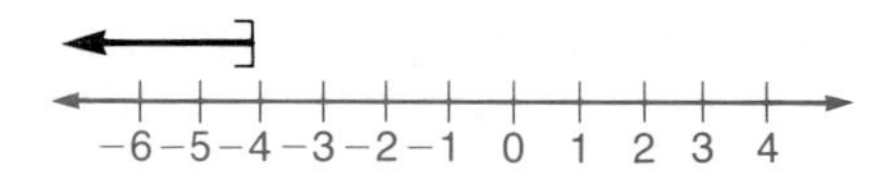

5. $4 > x$

7. $4 \leq x$

9. $x > 3$

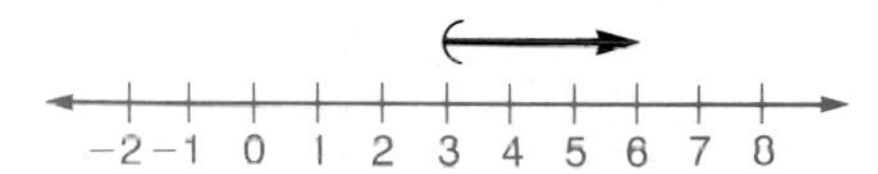

11. $x < -2$

13. $x \geq 1$

15. $x > 2$

17. $x \leq 8\frac{2}{3}$

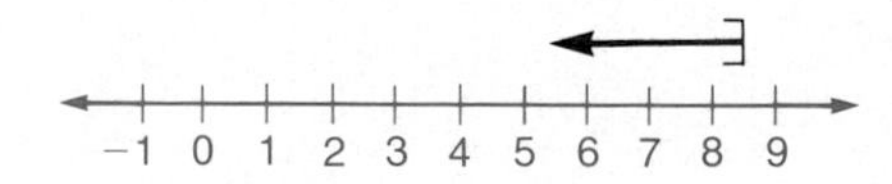

19. $x < 6$

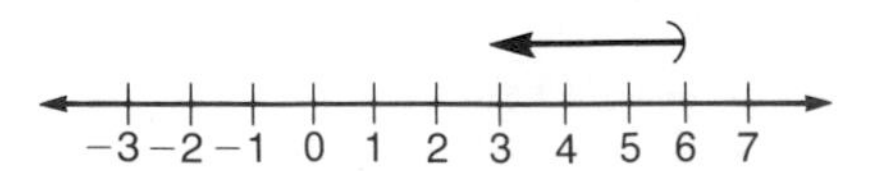

1.

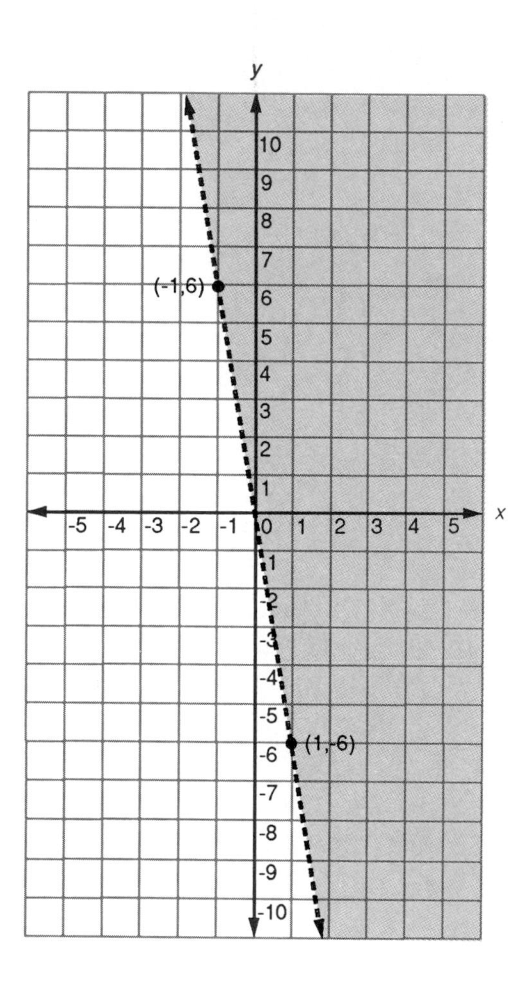

3.

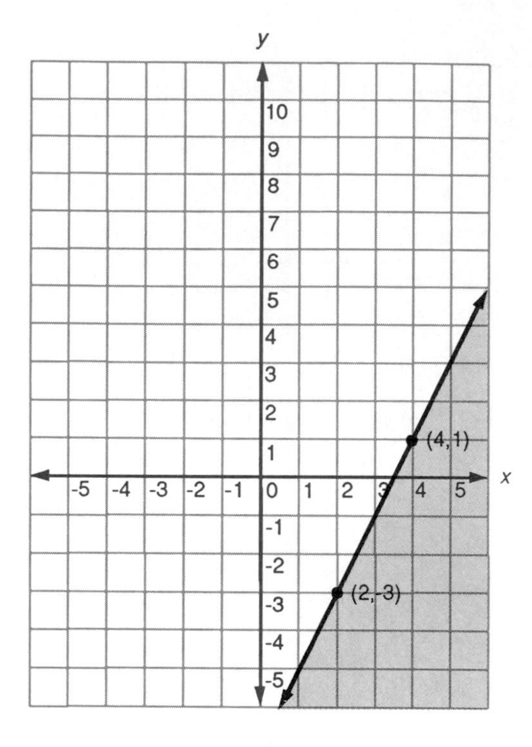

5.

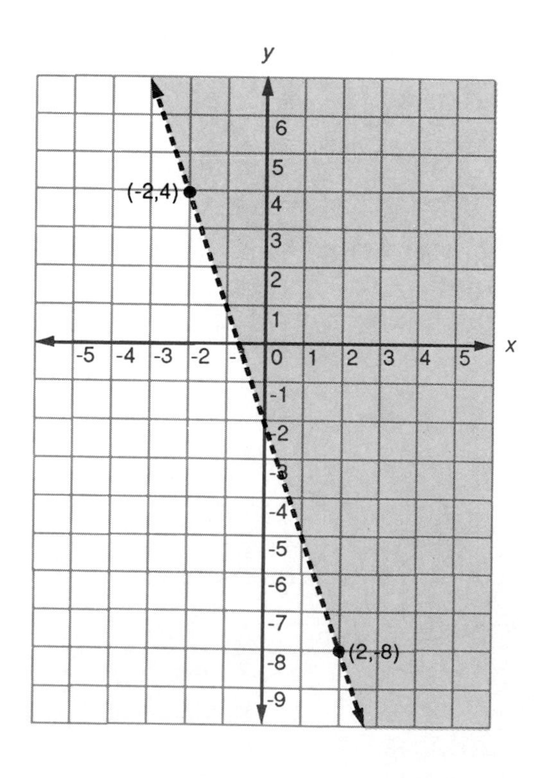

7.

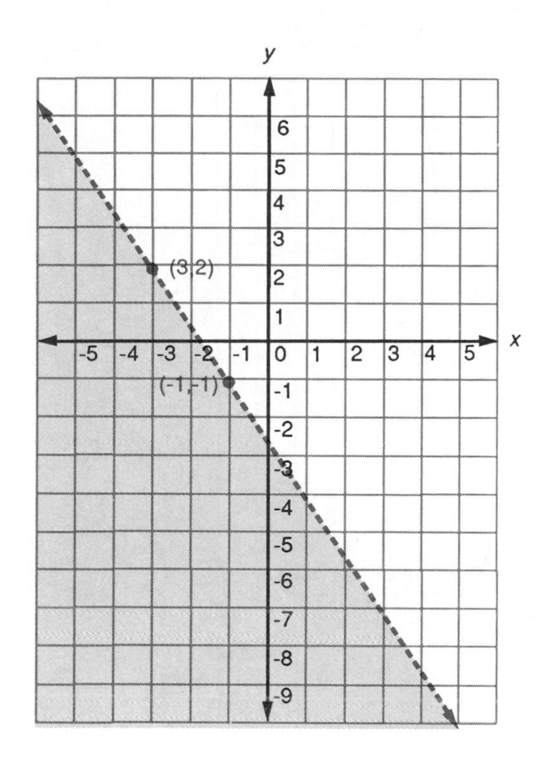

9.

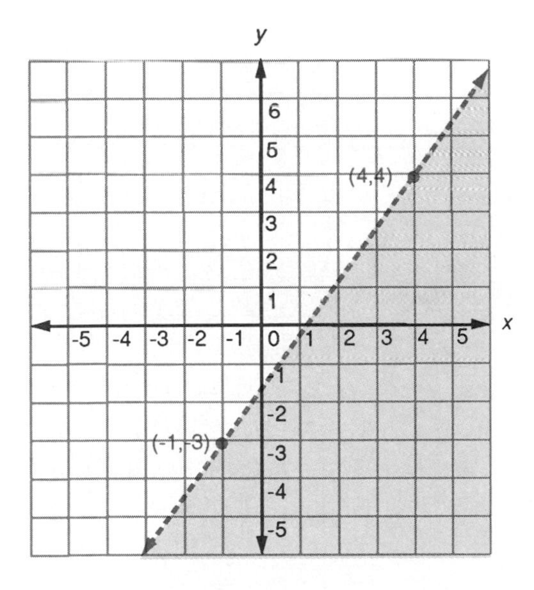

Application Exercises/Section 12.6

1. 90 **3.** 14 feet **5.** 15 inches **7.** $613.64

Appendix A

1. 1.41 **3.** 8.37 **5.** 5.66

Index

Boldface page numbers refer to boxed material.

G

I

L

M

O

P

Q

R

S